A DICTIONARY OF THE ENVIRONMENT

With the assistance of:
Ailsa Allaby

Dr G. Browning, Harwell Atomic Energy Research
Establishment

Dr M. D. Hooper, Nature Conservancy Council

John Macadam

Margaret Palmer

Professor F. Roberts, Brunel University and
Harwell Atomic Energy Research Establishment

Professor R. S. Scorer, Imperial College,
London

Professor E. K. Walton, St Andrews University

A DICTIONARY OF THE ENVIRONMENT

Michael Allaby

This book is sold subject to the
standard conditions of the Net Book Agreement.

First published 1977 by
THE MACMILLAN PRESS LTD
London and Basingstoke
Associated companies in New York Dublin
Melbourne Johannesburg and Madras

British Library Cataloguing in Publication Data

Allaby, Michael
    Dictionary of the environment.
    1. Ecology-Dictionaries
    I. Title
    301.31'03   QH540.4

    ISBN 0-333-19754-2

Printed in Great Britain by
The Camelot Press Ltd, Southampton

INTRODUCTION

A dictionary of a living language does not seek to lay down arbitrary laws dictating the sense in which words and phrases are to be used. It is a record, no more, of the way particular expressions are being used at a particular time. It is as alive, or as dead, as the language it records. In most scientific disciplines, developments are occurring so rapidly, new terms are being coined – some unnecessarily – at such a rate, that the collection and arrangement of them presents very real problems. Nowhere is this more true than in the environmental sciences. Still emerging, they do not even have a name of their own to describe them, for 'ecology', the word most commonly used, has a special limited meaning of its own. The professional ecologist today must be part botanist, zoologist, chemist, physicist, mathematician, geologist, meteorologist, statistician, geographer and, if he finds himself drawn into the wider environmental movement, economist and philosopher as well. He will find himself confronted with words drawn from all of these disciplines, and probably more.

In this dictionary I have attempted to draw from as wide a range of subject areas as I can but, inevitably, the final decision, the allocation of priorities, is mine and it is arbitrary. There are omissions, some of them glaring. Economics is covered in a somewhat cursory fashion, some subjects are barely mentioned. The fault is mine, and that of the book itself, which stubbornly refuses to accommodate more words to a page, or to grow pages (so adding to its biomass?).

No one can know everything, and in most cases I have returned for definitions to the disciplines from whence the terms are taken, so that they are defined in the sense in which they are used within that discipline. For this I must thank my collaborators, Margaret Palmer, John Macadam, Ailsa Allaby, Richard Scorer and Gavin Browning. I must thank, too, Dr Max Hooper, Professor F. Roberts, and Professor E. K. Walton FRSE, for undertaking so willingly the dreary task of checking the completed entries.

I have not used extensive internal cross-referencing, on the assumption that any term unfamiliar to the reader will have an entry of its own. *Cf.* and *see* are used only where such a cross reference might help the user. Such abbreviations as 'Geol.', 'Biol.' and 'Bot.' are, I hope, self-explanatory.

Michael Allaby
Wadebridge, Cornwall

A

A. Ampere.

a. Atto-.

aa. Hawaiian term describing basaltic lava with a rough, blocky surface.

abaca. Manila hemp (*Musa textilis*). *See Musa.*

abaxial. *See* dorsal.

abiocoen. All the non-living components of the environment.

abioseston. *See* Seston.

abiotic. Non-biological. *See* biotic factors.

ablation. The removal of a surface layer, particularly used for the melting and evaporation of the surface of ice, and also for the removal of loose surface material by the wind (deflation).

abscisic acid. Abscisin.

abscisin (dormin, abscisic acid). An auxin which induces leaf-fall and dormancy in seeds and buds, probably by inhibiting the synthesis of nucleic acid and protein.

absolute age. The age of a rock, mineral or fossil in years, determined as a radiometric age or by counting varves. Radiometric dating involves experimental errors, so such dates are usually quoted with a plus or minus error.

absolute humidity. The amount of water vapour present in a unit mass of air, usually expressed in grams of water per kilogram of air. Also called the humidity mixing rate.

absorbate. *See* absorption.

absorbent. *See* absorption.

absorbing duct. Tube used in a ventilator system to attenuate sound waves while offering low resistance to a continuous flow of air.

1

absorption. Process in which one material (the absorbent) takes up and retains another (the absorbate) to form an homogenous solution. Absorption may also refer to substances becoming attached to a solid surface by physical forces, but see adsorption. In air, it may refer to the incorporation of a gaseous component of the air into solid surfaces (also called adsorption) e.g. absorption of sulphur dioxide (SO_2) by stone, vegetation, particulate aerosol etc., or the absorption of radiation passing through the air by aerosols or a gaseous component of the air (e.g. of ultra-violet radiation by ozone (O_3) or of infra-red by carbon dioxide (CO_2) or water vapour.) Absorption also occurs in the ocean. The absorption of gases depends on humidity, state of vegetation, temperature and various physical laws. Absorption of light, etc., by air or water may be expressed as the path length in which the intensity is reduced by the factor e ($\simeq 2.73$) or by the reduction of intensity in unit path length.

absorption coefficient (acoustics). If a surface is exposed to a field of sound, the ratio of the sound energy absorbed by the surface to the total sound energy that strikes it. An absorption coefficient of 1 would mean that all of the sound energy was absorbed. *See* anechoic.

absorption tower. Structure, usually found in a chemical works, in which a liquid is made to absorb a gas, as in the production of sulphuric acid from sulphur dioxide/trioxide and water.

abstractive use. Of water, a use which removes water so that it is lost temporarily as a resource (e.g. in a cooling tower). A non-abstractive use (e.g. hydro-electric power generation) returns the water as surface water.

abyssal. The deepest region of the ocean, roughly below 1000 m. The term may be applied to the zone in lakes below the depth of effective penetration of light.

abyssal gap. A gap in a sill, ridge, or rise separating two abyssal plains through which the sea floor slopes from one plain to the other.

abyssal hill. Relatively small topographic feature of the deep ocean floor, ranging up to 1000 m high and a few kilometres wide.

abyssal plain. A large flat area of deep sea floor having slopes of less than about 1 : 1000.

abyssobenthos. *See* benthos.

Acacia. Genus of leguminous trees (also known as wattles) of tropical and sub-tropical origin (esp. Australasia) which may form dominant vegetation in arid areas. Many commercial products are derived from acacias, including dyes, perfumes, timber, etc.

acanthite. A major ore mineral of silver, acanthite, Ag_2S occurs in hydrothermal deposits, characteristically with lead, zinc, and copper minerals, which also contain silver by atomic substitution. Acanthite also occurs in supergene deposits. The other major ore mineral of silver is argentite. Nearly all the silver produced is a by-product from mining for lead, zinc and copper. Apart from its uses in photography, most industrial applications of silver utilize its high reflectivity and conductivity, as well as its resistance to organic corrodants.

Acanthocephala. Spiny-headed worms, a phylum of parasitic worms with affinities to the Nematoda. The larvae live in insects or crustacea, and the adults in the gut of vertebrates, to which they cling by means of a spiny proboscis. An example is *Echynorhynchus proteus*, the adult of which lives in ducks and the larva in freshwater shrimps. *See* parasitism.

Acanthodii. A group of extinct fishes, originating in the Silurian Period. They were small fishes, with fins supported by large spines, and were the first vertebrates known to have possessed jaws.

acaricide. Chemical, such as Derris and some organophosphorous, dinitro, and organochlorine compounds used to kill ticks and mites, Acarina.

Acarida. Acarina.

Acarina (Acarida). Mites and ticks, Arachnida with round bodies. Mites are very abundant in the soil, feeding on plant material and invertebrates. Some parasitic mites (e.g. 'red spider') damage crops and can be serious pests, and others cause diseases such as mange in animals. Ticks are blood suckers, some being vectors of diseases such as relapsing fever in man and fowls, Rocky Mountain spotted fever, and louping-ill in cattle and sheep.

acceleration. Rate of change of velocity with time. According to Newton's laws, acceleration × mass = force = rate of change of momentum. Momentum is a vector and motion in a curved path therefore requires the application of a force. In the atmosphere or ocean, vertical motion always requires horizontal acceleration, which results from buoyancy forces.

access agreement. In British planning, an agreement allowing the public access to privately owned land, being open country for open-air recreation. (As defined by the Countryside Commission).

access order. In British planning, an order allowing the public access to privately owned land, being open country where an access agreement is impracticable or is not adequately securing public access to the land for open-air recreation. (As defined by the Countryside Commission).

accessory (ecol.). Species which occurs in one-fourth to one-half of a community.

accessory mineral. A mineral occurring in small amounts in a rock and disregarded in the classification of that rock (which is based on essential minerals). Accessory minerals can yield evidence about the origin of the rock (e.g. the presence of metamorphic (see Metamorphism) minerals in a sandstone suggests a provenance , at least in part, from a metamorphic belt).

accidental species. A species which occurs in less than one-fourth of a stand.

acclimatisation. The process of adapting to abiotic environmental conditions, by phenotypic (*See* phenotype) rather than genetic variation.

accommodation (ecol.). The location of a population within an area or habitat.

accretion (meteor.). The attachment of airborne material to fixed or falling or flying objects. Ice accretion occurs on wires, hailstones or aircraft wings when the air contains supercooled cloud droplets, drizzle or rain, and is particularly dangerous on the rigging of ships in polar regions or on TV masts on hills in winter. Pollution accretion is exemplified by smoke deposition on window frames, ventilation intakes, etc., where the airflow is swift and curved.

Acer. Genus of maple trees and shrubs found in temperate regions. They yield charcoal, timber and maple sugar. (*A. saccharum*).

acetaldehyde. A direct oxidation product of ethyl alcohol which can be further oxidised to acetic acid. An important raw material for certain organic compounds.

acetate film. Non inflammable cinema film based on cellulose acetate.

acetic acid ($CH_3 . COOH$). The acid in vinegar, and an important raw material obtained from wood, acetylene or alcohol.

acetone. An important laboratory and industrial solvent. It is very volatile and is also miscible with water.

acetylcholine (ach). Substance released in minute amounts when impulses arrive at many nerve endings. It causes a contraction of muscle and is responsible for the passage of impulses across synapses in the ganglia of the autonomic nervous system. Its effects disappear rapidly after secretion because it is destroyed by the enzyme cholinisterase.

acetylene. A colourless, poisonous hydrocarbon gas, prepared historically in the

4

laboratory by the action of water on calcium carbide. It is used for welding, acetic acid synthesis, and as a starting material for many chemicals, e.g. PVC.

achene. A dry, one-seeded fruit which does not split open (e.g. buttercup fruit). Dispersal may be aided by wings (e.g. Sycamore), plumes (e.g. Old Man's Beard) or hooks (e.g. Wood Avens).

Achira (Canna edulis, C. indica). Plant cultivated for its starchy root, first domesticated in Peru some time prior to 2200 BC. Today *C. edulis* is still cultivated for its root and the tops are sometimes fed to cattle. It is also known as 'Queensland arrowroot'.

achondrite. Stony meteorite, without chondrules.

acicular. Having the form of needles, used especially for elongated crystals.

aciculilignosa. Needleleaf forest and bush. Evergreen, coniferous vegetation.

acid. (*a*) (Geol.) Acidic. (*b*) (Chem.) *See* pH. (*c*) (Colloquial) Lysergic acid diethylamide.

acid dipping. Immersion of a metal in a tank of suitable acid or acids to remove scale from and clean the surface. Often produces hazardous fumes and acid mists.

acid droplets. Minute liquid particles, emitted by certain industrial processes, which act as condensation nuclei. *See* sulphuric acid.

acidic (acid). (Geol.) A term applied to igneous rocks containing more than a high percentage (commonly set at 65%) of silica (SiO_2) in their chemical composition. Most of the silica is in the form of silicate minerals, such as feldspars, micas, and amphiboles, but the excess silica manifests itself in the presence of 10% or more free quartz. Granite, rhyolite, and obsidian are all acidic rocks. In petrology, acidic is contrasted with intermediate, basic and ultrabasic, but not with alkaline.

acid refractory. Furnace lining materials which are composed mainly of silica, designed to resist acid slags. *See* Bessemer process.

acid soot (acid smut). Particles of carbon, held together by water that is acidic through combination with sulphur trioxide. The carbon particles are emitted during combustion, and the soot particles are roughly 1 to 3 mm in diameter. Where oil-burning installations have metal chimneys, acid soot can acquire iron sulphate, which makes brown stains on materials and damages paintwork. Acid soot emissions can be reduced by using low-sulphur fuels, by reducing the air flow to minimise sulphur trioxide formation, by making flues airtight, by insulating chimneys, by raising the temperature, etc.

acoustic. Refers to properties or characteristics connected with sound, e.g. the acoustic qualities of an auditorium. It is not used to refer to people, where the term is acoustical (e.g. acoustical engineer).

acoustical. *See* acoustic.

acoustic reflex. The mechanism by which the mammalian ear protects itself against sounds that are too loud, by adjusting the connecting muscles that regulate the relative positions of the ossicles.

acquired character. Variation in an organism which appears as a response to environmental influence. *See* Lamarck.

Acrania (Cephalochordata). The lancelets, small, fish-like ciliary feeders, forming a small sub-phylum of the Chordata which may be similar to the ancestors of fish. They have poorly developed heads, no brain, bone or cartilage, and nephridia as excretory organs. *Branchiostoma* (*amphioxus*) is a living representative and *Jaymoytius* a Silurian form.

Acraniata. Invertebrata.

acre-foot. The volume of any substance required to cover one acre of a surface to a depth of one foot, equal to 43560 cubic feet, (1232.75 cu m).

Acrididae. The short-horned grasshoppers, whose antennae are shorter than their bodies. Some (e.g. the locust, *Locusta migratoria*) although commonly solitary, under certain conditions develop a gregarious and migratory form which causes incalculable harm to crops. *See* Orthoptera.

acrodont. *See* Thecodont.

acrosome. Projection on the head of a sperm which contains enzymes which play a part in the fusion of egg and sperm.

acrylic resins. A group of synthetic resins, obtained by polymerisation (*See* polymer) of acrylic acid-derived monomers. They are transparent, resistant to light, weak acids, alkalis and alcohols, but are attacked by oxidising acids, chlorinated hydrocarbons, ketones and esters. They are used widely in industry and consumer products.

ACTH (adreno-cortico-tropic hormone, cortico-tropin). A pituitary hormone secreted by the anterior lobe of the pituitary at the base of the brain in vertebrates. It stimulates the cortex of the adrenal glands to produce hormones such as cortisone.

actin. *See* actomyosin.

actinomorphic (radially symmetrical). Applied to animals (e.g. Coelenterata and Echinodermata) and flowers (e.g. buttercup) which have more than one plane of symmetry. Sessile animals are commonly actinomorphic. *Cf.* bilaterally symmetrical.

Actinomycetes. Group of bacteria with cells arranged in fine filaments. Important constituents of the soil, where they assist in the decomposition of organic matter. *Streptomyces griseus* produces the antibiotic streptomycin.

actinomycin. Antibiotic which blocks the synthesis of RNA by combining with DNA. It is produced by some Actinomycetes.

Actinopterygii. A large group of fishes containing the great majority of present-day bony fishes and many fossil forms. They are characterised by having the paired fins supported by horny finrays, with no skeletal axis. *See* Choanichthyes, Teleosti, Osteichthyes.

Actinozoa (Anthozoa). Class of marine Coelenterata including the sea anemones, stony corals, sea pens and 'Dead Men's Fingers' (*Alcyonium digitatum*). Some species are solitary, some colonial. The medusa stage typical of other Coelenterates is absent in the Actinozoa. *See* Hydrozoa, Scyphozoa.

activated alumina. Granular, porous form of aluminium oxide capable of absorbing (*See* absorption) water oil vapour or certain other substances from gases or liquids. Used in pollution control, chromatographic analysis, and as a catalyst.

activated carbon. Forms of carbon with a high adsorptive (*See* adsorption) capacity for gases, vapours and colloidal solids. The property is achieved by heating to 900°C with steam or carbon dioxide, giving a porous particle structure. Used for odour, fume and other pollution control.

activated carbon processes. Japanese processes for removing sulphur dioxide from flue gases. There are three processes: (a) Water washing, in which the gas is absorbed on dry activated charcoal and the charcoal washed with water to give dilute sulphuric acid or gypsum; (b) Gas desorption, in which the gas is absorbed dry and then desorbed (*See* desorption) to give sulphur dioxide; (c) Steam desorption, in which the gas is absorbed dry, then desorbed to give sulphur dioxide.

activated charcoal. Activated carbon.

activated manganese oxide process. Japanese process for removing sulphur dioxide from flue gases by dry absorption to produce ammonium sulphate.

activated sludge. The active material, consisting largely of protozoa and bacteria, used to purify sewage. When mixed with aerated sewage, the sludge organisms break down the organic matter in the sewage, using it as food, and multiply, thus producing more activated sludge.

active factors. The factors which supply energy and nutrient for the active operation of natural processes in plants.

active transport. The passage, accompanied by the expenditure of energy, of a substance from a region of low concentration to one of high concentration (i.e. against the concentration gradient). This usually occurs across cell membranes.

activity. The total flow of energy through a system in a unit of time.

actomyosin. A complex of the proteins actin and myosin which make up the major part of muscle. Contractions of muscle are probably accomplished by myosin molecules sliding along filaments of actin, using the energy from the breakdown of ATP.

actual vegetation. Vegetation that actually exists at the time of observation, regardless of the character, condition and stability of its constituent species.

Aculeata. Ants, bees and wasps. Hymenoptera most of which are parasitic. Gall wasps induce the formation of galls on oak and other plants. Many Parasitica (e.g. Ichneumons) lay their eggs in the eggs, larvae or pupae of other insects, and play an important role in controlling pests, esp. Lepidoptera.

adamantine. *See* lustre.

adaptation. (*a*) Evolutionary: The fitness of a structure, function or entire organism for life in a particular environment (e.g. the webbed feet of water birds). The process, brought about by natural selection, of becoming so fitted. (*b*) Physiological: The modification of an organism in response to environmental conditions, e.g. an increase in specific enzyme production by bacteria in response to the presence of certain substrates. (*c*) Sensory: A reduction in the excitability of a sense organ which is continuously stimulated. *See* adaptive radiation.

adaptive radiation. The evolution from primitive stock of divergent forms, each adapted to survive under different conditions (e.g. on the Galápagos Islands, the

14 species of 'Darwin's finches', each with a different mode of life, must all have evolved from an ancestral species which colonised the islands from the mainland).

adaxial. (Bot.) Ventral.

adenosine di (tri) phosphate. *See* ATP.

ADH (anti-diuretic hormone). A pituitary hormone secreted by the posterior lobe of the pituitary gland which stimulates the reabsorption of water by the mammalian kidney, so reducing the volume of urine. A form of diabetes is caused by its deficiency.

adiabatic. Occurring without a gain or loss of heat by the system involved.

adiabatic lapse rate. The rate of decrease of temperature with height of a parcel of air rising without exchange of heat (by mixing or conduction) with surrounding air, but taking at each height the ambient pressure. It is deduced from the equations of state, hydrostatic equilibrium, and adiabatic change. If, in the adiabatic ascent, the parcel has the same temperature as the surroundings, no buoyancy force will result from vertical displacement and the lapse rate of temperature in the surroundings is adiabatic, and neutral. A larger, or super-adiabatic lapse rate is unstable, and a smaller one stable. The adiabatic lapse rate for unsaturated air, usually called the dry adiabatic lapse rate, and denoted by Γ, has the value $(\gamma - 1)g/\gamma R \simeq 9.86°$C km^{-1}, where γ is the ratio of the specific heats of air ($\simeq 1.4$), g is gravity and R is the gas constant for air.

adipose tissue. *See* connective tissue.

adit. In mining, a horizontal or nearly horizontal opening from the surface to the ore. Adits are used for access and for draining mines.

adjustment. The behavioural response of organisms to a change in environmental conditions.

adobe. (N. American usage) Fine rock flour deposits produced by ice abrasion of recent glaciation and transported by wind to the site of deposition. Used for brick making, so that the term has also come to mean a sun-dried brick. *See* rammed earth.

ADP. *See* ATP.

adrenal gland. Hormone-producing organ of vertebrates. In mammals there is an adrenal gland adjacent to each kidney. Each consists of a central medulla, which secretes adrenaline and the similar hormone nor-adrenaline, and an outer cortex, which secretes sex hormones (e.g. androgen), glucocorticoids (e.g. cortisone) and mineralocorticoids (e.g. aldosterone) which regulate the salt and water balance of the body. The secretion of the cortical hormones is regulated by ACTH while the central medulla is controlled by the sympathetic nervous system.

adrenalin(e) (epinephrine). Hormone secreted by the adrenal gland (medulla) and produced at many nerve endings of the sympathetic nervous system (*See* autonomic nervous system) of vertebrates. The effects of this hormone are to prepare the animal for fight or flight by increasing the rate of heart beat, dilating the blood vessels of the muscles and brain while contracting those of the skin and viscera, increasing the level of blood sugar, erecting the hair, etc. Adrenaline is also found in some invertebrates.

adreno-cortico-tropic hormone. ACTH.

adsere. That part of a sere which precedes its development into another at any time before the climax stage is reached.

adsorption. The physical or chemical bonding of molecules of gas, liquid or dissolved substance to the external surface of a solid or the internal surface if the material is porous in a very thin layer. *See* absorption.

advanced gas-cooled reactor (agr). Nuclear reactor using enriched uranium dioxide as a fuel, gaseous carbon dioxide as coolant, and graphite as a moderator. Operating temperatures are higher than those in the earlier Magnox reactor being about 675°C.

advanced waste treatment. Any process for the treatment of waste water that follows other physical, chemical or biological treatments and aims to improve the quality of effluent prior to re-use or discharge. The term often refers to the removal of nitrate and phosphate plant nutrients. *See* eutrophication, primary treatment, secondary treatment.

advection. Transport by motion of the air, water or other fluid. Advection has the same general meaning as convection, but is used particularly to refer to horizontal transport by wind of something carried by the air (e.g. pollutants, heat, fog, etc.)

advection fog. *See* fog.

adventitious. Applied to roots or buds which arise in an unusual position (e.g. roots which grow from stems, buds produced elsewhere than in the axils of flowers).

adventive (casual) plant. An introduced or alien plant growing unaided by Man, but not permanently established.

AEC. Atomic Energy Commission (US).

aeolian deposit (eolian deposit). A sediment deposited after being carried by the wind.

aeon. Eon.

aeration. Any process where a substance becomes permeated with air or another gas. It is usually applied to aqueous liquids being brought into intimate contact with air by spraying, bubbling or agitating the liquid. Refers esp. to oxygen required by fish, and other aerobic aquatic organisms.

aerator. A device for introducing air into a liquid.

aerial plankton. Spores, bacteria and other micro-organisms floating in the air.

aeroallergen. Pollen and plant dust which cause hay fever and other allergic conditions.

aerobic. Living, using or active only in the presence of oxygen or air.

aerobic respiration. Process whereby organisms, using gaseous or dissolved oxygen, release energy by the chemical breakdown of food substances. *See* respiration.

aerobiosis. The functioning of biological processes that require oxygen.

aerodynamic drag. The force required to move a solid body through air. At low speeds, when flow is laminar, the drag is viscous and proportional to the speed. When the body leaves a turbulent wake, the drag is due to the creation of kinetic energy and is proportional to the square of the speed. At speeds greater than sound, energy of shock and sound waves is the predominant form of energy created and the drag is proportional to a higher power of the speed.

aerodynamic roughness. The aerodynamic roughness of a surface depends on the size of the roughness elements which cause the retardation of the air. The roughness height corresponds to the height of the wakes of the roughness, and below it the logarithmic profile does not apply.

aerogenerator. Machine with fast-moving wind-driven rotor blades used to generate mechanical power. *See* windmill, Darrieus generator, Grandpa's Knob generator, Savonius rotor. Historically, the power was used to grind corn or pump water from wells or dykes, conversion to electrical energy is now more common.

aerosol. Dispersion of solid or liquid particles of microscopic size in gaseous media. The particles are so small that their fall speeds are small compared with the vertical component of the air motion. Haze and cloud are the commonest atmospheric aerosols, fall speeds being fractions of 1 cm per second. Aerosol is used colloquially as an abbreviation for 'aerosol spray', or for liquid 'atomisers' which produce a liquid aerosol. *See* Rayleigh scattering.

aestidurilignosa. Mixed evergreen and deciduous hardwood forest.

aestilignosa. Broadleaf deciduous forest and bush in which the trees are leafless in winter.

aestivation. (*a*) (Zool.) The dormancy of certain animals (e.g. lungfish) during the dry season, the summer, or a prolonged drought. (*b*) (Bot.) The folding of parts of a flower bud. *See* hibernation.

aetiology. The science of the cause or origin of disease.

after-blow. In the Bessemer process (now rapidly becoming obsolete) for making steel, phosphorus is removed by continuing to blow air after the carbon has been consumed. It can be a cause of pollution from steel mills.

afterburner. In incinerators, a burner located so that the combustion gases are made to pass through its flames to remove smoke and smells.

after-ripening. The period of chemical and physical change undergone by the embryos of certain seeds (e.g. Hawthorn) after they apparently are fully developed, without which they will not germinate. A similar phenomenon may occur in bulbs and tubers.

aftershocks. A series of smaller shocks following an earthquake of large magnitude and occurring fairly close to the focus of the main shock.

Ag. Silver.

agar-agar. Gelatinous substance derived from certain seaweeds, used as a bacteriological culture medium, as a thickening agent in foodstuffs, and in pharmaceutical products.

Agaricaceae. Agarics.

agarics (Agaricaceae, gill fungi). A group of basidiomycete fungi characterised by the production of spores on gills (e.g. common mushrooms and toadstools).

agate. A form of chalcedony.

Agave. Genus of American plants. *A. americana* produces large quantities of sap which, fermented, gives *pulque*, the Mexican national drink from which *mescal* is distilled. Other species are cultivated for their fibre.

ageostrophic. A form of motion in air in which the horizontal pressure gradient force is not in balance with the deviating (coriolis) force due to the wind velocity. Ageostrophic motion cannot be deduced from the pressure field, and may be caused by friction, acceleration (either linear or due to curvature) or changing pressure field. It is necessarily associated with vertical motion and the creation of clouds and weather.

agglomerate. A rock composed of fragments of volcanic material with a preponderance of pieces greater than 2 cm in diameter. Some agglomerates are bedded and were sedimented out after ejection of the fragments from a vent. Others (vent-agglomerate) are vent fillings.

agglomeration. The gathering together of particles (e.g. smoke particles in air under the influence of ultrasonic radiation). *Cf.* flocculation.

aggradation. The raising of the level of the land surface by the accumulation of deposited material.

aggregate. Construction material made from particles, most commonly of crushed stone. For road construction (*see* pavement) UK regulations require particles to have a diameter between 0.075 mm and 38 mm, depending on the use to which they are put, and be well graded (*see* sorting). For surfacing layers the aggregate must bond well with bitumen.

Agnatha. The lampreys, hagfishes and their extinct relatives. A class, or sub-phylum, of primitive, jawless, fish-like vertebrates. *See* Cyclostomata.

agonistic behaviour. Threatening or otherwise aggressive behaviour of an animal toward another of the same species.

agora. In ancient Greek cities, a public open space at the centre of the city, of no special shape, used as a market place and meeting place. It developed into a place for sporting activities and eventually into the plaza, campo, piazza, etc. that survive in many modern Latin towns.

AGR. Advanced gas-cooled reactor.

agric. Depositional (B) soil horizon containing clay and humus. Formed by cultivation. *See* soil horizon.

agricultural revolution (Neolithic revolution). The introduction of agriculture to supplement hunting and gathering in Neolithic times, around 10000 to 8000 BC. This made food much more readily available and meant that less space was needed to support a given population, so it gave rise to a great increase in the human population. The term 'agricultural revolution' may also be applied to the introduction of mechanisation, enclosure and other improvements in agriculture that occurred much more recently (18th century in England).

agronomy. The study of rural economy and husbandry.

Agung, Mount. Mount Agung.

aid. (*a*) *A*rtificial *I*nsemination by *D*onor, used widely in livestock breeding. (*b*) Assistance given by developed to developing countries in economic or technical form. US AID is the US *A*gency for *I*nternational *D*evelopment, the major US organisation concerned with aid projects in developing countries.

air conditioning. The process of bringing atmospheric or other air to specified conditions of cleanness, temperature and humidity for use in buildings. In industrial countries this is often a prodigiously energy-intensive process. In poor countries it is more usually achieved by draughts, humidifiers and wind traps.

air, conservative properties of. Conservative properties of air.

air lock. Device to give access from one region or enclosed space to another without direct contact of the fluids in those regions. Used in control of radioactive contamination and for air/water transitions.

air mass. A mass of air extending over a large distance, say 1000 km upwards, possessing distinct characteristics of temperature, humidity and lapse rate. *See* adiabatic lapse rate.

air parcel. A theoretical concept useful in atmospheric physics. A body of air whose properties and transformations may be considered by application of the laws of physics and mechanics.

air pollution. Gaseous or aerosol material in the air not considered to be a normal constituent, or excess of a normal minor constituent such as sulphur dioxide, carbon monoxide, nitrogen dioxide, dust, etc. The definition is confused by the occurrence of natural pollutants, such as methane, hydrogen sulphide in boggy areas, dust in deserts, volcanic debris, etc.

air quality. Air quality is considered high when the pollution levels are low. This may be judged aesthetically, or by reference to medical, botanical, or material damage. Standards of quality are not absolute and have cultural variations. They are also confused by possible synergistic effects (*see* synergism) or effects of humidity, wind speed, time of year, duration of incidence of the pollutant, etc., which means that a single value of the concentration of a pollutant is not a complete specification of air quality in respect of that pollutant.

air shed. A concept, invented by analogy with 'watershed', used in crude computations of air pollution concentrations to be expected from a given catalogue of pollution sources, expected air movement, and chemical rates of change.

aircraft wake. Two parallel, contra-rotating vortices trailing approximately from the wing tips, which drive one another downwards together with a body of air surrounding them, containing the engine exhaust. The intensity of the vortices corresponds to the wing downwash which supports the aircraft and is reduced with increased speed. It is greatest behind a slow moving heavy aircraft, and on the approach to landing it must be avoided by other aircraft, particularly smaller ones which might be overturned in one of the vortices.

Aitken counter. Device which causes rapid condensation of droplets in air by rapid adiabatic expansion of it. It makes possible the counting of the number of condensation nuclei present in a unit volume by means of a microscope. The Aitken nuclei are those which produce a droplet during the expansion and correspond to those naturally available for cloud formation in rising air or fog. In a slow expansion only some of the Aitken nuclei form droplets. Named after the physicist John Aitken, of Edinburgh.

aitogenic. paratonic.

Al. aluminium.

alabaster. Massive, fine-grained gypsum used for carvings and vessels.

Alamogordo cattle. A herd of 32 Hereford cattle that was exposed to fallout, probably 10 to 15 miles northeast of ground zero, from the first atomic bomb test at Alamogordo, New Mexico, in July, 1945. Subsequently the animals were moved to Oak Ridge, Tennessee, where the effects of radiation were studied. The animals developed patches of grey hair on their backs and sides, hornlike growths apparently no deeper than the dermis but which took 2 to 3 times longer to heal when wounded, and 3 cows developed skin carcinoma. Their general health and reproductive efficiency were otherwise no different from those of 74 control cattle.

alanine. Amino acid with the formula $CH_3CH.(NH_2).COOH$. Molecular weight 89.1.

Albatross. Family of oceanic birds (Diomedeidae) which engages in dynamic soaring, exploiting the wind gradients near the sea surface and over wave crests to remain airborne without flapping flight. Since this generally leads to downwind drift, the largest albatrosses live in the southern hemisphere, where the pole may be circumscribed repeatedly without land obstacles. Albatrosses live by scavenging surface material and breed on steep oceanic islands. The Wandering Albatross (*Diomedea exulans*) which is the largest of all ocean birds, with a mature wingspan of more than 3 metres, has difficulty in getting airborne in the absence of wind.

albedo. The whiteness, or degree of reflection of incident light from an object or material. Smooth-surfaced materials often have a high albedo, reflecting most of the incident light. Snow and cloud surfaces have a high albedo, near to unity because most of the energy of the visible solar spectrum is reflected or scattered. Vegetation and the sea have a low albedo, nearer to zero, because they absorb a large fraction of the energy. Cloud is the chief cause of variations in the Earth's average albedo, which varies around a value of one-half.

Albian. A stage of the Cretaceous System.

albic. Applied to soil horizons from which clay and free iron oxides have been removed.

albinism. The lack of normal skin pigments caused by the absence of the enzyme tyrosinase, often in mammals due to an autosomal recessive gene. In a pure albino the eye appears pinkish. Most bird albinos are partial, with white patches on the plumage.

Alcedinidae (kingfishers). A family of brightly coloured, hole-nesting birds which dive for fish. They have large heads, short tails and long, sharp beaks. Along with the Hoopoe and Bee-eaters, they belong to the mainly tropical order Coraciiformes.

Alcidae. Auks, razorbills, guillemots, puffins, etc. Marine, fish-eating birds with webbed feet, short wings and heavy bodies. They come ashore only to breed, and swim under water using their wings. They often suffer heavy casualties after oil spillages.

alcoholic fermentation. The fermentation of ethyl alcohol from sugar (*see* carbohydrate) by yeasts.

aldehydes. Class of chemical compounds, mostly colourless, volatile and with suffocating vapours, intermediate between acids and alcohols.

aldrin. Persistent organochlorine pesticide similar to dieldrin. Its use in the UK is now very restricted. It is effective against many insects but poisonous to vertebrates. *See* cyclodiene insecticides.

aleurone grains. Large granules of protein found in the food storage organs of plants, notably in seeds.

Aleutian Low. A low pressure area in the neighbourhood of the Aleutian Islands (N. Pacific) which is evident on long-term-average pressure charts on account of the high frequency with which cyclones occur in that area.

alfisols. *See* soil classification.

Algae (pl.), (sing. Alga). Large group of plants which contain chlorophyll and which have a relatively simple body construction varying from a single cell to a multicellular, ribbon-like thallus. They are either aquatic (e.g. diatoms, seaweed) or live in damp places (e.g. the green stain on damp walls, etc.). Planktonic algae form the bases of marine and freshwater food chains. The bacterial culture medium agar-agar is extracted from seaweeds. *See* bacteria, Chlorophyta, Chrysophyta, Bacillariophyta, Charophyta, Cryptophyta, Cyanophyta, Euglenophyta, Phaeophyta, Pyrrophyta, Rhodophyta, Xanthophyta.

algal bloom (phytoplankton bloom). A sudden proliferation of microscopic algae in water bodies, stimulated by the input of nutrients (e.g. phosphates). *See* eutrophication.

aliphatic. Applied to organic compounds containing open chains of carbon atoms, and comprising the paraffins, olefins and acetylenes, their derivatives and substitution products. *Cf.* aromatic.

alkalic. Alkaline.

alkali-feldspar. *See* feldspar.

Alkali Inspectorate. The official body in Britain most concerned with the control of air pollution from industrial sources. Formed in 1880, in 1971 the name was changed to the Inspectorate of Alkali &c. Works.

alkaline (alkalic). (*a*) *See* pH. (*b*) (Geol.) Applied to igneous rocks containing an abundance of the alkaline metals, indicated by the presence of such minerals as sodium-rich pyroxenes and amphiboles and sodium and/or potassium-rich feldspars. In petrology, the opposite of alkaline is calc-alkaline, not acidic. Alkaline is also used to refer to igneous rocks rich in feldspathoids and to

17

minerals rich in sodium and/or potassium (e.g. alkali-feldspars).

alkalised alumina process. British process for removing sulphur dioxide from flue gases, using pellets of sodium aluminate, to produce sodium sulphite and sulphate. The pellets can be regenerated by heating with hydrogen, giving hydrogen sulphide from which the sulphur can be recovered. The process was invented in 1968.

alkaloids. Complex, basic organic compounds containing nitrogen, found in certain families of flowering plants (e.g. Ranunculaceae, Scophuloariaceae, Solanaceae, Umbelliferae). Important for their poisonous and medicinal qualities (e.g. morphine from the fruit of the opium poppy, nicotine from tobacco leaves, atropine from deadly nightshade, quinine from the bark of *Cinchona* spp., strychnine from the seeds of *Strychnos nux-vomica*) (*see* Strychnos).

alkylmercury. Highly toxic mercurial compound used as a fungicide and seed dressing. It is the cause of many accidental deaths and permanent injuries resulting from the consumption of dressed grain.

alkyl sulphonates. Surface-acting agents used in synthetic detergents. Those with a linear molecular structure are degraded fairly readily by micro-organisms. Those with a branched structure (e.g. alkyl benzene sulphonate) are stable and resist biodegradation, causing foaming in water into which they are discharged.

allantois. One of the embryonic membranes of Amniota. It is a large outgrowth from the hind gut of the embryo and is covered in a network of blood vessels. In reptiles and birds excretory products are stored in the allantoic cavity, and the allantoic blood vessels are responsible for gaseous exchange as they lie close to the shell. In mammals these blood vessels supply the placenta.

alleles. Allelomorphs.

allelochemistry. The chemistry of substances released by one population that affect another population.

allelomorphic genes. Allelomorphs.

allelomorphs (allelomorphic genes, alleles). The alternative forms of a gene, which lie at the same locus (relative position) on homologous chromosomes, and which pair during meiosis. They behave in a Mendelian manner, producing different effects on the same developmental process. *See* Mendel.

allelopathy. The influence which one living plant exerts upon another via chemical exudates.

Allen's rule. As the mean temperature of the environment decreases (e.g. with increase in latitude) the relative size of the appendages (e.g. limbs, tails, ears) of warm-blooded animals tends to decrease. This is correlated with the increased need to conserve heat.

allochthonous. (*a*) Applied to material which originated elsewhere (e.g. drifted plant debris on the bottom of a lake). (*b*) (Geol.) Applied to rocks whose major constituents have not been formed *in situ*. Contrasted with autochthonous.

allogeneic (allogenic). Having different genes. *cf.* isogenic.

allogenic. (*a*) (Geol.) Refers to those constituents of a rock which were formed previously and separate from the rock (e.g. the pebbles in a conglomerate). Opposite of authigenic (*b*) (Biol.) Allogeneic.

allogenic succession. A succession which takes place because of changes in the environment not produced by the plants themselves, but by external factors (e.g. alterations in drainage or climate). *Cf.* Autogenic succession.

allopatric. Refers to different species or subspecies whose areas of distribution do not overlap. *Cf.* sympatric.

allopolyploid. Organism with more than two haploid sets of chromosomes derived from two or more species. Cultivated wheats are probably allopolyploids.

allotetraploid (amphidiploid). Plant species produced when a diploid hybrid doubles its chromosome number. Unlike most diploid hybrids, allotetraploids are fertile. Artificially induced allotetraploidy is an important means of producing new varieties in agriculture. *See* endomitosis.

allotropy. The existence of an element in two or more physical forms (e.g. carbon as diamond and graphite; red and white phosphorous).

alloy. Metallic substance prepared by adding other metals to a major metallic ingredient to obtain certain required properties (e.g. bronze, brass, solder). Often used metaphorically for intimate non-metallic mixtures.

alloy steel. Any special type of steel made by adding other elements (e.g. chromium, nickel or cobalt) to ordinary carbon steel. Many steels with properties of corrosion resistance, great hardness, specific magnetic characteristics, etc. have been produced.

alluvium. Sediment transported and deposited by flowing water. *See* flood plain.

almanac. A catalogue of predictions. Some contain reliably computed astronomical predictions and tidal tables, others contain guesses with a variety of bases in experience, yet others are founded in fancy.

alpha. α

alpha bronze. A solid solution of tin in copper, the former usually forming 4 to 5%. Unlike phosphor bronze and gun metal, it is workable and is used for springs, turbine blades, coins, etc.

alpha diversity (niche diversification). Diversity resulting from competition between species that reduces the variation within particular species as they become more precisely adapted to the niches they occupy.

alpha particles. Heavy particles ($_2^4$He^{24}) produced by radioactive decay (e.g. decay of plutonium) with little penetrative power, but damaging when in contact with living tissue.

Alpine. (*a*) The parts of a mountain above the tree line, but below permanent snow. The plants and animals living there. (*b*) (Geol.) Refers to processes or products associated with the Alpine orogeny, which can be considered to span the time from the Triassic to the present.

Alpine orogeny. The orogeny (mountain-building) that culminated in the Miocene Period, forming the Alpine-Himalayan belt.

alternation of generations. The alternation of sexual and asexual reproduction in the life cycle of an organism. True alternation of generations occurs in plants which have a haploid gamete producing generation (gametophyte) alternating with a diploid, spore-producing generation (sporophyte). The gametophyte is dominant in mosses, liverworts and some algae, but in ferns it is an insignificant thallus, the familiar fern plant being the sporophyte. In flowering plants the gametophyte is microscopic, forming part of the pollen grain or ovule. What has been called a form of alternation of generations occurs in some animals (e.g. tapeworms, jellyfish) but in this case both generations are diploid.

alternative technology. Technology that aims to use resources sparingly and to do the minimum damage to the environment or to species inhabiting it, while permitting the greatest possible degree of personal control over the technology on the part of its user.

altimeter. An aneroid barometer calibrated to read height above some reference level whose barometric pressure (QFE) is set on the altimeter dial.

altitude. (*a*) Height above some reference level (e.g. ordnance datum, mean sea level) which is equipotential in the Earth's gravitational field. (*b*) In astronomy, a measure of the angular elevation above the horizon.

altocumulus. Cumulus clouds generated by upcurrents not having their origin at the ground or sea, but at some higher level. The cloud elements thus appear smaller from the ground than cumulus originating at the ground. Altocumulus is typically dappled.

altostratus. A high level layer of cloud not having dappled or fibrous features but of uniform appearance over a large part of the sky. Typically altostratus appears before and in the rain of a warm front of a cyclone.

altrices. Birds which are hatched naked or nearly so, and at first are unable to look after themselves (e.g. pigeon).

altruistic. Applied to traits selected for their contribution to the survival of the group rather than of the individual.

alumina. Aluminium trioxide (Al_2O_3).

aluminium (aluminum) (Al). A very light, ductile, metallic element used in many industries. Produced from bauxite by an energy-intensive electrolytic process, taking more energy to produce than steel from ironores. It is very resistant to corrosion due to the monomolecular layer of oxide formed on exposure to air and water. AW26.9815; At.No.13; S.G. 2.7; MP.659.7°C.

aluminous cement (high alumina cement). Portland cement containing 30 to 50% alumina, 30 to 50% lime and not more than 30 to 50% of iron oxide, silica, and other ingredients. It hardens rapidly and resists sea water. After extensive use in Britain in the 1960s it was found to deteriorate with time and has caused the collapse of a few large buildings, and led, in certain situations, to the closing of others pending investigation.

aluminum (Al). aluminium.

alveoli. The innumerable, small sacs in the lung through whose walls oxygen enters the bloodstream and carbon dioxide leaves. *See* bronchial diseases.

amalgam. A solution of a metal in mercury.

Amaranthus. Genus of temperate and tropical plants, some used as pot herbs, others as cereal grains in Asia.

Ambari hemp. Kenaf.

amber. Fossil resin. Amber is known from the Cretaceous onwards and often contains excellently preserved insects with only their soft inner tissues missing.

ambergris. A secretion of the sperm whale, at one time used extensively in perfumery, but now scarce as whale stocks decline. It has been largely discarded by the cosmetic industry.

ambient. Surrounding a phenomenon, present before a phenomenon. Ambient temperature is the temperature of the surrounding air; ambient turbulence is the eddy motion present not caused by the phenomenon, and is a cause of bumpiness or dispersion of pollution.

ambient noise. The background, or prevailing noise.

ambient quality standards. Environmental quality standards.

amenity. An abstract concept expressing those natural or man-made qualities of the environment from which Man derives pleasure, enrichment and satisfaction.

amensalism. The opposite of commensalism.

ametabola. Apterygota.

amethyst. Quartz made purple by slight impurities.

ametoecious (monoxenic) parasite. *See* parasitism.

amines. Compounds formed by replacing hydrogen atoms of ammonia (NH_3) by organic radicals (R). Amines are classified according to the number of hydrogen atoms so replaced, as primary (NH_2R), secondary (NHR_2) and tertiary (NR_3). The release of nitrites into the environment (either as nitrites or as nitrates reduced naturally by bacteria to nitrites) may lead to the random reaction with secondary amines to form nitrosamines.

amino acid. Organic compound containing a carboxyl group (COOH) and an amino group (NH_2) (e.g. glycine, CH_2NH_2COOH). About 30 amino acids are known. They are fundamental constituents of living matter because protein molecules are made up of many amino acid molecules combined together. Amino acids are synthesised by green plants and some bacteria, but some (e.g. phenylalanine) cannot be synthesised by animals and therefore are essential constituents of their diets. Proteins from specific plants may lack certain amino acids, so a vegetarian diet must include a wide range of plant products.

ammonia (NH_3). Liquid or gas used directly as a fertilizer or as the basis of ammonium (NH_4) fertilizer compounds. Used industrially in many processes, and as a refrigerant. It is produced naturally by the decomposition of animal urine. Its irritant vapours are immediately recognisable. *See* coal gas, Haber process.

ammonia process. Japanese process for removing sulphur dioxide from flue gases, using ammonia as the reagent in a wet scrubber to produce ammonium sulphate.

ammonia scrubbing Hixon. US process for removing sulphur dioxide from flue gases using a scrubbing tower in which sulphur dioxide is absorbed by an aqueous stream of ammonium hydroxide, bisulphite and sulphite, makeup ammonia being added at the inlet. The effluent goes from the scrubber to a separator where fly ash is removed and then to an acidulator, where ammonium sulphite is reacted with ammonium bisulphite to form sulphur dioxide and ammonium sulphate. The sulphur dioxide is recoverved and the ammonium sulphate decomposed thermally to ammonia and ammonium bisulphite, which are recycled.

ammonification. The biochemical process which produces ammoniacal nitrogen from organic compounds. *See* nitrification.

ammonifying bacteria. Those nitrifying bacteria (e.g. *Pseudomonas*) which break down the protein in the dead bodies of plants and animals into ammonia. *See* nitrogen cycle.

ammonium nitrate. Used as a fertilizer and as a component of explosives, weedkillers and insecticides. The main source of nitrate run-off from farm land resulting from excessive fertilizer use. In water it can be a pollutant and may cause eutrophication.

Ammonites. Group of extinct cephalopod molluscs. They were marine Mesozoic animals, similar to the present-day Pearly Nautilus, with shells curled in one plane and divided by internal septa.

Ammophila. Genus of grass of which *A. arenaria* (Marram grass) is often used to stabilize blowing sands. *See* sand dune.

amnion. Fluid-filled sac which is an outgrowth of the embryo of a reptile, bird or mammal. The amnion envelops the embryo and provides it with the aquatic environment necessary for its development.

Amniota. Reptiles, birds and mammals, the land-living vertebrate classes whose embryos develop an amnion.

Amoeba. Genus of minute, acellular animals (Protozoa) which move and ingest food by means of pseudopoda and thus are continually changing shape. Members of the genus *Amoeba* are free living, but some closely related genera are commensal or parasitic (e.g. *Entamoeba histolytica*, which causes amoebic dysentry). *See* parasitism.

amoebocytes. Cells which move by means of pseudopoda and are found in the tissues and body fluids of vertebrates and invertebrates. Many are phagocytic (e.g. some types of white blood cell in vertebrates).

amorphous. (Mineral) Noncrystalline.

ampere (a). The SI unit of electric current, defined in 1948 as the intensity of a constant current which, if maintained in two parallel, rectilinear conductors of infinite length, of negligible circular section and positioned one metre apart in vacuo, will produce between the conductors a force equal to 2×10^{-7} newton per metre of length.

Amphibia. Class of vertebrates including newts, salamanders, frogs and toads, which, during their evolution, have succeeded only partly in conquering the difficulties of life on land. The adults are aquatic or live in damp situations, and have moist, smooth skins, often used as well as lungs for respiratory purposes. Apart from a few viviparous species, most have to return to the water to breed, because the eggs have no shell to prevent dessication. The larvae have gills. The earliest (Carboniferous) amphibia resembled osteolepid fishes, from which they differed in having pentadactyl limbs instead of fins, and a middle ear apparatus. *See* respiration, Anura, Apoda, Urodela, Labyrinthodontia.

amphibole. A group of rock-forming silicate minerals with a very wide range of compositions. The general formula can be given as $X_{2-3}Y_5Z_8O_{22}(OH)_2$, where X is usually calcium, sodium or potassium, Y is usually magnesium, iron or aluminium, and Z is silicon or aluminium (with a maximum of 2Al). Hornblende, one of the amphiboles, has a range of compositions which includes the typical $Ca_2Mg_4Al Si_7Al O_{11}(OH)_2$. Amphiboles occur in igneous rocks, being characteristic of those of intermediate character, and in metamorphic rocks (*See* metamorphism).

amphidiploid. Allotetraploid.

amphimixis. True sexual reproduction, involving fertilisation. *Cf.* apomixis.

Amphineura. A class of marine Mollusca. The most familiar members of the group (chitons, coat-of-mail shells) are limpet-like, bearing eight calcareous plates, and live on seaweed in the intertidal zone.

amphioxus (lancelet). Small, rather eel-like animals, of the genera *Branchiostoma* and *Asymmetron* which live in clean sand in shallow water where there is a good flow of current. Amphioxus exhibit primitive chordate (*see* Chordata) characteristics, but although their general structure is related to that of vertebrates, they lack a true head and backbone. Thus any theory concerning the origin of the Vertebrata must rest heavily on the true nature of Amphioxus. The term Amphioxus was formerly confined to the genus *Branchiostoma*, but it is used now in respect of both genera. *See* Acrania.

Amphiphyte. Amphibious plants (e.g. Amphibious Bistort, *Polygonum amphibium*). The aquatic and terrestrial forms often differ considerably in vegetative features.

Amphipoda. Order of Crustacea including the freshwater shrimps and sandhoppers. Most are marine and have laterally compressed bodies. *See* Malacostraca.

amplitude. The peak, or maximum value in wave-type transmission.

amygdale. (Geol.) Vesicle.

amylases (diastases). Enzymes, widely distributed in animals and plants, which break down starch or glycogen to dextrin, maltose or glucose (e.g. ptyalin in human saliva).

Anabaena. Blue-green alga that is able to fix atmospheric nitrogen. Forms a symbiotic (*see* symbiosis) relationship with *Azolla* forms.

anabatic wind. An upslope wind generated when a slope is warmed (e.g. by sunshine). It is local and unstable.

anabiont. A perennial plant which fruits many times.

anabiosis. Resuscitation after apparent death (e.g. after dessication).

anabolism. Synthesis of complex molecules from simpler ones. *Cf.* catabolism.

anadromy. The migration of some fish (e.g. salmon) from the sea into freshwater for breeding.

anaerobic. Living or active only in the absence of oxygen, or air, dissolved in the medium (e.g. water) or otherwise available. Anaerobic bacteria can obtain oxygen from nitrates or sulphates, releasing nitrogen, nitrous oxide, ammonium or hydrogen sulphide, provided oxidisable material (e.g. vegetable waste, methanol, or other carbon compound) is available.

anaerobic biological treatment. Any treatment utilising anaerobic organisms to reduce organic matter in wastes.

anaerobic decomposition. Organic breakdown in the absence of air. Commonly applied to the digesting of sewage producing methane.

anaerobic respiration. Process whereby organisms release energy by the chemical breakdown of food substances, without consuming gaseous or dissolved oxygen. Examples are the breakdown of glycogen to lactic acid in the muscles of vertebrates, and the breakdown of sugar to ethyl alcohol and carbon dioxide by yeast (fermentation). Anaerobic respiration is less efficient than aerobic respiration Many organisms are able to respire anaerobically when the oxygen supply is insufficient for aerobic respiration, but some (e.g. a few bacteria) never use free oxygen. *See* respiration.

analogous. (Zool.) An organ of one species is analogous to that of another species when both organs have the same function (e.g. the wings of a bat and an insect are analogous, but this does not indicate a close evolutionary relationship because these structures did not originate in the same way). *Cf.* homologous.

analogue. (*a*) A physical model in which the properties of the original are replaced by the consistent, but physically different, properties of the model (e.g. using an electric spark travelling along a length of gunpowder fuse as an analogue of an impulse travelling along a nerve). *See* analogue computer. (*b*) (Meteor.) Analogue methods make predictions on the basis of the study of other (theoretical or actual) systems having analogous properties. In weather forecasting, the analogue could be a similar situation taken from historical records. (*c*) (Eng.) A model of a mechanical or electrical system whose main properties are governed by the same kind of mathematical equation. Thus waves in a channel are analogous to sound waves in a pipe, but are more readily studied visually.

analogue computer. A device used in modelling (see Model) that represents selected time-varying properties of the original by a set of time-varying voltages, the correspondence between original and model being dictated by a mathematical representation common to both. *See* digital computer.

ana-front. A front at which the air on one side is ascending. Used particularly for fronts at which the cold air is ascending, because this is unusual.

Anarcardiaceae. Family of tropical plants often grown for their fruits (*Anarcardium occidentale* is cashew; *Mangifera indica* is mango). *See* Rhus.

Anatidae (waterfowl). Swans, geese and ducks. Aquatic birds with webbed feet, long necks and broad, flat beaks edged with horny lamellae. The family Anatidae belongs to the order Anseriformes.

ancient monument. In Britain, any structure of historic, architectural, traditional, artistic or archaeological value (other than an ecclesiastical building or a building which is for the time being used for ecclesiastical purposes) which, in the public interest, is listed by the Minister of Public Building and Works as an 'Ancient Monument'.

ancient soil. An older, weathered zone lying beneath present surface deposits, a product of past processes and climates.

andesite. A fine-grained intermediate volcanic rock associated with continental crust. Andesites are the volcanic equivalent of diorite.

androecium. *See* flower.

androgen. Any male sex hormone (e.g. testosterone, produced by the interstitial cells of the testis). Natural androgens are steroids, responsible for the initiation and maintenance of many male sexual characteristics.

anechoic. Applied to surfaces that are almost totally sound absorbent (*see* absorption coefficent). An anechoic chamber gives almost free field conditions.

anemochore. A plant whose seeds, spores or other reproductive structures are dispersed by wind (e.g. thistles).

anemograph. *See* anemometer.

anemometer. Wind measuring instrument, usually employing a pitot tube directed by a vane or a rotor carrying three or four hemispherical cups, or a pressure plate deflected against a spring or gravity. An anemograph makes a record of the wind. *See* panemone.

anemophily. Wind pollination.

aneuploid. Applied to an organism which is genetically unbalanced because its chromosome number is not a multiple of the basic haploid number. *Cf.* euploid, polyploid, nullisomic, trisomic.

aneurin. Thiamin.

Angiospermae. The flowering plants, characterised by the possession of flowers, although these may be (e.g. grasses) relatively inconspicuous structures. The ovules, which become seeds after fertilisation, are enclosed in ovaries. The xylem contains true vessels. Angiospermae are divided into two classes, Monocotyled-oneae and Dicotyledoneae. *See* Spermophyta, Gymnospermae.

angle of indraught. The angle of inclination of the direction of a steady wind to the isobars called positive when the direction is towards low pressure. Angle of indraught is a measure of departure of the surface wind from geostrophic and is usually towards low pressure, at a greater angle over land than over sea because of greater friction. The angle may be zero or negative over the sea when there is a strong thermal wind.

angle of rest. The steepest angle, measured from the horizontal, at which a fragmental material remains stable. *See* talus.

Ångström (Å). Measure equivalent to one ten-millionth of a millimetre (mm^{-7}), or one-tenth of a nanometre.

angular frequency. (Acoustics). The frequency of sound expressed in radians per second, obtained by multiplying the frequency in Hertz by 2π.

angular momentum. The same as moment of momentum, therefore measured about a point or axis. Angular momentum is changed only by moment of forces. Since the angular momentum of a particle is proportional to distance from the axis, motion of air towards the pole causes the velocity of the air in the direction of the Earth's rotation (eastwards) to increase unless decreased by friction or movement towards higher pressure. Similarly, equator-ward displacement increases distance from the Earth's axis and so decreases eastward velocity. This is manifested in a wind from the east in tropical latitudes (Tradewinds).

angular unconformity. (Geol.) An unconformity marked by angular divergence between the older and younger rocks.

anhydrite. An evaporite mineral, calcium sulphate, $CaSO_4$, found in sedimentary rocks.

anhydrous. Applied to oxides, salts and other compounds to indicate absence of water of crystallisation or combined water. More generally, means containing no water.

animal. A living organism characterised chiefly by holozoic mode of feeding. Unlike most plants, most animals possess powers of locomotion and have compact bodies. Their cells lack chlorophyll and cell walls. The distinction between plant and animal kingdoms, which are now often regarded as part of a third kingdom, the Protista, becomes blurred in some simple organisms (e.g. Protozoa, slime moulds , unicellular algae).

animal unit. A unit of livestock based on the equivalent of a mature cow (approx. 1000 lb (454 kg) live weight). One cow is considered equal to one horse, one mule, five sheep, five swine or six goats.

anion. Negatively charged ion. In an electrolytic process, a potential gradient will cause anions to migrate towards the anode, the electrode at which oxidation occurs. The other (negative) electrode is the cathode.

anisogamy. heterogamy.

anisotropic. Having different optical or other physical properties in different directions (e.g. quartz, graphite.)

ankerite. A carbonate mineral with the formula $Ca(Mg, Fe) (CO_3)_2$.

Annelida (Annulata). A phylum of animals comprising the ringed or metamerically segmented worms. They possess a coelom, well developed blood and nervous systems, nephridia and, typically, bristles (chaetae). The two main classes are the Chaetopoda,which includes earthworms and ragworms, and the leeches (Hirudinea). In addition, there is a small class of marine worms, the Archiannelida, whose members usually lack bristles, also two other classes of marine worms the Echiuroidea and Sipunculoidea, which are often included in the phylum, although their members show few signs of segmentation.

annual. A plant which completes its life cycle from seed to seed in a single year. *Cf.* biennial, perennial, ephemeral.

annual cycle. One of the two cycles dominating the terrestrial environment and determined by obvious celestial mechanisms, in this case emphasised as a result of the inclination of the Earth's axis to the normal to the ecliptic.

annual rings (growth rings). Concentric rings of secondary wood evident in a cross section of the stem of a woody plant. The marked difference between the dense, small-celled late wood of one season and the wide-celled early wood of the next enables the age of a trunk to be estimated.

Annulata. Annelida.

anode. *See* anion.

anodising. The formation of a thin film of alumina on the surface of aluminium by making the treated object the anode in an appropriate acid electrolyte. This improves resistance to corrosion.

anomalous audibility. Due to wave fronts from a source of sound on the ground refracted at warm layers, usually above about 30 km in the air, where the increased velocity of sound refracts the rays downwards. Because of absorption at the ground and the upward refraction of rays due to a decrease of temperature upwards and/or wind shear, even a very loud source of sound is inaudible at 20 or 30 km distance, but may be audible again in the zone of anomalous audibility 100 to 150 km distant.

anomaly. Departure from some defined value (e.g. ambient or normal). Atmospheric anomalies include the temperature anomaly of rising warm air, and anomalous visibility due to refraction (superior image).

Anoplura (Siphunculata). Sucking lice. Wingless, parasitic insects (*Exopterygota*) with flattened bodies, which live on mammalian blood. The human body louse, *Pediculus humanus* is a vector of typhus. *See* parasitism, mallophaga.

anorthosite. A coarse-grained igneous rock, usually Precambrian, composed almost entirely of plagioclase feldspar.

anoxia (hypoxia). Deficiency of oxygen in tissues, in blood, or in a body of water.

antagonism. (*a*) Opposing effects produced by drugs, hormones, etc. on living systems (e.g. adrenalin accelerates heart beat while acetylcholine slows it. (*b*) Adverse effect of one organism on the growth of another. The interference may be due to the production of antibiotic substances or to competition for food, etc. (*c*) Action of opposing pairs or sets of muscles (e.g. biceps and triceps) one of which must contract while the other relaxes.

antarctic. The continent and nearby seas surrounding the South Pole. Living organisms associated therewith.

antecedent drainage. A drainage system maintaining its original direction across a line of localised uplift.

anthelion. A bright spot of white light at the same altitude as the sun, but at opposite azimuth, produced by reflection and refraction in hexagonal ice prisms in the air having their axes vertical.

anthelminthics. Drugs used against parasitic worms (Helminths).

anther. *See* flower.

antheridium. The male sex organ of Algae, Fungi, Bryophyta and Pteridophyta, which produces spermatozoids.

antherozoid. spermatozoid.

30

Anthocerotae. *See* Bryophyta.

anthocyanins. Glycoside pigments, dissolved in cell sap, which impart colour to petals, autumn leaves, fruits, etc. Their colour depends on the pH of the cell sap, being red in acidic and blue or violet in alkaline conditions.

Anthozoa. Actinozoa.

anthracene. Aromatic hydrocarbon obtained during coal distillation used as an intermediate in the production of dyestuffs.

anthracite. A coal of high rank with a vitreous lustre.

anthrax (splenic fever, malignant pustule, woolsorters' disease). An acute, specific, infectious, febrile disease of mammals (including Man) caused by *Bacillus anthracis*, whose spores may persist in virulent form for many years in soil or other contaminated material. It occurs throughout the world and has caused major epidemics in the past. It is transmitted to Man most commonly from infected wool or carcases of diseased herbivores, and may begin as a localised skin infection which can lead to fatal septicaemia if it becomes generalised, or as a pulmonary infection (woolsorters' disease) caused by the inhalation of spores, which proceeds rapidly and terminates fatally. It may also occur as an intestinal infection following the ingestion of contaminated food. The spores can be transmitted aerially under certain conditions, and the disease is one that has been considered as a weapon of war.

anthropic. Applied to surface soil layers. Similar to mollic, but with high phosphate content derived from long-term cultivation. *See* soil horizons plaggen, agric.

anthropic zone. The area of the world's surface entirely transformed or transformable by Man for his various activities.

anthropochore. Plant introduced to an area by Man.

Anthropoidea (Pithecoidea). A sub-order of primates, including Man, monkeys and apes (gibbons, and Orang-utan, Gorilla, Chimpanzee), characterised by the possession of large brains, forward-facing eyes and a high degree of manual dexterity.

antibiosis. A special case of antagonism in which an organism produces a substance injurious to other organisms. *See* antibiotic.

antibiotic. Substance given out by one species which has an antagonistic (*see* antagonism) effect on organisms belonging to different species. Some antibiotics

31

are of outstanding importance in medicine (e.g. penicillin, produced by the mould *Penicillium notatum*, inhibits the growth of many pathogenic bacteria, and streptomycin, produced by *Streptomyces griseus*, is effective against the tuberculosis bacillus, which is not affected by penicillin).

antibody. Protein produced by an animal in response to the presence in its tissues of a foreign substance, or antigen, usually a protein or carbohydrate. Thus, the presence of a parasite or its toxins will stimulate the formation of specific antibodies which will kill or damage the parasite or neutralise its toxins. Natural immunity to diseases occurs because antibodies remain in the blood after the parasite has disappeared.

anticline. A fold that is convex upwards. *Cf.* syncline.

anticlinorium. (Geol.) A compound anticline in which the limbs are folded. The term is used for large-scale structures.

anticyclone. A centre of high atmospheric sea level pressure. The air circulates in the direction relative to the Earth opposite to the vertical component of the Earth's rotation (anticyclonically). The circulation is cyclonic in space but slower than the Earth's rotation. In a cold anticyclone, the high pressure is due to cold air in the lowest 2 to 5 km, but in a warm anticyclone it is due to cold air at greater altitudes, usually above the tropopause.

anti-diuretic hormone. ADH.

Anseriformes. The geese, a sub-family of the Anatidae, whose most representative member is the Greylag Goose (*Anser anser*) from which the domestic goose is derived. The genus *Anser* comprises the grey geese (Bean Goose, *A. fabalis*; Pink-footed Goose, *A.f. brachyrhynchus*; and White-fronted Goose, *A. albifrons*). Other members of the sub-family are black geese (e.g. Barnacle Goose, *Branta leucopsis* and Canada Goose, *B. canadensis*). The Hawaiian Goose, or Nene (*B. sandvicensis*) became almost extinct but was saved by the breeding and rearing of captive birds, notably at the *Wildfowl Trust*.

antigen. Substance which stimulates the production of an antibody.

antigibberellin. Organic compound causing stunted growth in plants, or having opposite effects to those of the gibberellin auxins. Maleic hydrazide is an example.

antiknock additives. Compounds added to petroleum to prevent pre-ignition (knocking or pinking) in internal combustion engines. The most widely used is tetraethyl lead.

antistatic agents. Compounds which give sufficient conductivity to normally non-conducting surfaces to prevent the build-up of charges of static electricity.

antimetabolite. Substance that inhibits the action of a metabolite. Vitamins, amino acids, some proteins and other substances necessary to living organisms can be prevented from performing their metabolic function by an interaction between an antimetabolite and an enzyme. Antimetabolites are often related structurally to the nutrients or metabolites they inhibit.

antimony (Sb). Element whose compounds are used in metal used in letterpress printing, dyestuffs and lead-acid batteries. Prolonged exposure can lead to heart disease and dermatitis. AW 121.75; At. No. 51; SG 6.69; mp 630°C.

anti-natalist. Term applied to views or policies designed to reduce birth rates or to persons holding such views. Pro-natalist views or policies aim to increase the birth rate and so achieve an increase in human numbers.

antinode. (Acoustics). The point, line or surface where the amplitude of sound is at its maximum. The opposite of node, where a sound wave has an amplitude of zero.

anti-solar point. The point opposite the sun, in the direction of the observer's shadow.

antithrombin (heparin). *See* clotting of blood.

antitrades. Winds that blow above the tradewinds from a westerly point.

antitriptic wind. A wind blowing against significant drag at the surface, comparable with the force due to the pressure gradient, and opposite in direction to the wind. Examples are sea breezes early in the day, and winds in canyons blowing towards the lower pressure end, or katabatic winds, esp. in ravines.

antlers. Horns of deer which are equivalent to the horns of other members of the Ruminantia. In most species antlers are grown only by the male, but in Reindeer they are grown by both sexes. They grow as bony processes from the frontal bones of the skull which reach their full size quickly. Antlers are shed and grown anew each year, usually adding more branches at each regrowth. Antlers are employed in ritualized mating contests between rival stags, but not for combat with other species, when the forehooves are used. Reindeer used the widened part of their antlers to shovel snow.

Ant, White. Isoptera.

Anura (Salientia). Frogs and toads. An order of Amphibia characterized by the absence of a tail and the possession of long hind legs for jumping.

AONB. Area of Outstanding Natural Beauty.

apatetic coloration. Protective coloration which misleads an animal's predators, for instance by masking the outline (disruptive coloration) or producing the appearance of eyes where there are none.

apatite. A group of minerals, $Ca_5(OH, F, Cl)(PO_4)_3$, found in igneous rocks, metamorphosed (see Metamorphism) limestones and which is the main source of phosphate. Bones and teeth are formed from varieties of apatite.

apetalous. Having no petals (e.g. the flower of Marsh Marigold, *Caltha palustris*).

aphagia. Failure to eat, that can lead to death and can necessitate forced feeding in animals. Voracious eating, often of edible or inedible substances, proceeding past the point of normal satiation, is hyperphagia.

Aphaniptera (Siphonaptera). Fleas. An order of wingless, blood-sucking insects (*Endopterygota*) parasitic on mammals and birds. The adults have laterally compressed bodies and strong legs for jumping, and the larvae are grub-like. Most are specific to one or a few hosts. Some rat fleas transmit the plague bacillus from rat to Man, and the rabbit flea is the vector of myxomatosis. *See* parasitism.

aphanitic. Refers to an igneous rock in which the individual crystals of the ground-mass cannot be distinguished by the naked eye.

Aphidae. Aphididae.

Aphididae (N. American Aphidae). The aphids (Greenfly, etc.), members of the insect order Hemiptera (bugs). Their mouthparts are adapted for piercing and sucking plants. They are notable for wide distribution and prolific reproduction, and are of great economic importance as vectors of virus diseases (e.g. sugar beet yellows). The yearly life cycle of *Aphis rumicis* involves the production of many generations of wingless parthenogenetic females which produce young viviparously, the host plants being the spindle tree (*Euonymus europaeus*) and crops such as beans. In the autumn, oviparous females arise, mate with winged males, and lay eggs which overwinter in the spindle tree. New adults are distributed by air motion, becoming airborne on warm days. They exemplify the principle of survival by dispersion and often travel many kilometres when carried to several thousand metres by convection currents. *See* parthenogenesis.

aphids. Aphididae.

aphotic zone. The deeper parts of seas and lakes where light does not penetrate.

34

aphytic zone. The part of a lake floor that, by reason of its depth, lacks plants.

apical meristem. *See* meristem.

API scale. An arbitrary gravity scale adopted by the American Petroleum Institute for use with crude oil. API gravity values in degrees API or, more usually, just degrees, are related to the formula:

$$\text{Degrees API} = \frac{141.5 - 131.5}{\text{SG at } 60°\text{F}}$$

Hence a heavy crude has a numerically low API gravity. Most Middle Eastern crudes fall in the range 27.0 to 43.0° API.

apocrita. *See* Hymenoptera.

Apoda (Gymnophiona, Caecilians). An order of limbless, worm-like, burrowing Amphibia of tropical countries. Unlike other present-day Amphibia, some Apoda have small scales embedded in the skin.

apodemes. *See* endoskeleton.

Apodidae (Micropodidae) (swifts). Swallow-like birds, accomplished fliers, highly adapted for life in the air. Swifts fly for very long periods, catching insects in their wide mouths. They are able to roost on the wing, and have difficulty in taking off if grounded. Along with hummingbirds, swifts belong to the order Micropodiformes.

Apodiformes (Micropodiformes). *See* Apodidae.

apomixis. Reproduction by seeds formed without fertilisation. The usual form of reproduction in some plants (e.g. Dandelion). Can be taken to include parthenogenesis.

aposematic coloration (sematic coloration). Warning coloration. Conspicuous markings on an animal which is poisonous or distasteful to predators (e.g. the black and yellow markings on wasps and the poisonous amphibian *Salamandra maculosa*).

apparent mortality. Successive percentage mortality.

appetitive behaviour. Behaviour that increases the chances of an animal satisfying a need (e.g. a hungry animal moving around until food is found). The variable introductory phase of an instinctive pattern or sequence (as defined by W. H. Thorpe).

Appleton layer. The region around 200 km in the atmosphere where ionisation by solar radiation causes the refraction and reflection of radio waves. Named after Sir Edward Appleton (1892–1965) of Edinburgh, the inventor of radar.

application factor. The ratio between that concentration of a substance that produces a particular chronic response in an organism and that which causes death in 50% of the population within a given period of time (the LD_{50}). The application factor is used to determine the maximum safe concentration of substances for the organisms under consideration.

appressed. Applied to organs in plants that are pressed together but not united.

Apterygota (Ametabola). Primitive, wingless insects whose young closely resemble the adults. This sub-class includes the Collembola (spring-tails) and Thysanura (bristle-tails, e.g. the 'silver fish') which are of little economic importance. *see* Pterygota.

Aptian. A stage of the Cretaceous System.

aquaculture (pisciculture). The breeding and rearing of freshwater or marine fish in captivity. Fish farming or ranching.

aquafalfa. Ground where the water table is high.

aquamarine. Blue beryl of gem quality.

aquiclude. A geological rock or soil formation which is sufficiently porous to absorb water slowly, but which will not allow water to pass quickly enough to furnish a supply for a well or spring. Clays and shales are commonly regarded as aquicludes.

aquifer. A geological formation through which water can percolate, sometimes very slowly, for long distances. Springs and wells are charged from aquifers and the contamination of an aquifer may lead to the contamination of wells and springs over a wide area. Aquifers are described as 'confined' when they are overlain by aquicludes and as 'unconfined' when their upper surface is at ground level. *See* ground water, porosity, permeability, artesian basin.

aquiherbosa. Submerged aquatic vegetation, marsh and fen vegetation, and sphagniherbosa.

aquiprata. Plant communities in wet meadows.

aquitard. Aquiclude (US usage).

Ar. Argon.

arachidonic acid. Constituent of vitamin F. *See* vitamin.

Arachnida. A class of Arthropoda containing the spiders, mites, ticks, harvestmen, scorpions and king-crabs, all of which possess four pairs of walking legs, prehensile first appendages instead of antennae, and simple eyes. *See* Araneida, Acarina, Scorpionidea, Xiphosura, Pentastomida, Tardigrada.

Araneida (spiders). Arachnida whose bodies are divided into two parts, and which have abdominal silk glands and poison glands at the bases of their first appendages. Some spiders catch their prey by chasing it, others by using webs or trapdoors. They digest the prey externally with saliva, and feed by sucking up the resulting fluid.

arboreal. Pertaining to trees. An arborist is a person who studies trees; arboriculture is the cultivation of trees; arboretum is an area (often of parkland) dominated by trees. *See*-Etum.

arboriculture. *See* arboreal.

arborist. *See* arboreal.

arc furnace. Any furnace heated by an electric arc, rather than a fuel or external heat.

Archaean era. Precambrian.

Archaeornithes. *Archaeopteryx* and *Archeornis*, the earliest (Jurassic) birds, which had feathers, teeth, long tails containing vertebrae, and wings ending in three-clawed digits. They had many reptilian features and probably only limited powers of flight.

arch clouds. Wave clouds extending a considerable distance along a mountain range and remaining stationary with a wind blowing through them, having well defined upwind and downwind edges and giving the appearance of an arch in the sky (e.g. the Chinook arch in the Albertan Rockies, the Southern Arch in the New Zealand Southern Alps).

Archegoniatae. A group of plants comprising the Byrophyta and Pteridophyta, members of the Cryptogamia in which the female sex organs are archegonia.

archegonium. The female sex organ of Bryophyta, Pteridophyta and most Gymnospermae. An archegonium is a hollow, multicellular, flask-shaped organ, containing an egg cell.

Archiannelida. *See* Annelida.

Archimedes' principle. The buoyancy force on an object immersed in a fluid is equal to the weight of fluid displaced by it. Archimedes, a Greek engineer of the Alexandrian school (c. 250 BC) first appreciated that by means of the principle, the density of the material of an object of complex shape could be measured accurately as a multiple of the density of the fluid.

arctic. (*a*) Region around the North Pole, within the Arctic Circle (i.e. latitudes higher than 66°30 N). The Arctic Circle defines the region north of which there is at least one period of 24 hours in summer during which the sun does not set, and one similar period in winter during which it does not rise. The word 'arctic' is derived from the Greek *arktos*, 'the bear', referring to the constellation of The Bear. The climate is characterised by extreme fluctuations between summer and winter temperatures, permanent snow and ice in the uplands, and permafrost which may melt superficially in summer in the lowlands. (*b*) High latitude regions, which may or may not be inside the geographical Arctic Circle, from which tree growth is usually absent because of unfavourable conditions (e.g. shortness of the growing season). The plants and animals living there.

arctic brown earth. *See* Tundra soil.

arctic smoke. Steaming fog, produced when very cold air passes over warm water. Common off the coast of N. Norway, and on rivers warmed by power stations on cold mornings.

Arctogea (Metagea). *See* zoographical regions.

Area of Outstanding Natural Beauty (AONB). In British planning, an area, not being a national park, designated under the *1949 National Parks and Access to the Countryside Act* to preserve and enhance its natural beauty. AONB designation enables local planning authorities to operate more stringent policies of development control. (As defined by the Countryside Commission.)

Arecaceae. Palmae.

arenaceous. *See* arenite.

Arenigian. The oldest series of the Ordovician System in the UK. In Europe, the Tremadocian is considered to be Ordovician rather than Cambrian.

arenite. (Geol.) A sedimentary rock made up of grains of sand between 0.06 mm and 2.00 mm. Sediments formed mainly of arenite grains are termed arenaceous. *cf.* rudite, argillite.

arête. Sharp-edged ridge left between two corries, cwms or cirques as the ice hollows them out.

argentite. A mineral which is the form of silver sulphide, Ag_2S, stable above 180°C. Acanthite is the form stable below 180°C. These two minerals are the major source of silver. Argentite is found in hydrothermal deposits, characteristically with lead, zinc and copper minerals.

argillaceous (argillic). Term used in sense of 'Rich in clay minerals'. *Cf.* argillite.

argillite (**lutite**). Sedimentary rock with grains less than 0.06 mm in diameter; slightly metamorphosed and does not split easily. *Cf.* soil particles size, rudite, arenite.

arginine. Amino acid with the formula $NH_2C(:NH).NH(CH_2)_3.(NH_2)COOH$. Molecular weight 174.2.

argon (Ar). Element. An inert gas which comprises nearly one per cent of the atmosphere by weight. It has an At.No. of 18, and a radioactive isotope Ar^{14}, with a half-life of 30 days, which is formed in the air cooling systems of some nuclear reactors.

arid zone. A zone of latitude (15°–30°N and S) of very low rainfall containing most of the world's deserts. Although the arid zone is very dry, rain does occur there and either evaporates or sinks underground, so it is of little use to vegetation. An oasis occurs where the water table is close to the surface. Even in arid zones there are sometimes years in which cultivation or grazing is possible, but there is a danger of over-population at such times if modern aids are used. High populations cannot be supported in a long sequence of dry years and at such times overgrazing or tillage can cause erosion and the destruction of topsoil. *See* desert, drought, Sahel.

aridisols. *See* soil classification.

aril. An additional covering to the seed in some plants.

Aristotle's lantern. *See* Echinoidea.

arithmetic growth. Linear growth. *See* exponential growth

arithmetic mean. *See* mean.

arkose. A feldspar-rich sandstone. Strictly, a sandstone containing 25 % or more feldspar.

armorican. (Geol.) Usually understood to refer to processes and products of mountain-building in Europe in late Palaezoic times. Armorican was originally restricted geographically to Brittany and Massif Central, but most current usage is broader, and armorican and hercynian are used interchangeably.

aromatic. Applied to organic compounds derived from benzene and having closed rings of carbon atoms. *Cf.* aliphatic.

arrow worms. Chaetognatha.

arroyo (wadi). A steep-sided, flat floored stream channel, a landform of arid regions.

arsenic (As). Element. Exists in three allotrophic forms (see Allotropy): as ordinary grey crystalline arsenic (SG 5.727), as black metalloid arsenic (SG 4.5), and as yellow arsenic (SG 2.0). It occurs in elemental form, combined with sulphur as realgar (AS_2S_2) or orpiment (As_2S_3), with oxygen as white arsenic (As_2O_3), and with some metals. Its compounds are very poisonous and are used medicinally and in the elemental form arsenic is used in transistors. It was formerly used widely in insecticides. AW 74.9216; At.No. 33.

arsenopyrite. A mineral with the composition FeAsS, found in hydrothermal deposits. Arsenopyrite is a major ore of arsenic and was formerly mined extensively to provide insecticides against such pests as the cotton boll-weevil. Arsenic currently finds application in alloys for sporting ammunition, cable-sheathing and transistors. Arsenopyrite is a common gangue mineral of hypothermal deposits of gold and tin and tungsten ores.

Artemisia. Genus of Compositae common in arid areas, but some species cultivated elsewhere. *A. tridentata* is American sagebrush; *A. absinthium* is wormwood, used to flavour absinthe.

artesian basin. A syncline in which the level of ground water in the outcrops of an aquifer is sufficient to produce natural flow from wells tapping the confined part of the aquifer.

artesian well. Well bored into a confined aquifer which overflows the well-head, i.e. the piezometric level of the aquifer is above the level of the well-head.

Arthropoda. The largest phylum of animals, containing over 700 000 species, and the only major invertebrate phylum with members adapted for life on truly dry land. Arthropods are characterised by the possession of a jointed exoskeleton containing chitin, paired jointed limbs, a well developed head and a haemocoelic body cavity. *See* Crustacea, Myriapoda, Insecta, Arachnida, Onychophora, Tardigrada, Trilobita.

artificial formation. A vegetation pattern caused by the continuous activity of Man.

artificial insemination. The practice of introducing semen artificially into the female animals used for breeding purposes. The semen of a male with desirable hereditary qualities can be preserved under refrigeration and used to inseminate numerous females when optimum conditions occur.

artificial rain. Rain caused by an artificial stimulus such as common salt, which forms larger droplets because it is hygroscopic, dry ice (solid carbon dioxide) pellets, which sublime at $-72°C$ and so leave a trail of air cooled to near that temperature in which supercooled droplets freeze, or silver iodide (AgI) which has a crystal structure close enough to that of ice to initiate crystal growth in droplets supercooled to below about $-4°C$. In the latter cases the crystals formed grow by the Bergeson-Fuideisen mechanism; in all cases the larger particles grow by accretion of smaller, slower-falling cloud droplets. Although fallstreak holes have been made in layers of supercooled clouds and some cumulus appear to have been made to grow into showers more quickly, the record of rainmaking attempts is unimpressive and usually undetectable in terms of identifiable increases in rainfall. As a human resource it is an ineffective technological fix and from the practical viewpoint, technological fiction.

artificial recharge. The introduction of surface water into an underground aquifer, through recharge wells.

artificial reef. A reef in the sea made from solid waste materials, to provide cover for marine organisms. As a method of waste disposal it is effective. Its ecological advantage is more dubious, and its effects on commercial fisheries is negligible.

Artiodactyla (Paraxonia). The even-toed ungulates (hoofed animals) including cattle, sheep, pigs, deer, giraffes, hippopotamus and camels. Large, herbivorous, cloven-hoofed mammals, some of which chew the cud. *See* Perissodactyla, Ruminantia.

Arusha declaration. The declaration issued following an IUCN conference at Arusha, Tanzania, in 1961, which recognised the importance of the survival of wildlife in Africa 'as a source of wonder and inspiration (and as) an integral part of our national resources and of our future livelihood and well-being'.

As. Arsenic.

asbestos. The fibrous form of several silicate minerals. Chrysotile, a serpentine mineral, accounts for about 90% of the asbestiform minerals mined. Blue asbestos is natural sodium iron silicate, or crocidolite. Its use is forbidden in the UK, although it is found in demolition work. Asbestos fibres are non-inflammable and stable in many corrosive environments, they can be woven or bound by inert media, and they are used as electrical and thermal insulators and in tiles and roofing sheets. Prolonged exposure to asbestos fibres can cause asbestosis and cancers. *See* mesothelioma.

asbestosis. Dust disease caused by the inhalation of asbestos fibres. Each fibre that enters the lungs may scar the oxygen-absorbing lung tissue, reducing lung function. Chronic shortness of breath results.

Aschelminthes. A phylum used by some taxonomists which includes the Rotifera, Nematoda, Nematomorpha and some of the Gephyrea.

Ascidians. Sea squirts. *See* Urochordata.

Ascomycetes. Group of fungi characterised by the production of cylindrical or spherical spore sacs (asci) containing 8 spores. Some cause diseases of trees (e.g. Chestnut Blight) and powdery mildew diseases, while others produce flavour in cheese or are edible (e.g. truffles, morels). A few are used to produce important drugs (e.g. *Penicillium* gives penicillin and *Claviceps* gives ergotamine (see Ergot). The yeasts (Saccharomyces) used in baking and brewing are Ascomycetes.

ascorbic acid. Vitamin C. A water-soluble vitamin needed by Man and some animals for the forming of collagen and thus for the maintenance of capillary walls and the healing of wounds. Fresh vegetables and fruit (esp. citrus fruits) are important sources of the vitamin. Deficiency causes scurvy, bleeding in the mucous membranes, under the skin and into the joints.

asexual reproduction Reproduction not involving fertilisation. Includes spore formation and propagation by means of bulbs, cuttings, etc. in plants, and reproduction by fission or budding in animals. *See* Coelenterates.

ash. (*a*) Mineral content of a product that remains after complete combustion. (*b*) Pyroclastic fragments ejected by a volcano, and less than 4 mm in their long dimension.

ash, European. *Fraxinus excelsior*

Ashgillian. The youngest series of the Ordovician System in Europe.

asparagine. Amino acid with the formula $NH_2CO.CH_2CH.(NH_2).COOH$. Molecular weight 132.1.

aspartic. Amino acid with the formula $COOH.CH_2CH.(NH_2).COOH$. Molecular weight 133.1.

asphalt. A brown to black, solid or semi-solid bituminous substance (*see* bitumen) which occurs naturally but is also obtained as a residue during the refining of certain petroleums. Asphalt melts between 65°C and 100°C. It is used in the construction of roads, as a waterproofing component in roofing, in fungicides, and in paints.

assembly. The smallest community unit of plants or animals (e.g. a colony of the blackfly *Aphis fabae* on a broad bean plant).

assimilation. (*a*) The incorporation of food substances into the parts of a living organism. (*b*) The natural process whereby a water body purifies itself of organic pollution. This is accomplished largely by the action of micro-organisms.

association. (*a*) (Bot.). A stable plant community of definite floristic composition, dominated by particular species, and growing under uniform habitat conditions. The largest natural vegetation group recognised (e.g. oak-ash woodland) in some systems, but also applied to smaller units in Europe. *See* community. (*b*) (Stat.) The degree of dependence or independence between two or more variables, esp. in cases of simple two-fold classification into A or not A, B or not B. *See* correlation, regression.

assortative breeding. (mating). Mating which is non-random and involves the tendency of females and males with particular characteristics to breed together.

Asteroidea (star fishes). A class of predaceous Echinodermata whose members have flattened, star-shaped bodies with arms not sharply distinct from the central region of the body. They move by means of tube feet. They feed mainly on molluscs and some may cause losses on commercial oyster beds. *See* Ophiuroidea.

asthenosphere. (*a*) The less rigid part of the Earth's interior (i.e. the outer core). (*b*) The mantle and core. (*c*) The lower part of the crust, the sima where adjustment for isostasy is supposed to take place. (*d*) A zone of the Earth where earthquake shock-waves travel at much reduced speeds. This zone is also called the Low Velocity Zone. The top of this zone is from 70 to 150 km below the surface, and the base 200 to 360 km. Since the advent of the theory of plate tectonics, this fourth use has been the most common.

asulam. Translocated herbicide of the carbamate group used to control docks in grassland and orchards, and bracken in grassland, forest plantations and upland pastures.

asymmetrical fold. (Geol.) A fold in which one limb dips more steeply away from the axis than the other. Such a fold will have an inclined axial plane.

atmosphere. (*a*) The gaseous envelope surrounding a planet. The Earth's atmosphere consists of nitrogen (79.1%), oxygen (20.9%), by volume, with about 0.03% carbon dioxide and traces of the noble gases (argon, krypton, xenon, neon and helium) plus water vapour, traces of ammonia, organic matter, ozone, various salts and suspended solid particles. *See* evolution of the atmosphere. (*b*) Unit of pressure, defined as that pressure which will support a column of mercury 760 mm high at 0°C, sea-level, at a latitude of 45°. One normal atmosphere = 101.325 newtons per square metre.

atmospheric dispersion. The mechanism of dilution of gaseous or smoke pollution whereby the concentration is progressively decreased. The dilution rate is very variable and pollution incidents occur when the atmospheric dispersion is insufficiently effective. Atmospheric dispersion is a most important world-wide mechanism for the distribution of salts by rain and is relied upon for the removal of the products of combustion. It is not subject to more than very rough mathematical treatment because of its complexity and variability. The highest arts of mathematical analysis and computation have been employed to little practical effect and have generated little wisdom or knowledge.

atmospheric tides. These are produced by gravitational influence of the sun, and moon as in the case of the ocean, but their magnitude is negligible. A solar semi-diurnal tide is generated by the daily variations in atmospheric heating and consequent expansion. The wave travels round the world in 24h, with two crests 12h apart. It is of the order of 2 mb at the Equator, with crests at 10 and 22h, solar time. The gravitational field of the sun exerts a couple on this tide, accelerating the Earth's rotation.

at.no. Atomic number.

atomic energy. Energy, as heat and radiation, derived in a controlled fashion from nuclear fission or fusion. See nuclear reactor.

Atomic Energy Commission (AEC). US federal agency responsible for all aspects of atomic energy in the US.

atomic number (at.no., Z). The number of electrons rotating round the nucleus of the neutral atom of an element, or the number of protons in the nucleus.

atomic spectrum. The spectrum emitted by an excited atom due to changes within the atom. It is often used to identify trace metals in the environment.

atomic weight (AW). The relative atomic mass, being the ratio of the average mass per atom of a specified isotopic composition of an element to 1/12 of the mass of an atom of $^{12}_{6}C$.

atomisation. The dividing of a liquid into extremely small particles, so increasing its surface area.

ATP (adenosine triphosphate). A co-enzyme that provides a source of energy for activities such as muscular contraction and the synthesis of organic compounds in plants and animals. By removing a phosphate group from ATP, thus forming ADP (adenosine diphosphate), and transferring it to other substances, enzymes are able to make this energy available. ADP is subsequently rebuilt to ATP, using energy from catabolism or from light during photosynthesis.

ATPase. Enzyme that breaks down ATP to ADP.

attenuation. Loss of virulence in bacteria or other pathogenic micro-organisms.

atto- (a) Prefix used in conjunction with SI units to denote the unit $\times 10^{-18}$.

attribute. (Stat.) A qualitative characteristic of an individual (e.g. male or female). *Cf.* variable.

audio frequency. A sound frequency within the audible range, about 20 to 20 000 Hz.

audiogram. A graph, usually plotted by an audiometer which describes the response of a subject to sound as a function of sound frequency, usually measuring each ear separately.

audiometer. Instrument for producing an audiogram. The audiometer feeds calibrated pure sound tones into an earphone and records signals made by the subject that indicate whether or not the sounds have been heard.

auditory ossicles. *See* ear.

Audubon, John James (1785–1851). An early American conservationist, famous for his paintings of birds, after whom modern American conservation groups, most notably the Audubon Society, are named. He initiated the study of migration by ringing birds.

Audubon Society. American society concerned with conservation and named after J. J. Audubon. The *Audubon Society* and the *Sierra Club*, are two of the most influential conservation groups in the US.

aufeis. Layer upon layer of frozen water, discharged sequentially, as by a spring.

auger. Instrument for drilling and collecting cores of soil or surface deposits.

augite. A calcium magnesium aluminosilicate mineral which is one of the pyroxene group.

auks. *See* Alcidae.

aureole. The bright orange glow seen around the sun in haze or thin cloud. *See* Corona.

Aurochs (*Bos primigenius*). The species of wild cattle from which modern domestic cattle are descended. The aurochs became extinct in the 17th century. *See* Bovidae.

aurora borealis. 'Northern Lights', in curtain and flashing formations of white and coloured light in the ionised layers around 400 km above ground. Particles emitted from sunspots travel down the lines of force of the Earth's magnetic field and the lights appear to emanate from near the magnetic pole.

Austausch coefficient. The coefficient originally proposed by the German physicist Ludwig Prandtl (1875–1953) to represent the diffusive effect of eddies in a fluid according to the mixing length theory. *Austausch* (German) means 'exchange'.

Australian region (notogea). *See* zoographical regions.

autacoid. hormone.

autecology. The study of relationships between a single species and its environment. *Cf.* synecology, ecology.

authigenic. Refers to minerals which grew *in situ*, during or after deposition. Contrasted with allogenic.

autochthonous. (*a*) Applied to soil organisms whose activity is not affected by the addition of organic material to the soil. *Cf.* Zymogenous. (*b*) Applied to rocks or organic material in which the major constituents have been formed in situ. Opposite to allochthonous.

autoclave. Strong pressure vessel used for carrying out reactions at high pressures and sometimes at high temperatures.

autocoprophagy. Refection.

autoecious parasite. *See* parasitism.

autogamy. (*a*) (Bot). Self-fertilisation. (*b*) Paedogamy. A process occurring in some diatoms (*see* Bacillariophyta) and Protozoa in which the cell divides to form two gametes, which then reunite.

autogenic. autonomic.

autogenic succession. A succession which results from changes created by the organisms themselves. *Cf.* allogenic.

autoimmunity. Conditions in which the immune responses of an animal are directed against its own tissues. *See* immunity.

autolysis. The breakdown of animal and plant tissues which occurs after the death of the cells and is carried out by enzymes within these cells.

autonomic (autogenic). Applied to spontaneous movements (e.g. nutation) which occur in response to external stimuli. *See* paratonic.

autonomic nervous sytem. The sympathetic and parasympathetic nerve supply to the involuntary muscle (e.g. heart, blood vessels, gut) and glands of vertebrates, together with the sensory nerve fibres which pass impulses from internal sense organs to the central nervous system (brain and spinal cord). Some of the cell bodies of the autonomic nervous system are outside the central nervous system, situated in ganglia (e.g. the solar plexus). Where an organ is supplied by both sympathetic and parasympathetic systems, the two act antagonistically. For instance, the heart beat is accelerated by the sympathetic supply and slowed by the parasympathetic, whereas peristaltic movements of the gut are stimulated by the parasympathetic and inhibited by the sympathetic supply. Acetylcholine is released at endings of parasympathetic nerves, and adrenaline at many sympathctic nerve endings.

autopolyploid. Organism with more than two haploid sets of chromosomes all derived from the same species. *See* endomitosis.

autosome. A chromosome other than a sex chromosome.

autotomy. The automatic self-amputation of part of an animal, usually followed by its regeneration. e.g. by intense muscular contraction a lobster will break off a limb, and a lizard its tail, when seized by a predator.

autotrophic. Applied to organisms which produce their own organic constituents from inorganic compounds, utilising energy from sunlight or oxidation processes. Chlorophyll-containing plants are autotrophs which elaborate organic compounds from simple substances (carbon dioxide, water) during photosynthesis. A few bacteria are autotrophic. Those containing bacteriochlorophyll are phototrophic. Others, such as those which obtain energy by oxidising hydrogen sulphide, are chemotrophic. *Cf.* heterotrophic.

autumn. The season of decreasing length of day, of cooling weather, of cessation or reduction of vegetable growth in higher latitudes. For climatological purposes, it is defined arbitrarily as September, October and November in the Northern Hemisphere, but it is not without great variation from one year to another. *See* seasons.

auxins. Plant hormones (e.g. indole-3-acetic acid (IAA). Auxins are made in the growing tips of stems and roots, and move to other parts of the plant, where they regulate growth. They may, for instance, initiate cell division in cambium, increase the rate of cell enlargement causing curvature responses, initiate root formation in cuttings, inhibit bud development and fruit drop, or initiate flower formation. *See* geotropism, phototropism, hormone weed killers, cytokinin, gibberellin, abscisin, IAA.

auxin-type growth regulators. Hormone weed killers.

auxotroph. A strain of micro-organism which has nutritional requirements additional to those of the organism in its natural state. *See* prototroph.

avalanche. The most rapid form of mass-wasting. A catastrophic slide of rock debris and/or ice and snow.

average. (Stat.) A value summarising a set of values (e.g. geometric, or more usually, arithmetic mean). It is not always obvious what averages are significant. Usually they are defined for reference purposes and then called 'normal'. Atmospherically, average wind speed represents the run of the wind, but the average of the cube of the speed of the wind represents the power obtainable aerodynamically from a windmill, if it can make use of strong winds. The average day temperature is often defined as midway between the maximum and minimum, but this has many defects as an indicator of trends, even if it possesses historical advantages. No one would define the average wind in that way.

Aves (birds). Class of warm-blooded vertebrates which have feathers and forelimbs modified as wings. Birds have marked affinities to reptiles, and probably evolved from reptiles related to primitive dinosaurs. Some fossil forms (*Archaeopteryx* and *Hesperornis*) had teeth on the jaws. Birds lay large-yolked, hard-shelled eggs, and exhibit a high degree of parental care.

AW. Atomic weight.

awn. (Bot.) bristle-like projection from the tip of back of the glumes in some Graminae.

48

axenic culture. A pure culture. One which contains only a single kind of organism.

axerophtel. Vitamin A. *See* vitamin.

axial plane. (Geol.) The imaginary plane which divides a fold as symmetrically as possible, and which passes through the axis.

axial velocity. The component of velocity along an axis, usually the axis of symmetry of a motion system. Thus there is a velocity along the axes of the vortices behind an aircraft with a maximum towards the aircraft in the vortex cores. Swirling flow in a pipe or in a tornado has a velocity component along the axis which varies with the distance from it.

axillary. (Bot.) The angle between the leaf stalk of a plant and the stem.

axis. (*a*) *See* flower. (*b*) (Geol.) The median line between the limbs of a fold, along the trough of a syncline or the crest of an anticline. (*c*) (Meteor.) The direction of flow at the centre of a jetstream.

axon. The process of a nerve cell, usually an especially long one, that conducts impulses away from the nerve cell body. *See* dendrites.

Azolla. Genus of aquatic (floating) ferns forming a symbiotic (*see* symbiosis) relationship with the blue-green alga *Anabaena*, which can fix nitrogen and thus may be of significance in rice culture.

azotobacter. The most important of the nitrogen-fixing bacteria (*see* nitrogen fixation). It is aerobic and obtains its energy by breaking down carbohydrates in the soil. Along with other nitrogen-fixing bacteria it is responsible for the gradual increase in the nitrogen content of unmanured grassland. *See* nitrogen cycle, *Clostridium pasteurianum*.

B

B. (*a*) Billion (in the American sense of one thousand million, 10^9). Used in conjunction with brl or cf in figures for hydrocarbon reserves to signify billion barrels or billion cubic feet. (*b*) Boron.

Ba. Barium.

Bacillariophyceae. *See* Bacillariophyta.

Bacillariophyta (Bacillariophyceae, diatoms). Unicellular algae, some of which are colonial, either green or brownish (but all containing chlorophyll), with siliceous and often highly sculptured cell walls. Diatoms make up much of the producer level in marine and freshwater food chains and they have contributed to the formation of oil reserves. Deposits of diatomaceous earths were formed by the accumulation of diatom cell walls.

bacillus. Any rod-shaped bacterium. More specifically, a genus of bacteria which includes *Bacillus tetani*, responsible for lockjaw, and *Bacillus subtilis* a symbiotic inhabitant of the intestine of many animals, including Man, which synthesises vitamin B_{12}. *See* symbiosis.

Bacillus subtilis. Bacterium that produces a pectin-splitting enzyme that is very stable, enabling it to derive nutrients from wood. It produces a fructans (D-fructose polymer) which substitutes for starch as an energy store.

backcross. The crossing of a hybrid (heterozygote) with one of its parents, or with an organism genetically similar to one of its parents.

backing wind. A wind showing a cyclonic turning of its direction. As a front approaches from the west (N. Hemisphere) the wind backs from west to southwest or south. A wind that changes its direction in the opposite sense is a veering wind.

backscatter. Solar radiation is scattered in a forward (downbeam) direction by atmospheric aerosols. Backscatter is the dispersion of radiation in an upbeam direction, so tending to increase albedo. *See* Rayleigh scattering, Mie scattering.

bacteria (Schizomycophyta, Bacteriophyta). Group of ubiquitous microscopic organisms, mostly unicelluar, classified as plants or Protista. They may be rod-shaped (bacilli), spherical (cocci), roughly spiral (spirelli), or filamentous (Actinomycetes) and may possess flagella which enable them to move. They reproduce by simple fission, spore formation and sometimes by sexual means. A few are autotrophic either obtaining energy from oxidation processes or by photosynthesis if they contain bacteriochlorophyll. Many are saprophytic (*see* saprophyte) and play a vital role in the decomposition of organic matter, (*see* Carbon cycle, Nitrogen cycle). Many are parasites (*see* parasitism), causing animal diseases such as bubonic plague, tuberculosis, cholera and tetanus, and a few plant diseases such as soft rot of carrots. Some bacteria are symbiotic (e.g. *Bacillus subtilis*). *See* mycoplasmas, myxobacteria.

bacteriochlorophylls. Pigments, found in photosynthetic bacteria (see Photosynthesis) which trap light energy, thus enabling the bacteria to be autotrophic.

bacteriophages (phages). Viruses parasitic on bacteria. A bacteriophage consists largely of nucleic acid (usually DNA) which takes over the nucleic acid producing mechanisms of the bacterial cell, forcing it to make virus nucleic acid instead of its own (*see* parasitism). *See* lysogenic bacterium.

Bacteriophyta. Bacteria.

bacteriostat. Substance which retards or inhibits the growth of bacteria.

baffle. Plate, grating or refractory wall used especially to block, hinder or divert a flow or to redirect the passage of a substance.

bagasse. Crushed sugar cane left as a fibrous residue after the sugar has been extracted. Bagasse can be treated for use as a fertilizer, or for protein extraction. *See* novel protein foods.

bag filter. Device based on a filter made from woven or felted fabric, often tube-shaped, used to remove particulate matter from industrial waste gases. Bag filters may be up to ten metres long and a metre wide. They are closed at one end, the other end being connected to a gas inlet. More than 99 % of all particles may be extracted, provided the velocity of the gas is low. The material from which the filter is made is limited by operating temperatures. Natural fibres cannot be used above 90°C, nylon above 200°C, glass fibres, sometimes impregnated with graphite and siliconized for greater durability, above 260°C.

Bajocian. A Stage of the Jurassic System.

Balaenoidea (Mysticeti). Toothless whales which have large heads and baleen (whalebone) hanging from their palates, for straining shrimps and other small animals from the water. The group includes Rorquals, the Right Whale and the Blue Whale. *See* Cetacea.

balance of nature. *See* ecological balance.

ball lightning. An elusive form of electric discharge said to take the form of an incandescent ball moving at a rate comparable with air motion. The reality is confused by faked and unscientific photographs and descriptions of halluci-nations or dazzle.

balloons. Used widely either as pilot balloons, which are followed by theodolite to measure wind, or as sounding balloons carrying instruments which record and are recovered or transmit (radiosonde) or are tracked by radar. A typical sounding balloon carries a few kg to heights of up to 30 km occasionally, 20 km regularly. 'Constant level' balloons are made by enclosing the rubber in a silk net or sheath to give constant volume, otherwise the expansion of a rubber balloon maintains a fairly constant rate of rise of the order of 5 to 10 m sec^{-1}.

51

bamboo. *See Bambusaceae.*

Bambusaceae. The bamboos, a group of woody plants belonging to the grass family (Graminae) which often grow to a great height. They form a very important part of the ecosystem in tropical and sub-tropical areas.

banana. *See Musa*

banana republic. *See* One-crop economy.

band. A segment of the spectrum of wave frequencies, e.g. the long, medium or short wavebands of radio waves. Sound bands are measured in octaves.

banded ironstone. A Precambrian deposit of alternating fine bands, usually less than 1 mm thick, of chert and haematite. The cherts contain fossils of microscopic plants. Banded ironstones have been dated at between 3200 and 1700 million years old, at which time, it is believed, the atmosphere was far poorer in oxygen and richer in carbon dioxide than it is now. Unleached banded ironstone is known as taconite and contains 15–40% iron, but surface leaching of siliceous and carbonate minerals has produced what was considered a more desirable ore grading 55% or more iron. The taconite, after beneficiation, is pelleted and has proved to be a better raw material for smelting.

banner cloud. A wave cloud fixed over a mountain, extending downwind of the peak.

bar. (*a*) The meteorological unit of pressure, equal to 10^5 Nm^{-2} (10^6 dynes cm^{-2}), or 10^3 millibars (mb) which is roughly the sea level atmospheric pressure (usually taken to be 1013.2 mb). (*b*) A barrier that is wholly or partially submerged beneath the sea. Traditionally a navigation hazard. *See* point bar.

barban. Translocated herbicide of the carbamate group used to control wild oat and other seedling grasses in cereal crops, beans and sugar beet. Harmful to fish.

Barberry. *See Berberis vulgaris.*

Barcelona Convention. Convention drawn up in 1976 by nations bordering the Mediterranean, that forbids the discharge of a wide range of substances into the Mediterranean, including organosilicon compounds, all petroleum hydrocarbons, all radioactive wastes, and acids and alkalis from sources as yet unidentified.

barchan. Sand-dune, crescent-shaped in plan, with the horns of the crescent pointing downwind. The steeper slope is on the leeward side. Other spellings include barkhan, barkan, barkham and barchane.

barchane. *See* barchan.

barium (*Ba*). Element. A silvery-white, soft metal that tarnishes readily in air. It occurs as barytes and as barium carbonate. Its compounds resemble those of calcium, but are poisonous. The compounds are used in the manufacture of paints, glass and fireworks. AW 137.34; At. No. 56; SG 3.5; mp 710°C.

bark. The protective layer of dead cells outside the cork cambium (phellogen) in the older stem or root of a woody plant. Bark may consist of cork alone, or alternating layers of cork and dead cortex or phloem tissues. The term is often used more loosely to mean all the tissue outside the xylem.

barkan. *See* barchan.

barkham. *See* barchan.

barkhan. *See* barchan.

barklice. *See* Psocoptera.

barley. *Hordeum vulgare. See* Graminae.

barnacles. *See* Cirripedia.

baroclinic. Having a different distribution of pressure at different heights owing to horizontal variations of density. This is the normal state of the atmosphere.

barograph. A recording barometer, usually employing a pen activated by an anaeroid barometer writing on a cylinder that rotates, completing one rotation each week.

barometer. Instrument for measuring atmospheric pressure, usually of the mercury type, which measures the height of a column of mercury supported under a vacuum in a glass tube. Different designs use various methods for correcting for temperature, and gravitational strength (which varies with height and latitude). The anaeroid (without fluid) type is used in altimeters and barographs, but is not capable of the accuracy (0.1 mb) required for meteorological observations.

barotropic. Having the same pressure field at all heights and no horizontal density gradients. A barotropic fluid conserves the vertical component of vorticity. A barotropic model was used in early computed forecasts, but it had serious limitations.

barrage. An obstruction constructed across the flow of sea tides in order to hold

53

back water in a basin for subsequent release, usually in order to generate power by the flow of water in one direction, or in both directions. *See* tidal power.

barrel. (*a*) Container for liquids, used for storage or transport. (*b*) Brewing. Cask containing 36 Imperial Gallons (164 litres). (*c*) A volumetric unit of measurement used for oil. The most commonly used unit is equivalent to 42 US gallons (or 5.61 cu ft or 0.159 cu m). Depending on the specific gravity (or degrees API of the particular crude oil, between 6.5 and 7.5 barrels weigh one metric tonne. Barrel is abbreviated to brl, bbl or bl, and coupled to this is usually MM for million (10^6), B for billion (in this case 10^9) or T for trillion (10^{12}).

Barremian. *See* Neocomian.

barrier spit. A beach barrier connected to the mainland at one end and terminating in the open water. As it builds down-drift of the longshore currents it may develop a hook at the open end as currents are refracted shorewards. *See* baymouth barrier, bar, bayhead barrier.

baryte. A barium mineral, $BaSO_4$, chiefly found as gangue with galena in hydrothermal veins. Because of its high specific gravity (4.5), baryte is used to make drilling mud more dense. Baryte is also used in paints, and as an opaque material for radiography.

basal body. *See* centriole, cilia, flagellum.

basalt. A fine-grained basic igneous rock, consisting essentially of a calcium-rich plagioclase feldspar and a pyroxene, with or without olivine.

basaltic lava. Lava with the composition of basalt. Basaltic lava has the lowest viscosity of the lavas, and is the most common: 90% of lavas are basaltic.

basecourse. *See* pavement.

base exchange. Transfer of cations (positively charged ions) between an aqueous solution and a mineral. Principle has many important applications, e.g. in the softening of water when Ca^{2+} ions in the water are replaced by Na^+ ions in the softening agent. It is also used to improve soil fertility by the addition of solutions rich in a mineral deficient in the soil. e.g. Ca^{2+}, Mg^{2+}, Na^+, K^+, to replace another cation, say H^+, adsorbed on to the soil particle. The efficiency with which cations will replace other cations depends on such variables as the number of charges on the ion, the hydration property of the ion, its size and the relative concentrations of the ions. The replacement series in soil is usually: $Al >$ (will replace) $Ca > Mg > K > Na$.

base level. Lowest level to which stream can erode its channel: sea level, lake level, or the level of the main stream into which the tributary stream flows.

baselines. The tolerance levels of organisms to particular concentrations of elements. These vary from species to species.

base pairing. *See* DNA, RNA.

base saturation. When the cation exchange capacity of a soil is saturated with the exchangeable bases Ca, Mg, K or Na. Expressed as a percentage of the total cation exchange capacity.

base subsistence density. The density above which the survival of a human population is impossible.

basic. (*a*) A term applied to igneous rocks containing a relatively low (commonly set at 50%, or else in the range 44–52%) amount of silica (SiO_2) in their chemical composition. The silica is in the form of silicate minerals such as plagioclase feldspar, pyroxene, and olivine. Quartz is rarely present in basic rocks. Basalt, dolerite, and gabbro are basic rocks. In petrology, basic is contrasted with acidic, intermediate and ultrabasic. (*b*) (Chem.) Applied to substances that react with acids to form salts. *See* pH.

basic refractory. Heat resistant material containing a large proportion of metallic oxides, used as a furnace lining, e.g. dolomite and magnesite.

Basidia. *See* Basidiomycetes.

Basidiomycetes. Group of fungi in which spores are borne externally, usually in fours, on mother cells (basidia). The group includes mushrooms, puffballs, bracket fungus, most of the fungi of mycorrhizas, smuts and rusts, and dry rot fungus *Serpula* (*Merulius*) *lacrymans*. *See* agarics, rust fungi, smut fungi.

basket-of-eggs topography. A drumlin field.

Batesian mimicry (pseudaposematic coloration). The resemblance of a harmless animal to a poisonous, dangerous or distasteful one, which is often conspicuously marked (*see* aposematic coloration). This affords protection, as predators tend to avoid both (e.g. bee-hawk moths resemble bees and wasps). *See* Müllerian mimicry.

batholith. A large igneous intrusion with steeply dipping contacts and no apparent floor. Exposed batholiths may cover hundreds of thousands of square kilometres, and are associated with orogenic belts. Most batholiths are granitic.

Bathonian. A Stage of the Jurassic System.

bath plug vortex. The most homely model of a tornado or hurricane, demonstrating the conservation of angular momentum with spin increasing as fluid approaches the centre of the vortex. The shape of the funnel being a constant pressure surface is like that of a tornado cloud and resembles roughly the shape of the tropopause which is sucked down into the eye of a hurricane.

Batrachia. Amphibia.

bats. *See* Chiroptera.

Battersea gas washing process. Method of wet scrubbing (*see* wet scurbber) the flue gases at Battersea Power Station, London, to remove sulphur dioxide. The gases are washed with river water and chalk, using about 157 litres to each tonne of coal burned. The effluent is an almost saturated solution of calcium sulphate. The system is now used at other power stations. Its disadvantages are that the cooling of the plume causes a localised fallout of residual sulphur dioxide, the river is polluted, the recovered sulphur cannot be used, and the process causes corrosion and increases costs. The process was invented in 1930 and is still used at Battersea 'A' and Bankside power stations in London.

battery (electric). Device for storing electrical energy chemically so it can be used later as required.

battery (poultry). Most common source of hen eggs in most industrial countries. Semi industrialised system of husbandry where birds are confined in cages, fed and watered by partly automated systems and their eggs collected in channels below the cages, which commonly rise in several banks. One unit may house many thousands of birds. Source of polluting effluent, recycling or disposal of which is often difficult. An energy-intensive way of producing eggs.

bauxite. The most important ore of aluminium, mainly hydroxides. Also used as a filler in plastics and rubber, in catalysts, and as an abrasive.

bayhead barrier. A barrier enclosing a lagoon near the inland head of a bay. *See* barrier spit, baymouth barrier, bar, beach barrier.

baymouth barrier. A barrier across the mouth of a bay. *See* bayhead barrier, barrier spit, bar, beach barrier.

Be. beryllium.

beach barrier. A wide ridge of sand protruding above normal high tide level, parallel to the coast and separated from it by a lagoon or marsh. Two proposed modes of origin: (*a*) that barriers are formed by longshore drift (*b*) that barriers can be produced by wave action transporting sand shoreward and turbulence, due to surf, piling sand on to the barrier site. This is augmented by longshore drift. *See* barrier spit, baymouth barrier, bayhead barrier, bar.

beach drift. Zig-zag movement of material along beach platorm when waves swash obliquely on the shore. *See* littoral drift, longshore drift, littoral currents, longshore currents.

bear animalcules. *See* Tardigrada.

beat. (acoustic) A periodic increase and decrease in amplitude caused by the superposition of two tones of different frequencies, the beat frequency being the difference between the two frequencies.

Beaufort scale. A scale of wind force devised by Admiral Sir Francis Beaufort (1774–1857) originally related to the state of the sea, and adapted for use on land with a scale from 0 to 12, although forces above 8 are rather rare over land. It is described in terms of behaviour of trees, smoke, waves, umbrellas, ease of walking, etc., and facilitated and internationally agreed reporting of wind speed which has been largely superseded only since 1950 by instrumental means.

Beckmann thermometer. A type of mercury thermometer with a large bulb, which gives it high sensitivity over a limited range. Mercury can be added to or removed from the column for the measurement of different temperature ranges.

Beckman process. Process, designed before the Second World War, to treat cereal straw with sodium hydroxide (caustic soda), followed by washing, so producing a substance that can be used as a feed additive. Ruminants will eat straw, but the Beckman Process increases the amount of nutrient they derive from it. The process was not developed for on-farm use, mainly because of the dangerous chemicals that it used, but it is being considered further either as an off-farm process, or for use on-farm in a modified form.

bed. A layer of sediment thicker than 1 cm and distinguished from adjacent layers by its composition, structure or texture.

bedding plane. The surface separating successive layers of stratified rock. Such planes form planes of weakness.

bed load. The sand and gravel particles rolling, sliding or bouncing along the bed of a stream.

bedrock. Consolidated rock below loose superficial material such as soil, glacial drift, or alluvium.

Beech. *See Fagus sylvatica.*

beetles. *See* Coleoptera.

Beetroot. *See Beta vulgaris.*

behaviour. The manner in which organisms respond to stimuli, including the accommodation of a population, individual interactions within the population, reproductive behaviour. *See* ethology.

behaviourism. The name of a school of experimental psychology that holds that the proper subject for psychological investigation is the prediction and control of behaviour and no other, so dismissing such issues as consciousness, mind and freewill. Basic drives are defined operationally (e.g. hunger or thirst in terms of length of time without food or water).

beheaded river. *See* river capture.

bel. Ten decibels.

Belemnites (Belemnoidea). An extinct group of cephalopod molluscs (Decapoda) abundant in the Mesoic. The bullet-shaped internal shells are plentiful fossils.

Belemnoidea. Belemnites.

Belladonna. *See* Solanaceae.

Bénard cell. A convection cell best observed in shallow layers of viscous liquid (e.g. diluted aluminium paint in which the flow is revealed by the appearance of the flat metallic particles.) They are steady when the heating at the bottom, or cooling at the top, is just sufficient to maintain the motion. Otherwise, the heat is transferred entirely by conduction at lower heating rates, or the motion is unsteady at greater heating rates. The cell is usually rectangular or hexagonal. Named after H. Bénard, who described them in 1901, by Sir David Brunt.

beneficial use. A use of the environment, or some part of it, for the benefit of the human population (e.g. for recreation, agriculture, water storage, etc.) requiring protection from pollution or despoliation.

beneficiation. The separation of a valuable mineral from gangue and country rock to effect a low-cost artificial concentration of the ore. Beneficiation includes such processes as crushing, magnetic separation, and froth flotation.

Bengbauforschung (BGF) process. German process for removing sulphur dioxide from flue gases with 95 % efficiency, using a special form of coke in a moving bed reactor at 140°C to absorb sulphur dioxide, which can be recovered either by extraction with water to recover the reagent and sulphuric acid, or in a moving bed reactor at 600°C with hot sand, to recover the reagent and sulphuric acid or elemental sulphur.

Benioff zone. A plane or planar zone of earthquake foci (*see* focus) which dips inwards and downwards from the edge of continents.

Bennettitales. An order of extinct Gymnospermae prominent in the Mesozoic. They resembled Cycadales but unlike these had hermaphrodite cones.

benthic. *See* demersal benthos.

benthon. *See* Benthos

benthos (benthon). (*a*) The organisms living on or in the bottom sediment of a sea or lake anywhere between the high water mark down to the deepest level. (*b*) The bottom of a sea or lake, the deep zones of water. Abyssobenthos, organisms living on an ocean floor at great depths; Phytobenthos, the part of a lake or sea bottom covered with vegetation, the plants living there; Potamo-benthos, the organisms living on a river bottom; Geobenthos, that part of the bottom of a fresh water lake which does not support rooted vegetation.

bentonite. A very fine clay formed by the alteration of volcanic ash deposits. Bentonite is used in water-softeners, and in reducing seepages in mines and channels, and is also a major component of drilling mud.

benzene (C_6H_6). The simplest aromatic hydrocarbon, found in coal tar and used extensively as an industrial solvent and in laboratories. Also used in manufacture of styrene, lacquers, varnishes and paints. A highly inflammable, narcotic liquid that is also a carcinogen.

benzene hexachloride. *See* Lindane.

benzene ring. The representation of benzene as a hexagonal ring of carbon with alternate double and single bonds, as:

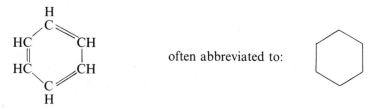

often abbreviated to:

benzo α pyrene. Carcinogen found in tars. Produced, *inter alia*, in the manufacture of asphalt, tar and other organic products.

Berberis vulgaris. Barberry, a hedgerow shrub occurring locally in small quantities in Britain. It harbours a stage in the life cycle of the fungus *Puccinia graminis* which causes black rust on wheat and other Graminae.

Bergeson-Fuideisen mechanism. The vapour pressure over supercooled water exceeds that over ice at the same temperature, so that if ice and water particles are present together in a cloud the droplets evaporate and the ice crystals grow. The Bergeson-Fuideisen mechanism is the explanation of the initiation of some kinds of rain, where the upper parts of clouds are frozen: it is supposed that after the initial growth the ice crystals grow by accretion as they fall through the cloud of droplets.

Bergmann's rule. As the mean temperature of the environment decreases (e.g. with increase in latitude) the body size of warm-blooded animal species or subspecies tends to increase.

bergschrund. A fissure formed during warm weather between the steep head wall of a cirque and the glacial ice filling the cirque.

berg winds. Warm winds blowing from the mountains, especially known in coastal Natal. Sometimes their temperature indicates recent adiabatic descent of air from above the height of the mountain. A type of föhn wind, not associated with rain or cloud.

beri-beri. Disease caused by deficiency of vitamin B_1 (thiamin) common in populations subsisting on a diet of polished rice.

Bermuda High. A high pressure region near Bermuda, evident in pressure charts averaged over long periods because of the frequent occurrence of anticyclones in that area. Part of the subtropical high pressure belt and the cause of the Sargasso Sea.

Berriasian. *See* Neocomian.

berry. A juicy fruit, usually containing many seeds, with a pericarp consisting of a skin (epicarp), a succulent mesocarp, and no stony endocarp surrounding the seeds. e.g. tomato, grape, orange, cucumber. *See* drupe.

beryl. The mineral beryllium aluminosilicate, $Be_3Al_2(SiO_3)_6$, and the major ore mineral of beryllium which is used to make very hard non-ferrous alloys. Gem-quality beryl is known as aquamarine if blue or emerald if green. Beryl is found in pegmatites.

berylliosis. A disease of the lungs caused by the inhalation of particles of beryllium oxide.

beryllium (Be). (Element) A hard, poisonous metallic element used in the production of corrosion-resistant alloys, in X-ray tubes and in the nuclear industry as a moderator. AW 9.0122; At. No. 4; SG 1.85; mp 128 %C. If inhaled, beryllium has been known to cause malignant growths in the lungs of workers and some residents near factories. Its use is now controlled strictly. The element is found in the atmosphere in minute traces. It is converted into a radioactive isotope in the stratosphere by cosmic rays.

Bessemer process. Steel-making process named after its inventor, Sir Henry Bessemer (1813–1898) by which much steel was made from about 1860 to the 1960's. A large vessel (the 'converter') is charged with 25 to 50 tons of molten pig iron and air is blown through holes in its base. The oxygen in the air produces iron oxide, which dissolves and then oxidises silicon, manganese and carbon impurities. The metallic oxides form a slag and the carbon is removed as carbon monoxide, some of which burns to form carbon dioxide. The vessel is tilted and while its contents are being poured into a large ladle, a manganese alloy is added which combines with the remaining iron oxide, so removing it. Other compounds are added to assist the de-oxidising of the steel, then anthracite is added to bring the carbon content to the desired level.

best practicable means. Concept applied to the control of air pollution in Britain by the Alkali Inspectorate, whereby the best possible level of control is attained that is technologically and economically feasible for the industrial plant in question.

beta-diversity (habitat diversification). Diversity resulting from competition between species leading to more precise adaptation to the habitat as a whole and thus narrowing the range of tolerance to other environmental factors.

beta (β) particles. Electrons emitted by radioactive decay, with moderate penetrative power and able to damage living tissue.

β-napthylamine. Powerful bladder carcinogen whose use is prohibited in UK.

***Beta vulgaris* (Chenopodiceae).** A species of very varied habitat, which includes the beetroot, sugar beet, chard, mangold-wurzel, and the wild sea beet from which all the cultivated forms were derived. Most forms are biennial and store sugar in a swollen root.

Betula (birch). (Betulaceae) A genus of trees and shrubs of north temperate regions. The birches extend to the northern limit of trees. The genus bears catkins and has hard wood which is used for making shoes, charcoal, etc.

Bewsey's method. Observation of the orientation of fallstreaks of ice cloud. If these lie at right angles to the wind direction at the level at which they occur, with the tail pointing in a direction backed from the wind, development is indicated. Typically the approach of a warm front is revealed and rain is to be expected. (Bewsey is a British meteorologist).

Bezoar (*Capra aegragus*). An animal with large, scimitar shaped horns, that inhabited Persia in prehistoric times and is believed to be the wild ancestor of domesticated goats.

bhc. *See* Lindane.

bias (Statis.). Systematic distortion as distinct from random error which may distort on one occasion but balances out overall.

biennial. A plant, (e.g. carrot) which completes its life cycle from seed to seed in two years. Food is stored in the first year and used to produce seeds in the second year. *See* annual, perennial, ephemeral.

biennial oscillation. An oscillation with a period of two years. It has been claimed that there is a 26 month oscillation of high altitude low latitude winds, but like all oscillations with no astronomical base, their reality is doubtful.

big bang. An explosion producing a pressure oscillation of the order of a millibar or more at a distant point on the Earth. e.g. the eruption of Krakatoa in 1883. Large nuclear fusion explosions are comparable.

bilaterally symmetrical. Applied to animals and flowers with only one plane of symmetry. Most freely-moving animals have similar right and left halves. Bilaterally symmetrical flowers (e.g. Sweet Pea, *Lathyrus odoratus*) are usually called zygomorphic.

bile. Fluid secreted by the liver of vertebrates, stored (in many) in the gall bladder and poured via the bile duct on to the food in the duodenum. Bile salts (sodium bicarbonate, glycocholate and taurocholate) reduce the acidity of the intestinal contents, activate the pancreatic enzyme lipase and emulsify fats. Bile pigments are the breakdown products of haemoglobin.

bilharzia. *See* schistoma.

biliproteins. phycobilins.

billow clouds. Clouds arranged in several parallel bars in close proximity in a layer. Caused by overturning due to convection, the cloud being cooled on top by radiation to space, the arrangement being due to wind shear. Alternatively due to overturning of unstable waves due to the generation of shear at a very stable layer when it is tilted, as in flow over a mountain.

binaural. Applied to the use of the ability to hear with two ears for orientation, or the electronic simulation of this.

binocular vision. An arrangement in which both eyes face forward, imparting stereoscopic vision. Found in primates and predators, to whom distance judging is particularly important.

binomial nomenclature (binomial system). The present method of naming organisms, using pairs of Latin words (*see* Linnaeus). The first word denotes the genus and is shared by the organisms closest relatives. The second denotes the species. The name of the person responsible for naming and describing the species should follow, e.g. the Heath Violet is *Viola canina* Linnaeus. If varieties or sub-species have been described a third Latin name is added, e.g. *Viola canina* ssp. *canina* and *Viola canina* ssp. *montana*.

bio-assay. The quantitative measurement, under standardised conditions, of the effects of a substance on an organism or part of an organism.

biochemical oxygen demand (BOD). The amount of oxygen used for biochemical oxidation by a unit volume of water at a given temperature and for a given time. This is used as a measure of the degree of pollution. The more organic matter the sample contains, the more oxygen is used by its micro-organisms.

biochore. A subdivision of biocycle, comprising a group of similar biotopes, for instance desert (including sandy and stony desert biotopes) or forest (including coniferous, deciduous and other forest biotopes).

biocide. Any agent that kills living organisms. Sometimes used as a synonym for pesticide.

bioclastic. Composed of fragmental organic remains. *See* clastic.

bioclimatology. The science of the relationship of life to climate.

biocoen. All the living components of the environment.

biocoenology. Synecology.

biocoenosis. (*a*) (Biol.) A community of organisms occupying a biotope (*b*) (Geol.) An assemblage of fossils consisting of the remains of organisms living together. Biocoenosis is contrasted with thanatocoenosis. (*c*) Mutualism between plants and animals.

biocontent. The total energy content of an organism or community.

biocontrol. *See* biological control.

biocycles. Subdivisions of the biosphere. Those usually recognised are land, sea and freshwaters. *See* biochore.

biodegradation. The breakdown of substances by micro-organisms (mainly aerobic bacteria). Many manufactured substances are readily biodegradable, but others (e.g. organochlorine insecticides and 'hard' detergents) are much more resistant to bacterial action. *See* activated sludge, alkyl sulphonates.

biodynamic farming. Farming according to principles laid down by Rudolf Steiner. These are similar in many ways to organic farming principles, but in addition relate farm operations to phases of the moon and make use of various preparations in very small quantities.

bioengineering. The employment of biochemical processes on an industrial scale, especially to produce bulk foodstuffs for humans or livestock by the recycling of wastes. *See* novel protein foods.

biogenetic law. *See* palingenesis.

biogeochemical anomaly. Concentration of a minor or trace element in the soil determined by chemical analysis of plants. *Cf.* geobotanical anomaly.

biogeochemical cycles. The circulation of elements within ecosystems, *See* carbon, nitrogen and phosphorus cycles.

biogeosphere. The outer part of the earth's crust (lithosphere) as far down as life exists.

biological amplification (biological magnification). The concentration of a persistent substance (e.g. organochlorine insecticide) by the organisms of a food chain so that at each successive trophic level the amount of the substance relative to the biomass is increased.

biological bench-mark. The use of plant or animal species to measure pollution (e.g. lichens to measure sulphur dioxide) based on assessments (bench-marks) of population level and fitness, against which changes can be evaluated. *See* biological indicators, biological monitoring.

biological control (biocontrol). The control of pests by the use of living organisms. The dramatic reduction in the rabbit population of Australia by the introduction of the myxomatosis virus, and the control of the citrus scale insect in California by the introduction of an Australian ladybird (ladybug) beetle (foiled later by the use of insecticides) were successful examples. An advantage of this type of pest control is that, once established, the control is self-perpetuating. It is also specific, and does not pollute the environment. Another technique of biological control involves the introduction of large numbers of sterile (ir-radiated) males, resulting in the production of infertile eggs. This has been successful in some parts of the U.S.A. in the control of the screw worm fly (*Callitroga macellaria*).

biological husbandry. Organic farming.

biological indicators. Micro-organisms (e.g. some which migrate to regions of high or low oxygen concentration) which are used as indicators of chemical activity. *See* biological bench-mark, biological monitoring.

biological magnification. Biological amplification.

biological monitoring. The direct measurement of changes in the biological status of a habitat based on evaluations of the number and distribution of individuals and species before and after a change. *See* biological bench-mark, biological indicators.

biological shield. A thick (usually 3 to 4 metres) wall surrounding the core of a nuclear reactor designed to absorb neutrons and gamma radiation for the protection of personnel. The shield is usually made from concrete.

biologic specialisation. Physiologic specialisation.

bioluminescence. The production of light of various colours by living organisms (e.g. some bacteria and fungi, glow-worms, many marine animals such as fishes, Cephalopoda and Protozoa). Luminescence is produced by a biochemical reaction, catalysed by an enzyme, and produces no heat. In some animals (e.g. glow-worm) the light is used as a mating signal. In others it may be a protective device and in deep sea forms luminous organs may serve as lanterns.

biomass. The total weight of the organisms constituting a given trophic level (*see* food chain) or population, or inhabiting a defined area.

biome. A major ecological community of organisms, occupying a large area (e.g. tropical rainforest). *See* formation.

biometeorology. The science of the relationship of life to weather.

bioplex. Abbreviation of 'biological complex'. Biological system where the waste products of each stage are used as raw materials in a succeeding stage, the whole system forming a cycle.

bioseston. *See* Seston.

biosphere. The part of the earth and its atmosphere in which organisms live.

biostimulants. Substances which stimulate the growth of aquatic plants. *See* eutrophication.

biostratigraphic zone. *See* index fossil.

biosystematics. The study of the biology of populations, especially in relation to their evolution, variation, reproductive behaviour and breeding systems.

biota. The flora and fauna of an area.

biotic factors. Influences on the environment which are the result of the activities of living organisms.

biotic index. A rating used in classifying freshwater bodies (mainly rivers and streams) according to the type of invertebrate community present in the water. The biotic index is largely an indication of the amount of dissolved oxygen present, and this in turn is a measure of the level of organic pollution. Very clean water, holding a wide variety of species, including stonefly and may-fly nymphs, both of which are very sensitive to lack of dissolved oxygen, has a high biotic index. As pollution increases, oxygen levels decrease, and the more sensitive species disappear. Badly polluted water, in which only a few tolerant species (e.g. red midge larvae and annelid worms) can survive, together with a few animals which breathe air at the surface, has a low biotic index.

biotic potential. An estimate of the maximum rate of increase of a species in the absence of competition from predators, parasites, etc. This is usually enormous, but in fact is never achieved in natural surroundings because of the action of natural selection.

biotic province. A major ecological region of a continent (e.g. the Hudsonian province which covers most of Alaska and Canada).

biotic pyramid. The trophic levels within a stable food chain described graphically. Since a high proportion of the energy derived from food is used to sustain the metabolism of the organism that consumes it, the number of individuals at each tropic level decreases (e.g. one lion kills about 50 zebra a year). Thus a graphic description of the trophic levels in an ecosystem will resemble a pyramid, with primary producers at the base, then primary consumers (herbivores), secondary consumers and tertiary consumers. Animals at higher levels are often larger.

biotic succession. That component of a succession in which the composition of communities is related to interactions among different species.

biotin. A water-soluble nitrogenous acid that is classified as one of the B group of vitamins, although at one time it was known as vitamin H. Biotin is involved in the formation of fats and utilisation of carbon dioxide in many animals and it may be essential for Man. *See* vitamins.

biotite. An iron-rich mica mineral.

biotope. A habitat which is uniform in its main climatic, soil and biotic conditions. (e.g. a sandy desert).

biotype. A naturally occurring population consisting of individuals with the same genetic make-up.

bipinnate. Double pinnate. Applied to plants bearing compound leaves with more than three leaflets arranged in two rows on a single stalk, the leaflets then being further sub-divided.

birch. *See Betula.*

bird. *See* Aves.

bird ringing (banding). Method of marking birds with numbered light metal leg rings, to obtain information about their movements, life span, etc. In Britain, the main scheme is run by the *British Trust for Ornithology*, which trains people to catch birds (by means of mist nets, etc.) make observations on age, weight, etc., and ring them before release. About 500000 birds are ringed in Britain each year. The recovery rate is about 11000 per year. *See* duck decoy pond.

bird song. A form of communication between birds, developed strongly in passerines. The elaborate song of a male bird is used mainly to mark its territory and is often delivered from a vantage point (song post). Song is often the only safe means of separating some closely related species of warbler (e.g. Willow Warbler and Chiff Chaff) in the field. Sub-song is an extremely quiet form of song, the significance of which is not clear.

birth rate. The number of new births within a population in a given period of time, usually one year.

Bischoff process. German process for removing sulphur dioxide from flue gases with up to 90 % efficiency by injecting dry calcium oxide (CaO) into the scrubber.

bisexual. *See* hermaphrodite.

Bishop's ring. A brown ring, about 15 % in radius, rather rarely seen around the sun in haze of nearly uniform particle size.

bit. The unit of information, a contraction of '*bi*nary dig*it*', used in information theory and in modelling. *See* model, digital computer.

bittern. The liquid remaining after the crystallisation of sodium chloride from sea water; a source of iodine, bromine and magnesium.

bitumen. General geological term for various solid and semi-solid hydro-carbons. The term was defined in 1912 by the *American Society for Testing Materials* (ASTM) to include all hydrocarbons soluble in carbon disulphide, but current usage is generally more restrictive.

bituminous coal. A coal between lignite and anthracite in rank. Bituminous coals burn with a smoky luminous flame.

bivoltine. Applied to organisms which produce two generations in a year.

black body. (*a*) An ideal body which would, if it existed, absorb all and reflect none of the radiation falling upon it. The ideal absorber of solar energy. (*b*) A body radiating the spectrum appropriate to its temperature and by which its temperature may be measured. Thus daylight due to the sun is black body radiation at 3000°K. Snow and cloud are black bodies radiating in the infra-red.

Black Death. A plague prevalent in the 14th century in Europe, reducing the population by between one and two-thirds of its former value. Probably occasioned by less favourable climate and the impossibility of sustaining the growth and unprecedented population of the previous century.

black diamond (carbonado). A crystalline form of carbon, similar in some ways to diamond, a valuable abrasive because of its great hardness.

black earth. *See* Chernozem.

black frost. Clear ice not possessing the familiar whiteness of hoar frost, snow, etc. Sometimes occurring on wet roads at night. Refers to supercooled rain which accumulates on rigging, often with disastrous consequences in Arctic seas.

black mustard. *See* Brassica.

black rust. *See* Berberis vulgaris.

black sand. A coarse sedimentary concentrate of heavy minerals formed by water and/or wind action. The heavy minerals are commonly a mixture of magnetite, ilmenite, and haematite, with lesser amounts of rutile, cassiterite, gold, monazite, zircon, chromite, garnet and the ferromagnesian minerals. Black sands are placer deposits.

black smoke. Smoke produced by carbon particles released by hydrocarbon cracking followed by sudden cooling. *See* Brown smoke.

bladderworm (cysticercus). A larval stage of some tapeworms. It inhabits the tissues of an intermediate host and consists of a bladder containing an inverted scolex ('head'). This everts when eaten along with the tissues of the intermediate host by the definitive host. Bladderworms of the hydatid worm *Echinococcus granulosus*, whose adult stage infests carniverous mammals such as dogs, form large multiple cysts, which can cause great damage to their hosts, including Man.

blanket bog. *See* bog.

blastema. Mass of undifferentiated cells which will develop into an organ in an animal. The regeneration of a lost part often begins with the development of a blastema.

blast furnace. Furnace for the production of iron from iron ore. It is constructed of refractory bricks contained by steel plates. A charge of ore, coke and limestone $(CaCO_3)$ is introduced from above and sometimes sinter pellets. Molten iron and slag are removed from the bottom of the furnace. The product is pig-iron or cast iron.

blastula. An early stage of an animal embryo, usually a hollow ball of cells.

bleaching powder. Chlorinated lime (i.e. calcium hydroxide more or less saturated with chlorine). One of a number of chemical agents that will bleach or destroy natural colour, so rendering substances white, but has largely been displaced by elemental chlorine and especially by sodium hypochlorite.

bleicherde. Grey, leached layer of a Podzol. *See* soil horizons.

blepharoplast. *See* centiole, cillia, flagellum.

blizzard. A storm of snow blown up from the surface. Often used to refer to any snowstorm accompanied by strong wind.

block heating. *See* district heating.

blocking. The mechanism whereby an anticyclone remains stationary and cyclones move around it. It appears to block storms in their tracks.

blood. *See* blood corpuscles, blood groups, blood plasma, blood platelets, blood serum.

blood corpuscles. Cells which circulate in the blood of vertebrates or in-vertebrates. Erythrocytes (red blood cells), present in vertebrate blood, contain oxygen-carrying haemoglobin. In Man, they are biconcave discs without nuclei, about 5 million per cu mm of blood being the normal count; Leucocytes (white blood cells) are present in vertebrates and invertebrates. The three types in mammals are: Lymphocytes, small amoeboid but non-phagocytic cells, which make or carry antibodies; Monocytes, large phagocytic cells which enter the tissues and ingest invading micro-organisms; Polymorphs (polymorphonuclear leucocytes, granulocytes), phagocytic cells with lobed nuclei and granular cytoplasm, which also enter tissues to attack invaders.

blood groups. A group of people or animals bearing the same antigens on their red blood cells. If blood from different groups is mixed the red blood cells clump together because the antigens (agglutinogens) react with agglutinins in the plasma. In Man, there are four main blood groups, A, B, AB and O (the latter contains neither agglutinogen A or B). The proportions of the groups are not constant in populations throughout the world. If agglutinogens A or B are absent, the corresponding agglutinins are present. In a blood transfusion the corpuscles of the transfused blood must be compatible with the plasma of the patient, but little harm is done if the transfused plasma contains agglutinins incompatible with the patient's corpuscles. Other blood antigens, including the Rh (Rhesus) Factor occur, but their related antibodies are not normally present in plasma. If a Rh-negative woman is carrying a Rh-positive foetus and foetal blood crosses the placenta, antibody against the Rh factor appears in the mother's blood. In a subsequent pregnancy, antibody can cross the placenta and, if present in sufficient quantities, may damage a Rh-positive foetus.

blood plasma. The liquid part of blood. In vertebrates it contains proteins (e.g. fibrinogen, responsible for clotting), dissolved food substances (e.g. glucose), inorganic salts (e.g. sodium chloride and bicarbonate) and excretory products.

blood platelets (thrombocytes). Minute bodies, possible cell fragments, present in the blood of mammals. They produce an enzyme activator, thrombokinase, responsible for initiating the clotting of blood.

blood serum. The liquid which separates from clotted blood. It is similar to plasma (*see* blood plasma) but is without the ability to clot.

blood sugar. *See* carbohydrate.

bloodworms. (*a*) The aquatic larvae of certain non-biting (chironomid) midges, which contain haemoglobin and are tolerant of organic pollution.

(*b*) Sludgeworms or river worms. Mud-dwelling aquatic oligochaete worms (e.g. *Tubifex*), which contain haemoglobin and are tolerant of organic pollution. (*c*) Red polychaete worms which occur on muddy shores.

blow-by. Gases which blow past the piston rings and into the crankcase of an internal combustion engine, from which they are removed via a tube, and fresh air introduced into the crankcase through a filter. Crankcase emissions are now controlled in California, Australia and elsewhere, the gases being carried back to the intake manifold by the flow caused by the low (less than 1 atmosphere) pressure in the induction system. Uncontrolled crankcase emissions (unburnt fuel and exhaust gases) are a major source of hydrocarbon emissions.

blowdown. An occasion when a substantial but limited area of trees is blown down in a short time. They are rare, and most occur on the lee side of mountains and are due to the strong winds in the trough of lee waves.

blowing-down. The operation of discharging part of the water in a boiler while still under steam, to remove concentrated solids.

blow-out. A deflation basin excavated in sand or other easily eroded regolith. Blow-outs can be initiated in partly stabilised sand-dunes by the killing of marram and other binding agents by people's feet.

blue asbestos. *See* asbestos.

Blue-green algae. *See* Cyanophyta.

blue ground. Altered and brecciated (*see* breccia) igneous rock occurring in pipes within cratons. Blue ground is composed mainly of the ultrabasic rock kimberlite, with a range of xenoliths including eclogite. Some blue grounds contain diamonds.

blue-john. A decorative variety of fluorite consisting of alternating bands of blue and yellow crystals.

Blueprint for Survival. Document outlining the reasons for the non-sustainability of economic growth based on a constant expansion of industrial activity and resource and energy consumption, together with suggested strategies whereby a 'no-growth' society might be created. *A Blueprint for Survival*, by Edward Goldsmith, Robert Allen, Michael Allaby, John Davoll and Sam Lawrence was first published in January 1972 as a special issue of *The Ecologist* magazine, and subsequently in book form. It was translated into many languages and attracted much publicity.

blue sky. Rayleigh scattering of light by particles in the atmosphere that are

71

Blue whale

small compared to the wavelength of light. The blue sky is strongly polarised in directions seen at right angles to the sun, indicating that it is single scattering. Blue sky is seen in shadows when the air is viewed from above, e.g. from an aircraft or in dark sided mountain valleys.

Blue whale. *See* Balaenoidea.

BOD. *See* biochemical oxygen demand.

boehmite. One of the major ore minerals of aluminium, boehmite, with the formula AlO(OH), is one of the main constituents of bauxite and laterite.

bog. An area of wet acid peat in which grow characteristic plants, such as *Sphagnum* mosses, sundews (*Drosera* spp.) and Bog Myrtle (*Myrica gale*). Bogs form on badly drained ground, the lack of oxygen in the waterlogged soil preventing decomposition of dead plants, and may cover large areas as blanket bog. Raised bog may develop over valley bog or over fen.

boghead coal. A sapropelic coal rich in kerogen. The organic material in boghead coals appears to be mostly finely divided algal and fungal matter.

bog soil. Brown or black peaty material over buried peat (or over mottled mineral soil: Half bog). May be metres thick. Accumulation due to slow decomposition of organic matter in waterloggged conditions. *See* gley.

boiler. Pressure vessel designed to produce vapour from liquid by the application of heat.

boiling-water reactor (BWR). A light-water nuclear reactor using enriched uranium as a fuel, boiling light water as a coolant and light water as a moderator. Steam from the reactor is fed directly to turbines and then back into the reactor.

bolson. A low area or basin completely surrounded by higher ground. A bolson drains centripetally into a playa.

bolster eddy. An eddy sometimes found at the foot of a cliff or steep slope with air rotating in the corner driven by an upslope wind.

bomb calorimeter. Apparatus in which the heat produced by a reaction is measured, the reaction being made to take place in a closed vessel. Used in analysis to determine the calorific value of a material.

bombs, volcanic. Large pieces of molten lava ejected into the air. During flight and on impact, these develop characteristic forms giving rise to such descriptive terms as cow-dung bombs, bread-crust bombs, ribbon bombs and spindle bombs.

Bombyx mori. *See* Lepidoptera.

bone. A vertebrate skeletal tissue (*see* also cartilage) consisting of cells regularly arranged in a matrix of collagen fibres and bone salt (a complex compound containing calcium and phosphate). Channels permeating the matrix connect the cells and contain blood vessels and nerves.

bone conduction. The means by which sound can reach the inner ear and be heard without travelling through the air in the ear canal.

booklice. *See* Psocoptera.

bora. A cold katabatic wind from highlands, characteristic of Trieste and N. Yugoslavia.

Bordeaux mixture. Mixture of copper sulphate, quick lime and water long used as a fungicide in European vineyards. This, or similar mixtures containing inorganic copper, are used to control potato blight, fruit scab and other diseases of horticultural crops.

boreal forest. Coniferous forests of northern cold winter climates.

bornite. A copper iron sulphide mineral, Cu_5FeS_4. Bornite is a major ore of copper found in hydrothermal veins and in the zone of secondary enrichment.

boron (B). Element. A brown powder (SG 2.37) or yellow crystals (SG 2.34). Occurs as borax and boric acid. Used for hardening steel and in the production of enamels and glasses. It is used in the control rods of nuclear reactors because of its property of absorbing slow neutrons. It is an essential micronutrient for plants. mp 2300°C.

boss. The term is either synonomous with stock (North American usage) or else refers to a steep-walled stock which is roughly circular in plan. Some small bosses may be volcanic plugs.

Botanical Society of the British Isles (BSBI). The senior botanical society of Britain, founded in 1836, and responsible for the production of works such as the *Atlas of the British Flora.*

botryoidal. Having a form resembling a bunch of grapes.

bottle gas. lpg.

bottom load. *See* bed load.

bottom sets. The gently sloping distal sediment layers of a delta. The sediment is fine-grained.

botulism. A disease produced by the bacterium *Clostridium botulinum*, which causes many wild-fowl deaths during hot summers when water levels drop. The disease also affects other animals, including Man.

boulder. A block of rock more than 25 cm in diameter.

boulder clay. An unsorted, and usually unstratified, glacial deposit of rock-flour, often containing coarser, ice-transported material ranging in size from sand to boulders. Also called till, and one of the components of glacial drift.

boundary layer. The layer of fluid moving more slowly than the main stream because of closeness to a rigid boundary, or more generally applied to other properties of the fluid than velocity, e.g. pollution, temperature, etc. The layer partaking significantly of properties caused by the presence of the boundary. The boundary layer equations for the motion of a fluid are simplified versions of the fall equations but are approximately correct only close to the boundary.

Bovidae. Cattle, sheep, goats, etc., a large family of ruminant Artiodactyla with four-chambered stomachs and horns (which are not shed, *see* antlers).

BP (B.P.). Years before the present. O BP is taken conventionally to be 1950 AD.

Br. Bromine.

Brachiopoda (lamp shells). A phylum of animals superficially resembling bivalve molluscs with a shell of two valves, dorsal and ventral. Present-day Brachiopods are marine, solitary and usually attached to the bottom. Plentiful in Palaeozoic and Mesozoic times, but now a small group.

Brachycera (short horned flies). A sub-order of Diptera including horse flies and gad flies (Tabadinae), the females of which are blood suckers; robber flies, which catch insects in the air; and bee flies, the adults of which hover and feed in flowers.

Bracken (*Pteridium aquilinum*). A widespread fern which is often a troublesome weed on acid grassland and heaths because it is not eaten by most sheep or rabbits and recovers very fast after burning. It is the commonest dominant in the field layer of woods on acid soil in Britain.

brackish. Term used to describe water which contains some salt in solution, but less than is contained in sea water.

bract. A modified leaf beneath a flower.

bracteole. Small or secondary bract.

braided stream. Wide, shallow stream which repeatedly divides between islets of alluvium and rejoins.

brain. An anterior enlargement of the central nervous system in an animal. The development of a brain is correlated with the aggregation of sense organs at the anterior end. *See* cerebral hemispheres, cerebellum, medulla.

Branchiopoda. A sub-class of the Crustacea, including the Cladocera (water fleas, e.g. *Daphnia*) brine shrimps and fairy shrimps. Most live in fresh water and commonly reproduce by parthenogenesis. They have leaf-like thoracic appendages.

Branchiura (carp lice). A small group of Crustacea which are temporary ectoparasites (*see* parasitism) of freshwater and marine fishes. Branchiura have sucking mouthparts and a large, flattened head. They swim well by means of four pairs of thoracic limbs.

brass. Term for a series of alloys consisting basically of copper and zinc.

Brassica. Genus of plants, belonging to the family Cruciferae, many of which are cultivated for the stems, roots, leaves or seeds. *B. nigra* is the black mustard, *B. oleracea* is the cabbage, *B. campestris* is the turnip and *B. napus* is rape, grown for its seeds from which oil is extracted.

Brassicaceae. *See* Cruciferae.

braunerde. *See* brown forest soil.

breakwater. A wall built offshore to protect a beach or harbour from wave action.

bream zone. *See* river zones.

breccia. A rock composed of angular fragments of pre-existing rock predominantly over 2 mm in diameter. Breccias can be of sedimentary origin (implying little transport) or tectonic (fault-breccia), or volcanic (vent-breccia).

breeder reactor. Liquid metal fast breeder reactor.

breeze. Land and sea breezes are caused by temperature differences usually of diurnal origin. Similar breezes occur inland which are not part of the general motion of air over a large area. The name given to a welcome or refreshing wind of moderate or light strength.

brigalow forest. Forest in which the dominant species is the brigalow (*Acacia harpophylla*), covering large areas in Australia.

brimstone. Sulphur.

brine shrimps. *See* Branchiopoda.

Bristle-cone pine. *Pinus aristata*, a living specimen of which has been dated by dendrochronology as being 4600 years old. This is the oldest living organism so far dated, the oldest Redwood, *Sequoia sempervirens* so far dated is 3212 years old. *See* Taxodiaceae.

bristle-tails (Thysanura). *See* Apterygota.

bristle worms. (Polychaeta) *See* Annelida.

British Thermal Unit (BThU). The quantity of heat absorbed by one pound of water during an increase of temperature of one degree Fahrenheit. Use of this term is widespread in the USA. The therm is more popular in Britain, although the 'joule' and multiples thereof is rapidly gaining ground in both countries.

British Trust for Ornithology (BTO). Society founded in 1932 to further the scientific study of birds in the field. Its work includes running the bird ringing scheme and organising counts of species, e.g. herons.

brittle stars. *See* Ophiuroidea.

brl. Barrel, a volumetric unit used for crude oil.

broadleaved trees. Trees belonging to the Dicotyledonae. Some broadleaved trees are evergreen (e.g. Holly) but most British species are deciduous. *See* conifers, hardwoods.

Brocken spectre. A glory observed on the Brocken, a mountain in E. Germany.

broiler. System for semi-industrial production of poultry meat. Birds are housed indoors but not caged, and fed and watered by semi-automated systems. Stocking densities are usually very high. Effluent disposal often presents problems.

bromine (Br). Element that is liquid at room temperature, is brownish red and corrosive. It is a raw material in the manufacture of pharmaceuticals and dyestuffs, in the production of antiknock compounds, as a catalyst and it is a component of photographic materials. AW 79.909; At.No. 35. bp 58.8°C.

bronchial diseases. Diseases of the bronchial system involving the excessive production of mucus stimulated by smoking or pollution. They are chronic because the system becomes adapted to and dependent on the stimulant for the mucus production and cleansing of the system by coughing because of the increased viscosity. Infections are less easily prevented in a bronchitic lung. Coughing can cause structural damage to the alveoli (air sacs) causing emphysema in which the damaged part is filled with mucus and is permanently useless.

bronze. General term for alloys of copper and tin.

brood parasite. *See* parasitism.

Brown algae. Phaeophyta.

brown coal. *See* lignite, coal.

brown earth. *See* brown forest soil.

brown forest soil (brown earth, braunerde). A dark brown, friable soil with no visible layering, though lighter soil below surface. Organic matter gradually decreases down through well aerated soil to a calcareous parent material whose high calcium carbonate content helps to retard leaching. High agricultural potential. *See* soil horizons, soil classification.

Brownian movement. The random movement of small, suspended particles due to unbalanced impacts with molecules of the medium.

brown podzolic (podsolic) soil. Acid forest soil with surface litter layer over dark, greyish brown organic and mineral soil, with pale leached layer beneath. *See* grey brown podzolic soil, soil horizons, soil classification.

brown ores. Soils leached (*see* leaching) of the more soluble minerals but retaining rich amounts of iron oxides and hydroxides, particularly goethite. Such ores have declined in importance with the increased mining of banded ironstones, but may constitute a major resource for the distant future.

brown smoke. Smoke produced by volatile tarry substances emitting from coal burned at low temperatures. *See* black smoke.

brown soils. *See* chestnut and brown soils.

brunizem. *See* prairie soil.

brush discharge. An almost continuous release of electric charge to the air from prominent objects when a strong electric field, e.g. due to a thundercloud, is present.

bryocole. An animal which lives in moss (e.g. some Tardigrada).

Bryophyta. The liverworts (Hepaticae), hornworts (Anthocerotae) and mosses (Musci). Small plants, mostly terrestrial (although many prefer damp surroundings), attached by rhizoids. Mosses and some liverworts have stems and leaves, the rest are thalloid. The plant is a gametophyte and liberates motile male gametes which fertilise solitary egg cells housed in flask-shaped organs (Archegonia). A capsule (sporophyte), dependent on the gametophyte, grows out, and produces spores which give rise to new plants.

Bryozoa. *See* Polyzoa.

BThU. *See* British Thermal Unit.

bubble bursting. The chief mechanism for the projection of tiny water droplets from the sea surface, which are quickly dried and remain in the air as sea salt aerosol. Sea spray particles are too large to become dry before falling back, but the air entrapped in whitecaps and breakers generates bubbles which burst on reaching the surface.

budding. (*a*) A type of asexual reproduction in which new cells are formed as outgrowths of a parent cell (e.g. in yeasts) or new multi-cellular animals develop as outgrowths of the parent's body (e.g. in coelenterates such as Hydra) (*see* gemmation); (*b*) Grafting a bud on to a plant.

buffered solution. A solution (e.g. sea water and many fluids found in animal and plant bodies) which resists changes in pH if acid or alkali is added.

bug. *See* Hemiptera, Coleoptera.

bulb. A modified underground shoot with a much shortened stem bearing fleshy, food-storing leaf bases or scale leaves (e.g. onion, daffodil). A bulb is an organ of perennation and vegetative propagation, enclosing the next year's bud.

bumpiness. The rough ride experienced by aircraft travelling through the up and down currents of atmospheric convection. A glider or balloon may avoid it by remaining in an upcurrent. A fast aircraft may experience only rapid vibrations.

bunodont teeth. Cheek teeth which have separate conical elevations (cusps) on the crown. This type of tooth is typical of mammals with a mixed diet (e.g. pigs, Man). *See* selenodont, lophodont.

Bunter. The name of a Stage in European Lower Triassic System. Most of the reservoir-rocks in the southern North Sea gas-fields are Bunter sandstones.

buoyancy. The upward force on a body (including a body of fluid) immersed in a fluid. *See* Archimedes' Principle.

burette. A graduated glass column with a stopcock at the lower end. Used in chemical analysis for titration.

burn up. (*a*) The quantity of fissile material destroyed in a reactor by fission or by neutron capture, as a percentage of the original quantity present. (*b*) The heat obtained per unit mass of fuel.

bush layer. *See* layer.

buster, southerly. The name given to a burst of air of polar origin entering Australia behind a cold front with a strong wind.

butadiene. A major hydrocarbon raw material for the production of butadienne/styrene rubbers, water-based latex paints, etc.

butte. Formed when horizontal bedding, capped with resistant rock, has been worn away at the sides to form an isolated, flat-topped hill. *See* Mesa, Kopje.

Buys Ballot's law. The wind blows with low pressure on the left in the northern hemisphere and on the right in the southern hemisphere. Not applicable close to the Equator or violent squalls. *See* Coriolis. Buys Ballot (1817–1890) was a Dutch meteorologist.

bwr. Boiling water reactor.

b.y. Billions of years (a billion is one thousand million).

by-pass valve. A valve so arranged as to cause the fluid which it controls to flow past some part of its normal path. e.g. to allow a liquid to avoid a filter through which it usually passes.

byssinosis. A disabling lung disease caused by the inhalation of cotton dust during cotton manufacture.

C

c. (*a*) Carbon   (*b*) Coulomb.

c. Centi-.

Ca. Calcium.

cabbage. *See* Brassica.

Cactaceae (cacti). A family of dicotyledonous xerophytic plants found mainly in tropical America. The leaves are usually much reduced. The fleshy, water-storing stems are covered in a thick cuticle and bear spines which retard transpiration, promote the formation of dew at their tips, and protect the plant from the heat of the sun and from grazing. *Opuntia* bears edible fruit and is grown as the food plant of the cochineal beetle. A few cacti provide timber and some are used as hedge plants.

cacti. *See* Cactaceae.

Cactoblastis cactorum. *See* Lepidoptera, biological control.

caddis flies. *See* Trichoptera.

cadmium (Cd). Element. Soft, silvery-white metal, found associated with zinc ores and in the rare mineral greenockite. Used in the manufacture of fusible alloys, for electroplating, for making control rods in nuclear reactors. Cadmium pollution was blamed for the Japanese 'itai-itai' disease. AW 112.40; At.No. 48; SG 8.642; mp 320.9°C

caecilians. *See* Apoda.

caenogenetic. Applied to a special feature present in the embryo of young stage of an animal which adapts that particular stage to its environment, but which is not present in the adult (e.g. embryonic membranes of a vertebrate).

caesium (Cs). Element. A silvery-white metal that is highly reactive. Compounds are very rare. It is used in photo-electric cells and as a catalyst. AW 132.905; At.No. 55; SG 1.87; mp 28.5°C.

caffeine. An alkaloid derived from tea and coffee, being the stimulant ingredient of these drinks. Also produced synthetically.

Cainozoic. Cenozoic.

Cairngorm. Quartz made brown by slight impurities.

Cajon. Box canyon of American Southwest.

cake-urchins. *See* Echinoidea.

calc-alkaline. A term applied to igneous rocks in which the dominant feldspar is calcium-rich. Such rocks tend to contain calcium-rich ferromagnesian minerals (e.g. hornblende and augite). The opposite of calc-alkaline is alkaline.

calcareous. Containing calcium carbonate.

calcic. Applied to soil horizon of secondary carbonate accumulation, not less than 15cm thick and with calcium carbonate equivalent of more than 15%. *See* soil horizon.

calcicole. Plant that grows best on calcareous (chalk and limestone) soils (e.g. Old Man's Beard (*Clematis vitalba*), Rockrose (*Helianthemum chamaecistus*), Wayfaring Tree (*Viburnum lantana*). *See* calcifuge.

calciferol. Vitamin D. *See* vitamin.

calcifuge. Plant that grows best on acid soils. e.g. Ling (*Calluna vulgaris*), Bracken (*Pteridium aquilinum*). *See* calcicole.

calcimorphic soils. Soils containing excess limey material.

calcination. Strong heating in air, e.g. to convert metals to their oxides.

calcite. A very abundant carbonate mineral, with the formula $CaCO_3$. Calcite is the main constituent of limestones, marbles, and carbonatites, and is also a gangue mineral in some hydrothermal deposits. Calcite is an important cement in many sedimentary rocks.

calcitonin (thyrocalcitonin). Vertebrate hormone, secreted by the thyroid in mammals, which regulates the level of calcium in the blood by reducing its release from bone. *See* parathyroid.

calcium (Ca). Element. A soft white metal which tarnishes quickly in air. It occurs very widely as calcium carbonate and has many industrial uses. It is an *calcium carbonate (CaCO₃) (chalk, limestone)*

essential nutrient for plants and animals (*see* macronutrient). AW 40.08; At.No. 20; SG 1.55; mp 845°C.

calcium carbonate (CaCO$_3$) (chalk, limestone). Common mineral, widely distributed. Used as a filler for paints, rubbers and plastics, as a pigment in whiting, in the manufacture of putty, as an ingredient in polishes, in medicine.

calcium chloride (CaCl$_2$). A drying agent, often used in laboratories in U-shaped tubes to remove water from a gas stream.

calcium cycle. The circulation of calcium. Plants take up calcium from the soil, and it passes through the trophic levels of the ecosystem, returning to the soil. Losses by leaching enter aquatic ecosystems, and are made good in terrestrial ecosystems by the capture of calcium lost in surface water or migrating organisms from neighbouring ecosystems, or from underlying rock. Chalk and limestone rocks are predominantly of biological origin, being the accumulated insoluble calcium residues of former marine organisms. But *see* oolite.

calcium hydroxide (slaked lime). Can be produced by the action of water on calcium oxide. Used in sugar purification, glass production and in the manufacture of plasters and mortars; also agriculturally as lime.

calcium oxide (quicklime). A chemical of very wide industrial use, having a high affinity for reaction with water to give calcium hydroxide (slaked lime). Used in metallurgy, paper manufacture, petroleum processing and in the food industries.

caldera. A large (several kilometres in diameter) pit-like depression within a surrounding wall of volcanic material. Calderas are formed by explosive eruptions or by collapse caused by the removal of magma through subterranean conduits, or by ring-fracture or stoping.

caliche. *See* hardpan.

Callovian. A Stage of the Jurassic System.

Callunetum. A community of plants dominated by *Calluna vulgaris* (ling, heather).

caloric requirement. The amount of energy needed for an animal to maintain its normal functions. The oxidation of food provides this energy. When oxidised, 1g of carbohydrate (e.g. glucose) liberates 3.74 Calories (3740 Calories); 1g of protein liberates 4.1 Calories and 1g of fat liberates 9.3 Calories. The minimum daily caloric intake for a human being varies widely according to age, sex and occupation, but the average is approximately 2400 Calories.

calorie. A gram calorie is the amount of heat required to raise the temperature of 1 gram of water through 1°C at 15°C. A Calorie, or kilocalorie, is 10^3 gram calories and is the unit in which the energy value of food is measured, although this is now being replaced by the joule.

calorific value. The number of units of heat obtained by complete combustion of unit mass of a substance. This is the most important characteristic of a waste (municipal or industrial) that is to be disposed of by burning. Generally quoted in BThU/lb; the gross value is when the heat obtained by condensing water vapour is included, otherwise it is the net calorific value.

calyx. *See* flower.

cambic. Applied to a changed soil horizon either in structure or mineral content.

cambium. *See* meristem.

Cambrian. Refers to the oldest period of the Palaeozoic Era, usually taken as beginning some time between 570 and 610 million years before the present. Cambrian also refers to the rocks formed during the Cambrian Period: these are called the Cambrian System, and in the UK are divided into 4 Series, the oldest being the Comley Series, next the St. David's Series, then the Merioneth Series and finally the Tremadoc Series. These 4 Series were probably laid down in 70 to 80 million years. Geologists in most other countries put the Tremadoc Series into the Ordovician, so making the Cambrian shorter. The Cambrian is zoned by use of trilobites.

Camellia (Thea). *Camellia sinensis* is the tea plant, a small tree or shrub (family Theaceae) whose leaves contain caffeine. The plant is extensively cultivated in India, China, Japan, etc., and the young shoots are picked, fermented (except for green tea) and dried.

Campanian. *See* Senonian.

campanulate. Bell-shaped (esp. of flowers).

Campbell-Stokes sunshine recorder. Instrument which records sunshine by focussing it through a glass sphere on to a sensitive chart.

Canadian deuterium — uranium reactor (Candu). A nuclear reactor that uses natural uranium as a fuel, heavy water as a moderator, and as a coolant heavy water held under a pressure of 90 atmospheres inside tubes which also contain the fuel. The tubes run through a tank containing the moderator. A Candu reactor is working at Pickering, Ontario.

candela (cd). The SI unit of luminous intensity, being the intensity in a horizontal direction of a surface of 1/600000 square metre of a black body at the temperature of freezing platinum under a pressure of 101325 Nm^{-2} (newtons per square metre).

Candu. Canadian deuterium uranium reactor.

Cannabiaceae. (Cannabidaceae). A family of dicotyledonous plants containing only two genera, *Humulus* and *Cannabis*. *Humulus lupulus* (Hop) is a perennial climbing herb widely cultivated for its cone-like fruit used to flavour beer. *Cannabis sativa* (Hemp) is cultivated in temperate and tropical regions for its fibre and the drug (ganja, charas, bhang, pot) contained in its resin.

Cannabidaceae. *See* Cannabiaceae.

cannel coal. A sapropelic coal rich in kerogen. The organic material appears to be finely divided vegetable matter, spores, algae and fungi.

canopy cover. The percentage of the ground that is covered when a polygon drawn about the extremities of the undisturbed canopy of each plant is projected upon the ground and all such projections in a given area are added together.

cantharophily. Pollination by beetles.

CAP. Common Agricultural Policy.

cap cloud. A cloud which remains stationary on or over a mountain peak in a strong wind. *See* wave cloud.

capillary flow. (*a*) The ascent or descent of liquids within tubes of very small diameter due to the relative attraction between the molecules of the liquid and the molecules of the material composing the tube. (*b*) Movement of water upwards through soil spaces above the water table through pore-surface attraction. The smaller the pore spaces the greater the height to which water will rise.

capillary moisture. Water held in pores around soil particles by surface tension forces. *See* hygroscopic moisture, field capacity.

Caprimulgidae (nightjars). Nocturnal birds (Caprimulgiformes) which feed on insects taken on the wing. Eggs are laid on bare ground. Hibernation occurs in an American species.

cap-rock. The relatively impermeable upper seal for an underground reservoir containing crude oil or natural gas.

capsicum. *See* Solanaceae.

Capsidae (capsid bugs). A large family of plant bugs (Heteroptera), some of which are serious pests in orchards (e.g. the Apple Capsid, *Plesiocoris rugicollis*). A few are predatory and play an important role in the control of pests (e.g. *Blepharidopterus angulatus*, which feeds on the red spider mite).

capsid bugs. *See* Capsidae.

capsule. (*a*) A dry fruit formed from more than one carpel, which opens to liberate its seeds (e.g. poppy). (*b*) The swollen terminal portion of the sporangium of a moss or liverwort inside which the spores are formed. (*c*) An envelope of connective tissue, gelatinous material, etc., surrounding an animal organ, egg, bacterial cell, etc.

capture of droplets. This occurs when a large drop, i.e. a crystal or hailstone, falls faster than smaller cloud droplets and grows by accretion. Also significant on some trees, e.g. pine, whose leaves of small dimension accrete particles from cloud when it is present and materially add to the rainfall of the area. Sometimes a raingauge under a tree collects more than one in the open. *See* rime.

Caradocian. The fourth oldest Series of the Ordovician System in the UK.

carat. Unit of weight for jewels, equal to 200 mg. Also a standard of purity for gold alloys, 24 carat corresponding to pure gold, 22 carats to 22 parts of gold and 2 parts of alloying metal, etc.

carbamate (group of pesticides). Group of organic compounds with a wide range of biological activity, used as herbicides, fungicides or insecticides. They have a low toxicity to mammals and persist in the soil for only a short time. They include carbamates (e.g. Chlorpropham, Carbaryl, Barban, Asulam) also thiocarbamates and dithiocarbamates (e.g. Diallate, Triallate, Zineb, Maneb, Metham-sodium). *See* under individual headings.

carbaryl (sevin). Contact insecticide, earth-worm killer and growth regulator of the carbamate group, used to kill earthworms in turf and to control insects such as winter moth caterpillars and earwigs. Also used for fruit thinning in apples. Harmful to bees and fish, but not very toxic to mammals.

carbohydrate. Organic compounds with the general formula $C_x(H_2O)_y$. Glucose (dextrose) is a 6-carbon-atom (hexose) monosaccharide (simple sugar). It is made during photosynthesis and is the principal energy source for metabolic processes in plants and animals. It is the 'blood sugar' of vertebrates. Galactose is also a hexose sugar, differing only slightly in configuration from glucose. It is a constituent of many mucilaginous compounds and pectic substances. Ribose is a

5-carbon-atom (pentose) monosaccharide, and important constituent of nucleic acids. Sucrose (cane sugar) is a 12-carbon-atom disaccharide formed by the combination of the (6-carbon-atom) monosaccharides glucose and fructose. It is found in plants but not usually in animal tissues. Lactose (milk sugar) is a 12-carbon-atom disaccharide formed by combination of the monosaccharides glucose and galactose. It is a constituent of mammalian milk. Maltose (malt sugar) is a 12-carbon-atom disaccharide, each molecule a compound of two glucose molecules. It is formed as a breakdown product during the digestion of starch and occurs in germinating seeds. Starch is a polysaccharide, a combination of many monosaccharide molecules. It is made during photosynthesis and stored as starch grains in many plants. Glycogen (animal starch) is a polysaccharide made up of glucose units. It is stored by fungi and animals, notably in the liver and muscles of vertebrates. Cellulose is a long-chain polysaccharide made up of glucose units. It is a fundamental constituent of plant cell walls. Pectin is an acid polysaccharide, also a constituent of plant cell walls.

carbon (C). Element. Occurs in several allotropic forms (*see* allotropy) including diamond and graphite and as amorphous carbon (lampblack etc.). Owing to its valency of 4, carbon forms compounds with many other elements and is able to form very large molecules that are the chemical basis of life. AW 12.011. At. No. 6. mp 3550°C. SG: diamond 3.51; graphite 2.25; amorphous carbon 1.8 to 2.1.

carbon-14 dating. *See* radiometric age.

carbonaceous. (*a*) Coaly. (*b*) Rich in organic compounds and therefore a potential source rock for hydrocarbons.

carbon, activated. *see* activated carbon.

carbonado. *See* black diamond.

carbonate minerals. Minerals containing the carbonate group, CO_3. Since a limited substitution can occur between Ca, Mg, Fe^{2+}, Mn and Zn, many minerals exist with compositions intermediate between calcite, $CaCO_3$; dolomite $CaMg(CO_3)_2$; siderite, $FeCO_3$; ankerite, $FeMg(CO_3)_2$; magnesite, $MgCO_3$; rhodocrosite, $MnCO_3$; and smithsonite $ZnCO_3$. Other carbonate minerals include aragonite, $CaCO_3$ (a polymorph of calcite), and malachite, $CuCO_3.Cu(OH)_2$. Many living organisms produce hard parts of aragonite, or of calcite with differing amounts of magnesium substituted for calcium: after death many varieties change to a more stable form or suffer preferential leaching.

carbonation. *See* solution.

carbonatite. An intrusive or extrusive rock composed mainly of one or more of the carbonate minerals, calcite, dolomite, or ankerite, and typically a very rare associate of alkaline igneous intrustions. Carbonatites have suites of minor elements dissimilar to that of limestones, and some are sources of niobium and tantalum ores. Niobium (called columbium in the USA) finds increasing use in nuclear reactors, and is almost wholly produced from carbonatites.

carbon black. Finely divided forms of carbon made by the incomplete combustion or thermal decomposition of various hydrocarbons. The fine particles are a severe pollution nuisance in areas where they are produced industrially. The form of carbon used in printing ink is carbon black. It has been proposed for use in control of weather and climate by covering snow surfaces with carbon black to cause their melting by absorption of sunshine, but this is unlikely to be practical because the carbon black itself would be covered very readily by fresh falls of snow.

carbon cycle. The circulation of carbon through ecosystems. Carbon atoms from carbon dioxide are incorporated into organic compounds formed by green plants during photosynthesis. These compounds are eventually oxidised during respiration by the plants which made them, or by herbivores, carnivores, and saprophytes, thus releasing carbon dioxide for further photosynthesis.

carbon dioxide (CO_2). A minor constituent of the air, comprising about 0.4% of the atmosphere almost entirely of biological origin. CO_2 is essential to living systems, released by respiration (and to a much lesser extent by volcanic activity) and removed from the atmosphere by photosynthesis in green plants. The atmospheric content has increased by about one-fifth since the burning of coal and oil began on a large scale, but although there has been a theoretical increase of world temperature as a result of the enhanced greenhouse effect no effect has been observable. Atmospheric CO_2 varies by a small percentage with the seasons and the ocean contains many times the amount of the gas that is in the atmosphere.

Carboniferous. Refers to the fifth oldest Period of the Palaeozoic Era, usually taken as beginning some time between 345 and 350 million years before the present. Carboniferous also refers to the rocks formed during the Carboniferous Period: these are called the Carboniferous System, which in Europe is divided into two Series, the Lower (or Dinantian) and Upper (or Silesian). These Series are sub-divided into Stages: the Dinantian into Tournaisian, Visean and Namurian, and the Silesian into Westphalian and Stephanian. The Carboniferous Period lasted about 90 million years. In North America Carboniferous is replaced by Mississipian and Pennsylvanian, which are normally ranked as Systems.

carbon monoxide (CO). Colourless gas found in trace quantities in the natural atmosphere and produced by incomplete combustion, notably in motor cars and cigarettes. Able to form a stable compound with blood haemoglobin. Was a *carbon tetrachloride (CCl_4)*

means of suicide in kitchen ovens when domestic gas was coal gas supplemented by water gas. CO is not a component of natural gas. CO is harmless in small doses. The natural level in the blood is about 0.5% saturation due to the decomposition of old red blood corpuscles. In dense traffic it seldom rises above 2% or 3%, which produces less effect than half a pint of beer. Smokers frequently achieve levels between 4% and 8% saturation of the blood and this produces a slight intoxicating sensation similar to alcohol, which may be one of the attractions of smoking. Chain smokers achieve 9% or 10% frequently. No permanent harm is known except in patients with overloaded or diseased pulmonary or cardiac function. Although traffic fumes have other dangers, that from CO is less than from conversation with a smoker in a closed space, such as a car. Haldane raised his blood saturation by CO to 40% with acute anoxaemia symptoms, but no permanent effects.

carbon tetrachloride (CCl_4). Widely used industrial solvent, best known as a dry cleaning agent. An irritant whose vapour may cause dizziness and headache. Prolonged exposure may lead to liver and kidney damage. A carcinogen.

carboxyhaemoglobin (COHb). The result of the combination of carbon monoxide with blood haemoglobin. While haemoglobin has an affinity for oxygen, its affinity for carbon monoxide is greater.

carboxylic acid. Organic acids in which a carbon fragment is attached to a carboxyl (CO.OH) group. The carboxylic acids include formic acid (HCO.OH), acetic acid (CH_3COOH), proprionic acid (CH_3CH_2COOH), butyric acid (CH (CH_2) COOH), valeric acid ($CH_3(CH_2)_3COOH$), lauric acid ($CH_3(CH_2)_{10}COOH$), myristic acid ($CH_3(CH_2)_{12}COOH$), palmitic acid ($CH_3(CH_2)_{14}COOH$), stearic acid ($CH_3(CH_2)_{16}COOH$), and benzoic, phthalic and terephthalic acids formed from benzene rings joined to carboxyl groups.

carboy. Large glass bottle-like vessel, often enclosed in wicker or steel basket. Used for storing chemicals, but its use is decreasing in industry.

carcinogen. Cancer-inducing substance, e.g. some hydrocarbons such as benzpyrene, asbestos, vinyl chloride, cutting oils and radio-active substances.

carcinogenic. *See* carcinogen.

carcinoma. Cancer.

cardiac muscle. *See* muscle.

Carica Papaya (**Pawpaw**). (Caricaceae) A small tree cultivated in warm countries for its edible fruit. The milky juice is used in digestive salts, and meat is tenderized by being wrapped in the leaves, which brings about partial digestion.

88

carnallite. An evaporite mineral, with the composition $KCl.MgCl_2.6H_2O$, which is one of the major sources of potassium, an essential fertilizer. *See* sylvite.

carnelian. A form of chalcedony.

Carnivora. *See* carnivore.

carnivore. (*a*) A flesh-eating animal or plant (e.g. sundew). A secondary consumer in a food chain (*b*) A member of the order Carnivora, largely predatory placental mammals, including dogs, cats, bears, otters, stoats, badgers, mongoose, pandas, raccoons and hyaenas. Carnivora have large canine teeth and sharp molars and premolars. *See* Pinnipedia.

Carnot Theorem. No engine operating between two given temperatures can be more efficient than a perfectly reversible engine operating between the same temperatures.

carnotite. A mineral which is a hydrated oxide of uranium and vanadium with the formula $K_2(UO_2)_2(VO_4)_2.3H_2O$, and is found in sedimentary rocks. Carnotite is a major ore mineral of both uranium and vanadium.

carotene. *See* carotenoids.

carotenoids. Group of orange, yellow and red pigments, including carotene and xanthophylls, found in plants (e.g. in carrot roots, some fruits and flowers, blue-green and yellow-green algae, some bacteria and fungi). They absorb light and can therefore assist in photosynthesis. Carotene is made into vitamin A by vertebrates.

carpel. *See* flower.

***Carpinus betulus* (Hornbeam).** (Carpinaceae). A tree of northern temperate climates, native to southern England, which bears bunches of nutlets, each with a leaf-like, three-lobed wing. Hornbeam is often dominant as a coppiced shrub in oakwoods in south east England. The timber is not much used.

carp lice. Branchiura.

carr. Fen woodland dominated by plants such as alder (*Alnus*) and willows (*Salix*).

carrying capacity. (*a*) (Ecol.). The maximum number of species which an area can support during the harshest part of the year, or the maximum biomass which it can maintain indefinitely. (*b*) Grazing capacity. The maximum number of grazing animals that an area can support without deterioration. (*c*) Planning.

The level of use, at a given level of management, which a natural or man-made resource can sustain without an unacceptable degree of deterioration of the character and quality of the resource. e.g. ecological capacity: the maximum level/of recreation use, in terms of numbers and activities that can be accommodated before a decline in ecological value, assessed from the ecological viewpoint.

cartel. An association of independent undertakings in the same or related branches of industry which aims to improve or stabilise conditions of production or sale for the mutual benefit of its members. To be strong, a cartel must control a large proportion (usually 75 % or more) of the production of a commodity and also its distribution.

cartilage. Flexible vertebrate skeletal tissue consisting of groups of cells lying in a matrix containing collagen. It occurs on the ends of bones, in the pinna of the ear, and in intervertebral discs, and forms most of the embryonic skeleton. *See* Chondrichthyes.

Caryophyllaceae. A widespread family of dicotyledonous, mainly herbaceous plants. The British species include the campions, pinks, chickweeds, stitchworts, pearlworts and sand spurreys.

caryopsis. The fruit of grasses. An achene with the ovary wall fused to the seed coat.

cascade. A repetitive system for separation or purification, giving a gradual increase in the separation of components as the process is repeated. The principle is used in many industrial processes, including the separation of nuclear isotopes, including uranium.

case-hardening. Method for producing a hard surface layer on a non-hardening type of steel by heating the steel in contact with certain carbon compounds, from which carbon is absorbed producing a thin layer of high carbon steel, which can then be heat-treated.

casein. The chief protein of milk, also used in the production of paints, glues and foodstuffs.

cassava. *See* Manihot.

cassiterite. A mineral with the formula SnO_2, and the major source of tin. Cassiterite is found in acid igneous rocks (particularly pegmatites) and contact metamorphic zones and in hydro-thermal deposits, but much of the world's production comes from placer deposits. The major use of tin is in the form of tin-plate in which a thin layer of tin acts as an anti-corrosion coating for mild steel, but tin is also used for a great variety of alloys (e.g. solder, bearing-metal, gun-metal, bronze, bell-metal and pewter) as well as in chemicals.

caste. A type of individual anatomically specialised for particular function within an insect colony. In a honey bee colony there are three castes: fertile females (queens), sterile females (workers) and males (drones). *See* polymorphism.

castellatus. Turret shaped cumulus clouds often present in lines. A very unstable form of altocumulus, individual turrets having lifetimes of the order of 5 to 10 minutes.

casting. The making of a pottery or metal object by pouring material into a mould. The mould is usually of plaster for clay casting and of sand for iron casting.

castles in the air. *See Fata Morgana.*

casual plant. *See* adventive plant.

catabolism (katabolism). The breaking down by organisms of complex molecules into simpler ones, with the liberation of energy. *See* anabolism.

catadromy, (katadromy). The migration of some fish (e.g. freshwater eel) from the rivers down to the sea for spawning.

catalase. An iron-containing enzyme which breaks down hydrogen peroxide to oxygen and water.

catalysis. A process in which the rate of a reaction is altered by the presence of an added substance which remains unchanged at the end of the reaction. Catalysts are usually employed to accelerate reactions (positive catalysts); retarding (negative) catalysts are also used.

catalyst. Substance whose presence changes the rate of a chemical reaction without itself undergoing permanent change in its composition. *See* catalysis.

catalytic cracking. The breaking of carbon – carbon bonds with the aid of a catalyst; an essential process in the refining of petroleum. *See* cracking.

catalytic oxidation. Japanese dry process for removing sulphur dioxide from flue gases, producing ammonium sulphate.

catalytic reactions. Chemical reactions in which the amount of one of the substances involved, the catalyst, is not decreased, although its presence is essential for the reaction to occur. *See* NO$_x$.

catalytic reforming. The use of heat, pressure and a catalyst to isomerise (*see* isomer) hydrocarbon molecules, thus varying their properties for specific use.

catarobic. Applied to a body of water in which organic matter is decomposing fairly slowly, and oxygen is not being used in sufficient quantities to prevent the activity of aerobic organisms. *See* mesosaprobic, polysaprobic, oligosaprobic.

catastrophism. The belief that the history of Earth and life consists of a series of discontinuous events of a scale of magnitude higher than those operating today. Although catastrophic processes are important, most geologists today are adherents of Uniformitarianism, and most biologists of evolution.

catchment area. A drainage system into which water flows to a particular location such as a main river system or a lake. In Britain, Water Authorities are now based on main catchment areas. *See* watershed.

catena. (*a*) A group of soils showing variations in type due to differences in topography or drainage although derived from uniform or similar parent material. (*b*) Also used in the sense of a diagram of the above.

cathode. *See* anion.

cation. Positively charged ion. *See* anion.

catkin. A long, usually tassel-like inflorescence which contains either male or female flowers (e.g. Hazel, Birch, Poplar, Willow, Oak).

cattle. *See* Bovidae.

caudata. Urodela.

caulocarpus. Having a fruit-bearing stem, as, for instance, in *Theobroma* (cocoa) and *Coffea* (coffee).

Cause's Principle (competitive exclusion principle). The law stating that two species with identical ecological requirements cannot exist together in the same habitat unless there is a superabundance of environmental resources including food.

caustic scrubbing. Process for removing sulphur dioxide from flue gases, using a sodium hydroxide (caustic soda) solution with which the sulphur dioxide reacts to produce sodium sulphite and bisulphite, which can be discharged into a stream, after the addition of lime has caused the precipitation of the pollutants as calcium sulphate. The process is used in the USA.

caustic soda process. Japanese process for removing sulphur dioxide from flue gases, using a wet scrubber, and producing sodium sulphite.

CCC. *See* chlormequat.

Cd. cadmium.

cd. candela.

celestine. celestite.

celestite, (celestine). The mineral strontium sulphate ($SrSO_4$), found chiefly in sedimentary rocks, but also in hydrothermal deposits. Celestite is the major source of strontium, which imparts a crimson colour to fireworks, and also finds uses in sugar-refining.

cell. A minute unit of living matter (protoplasm) bounded by a thin membrane (plasma-membrane) and, in plants, surrounded by a cell wall usually made of cellulose. The protoplasm of most cells is divided into a nucleus and the cytoplasm, which contains various inclusions (plastids, vacuoles, mitochondria, ribosomes, centrioles, lysosomes, endoplasmic reticulum, Golgi apparatus). Many micro-organisms consist of a single cell. The cells of multicellular organisms (which in sexually reproducing forms are all derived from a single cell, the zygote) are specialised for various functions and vary enormously in structure. *See* syncytium, plasmodium, coenocyte, protoplast.

cell division. *See* mitosis, meiosis.

cell membrane. *See* plasma-membrane.

cellophane. A strong, flexible, transparent film made from cellulose nitrate or acetate and used as a wrapping material.

cell theory. The theory, put forward in 1838–39 by Schleiden and Schwann, which proposed that all organisms are composed of cells and their products, and that reproduction and growth are the results of cell division.

cellular convection. Convection in a layer of uniform thickness. Best exemplified in fluids of large viscosity, often called Bénard cell convection.

cellular respiration (tissue respiration). *See* respiration.

cellulose. *See* carbohydrate.

cell wall. The outer supporting layer of a plant cell, made by the protoplast and consisting (in higher plants, many algae and a few fungi) largely of cellulose. The long-chain cellulose molecules are arranged in bundles, forming a network of strands separated by other polysaccharides including pectin. A middle lamella of pectic material forms across the centre of a dividing cell, and the wall is built on this from each side. It is subsequently thickened in most cells, but minute pits are left through which pass fine threads of cytoplasm (plasmodemata) connecting adjacent cells. Further thickening of the wall may occur, incorporating strengthening deposits of lignin on schlerenchyma, vessels, and tracheids, or waterproofing deposits of cutin on epidermal cells and suberin on cork cells.

Celsius, Anders (1701–1744). Swedish astronomer who devised the centigrade scale of temperature.

cement. (*a*) (Geol.) The mineral matter binding the fragments together in sedimentary rock. Common elements are silica (as quartz or chalcedony) calcite, dolomite, haematite, limonite, siderite, and gypsum. (*b*) (Construction) A powder which sets hard after having been mixed with water. The chemistry of hardening is not well understood. Portland cement is the most commonly used cement. Cements contain burnt lime (quicklime CaO) with silica (SiO_2), alumina (Al_2O_3), with small amounts of iron (Fe_2O_3) and often magnesium oxide (MgO). Gypsum is usually added to slow the hardening process. Where these minerals occur in a natural rock in the required proportion, the rock is called cement (or limestone) rock. If the limestone clinker is ground more finely than for Portland the result is a rapid hardening cement. High alumina cement (Aluminous cement) is made by melting bauxite ($Al_2O_3.2H_2O$) and limestone and grinding the clinker without gypsum, to provide a cement which reaches its maximum hardness very quickly. Sulphate-resistant cement has a higher (not less than 3.5%) content of tricalcium aluminate ($3CaO.Al_2O_3$), hydrated aluminate compounds being very susceptible to sulphate attack. Glass fibre reinforced cement is a mixture of glass fibres and cement which can be cast, pressed or injection moulded for tunnel linings, window frames etc. *See* concrete, mortar.

cementation. Increasing the carbon content of steel by heating it in contact with carbon compounds, as in case-hardening.

cement kiln. A long, slightly inclined, rotary chamber in which the slurry is calcined (*see* calcination) in the manufacture of Portland cement.

cement rock. *See* cement.

Cenomanian. A Stage of the Cretaceous System.

Cenozoic. The Era of geological time following the Mesozoic. In some usages the Cenozoic Era was followed by the Quaternary Era (i.e. Cenozoic is synonomous with Tertiary), and in others (e.g. that of the US Geological Survey) the Cenozoic continues to the present day (i.e. the Cenozoic is comprised of the Tertiary and Quaternary, both of which rank as Periods). Cenozoic also refers to rocks formed during the Cenozoic Era.

census (stat.). A complete enumeration of a whole population with respect to specific variables, as distinct from sampling.

centi-(c) Prefix used in conjunction with SI units to denote the unit $\times$ 10^{-2}.

central eruption. A volcanic eruption from a central vent, as distinct from an eruption from a fissure. Central eruptions form cones of two major varieties: shield volcanoes and stratovolcanoes.

centres of diversity. *See* Vavilov, Nikolai Ivanovich.

centres of origin. *See* Vavilov, Nikolai Ivanovich.

centrifugal force. The inertial force directed away from the centre of curvature evident in a body made to move in a curved path.

centrioles. A body situated in the centrosome and revealed by electron microscopy to be very similar in structure to the kinetosomes found at the bases of cilia and flagella. Each centriole is a cylinder, the walls of which are formed by a ring of nine groups of fibres. Centrioles are self-duplicating and form the poles of the spindle during cell division.

centrosome. A small region of cytoplasm, present in animal cells, situated near the nucleus, and containing a pair (usually) of centrioles. 'Centrosome' is sometimes used as a synonym for 'centriole'.

centipedes. *See* Myriopoda.

central nervous system (CNS). Nervous tissue which co-ordinates an animal's activities. In most invertebrates the CNS consists of solid, ventral nerve chords bearing swellings or ganglia. In vertebrates it is hollow and situated dorsally, inside the skull and vertebral column, and consists of the brain and spinal cord. Nerve cell bodies are aggregated in a CNS and many junctions (synapses) between nerve cells are present, so that impulses coming in from sense organs and going out to effector organs can be finely co-ordinated. In higher animals the brain is able to store information and is the seat of learning. In vertebrates the outer layer of the CNS consists of white matter, composed largely of nerve fibres, and the inner layer (also the cerebral cortex) is the grey matter, made up mainly of nerve cell bodies. *See* autonomic (sympathetic and parasympathetic) nervous system, peripheral nervous system, nerve net, nerve cell.

centromere. Spindle attachment.

cephalic index. The breadth of the human head expressed as a percentage of the length (front to back).

Cephalochordata. *See* Acrania.

Cephalopoda (Siphonopoda). Octopuses, squids, cuttle fish, nautilus, ammonites, etc. A class of molluscs characterised by the possession of a well-developed head bearing tentacles, complex eyes and a shell which is often reduced and internal. *See* Decapoda, Octopoda.

CEQ. Council on Environmental Quality.

ceratostomella ulmi. The fungus which causes Dutch Elm disease, spread by the Elm Bark Beetle, *Scolytus scolytus*. *See* Ulmus.

cerebellum. Part of the hind brain of vertebrates, particularly well developed in birds and mammals, which co-ordinates complex muscular movements.

cerebral cortex. *See* cerebral hemispheres.

cerebral hemispheres (cerebrum). A pair of swellings at the anterior end of the vertebrate brain, concerned largely with the sense of smell in primitive forms, but very well developed in birds and mammals, and responsible for the general co-ordination of the activities of the animal. In higher vertebrates the outer layer, the cerebral cortex, consists of grey matter and is extensive and much folded. *See* central nervous system.

cerebrum. *See* cerebral hemispheres.

cesspool. Storage tank for sewage that must be emptied at intervals. *See* septic tank.

Cestoda (tapeworms). A class of flatworms (Platyhelminthes) the adults of which parasitize terrestrial and aquatic vertebrates. The adult tapeworm has no gut or complex sense organs, but possesses a thick cuticle and hooks and/or suckers at one end to attach it to the host's gut. Most produce a chain of proglottides (segments) each containing reproductive organs which gives the characteristic ribbon-like appearance. The larval stage is passed inside one or more intermediate hosts (e.g. rabbit, sheep, man or flea for various dog tapeworms). *See* bladderworm.

Cetacea. Whales, dolphins and porpoises. An order of aquatic placental mammals without hair or hind limbs, but possessing blubber, front limbs modified as paddles and a transverse tail fin. The hunting of whales for their oil, spermaceti (used in the manufacture of perfumes) and baleen (whalebone) has threatened some species with extinction. The catch is now limited by international conventions. *See* Balaenoidea, Odontoceti.

cf. Cubic feet. Used in measurements of hydrocarbon reserves.

CFBR 1. Commercial Fast Breeder Reactor No. 1. *See* nuclear reactor.

CFC. Chlorofluorocarbons. *See* chlorofluoromethanes.

CFM. Chlorofluoromethanes.

Chaetae *See* Annelida.

Chaetognatha (arrow worms). A small phylum of pelagic animals with transparent, elongated bodies, often found in swarms in marine plankton (e.g. *Sagitta*).

Chaetopoda. A class of Annelida whose members possess bristles (chaetae). *See* Oligochaeta, Polychaeta.

chain reaction. A mechanism in which its operation causes its continued operation. Thus when a raindrop, having grown by accretion of cloud droplets to about 5 mm in diameter is broken by aerodynamic forces into several smaller droplets, each of which grows by accretion, rain is produced by a chain reaction.

chalcedony. A cryptocrystalline variety of silica. Some varieties of chalcedony are semi-precious, such as agate, onyx, carnelian and jasper. Chalcedony occurs in sediments, as gangue in hydrothermal deposits, and filling amygdales.

chalcocite. A copper sulphide mineral, Cu_2S. Chalcocite is a major ore of copper and is found mainly in the zone of secondary enrichment.

chalcopyrite. A copper iron sulphide mineral, $CuFeS_2$. Chalcopyrite is a widely occurring mineral but is mined mainly from hydrothermal deposits (e.g. porphyry coppers) and from stratiform deposits confined to particular horizons in organic-rich sedimentary rock and metamorphosed sediments. The latter deposits show evidence both of deposition of copper minerals from sea-water during sedimentation and of metasomatism during early diagenesis.

chalk. (*a*) *See* calcium carbonate (*b*) A very fine-grained, friable, pure limestone forming extensive deposits of Cretaceous and Lower Tertiary age. Most chalks are white.

Chamaeophyte. *See* Raunkiaer's Life Forms.

change of state. Transformation from one to another of the three states of matter: gas, liquid and solid.

chaparral. Type of vegetation dominated by shrubs, with small, broad, hard evergreen leaves, found in areas with a Mediterranean climate.

char. General term for the solid product of pyrolysis or carbonisation of an organic material. The char from the pyrolysis of municipal waste is used as a fuel.

characteristic impedence. A measure of the quality of a substance through which sound travels, expressed as the ratio of the effective sound pressure at a point to the effective particle velocity, and is equal to the product of the density of the substance and the speed of the sound within it, the result being a value in 'Rayls'. Air at 20°C and a pressure of 1000mb has a characteristic impedence of 408 Rayls.

characteristic species. Species which, almost without exception, are localised within a given association. They provide the most reliable floristic expression of the ecology of the group.

Charadriidae (waders). A family of small or medium-sized birds with long legs, wings and bills. Most are shore-birds which nest on the ground and are highly migratory, often occurring in flocks on passage (e.g. curlews, snipe, sandpipers, plovers, the Avocet, and the Woodcock).

charadriiformes. A large order of swimming or wading birds including the waders (Charadriidae), Auks (Alcidae), gulls and terns (Laridae), skuas, oystercatchers and the Stone Curlew.

charcoal. Impure carbon obtained by heating carbonaceous substances, e.g. wood, with little or no oxygen. It is used as a deodorant, decolourant, absorbant and catalyst.

charcoal, activated. *See* activated carbon.

chard. *See Beta vulgaris.*

charophyta (stoneworts). A group of algae found in still or slow-moving fresh or brackish water. The filamentous thallus bears whorls of branches and the plant becomes heavily encrusted with lime (e.g. Chara).

chelate. Chemical term referring to a compound having a ring structure, where a metallic ion is attached by coordinate bonds to two or more atoms in the ring.

Chelonia (testudines). Turtles, terrapins and tortoises. An order of toothless reptiles in which the body is encased in a 'shell' consisting of two plates, with the thoracic vertebrae and ribs fused to the dorsal one in most groups.

chemical engineering. Originally the branch of engineering concerned with the manufacture of chemical products on an industrial scale. Many other products and industries are now within the compass of chemical engineering practice: in general the process industries covering a wide range of manufacture, including the conversion of industrial raw materials and production of foodstuffs.

chemical oxygen demand (COD). The weight of oxygen taken up by the organic matter in a sample of water, expressed as parts per million of oxygen taken up from a solution of boiling potassium dichromate in two hours. The test is used to assess the strength of sewage and trade wastes.

chemoautotrophic (chemosynthetic). Applied to organisms which produce organic material from inorganic compounds, using simple inorganic reactions as a source of energy. e.g. *Thiobacillus* obtains energy by oxidising hydrogen sulphide to sulphur and other bacteria utilise energy from the oxidation of ferrous salts to their ferric form. *See* autotrophic.

chemoreceptor. A sense organ which is sensitive to the presence of chemical substances (e.g. insect antennae, vertebrate taste buds).

chemosynthetic. *See* chemoautotrophic.

chemotaxis. *See* taxis.

chemotrophic. Applied to organisms which obtain energy from any source other than light. The energy may be obtained from simple inorganic reactions (*see* chemoautotrophic) or from the oxidation of organic material by heterotrophic organisms. *See* phototrophic.

chemotropism. (*a*) In zoology, once a synonym for chemotaxis (*see* taxis). (*b*) In botany, a growth response in which the stimulus is a chemical concentration gradient (e.g. the growth of a pollen tube through the stigma and style in response to substances contained in the ovule or ovary wall).

Chenopodiaceae. A family of dicotyledonous mostly herbaceous plants, mainly of arid regions, and frequently adapted to tolerate salty soil (halophytic). The leaves are often reduced, fleshy or hairy. British species include Fat Hen (*Chenopodium album*), oraches (*Atriplex* ssp.), Sea Purslane (*Halimione portulacoides*) and samphires (*Salicornia* spp.). Useful members of the family are *Beta vulgaris*, Spinach (*Spinacia oleracea*) and some species of *Chenopodium* grown for leaves and seeds.

chernozem (black earth). Grassland soil found in sub-humid cool to temperate areas where a humus remains near the surface and a black or brown surface layer grades downward through lighter coloured layer to a layer where lime has accumulated. Natural vegetation of Chernozem is tall grasses; it is the soil of the Russian Steppes, North American Great Plains and Prairies and the Argentinian Pampas. *See* soil classification.

chestnut soils & brown soils. Similar to Chernozem soils, but occur in warmer areas of lower rainfall. Natural vegetation grasses shorter than those on Chernozem. *See* soil classification.

chert. Cryptocrystalline silica, usually occurring as nodules or as a replacement material in sedimentary rocks. *See* flint.

chi γ

Chilopoda. *See* Myriopoda.

Chimaera (Chimera). (*a*) An organism composed of a mixture of cells with different genetic compositions. This results from chromosome changes in part of the organism during development or from grafting (plants then being graft hybrids). *See* mosaic. (*b*) A genus of cartilagenous fishes (King of the Herrings, Rabbit Fish), belonging to the largely fossil group Holocephali whose members have large, flat crushing teeth and the upper jaw fused to the skull.

Chimera. *See* Chimaera.

chimney. Vertical passage used to disperse a gas stream into the atmosphere. Most often used to cool combustion products from domestic fires and from a wide range of industrial processes, e.g. power stations, cement works. Typically made from brick or steel and movement in them may be activated by an induced-draught fan, although the gas stream, by virtue of being above the temperature of the surroundings, tends to produce a natural draught.

china clay. Kaolin.

china man. *See Homo.*

chinampas. Mexican style of agriculture based on irrigation canals constructed on a rectangular grid pattern with beds to which the silt from periodic dredging of the canals is added, together with organic wastes of all kinds, including water weeds. The water is flushed out from fresh water springs whose water is brought to the chinampas by aquaducts, so preventing salination. The system was developed in Aztec times, and was highly productive.

Chinook. A warm wind from the west in Alberta, Canada, which descends from the Rocky Mountains and displaces cold air masses in winter. *See* Föhn.

Chiroptera (bats). A very large order of insectivorous or fruit-eating placental mammals, with fore limbs modified to form wings. The wing membrane is supported by the elongated bones of the 'arm' and second to fifth digits.

chi squared (X^2) test. (Statistical). Widely used to test agreement between a set of observations and a set of hypothetical figures.

chitin. A tough, flexible substance which forms much of the exoskeleton on Arthropoda and the bristles of Annelida. Its long chain molecules are partly polysaccharide but also contain nitrogen. The cell walls of many fungi contain a similar substance.

chloracne. Disfiguring skin rash caused by exposure to TCDD. The rash is difficult to treat and has been known to persist for 15 years. Workers exposed to TCDD at the Bolsover, Derbyshire, UK, factory of Coalite and Chemical Products Ltd in 1968 were free from chloracne in four years. Many cases occurred at Seveso, Italy. *See* TCDD, Seveso.

chlorates. Salts containing the chlorate ion; very strong oxidising agents; like perchlorates they can form explosive mixtures with organic materials.

chlordane. Persistent chlorinated hydrocarbon earthworm killer used in turf. Because of its side effects, usage in the UK is now limited. *See* cyclodiene insecticides.

chlorinated hydrocarbon. *See* organochlorine.

chlorination. (*a*) The process of introducing one or more chlorine atoms into a compound, often using gaseous chlorine. (*b*) The application of chlorine to water, sewage or industrial wastes for disinfection or other biological or chemical results.

chlorine (Cl). Element. Essential precursor of many industrial chemicals. An extremely toxic gas, combining with water (as in the lungs) to produce hydrochloric acid (HCl) and hypochlorite. Also used as a bleach and in water purification. Used as a weapon in World War I. It occurs naturally as common salt (halite) and as chlorides of other metals and is manufactured almost entirely by the electrolysis of brine. AW 35.453. At. No. 17.

chlorine demand. The amount of chlorine needed to kill all the pathogens in a sample of water.

chlorite. A group of silicate minerals closely related to micas, which are found in low-grade metamorphic rocks, as an alteration product of igneous rocks, and in sediments.

chlormequat (CCC). A growth regulator used to shorten and strengthen straw in oats and wheat.

chlorofluorocarbons (CFC). *See* chlorofluoromethanes.

chlorofluoromethanes (CFM) (chlorofluorocarbons, CFC). Compounds of carbon similar to carbon tetrachloride (CCl_4) or methane (CH_4) but containing some chlorine and some fluorine. They are non-poisonous and inert at ordinary temperatures and easily liquifiable under pressure, which makes them excellent refrigerants, solvents, foam-makers, etc. They are exceptionally suitable for atomisers (aerosol sprays) for medical and domestic purposes. They are not at present known to occur naturally and are unlikely to do so because of the special conditions required for their synthesis. They are biodegradable only very slowly if at all and their decomposition in the stratosphere by UV is the most likely sink, the greater the proportion of fluorine the higher the altitude of decomposition. CFMs containing hydrogen e.g. $CHClF_2$ are relatively quickly decomposed in the troposphere, as are those containing no fluorine, e.g. $CHCl_3$, which is also relatively plentiful in nature.

chloroform ($CHCl_3$). Simple chlorinated hydrocarbon solvent used in plastic, rubber and resin industries. It is an anaesthetic, but has adverse effects on health and may be a carcinogen.

chlorophenotone. *See* DDT.

Chlorophyceae. *See* Chlorophyta.

chlorophyll. A green pigment, present in algae and higher plants, which absorbs light energy used in photosynthesis. Except in blue-green algae, chlorophyll is confined to chloroplasts. Several types of chlorophyll occur, but all contain magnesium and iron. Some plants (e.g. brown and red algae, the Copper Beech tree) contain additional pigments which mask the green of their chlorophyll. *See* Bacteriochlorophyll.

Chlorophyta (Chlorophyceae, Isokontae). The green algae. The largest and most diverse division of algae, occurring in fresh and salt water and in damp places on land. Some are microscopic, often able to move by means of flagella, and occur as single cells or colonies. Others are filamentous (e.g. *Spirogyra*) or have a flattened thallus (e.g. *Ulva*, the Sea Lettuce).

chloroplast. A cytoplasmic body (plastid) of plant cells which contains chlorophyll and is the site of photosynthesis. Each chloroplast contains numerous grana, in which chlorophyll molecules lie within envelopes stacked like piles of coins. Here the 'light' reactions of photosynthesis take place. The subsequent 'dark' reactions go on outside the grana, in the surrounding stroma, which is bounded by the membrane of the chloroplast.

chlorosis. A yellowing of the normally green parts of plants, which occurs when chlorophyll formation is prevented (e.g. by lack of light, iron or magnesium).

chlorpropham. *See* CIPC.

Choanichthyes. A group of mainly extinct bony fishes, comprising the lungfishes (Dipnoi) and Crossopterygii. They are characterised by having nostrils opening into the mouth, and fleshy, lobed pairs of fins with a central bony axis. Amphibia are believed to have evolved from the Devonian Choanichthyes.

choline. A nitrogenous alcohol once believed to be necessary in the diet of many animals and so classed as a vitamin. Provided the diet contains adequate amounts of methionine and vitamine B_{12} (cobalamine) it is probable that it is not necessary, so placing in doubt its status as a vitamin, although it may be valuable in marginal diets. Where it is necessary, its absence leads to haemorrhage of the kidneys and the excessive deposition of fat in the liver. Choline is an important constituent of lecithin. *See* vitamin.

chondrichthyes (Elasmobranchii). The cartilaginous fishes, including sharks, dogfish, skates, rays, chimaeras, and many extinct forms. They are almost all marine. They have no bones, but their cartilage may be calcified. *See* Osteichthyes, Holocephali.

chondriosomes. Mitochondria.

chondrites. A type of stony meteorite usually containing small, spherical mineral bodies, called chondrules.

chondrule. *See* chondrite.

Chordata. A phylum of animals characterised by the possession, at least in the early stages of development, of a notochord (a longtitudinal supporting rod of vacuolated cells enclosed in a firm sheath, lying just below the CNS), 'gill pouches' and a hollow dorsal nerve cord. The phylum includes the vertebrates (Craniata) and the more primitive Protochordata (Acraniata, Urochordata, Hemichordata). *See* vacuole.

chorion. (*a*) One of the embryonic membranes of amniota. In mammals it forms much of the placenta. (*b*) A non-cellular envelope surrounding some ova (e.g. insect eggs).

chorology. The study of areas and their development.

chromatid. One of a pair of strands formed by the duplication of a chromosome.

chromatin. The nucleoprotein which makes up the chromosomes.

chromatography. Technique of chemical analysis using different rates of movement of the material being analysed over a base material. Comparison with known substances allows identification. Widely known as a method for analysing small quantities of chlorinated hydrocarbons. Main types are: 1. partition chromatography, which involves selective solution of the material between two solvents; 2. paper chromatography where materials travel at varying rates along a piece of paper one end of which is immersed in a solvent. *See also* gas chromatography.

chromatophore. (*a*) (Bot.) A pigmented plastid or thykaloid. (*b*) (Zool.) A pigment cell. Some animals (e.g. prawns, flounder, minnow, frogs, chameleons) are able to change colour by the concentration or spreading of the pigment granules within the chromatophores. This is under the control of hormones, and in fish and chameleons it is also controlled by the nervous system. Cuttle-fish are able to change colour rapidly because the chromatophores can be dilated by the contraction of radiating muscle fibres attached to the surface of the cells. *See* melanins.

chromic acid. Bright orange caustic material that reacts with organic matter. Used widely in laboratories as the strongest cleaning agent for glassware, it is also used in ceramic glazes, the rubber and textile industries, chromium plating, as an oxidising agent, and in general for de-greasing.

chromite. A mineral of the spinel group, with the formula $FeCr_2O_4$, which is the major ore-mineral of chromium. Chromite ore occurs as layers in ultrabasic igneous bodies, and in placer deposits, and more rarely in hydrothermal veins. Most chromium is used in alloy steel rather than as chromium-plate.

chromium (Cr). A hard white metal. It occurs as chrome iron ore (chromite) and is used in the manufacture of stainless steel and for chromium plating. AW 51.996; At. No. 24; SG 6.92; mp 1890°C.

chromophores. Functional groups attached to hydrocarbon radicals that produce the colours in dyestuffs. More generally, any thing or process producing colour in a substance.

chromosomes. Strands of genetic material which become obvious during cell division. In animals and most plants they are situated in the nucleus and consist of DNA and protein, but in bacteria, blue-green algae and viruses they are made up of nucleic acid only (RNA in some viruses). Each species has a characteristic number of chromosomes (46 in man) which are present in pairs, the members of which look alike (except for the sex chromosomes), and associate during meiosis. These pairs are homologous chromosomes. Gametes and the cells of gametophytes contain the haploid number of chromosomes, having only one of each pair. Most other cells have the full (diploid) number. Each chromosome consists of a linear series of genes, the constant positions (loci) of which have been mapped in some species. All the body cells of an organism have an identical set of genes because the DNA in each chromosome is exactly replicated and during mitosis the identical halves of the chromosomes separate and form the daughter nuclei. *See* genes, megachromosomes, sex chromosome.

chrysalis. Pupa.

Chrysophyceae. *See* Chrysophyta.

Chrysophyta (Chrysophyceae). The golden-brown and orange-yellow algae. A diverse group of microscopic algae, inhabiting fresh and salt water, many being planktonic. They contain carotenoid pigments and may be unicellular, colonial, filamentous or amoeboid.

chrysotile. A fibrous serpentine worked for the manufacture of asbestos.

chylocaulus. Applied to plants with stems which are succulent (e.g. cactus) *See* xerophyte.

chylophyllous. Applied to plants with leaves which are succulent (e.g. stonecrops, *Sedum*). *See* xerophyte.

Ciconiiformes. An order of birds including the herons, bitterns, storks, spoonbills and flamingos. They are large birds, with long necks, legs and bills, most of which feed on fish and other aquatic organisms in shallow water.

cilia (sing. cilium). Many short threads of cytoplasm which project from the surface of cells and beat in a regular sequence in a constant direction. They occur in many Protozoa and on ciliated epithelium in many multicellular animals (e.g. in respiratory passages of land animals). They have the effect of either moving fluid past the cell or of rowing the cell through a liquid medium. Many animals (e.g. bivalve molluscs, amphioxus) feed by filtering off organic particles from a stream of water passed through the animal by the action of cilia. True cilia occur in very few plants (e.g. male gametes of Cycads). A cilium contains two central filaments surrounded by a ring of 9 double ones, and it is anchored to a body (kinetosome) at its base. *See* flagella.

ciliary feeder. *See* cilia.

Ciliophora. A class of Protozoa whose members move by means of cilia and possess a double nucleus (e.g. Paramecium, which belong to the sub-class Ciliata). *See* Infusoria.

Cinchona. *See* Rubiaceae.

cinders. (Geol.) Scoria between 4 and 32 mm in diameter.

cinnabar. The mineral mercuric sulphide, HgS, and the principal ore of mercury found in near-surface hydrothermal deposits. Native mercury is commonly associated with cinnabar. Mercury is used in detonators, in various scientific instruments and electrical switches and in dentistry.

CIPC (chlorpropham). Herbicide of the carbamate group, used to control germinating weeds and to prevent sprouting in stored potatoes.

circadian rhythm (diurnal rhythm). Rhythmic changes with a periodicity of approximately 24 hours which occur in plants and animals even when they are isolated from daily changes in the environment (e.g. sleep patterns and leaf movements).

circle of inertia. The circular track along which a theoretical particle would travel on a frictionless rotating earth if it were subjected to no horizontal forces.

circulation (meteor.). The flow around a closed circuit drawn in a fluid. Expressed mathematically by the formula $\oint v.ds$, v being the velocity and ds the element of arc of the circuit. Circulation adds kinetic energy without effecting any bulk transport.

circumferential velocity. Tangential velocity.

circumhorizontal arc. A brightly coloured horizontal spectrum of colours, with red at the top occurring at an altitude below 32°, and below the sun; seen when the sun is a little more than 58° above the horizon, and caused by the sun's rays entering a vertical face and emerging from a horizontal face of hexagonal ice crystals with vertical axes.

circumnutation. Nutation.

circumpolar vortex. The system of westerly winds in high latitudes, particularly at altitudes between 1000 and 12000 m, which circulate air around the pole.

circumscribed halo. A halo seen around the sun which is tangent to the 22° halo at the top and bottom, but outside it at the sides. It is roughly elliptical when the altitude of the sun is high, but has a sagging shape at low altitudes. Caused by the passage of the sun's rays through hexagonal crystals with horizontal axes.

circumzenithed arc. A brightly coloured circular arc, with red at the bottom, seen above 58° above the sun, when the sun is below 32°, centred on the zenith and caused by the sun's rays entering the horizontal top and emerging from a vertical side of hexagonal ice crystals with vertical axes.

cirque, corrie, cwm. Steep bowl-shape hollowed out by glacial ice in mountainous regions; often forms the head of a valley.

Cirripedia (barnacles). Sedentary, aquatic Crustacea, usually with a body enclosed in calcareous plates, and with feathery appendages used for gathering food particles from the water. Barnacles have free swimming larvae. Some species are parasitic. *See* parasitism.

cirrocumulus. A contradiction in terms, meaning a fibrous heap cloud. Used to describe altocumulus when it occurs at the same levels as fibrous-looking ice cloud and, perhaps because of the height, appears to be composed of small puffs.

cirrostratus. A layer of fibrous-looking cloud covering a large part of the sky.

cirrus. Cloud with fibrous appearance, always composed of ice crystals, compact masses of cloud being drawn out into 'fibres' as the crystals fall through the air without evaporation.

cistine. Amino acid with the formula $(HOOC.CH(NH_2)CH_2S)_2$. Molecular weight 240.3.

cistron. A length of DNA upon which a molecule of messenger RNA is built. The messenger RNA determines the sequence of amino acids which go to form a particular protein. A cistron is usually regarded as being equivalent to a gene.

citrate processes. Used in the USA to remove sulphur dioxide from flue gases. The gas is contacted with an aqueous solution of sodium citrate with which it reacts to form a bisulphite-citrate complex, which is reacted with hydrogen sulphide to form free sulphur and regenerate citric acid. About two-thirds of the recovered sulphur is used to produce the hydrogen sulphide by reduction with natural gas and steam over an alumina catalayst. The process is 90 to 95% efficient.

citric acid cycle (Krebs' cycle). Part of the process of aerobic respiration. A series of enzyme-controlled reactions during which pyruvic acid (formed in glycolysis) is broken down to carbon dioxide and water and ATP is built up. The reactions occur in the mitochondria. *See* diagram.

The Citric Acid (Krebs') Cycle

CH3.COOH(acetic acid) + COOH.CO.CH2.COOH (oxaloacetic acid)
(1)
COOH.C(OH).CH2.COOH (citric acid)
CH2.COOH
(2)
− H2O
COOH.C:CH.COOH (cis-asconitic acid)
CH2.COOH
(3)
+ H2O
COOH.CH.CH(OH).COOH (isocitric acid)
CH2.COOH
(4)
− 2H
COOH.CH.CO.COOH (oxalosuccinic acid)
CH2.COOH
(5)
CO2 + COOH.CH2.CH2.CO.COOH (alphaketoglutaric acid)
(6)
+ H2O − 2H
CO2 + COOH.CH2.COOH (succinic acid)
(7)
− 2H
COOH.CH.CH.COOH (fumaric acid)
(8)
+ H2O
COOH.CH(OH).CH2.COOH (malic acid)
(9)
− 2H
COOH.CO.CH2.COOH (oxaloacetic acid)

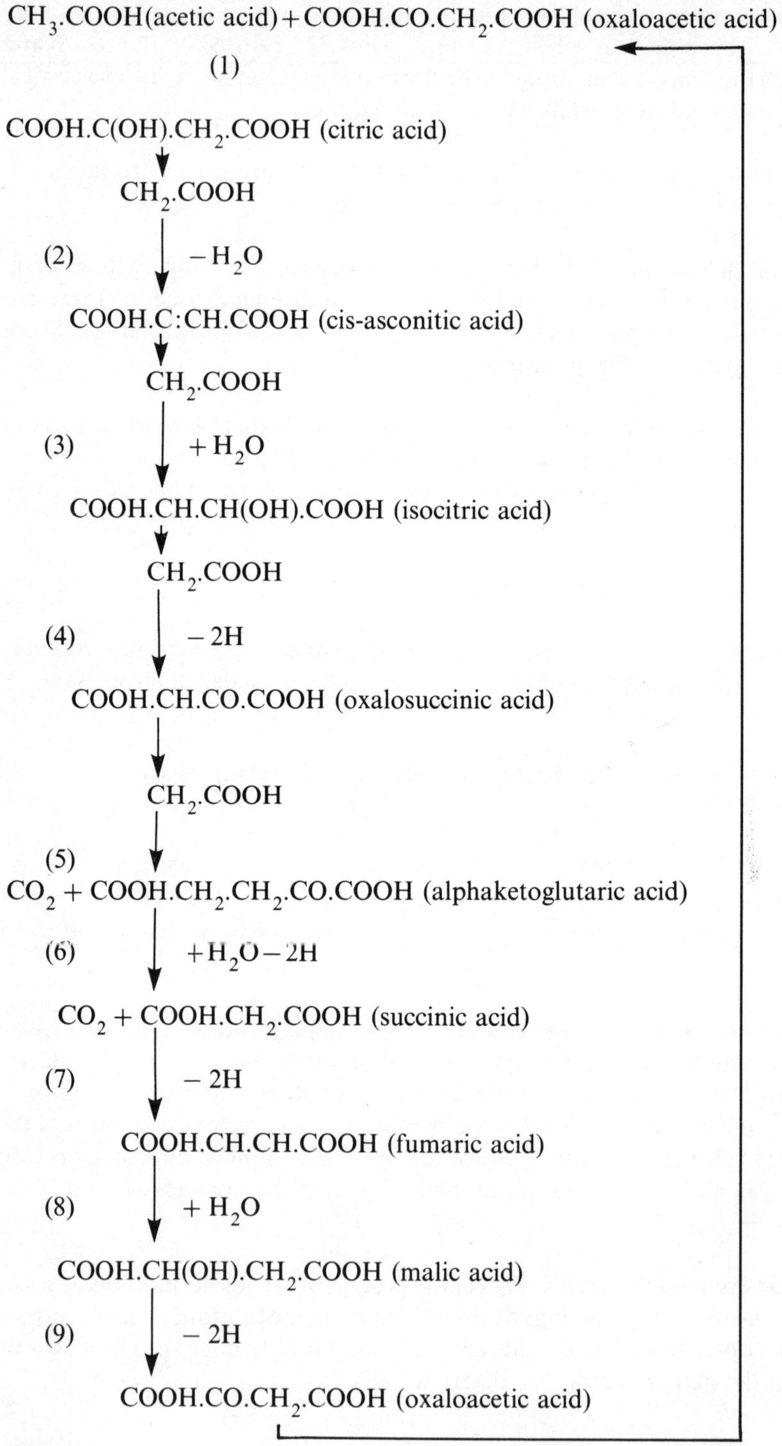

citrine. Quartz made yellow by slight impurities.

citrus. A genus of trees and shrubs (Family: Rutaceae) including the lemon, lime, orange, and grapefruit, widely cultivated in warm countries. The fruit is a berry with a leathery outer layer and flesh consisting of large cells growing out from the inner layer of the pericarp.

city climate. Changes in climate due to the influences of the city are sometimes beneficial as in the reduction of humidity due to rain running off roofs and paved surfaces which dry quickly and increase of night temperature due to use of fuel and trapping of air in streets, reduction of wind strength due to shading by small buildings, but are sometimes adverse, as in the production of pollution and consequent reduction in sunshine and lowering of temperature and generation of fogs or in the increase of wind caused by tall buildings.

city farm. A theoretical concept in which all the wastes from an urban community are processed into forms that can be fed to fish or to livestock. *See* bioplex.

Cl. Chlorine.

cladistic. Term applied to plants to indicate recent origin from a common ancestor.

Cladocera (water fleas). *See* Branchiopoda.

Cladode (Phylloclade). A stem modified to resemble a leaf in appearance and function, (e.g. Asparagus, Butcher's Broom).

clan. A group of plants composed of secondary species in local, or restricted, small, scattered areas. They are more or less permanent features of climax communities or of consocies which exist for long periods. A clan is a small community, of subordinate importance, but of distinctive character. Often it is the product of vegetative propagation.

clarification. Removing turbidity and suspended solids by settling. Chemicals can accelerate this process through coagulation.

clarke. Unit used to express the average percentage of an element in the Earth's crust. The clarke for the two most abundant elements, oxygen and silicon, is 46.60 and 27.72 respectively, and that for tin is 0.004.

clarke of concentration. The ratio of the amount of a particular element in a mineral or ore compared to its average crustal abundance. Thus an ore is a deposit in which the concentration clarke of the desired element is sufficiently high to make extraction profitable. Tin is economically worked at about 1% metal content, or a concentration clarke of 250.

class. *See* classification.

classification. Biological classification is based mainly on structural criteria and arranges organisms in a hierarchy of groups which reflect evolutionary relationships. Usually the smallest group is the species, although sub-species and varieties may be recognised. A species is generally regarded as a group of organisms which resemble each other to a greater degree than members of other groups and which form a reproductively isolated group that will not normally breed with members of another group. The common names of plants and animals often denote species. Similar species are grouped into genera, genera into families, families into orders, orders into classes, classes into phyla (for animals) or divisions (for plants) and these into kingdoms (Plant, Animal, Protista, etc.). The systematic position of the European Crested Newt can be given thus:

Kingdom	Animalia
Sub-kingdom	Metazoa
Phylum	Chordata
Sub-phylum	Craniata (Vertebrata)
Class	Amphibia
Order	Urodela
Family	Salamandridae
Genus	*Triturus*
Species	*cristatus*

The system is flexible and as knowledge increases organisms are continually being re-classified. *See* binomial nomenclature.

clastic. Descriptive term for sediments made from fragments of parent rocks or minerals that have been deposited by mechanical transport, sediments formed from organic remains are termed bioclastic.

clathrate. A type of compound in which one kind of molecule is enclosed within the structure of another without there being any direct bonding between the two.

Claus kiln. Oil refinery unit for recovering sulphur from gases rich in hydrogen sulphide. The hydrogen sulphide (H_2S) is burned in an insufficient supply of oxygen, giving the reaction:

$$2H_2S + O_2 \rightarrow 2H_2O + 2S$$

claviceps. *See* ergot.

clay. Sedimentary particles, mostly clay minerals, less than 0.004 mm across.

clay minerals. Stable secondary minerals formed by the weathering of some primary minerals. Form bulk of silicate material in soil and many sedimentary rocks. Have characteristic layered crystal structure, average grain size less than 0.002 mm in diameter. e.g. kaolinite, illite, montmorillonite, chlorite.

claypan. *See* hardpan.

clear air turbulence. The name given to small scale motion in clear air which causes bumpiness to aircraft with special reference to such motion above cloud base where convection from the surface is inoperative. It can be detected by radar and forward scattering of radio waves mainly because it is usually produced by unstable waves on stably stratified layers (*See* billow clouds). It is notably associated in its more intense forms with jet streams, particularly the exit region, and mountain waves.

cleavage. (*a*) (Biol.) *See* segmentation. (*b*) (Geol.) 1. The tendency for a crystal to split along planes determined by the crystal structure. 2. The tendency for a metamorphic rock to split along closely-spaced parallel planes, which are not usually parallel to bedding planes. Rock-cleavage is developed by pressure and is usually accompanied by at least a partial recrystallisation of the rock.

cleidoic egg. The egg of a land animal (e.g. bird, reptile, insect). The shell isolates it from the environment and allows only for gaseous exchange and a limited loss of water.

cleistogamy. Fertilisation which occurs inside an unopened flower. e.g. Wood Sorrel (*Oxalis*) and violets produce, after the showy flowers, small unattractive flowers which never open and so are capable only of self-pollination.

Clematis vitalba (**Traveller's Joy, Old Man's Beard**). (Ranunculaceae) A perennial woody climber bearing large heads of plumed achenes. It grows in hedgerows and wood margins and its presence indicates that the soil is calcareous. *See* calcicole.

clepto-parasite. *See* parasitism.

climate. The long term or integrated manifestation of weather.

climatic climax. The climax produced by a climate that is local and different from the climate of the area in general.

climatological station. A meteorological recording station whose purpose is to obtain records over a period of many years for the purpose of the study of climate.

climatological statistics. Statistics which describe the weather averaged over many years.

climatology. The study of climates over long periods of time.

climax. A community of organisms that is in equilibrium with existing environmental conditions, and forms the final stage in the natural succession. Oak woodland is the climax in much of lowland Britain, but on upland limestone it is ash woodland. *See* plagioclimax, preclimax, postclimax, subclimax, serclimax, proclimax.

cline. A gradation of structural differences among the members of a species which is correlated with ecological or geographical distribution. Populations at each end of a cline may differ substantially from one another, e.g. the Coal Tit (*Parus ater*) has 21 recognised sub-species which blend into one another and are distributed over Europe and parts of North Africa and Asia.

clinometer. Instrument for measuring angles in a vertical plane, by means of a pendulum or spirit level. Used chiefly for measuring the dip of such geological features as bedding planes, fault planes, joints and cleavage.

clint. A ridge in a limestone rock surface. Clints are separated by grikes.

clisere. A series of climaxes set in motion by a major change in climate (e.g. glaciation).

clone. A group of organisms of identical genetic constitution (unless mutation occurs) produced from a single organism by asexual reproduction, parthenogenesis or apomixis. *See* Ramet, Ortet.

closed community. A community in which colonisation is precluded because all the niches are well occupied.

Clostridium pasteurianum. An anaerobic, nitrogen-fixing bacterium responsible, along with other N-fixing bacteria such as *Azotobacter*, for the gradual increase in the nitrogen content of unmanured grassland.

clotting (coagulation) of blood. The solidifying of blood which normally occurs on injury, thus sealing the wound. In vertebrates the injured tissues and blood platelets release thrombokinase, which activates prothrombin, so forming the active enzyme thrombin. This acts on the soluble blood protein fibrinogen, converting it to a dense network of fine threads of fibrin, which entangles the corpuscles, forming a clot. Normally blood does not clot because heparin (antithrombin) neutralises thrombin and prevents the conversion of prothrombin to thrombin. Thrombokinase renders heparin inactive.

cloud. Water present in the atmosphere in the form of suspended droplets that have condensed on condensation nuclei. The formation of cloud requires that moist air be cooled, and the main cooling mechanism is decompression caused by the upward movement of air, the base of the cloud being determined by the limit of vertical convection. This process forms cumulus cloud in unstable air. In stable air, heat may be lost by radiation until the water vapour condenses into stratus cloud. Continuous layers of cumulus-type clouds may result from turbulence.

cloud amount. Usually measured in oktas, or eighths of the sky occupied by cloud, and reported thus by a single numeral, 9 indicating that the sky is obscured by fog.

cloud base. The height of the lowest part of a cloud or layer thereof, usually with special reference to low cloud, e.g. scud or lifting fog, a layer of stratocumulus, or cumulus for which the base represents the condensation level.

cloud bow. An arc caused like a rainbow in spherical cloud droplets.

cloud chamber. A vessel in which air may be expanded adiabatically to produce cloud. In physics laboratories the Aitken nuclei are removed by repeated condensation and sedimentation. The tracks of alpha particles can then be made visible because the ionisation produces condensation nuclei. The cloud chamber was the first instrument to be used to observe tracks of particles caused by radioactivity.

cloud generation. Almost entirely caused in the natural clouds by the adiabatic cooling due to ascent of air. Exceptions are steaming fog, radiation fog, aircraft trails.

cloud searchlight. Used to point a light beam vertically and measure cloud height by observing the elevation of the illuminated spot. Can be used by day, employing audio-frequency modulation or monochromatic source.

cloud seeding. *See* artificial rain. The introduction of dry ice, sea salt, or silver iodide into clouds in order to promote rainfall.

cloud shadows. The major cause of the reduction of sunshine and the unknown quantity in all physical theories of climatic change.

cloud streets. Clouds arranged in long parallel lines usually along the direction of the wind shear, which frequently differs little from that of the wind. They are caused when convection is limited to a uniformly deep layer over a wide area.

clubmosses. *See* Lycopodiales.

Club of Rome. An 'invisible college' founded in April, 1968, to foster under-standing of the varied but interdependent components that make up the global system in which we all live; to bring that new understanding to the attention of policy-makers and the public; and so to promote new policy initiatives and action. The club consisted initially of 30 scientists, educators, economists and humanists and civil servants from 10 countries, led by Dr Aurelio Peccei. Membership of the club is restricted and cannot exceed 100, but it grew quickly to a membership of 70 from 25 countries. The club initiated a project on the *Predicament of Mankind*, the best known result of which was the computer modelling of the world conducted at the Massachusetts Institute of Technology, funded by the Volkswagen Foundation, and published as *The Limits to Growth* in 1972.

Clupeidae. A family of bony fishes, including the herring (*Clupea*), sprat, pilchard, anchovy and shad, important items of Man's diet.

Cnidaria. Coelenterata.

cnidoblast. Nematoblast.

CNS. *See* central nervous system.

Co. Cobalt.

coagulation. Process of converting a finely divided or colloidally dispersed suspension of a solid into particles of such a size that reasonably rapid settling takes place. *See also* flocculation, colloid, clotting of blood.

coal. Carbonaceous fossil fuel of world-wide distribution, occurring in stratified accumulations. Coals form two series: (*a*) the humic coals, which with an increase in percentage carbon content (known as 'rank') pass from peat to lignite to bituminous coal to anthracite; and (*b*) the sapropelic coals such as cannel coal, boghead coal and oil shale. These consist of finely divided vegetable material such as spore cases, algae and fungal matter. Coal is the most abundant of fossil fuels, with fewer intrinsic dangers than nuclear fuels (*See* nuclear reactor), but it must be mined from the ground, either by open-cast mining or by pit mining, often in dangerous conditions.

coalescence. The formation of large drops out of smaller ones. Cloud droplets do not coalesce unless there is a fair spread of sizes causing collisions due to differential fall speeds. It is thought not to be a significant factor in rain generation among droplets all less than 20 microns in diameter.

coal gas. Mixture of combustible gases obtained by the destructive distillation of

114

coal, a process that produces a heavily contaminated waste gas and waste water containing a high content of organic substances. By-products of gas manufacture are coke, ammonia and many organic compounds. The process has declined in importance during the 20th century. *See* coal tar.

coal tar. A black, viscous liquid obtained by the destructive distillation of coal. It is a major raw material for pharmaceuticals, dyes, solvents, and other organic compounds. It contains several organic carcinogens as do many of the products of heating coal or oils, e.g. cutting oils. *See* coal gas.

coastal effects. Changes in weather at a coast may be caused by temperature differences, altitude differences, or roughness differences. The weather at a coast may exhibit increased wind due to cliffs. The vegetation may be affected by sea spray and snow may lie for shorter time because of salt or warmer sea air. Fogs may be more prevalent than in open sea due to cold upwelling water.

cobalamine (vitamin B$_{12}$). A cobalt-containing vitamin required by many organisms, including Man, for normal cell division. It is synthesised by the bacterium *Bacillus subtilis*, an inhabitant of the intestines in many animals. Deficiency in Man causes pernicious anaemia.

cobalt (Co). Element. A hard, silvery-white magnetic metal, which occurs combined with sulphur and arsenic. It is used in many alloys and also to produce a blue colour in glass and ceramics. It is an essential micronutrient. AW 58.9332; At. No. 27; SG 8.9; mp 1480°C.

cobalt 60. A radioactive isotope of cobalt used as a tracer, an X-ray source and in radiation therapy. A fallout product of nuclear explosions. It is also the medium of the 'Doomsday machine', an atomic weapon exploded in a casing of cobalt 59, the stable isotope, which is converted to cobalt 60 in the explosion to become a toxic product of a 'dirty' bomb, spread widely by fallout.

cobaltite. One of the major ore minerals of cobalt, cobalt arsenic sulphide, CoAsS, found in hydrothermal veins. The other major sources of cobalt are linnaeite, pyrite and cobaltiferous laterite.

cob nut. *See Corylus Avellana.*

coccidiosis. A disease of mammals and birds, often endemic in wild populations, caused by parasitic Protozoa (Sporozoa) which invade internal organs such as the liver.

Coccinellidae (ladybirds). *See* Coleoptera.

cochlea. Small snail-shaped structure forming the inner ear in mammals, containing hair cells in a fluid which move in response to vibrations transmitted by the ossicles of the middle ear. The hair cells transmit impulses to the brain.

Cockles. *See* Lamellibranchiata.

cocktail party effect. The faculty of concentrating on, and so hearing, one sound amid a background of similar sounds.

coconut palm. *See Cocos nucifera.*

Cocos Nucifera **(Coconut Palm).** A characteristic plant of tropical marine islands, with a large, single-seeded fruit (drupe) capable of floating for long periods and remaining viable. The plant provides many of the basic necessities of life for inhabitants of tropical areas, and its products are widely exported. The fruit provides food for humans and cattle and oil used in the manufacture of soap and margarine. Its outer layer yields valuable fibre (coir). The wood from the stem is useful timber. The leaves are woven for thatching, baskets, etc., and their stalks are used for stakes. The apical bud is edible, and the flower stalk is tapped for its juice, which yields sugar or is fermented to provide an alcoholic drink or vinegar.

COD. *See* chemical oxygen demand.

code, genetic. *See* RNA, DNA.

co-dominant. Species that occurs as a frequency dominant within the dominance area of another species.

codon. A sequence of three adjacent nucleotides in DNA or the corresponding RNA which specifies a particular amino acid.

Coelacanthini. *See* Crossopterygii.

Coelenterata (Cnidaria). Phylum of animals that includes sea anenomes, corals, jelly fish, comb jellies, etc. They are all aquatic, mostly marine, and usually radially symmetrical. They have simple, two-layered bodies with only one opening to the gut (coelenteron) and characteristically bear sting cells (nematoblasts) on tentacles. There are two basic types of individual, the hydroid and the medusoid. Colonial forms are common. *See* Actinozoa, Hydrozoa, Scyphozoa (three classes comprising the Cnidaria) and Ctenophora. *See* Diploblastica.

Coelenterates. *See* asexual reproduction.

Coelenteron. *See* Coelenterata.

Coelom. The main body cavity of animals. It is large in Vertebrata, Echinodermata and many Annelida and in these forms the cavity around the gut. In Arthropoda and Mollusca it is small, the main body cavity being a haemocoel, an expanded part of the blood system.

coelomic. Applied to structures and, especially, fluid contained in the coelom of animals.

coen. (*a*) All the components of an environment. (*b*) An abbreviation for coenosis.

coenocline. A sequence of natural communities associated with an environmental gradient.

coenocyte. A mass of protoplasm containing many nuclei, found in many fungi and some green algae. *See* Syncytium, Plasmodium.

coenosis. A random assemblage of organisms, which have common ecological requirements, as distinct from a community.

coenosite. A commensal (*see* commensalism).

coenospecies. A group of species amongst which fertile hybrids may occur.

coenzyme. An organic substance (e.g. ATP, cytochrome) which plays a part in an enzyme controlled reaction or group of reactions without being used up. Many coenzymes are continually changed and then reconstituted by enzymes (e.g. ATP is converted to ADP, which is then rebuilt to ATP; cytochrome is oxidised then reduced again).

COHb. *See* carboxyhaemoglobin.

coincidence (acoustic). When a solid object is exposed to air movement, including sound waves, it will tend to be displaced and if held under tension (e.g. the centre of the structure held in position by the remainder of the structure surrounding it) will tend to oscillate as it returns to its original position. This oscillation will have a wave length characteristic of the structure and bending waves will move within it. If the length of these waves coincides with the length of sound waves as these strike the structure, the sound will be transmitted with great efficiency through the material. There is a critical frequency below which coincidence cannot occur and above which sound insulation is reduced. The critical frequency is determined by the flexibility of the structure and its thickness so that the effectiveness of sound insulation increases the thicker and more rigid it is made.

coir. *See Cocos Nucifera.*

coke. Porous, brittle, solid fuel containing about 80% carbon. Made in coke ovens from coal and also obtained as a residue in the manufacture of coal gas.

col. (*a*) The region of small pressure gradient between two high or two low pressure regions; a region of light winds, often mistaken by birds ready to migrate for the middle of an anticyclone whose fair weather endures over a large area. (*b*) A saddle-back pass between adjacent mountain peaks.

cold-bloodedness. Poikilothermy.

cold front. The front boundary of an advancing cold air mass, at which the cold air usually undercuts the adjacent warm air and generates rain. It is accompanied by a trough of low pressure. Called kataor ana-front if the cold air is significantly descending or ascending, giving a pronounced clearance of cloud in the former case.

cold low. Low pressure centre generated by convection clouds, the upper air being colder than nearby.

cold pool. The cold mass of air causing a cold low.

cold trough. A trough of low pressure with cold air aloft stimulating vigorous convection clouds.

Coleoptera (beetles). One of the largest orders of insects, (Endopterygota), with a very wide distribution. The forewings from horny coverings (elytra) to the hind part of the body, and the hind wings are membranous, reduced or absent. Some beetle larvae are active and predaceous, some are caterpillar-like, others are maggot-like. Serious pests amongst beetles include the Cotton Boll Weevil (*Anthonomus grandis*), the Death Watch Beetle (*Xestobium rufovillosum*) and wireworms (the larvae of certain click beetles). The ladybirds (Coccinellidae) are carnivorous and of great importance in the natural control of aphids and scale insects. *Novius cardinalis* is a Coccinellid beetle used in the biological control of the citrus scale insect.

Coliform bacteria. A group of bacteria which are normally abundant in the intestinal tracts of Man and other warm blooded animals, and are used as indicators (being measured in the number of individuals found per millilitre of water) when testing the sanitary quality of water.

collagen. A fibrous protein forming one of the main skeletal substances in many animals. In vertebrates it is a component of bone, cartilage, and connective tissue and tendon is almost pure collagen. When boiled, collagen yields gelatin. *See* ascorbic acid.

collembola. *See* Apterygota.

collenchyma. A type of supporting tissue found in growing parts of plants (e.g. leaves and young stems). It consists of living cells in which parts of the walls, usually the corners, are thickened with cellulose. *See* sclerenchyma.

colloid. Substance present in a solution in a colloidal state. Particles dispersed in the medium are larger than those in a true solution, cannot pass through a semi-permeable membrane, and give rise to negligible osmotic pressure.

colluvium. Deposit moving downslope by mass-wasting or creep.

colostrum. The milk produced during the first few days after a mammal has given birth. It is rich in proteins and antibodies.

colour changes. *See* chromatophore.

colour phase. An unusual but regularly-occuring colour variety of plant or animal (e.g. the white variety of Sweet Violet).

Columbiformes. An order of birds which includes the pigeons and doves, and sandgrouse. They are rather heavy, grain- or fruit-eating birds, mostly good fliers. The domestic pigeon is descended from the Rock Dove, found wild on mountains and sea cliffs. The Dodo was a large, terrestial member of the order which lived in Mauritius and was exterminated by Man in the 17th century.

columnar joints. Joints which split up an igneous body, usually a sill or lava-flow, into columns which commonly have a hexagonal cross-section.

combe rock, head, coombe rock. An unsorted earthy deposit containing angular fragments, formed as a result of solifluction in periglacial conditions. The term combe rock is chiefly used in south-east England, the synonym head in south-west England and the USA.

combined sewer. A sewer pipe designed to carry household liquid waste (sullage water, sewage and storm-water), as distinct from a sewerage system in which rain water and storm water are carried separately.

Comley series. The oldest Series of the Cambrian System.

combustion. Reaction at high temperature of a substance with oxygen, to produce heat.

combustion air. In incinerators the air introduced into the primary chamber through the fuel bed by natural, induced or forced draught.

commensalism. A close association between organisms of different species during which only one partner receives benefit, but the other is not harmed. The association may be one in which the partners merely share a home (e.g. Man and House Martin) or it may be much closer (e.g. between a shark and a sucking-fish (remora), which is transported, protected and gathers superfluous food from the shark). *See* parasitism, symbiosis.

commercial fast breeder reactor (CFBR 1). Fast breeder reactor planned for commercial use in the UK. *See* nuclear reactor.

comminutor. Machine for crushing coarse material to a finer particle size.

commiscum. Species, i.e., all individuals that can exchange genes successfully.

Committee for Environmental Conservation (CoEnCo). A body, formed in 1969, made up of representatives of the major national conservation bodies, covering interests in wildlife, archaeology, architecture, outdoor recreation, amenity, etc.

Common Agricultural Policy (CAP). The framework on which is built the agricultural policy of the EEC. Its aim is to facilitate trade in food products by establishing Common Prices throughout the Community, measured in 'units of account', whose value against national currencies of members is fixed, rather than floating, to prevent distortions in trade caused by the revaluing of currencies. The CAP guarantees prices paid to producers for many commodities, provides grants for many capital schemes, and defines trading relationships in food commodities with non-member states.

common land. Land subject to 'rights of common' (e.g. rights of pasture) whether these rights are exercisable at all times or only during limited periods, and waste lands of a manor not subject to 'rights of common', provided that such land is registered as common land under the *Commons Registration Act* 1965. The public have no *'de jure'* rights of access to common land under common law. Where such rights exist they do so by virtue of specific grant, covenant or statute. The *Law of Property Act* 1925, S. 193, gave rights of access to the public over all commons wholly or partly in urban areas; in rural areas the owner of the common may, if he wishes, apply the benefits of this section to the land by deed.

common prices. *See* Common Agricultural Policy.

commons, tragedy of the. *See* Tragedy of the Commons.

community. Any naturally occurring group of organisms occupying a common environment. The term is a general one, covering various sized groups, e.g. assemblies, consociations, societies. *See association.*

compensation. The excess of some factors that ensures the survival of species that otherwise would be limited by a shortage of other factors.

compensation point. (*a*) For light. The light intensity at which the rates of respiration and photosynthesis in a plant are equal, so that there is no net absorption or evolution of carbon dioxide or oxygen. (*b*) For carbon dioxide. The concentration of carbon dioxide in the atmosphere surrounding a plant at which the rates of respiration and photosynthesis are equal, so there is no net absorption or evolution of carbon dioxide or oxygen. (*c*) Depth in water. The depth in a sea or lake below which, because of low light intensities, plants use up more organic matter in respiration than they make during photosynthesis.

competent bed. A layer of rock which is relatively strong and which flexes rather than flows during folding. Such beds, in a varied sequence, determine to a large extent the style of deformation. The incompetent beds passively accommodate themselves.

competition. The struggle for existence that results when two or more species have requirements in excess of the available supply.

competition curve. The relationship between parasitism and the area traversed by a parasite population, parasitism being expressed as a percentage. As parasite density increases, the curve rises toward 100%.

competitive exclusion principle. *See* Cause's Principle.

complemental males. Males, often reduced except for their reproductive organs, which live attached to female or hermaphrodite animals of the same species (e.g. angler fish, some stalked barnacles).

complex. The situation that occurs where several communities meet. Each community is characterised by its own dominants, sub-dominants, etc. and these may or may not be different for each. Several different stages in a sere may be found, for example. The communities are related to one another by certain species that are shared.

compositae. A very large and widespread family of mainly herbaceous, dicoty-ledonous flowering plants, which includes dandelions, daisies, thistles, etc. The flowers are arranged in heads, often with a disc of tubular florets surrounded by strap-like ray florets. The massing of the flowers makes them very attractive to insects, which pollinate many at a single visit. The fruit is a cypsela, and often bears a parachute of hairs which aids dispersal. The family contains few plants useful to man, other than decorative garden plants, but many troublesome weeds.

composite volcano. Stratovolcano.

composting. The planned decomposition of organic material, e.g. vegetable or municipal wastes, to produce a soil conditioner. *See* fertiliser.

compound interest. Exponential growth.

compression. (*a*) Descending air is subjected to compression by the increasing pressure, and this causes warming which evaporates cloud, and makes the air drier.(*b*)The compaction of refuse under pressure to make large blocks, which are then bound with bitumenized fabric, plastic or sheet metal, and used for land filling or reclamation.

conchoidal. Refers to a fracture consisting of a curved surface, often with concentric waves.

concrete. Structural material made by mixing Portland cement with sand and crushed stone or gravel and water in predetermined proportions. In large structures it is often reinforced with steel rods to improve tensile properties.

concretion. (*a*) A solid mass of matter formed by the cohesion or coalescence of its constituent particles. (*b*) A localised rock body in a sedimentary or pyroclastic rock. The body is harder than the surrounding rock and is authigenic.

condensate. Hydrocarbons present in a reservoir as a gas, but which separate as a liquid when brought to the surface. Condensates resemble petrol.

condensation. (*a*) (Chem.) A chemical change in which two or more molecules react with the elimination of water or some other simple substance. (*b*) The change of a vapour into a liquid, which occurs when the pressure of the vapour becomes equal to the maximum vapour pressure of the liquid at that temperature (i.e. the medium becomes saturated).

condensation level. The level to which a parcel of unsaturated air must ascend to become saturated. The cloud base in the case of convection cloud.

condensation nucleus. Liquid droplets are not formed if air is cooled beyond saturation (below the dew point) unless nuclei are present. In air there are normally at least 10 to 100 nuclei present, and often 1000, per cubic centimetre, particularly near cities. Therefore the air cannot become supersaturated. *See* supersaturation, ice nuclei, cloud chamber.

condensation sampling. Gas sampling technique in which gas is passed through tubes immersed in refrigerants, trapping various fractions of the gas mixture, depending on the refrigerant solution.

condensation trail (contrail). A trail of cloud formed by mixture of hot aircraft exhaust with cold ambient air. Trails persist only if they become frozen and the air is not unsaturated for ice. (*See* ice evaporation level).

conditional instability. The instability of cold air which comes into contact with a warmer land or sea surface, leading to penetrative vertical movements of air and the formation of cumulus and cumulonimbus clouds and turbulence. The situation is common in low latitudes and in temperate regions it occurs over land in summer and over sea in winter. The lapse rate is intermediate between the dry and saturated adiabatic lapse rates. This is conditionally unstable in that it is stable for unsaturated air but unstable for saturated air. If unsaturated air receives water evaporating from the surface, then as it approaches saturation its stability will decrease and it will be potentially unstable.

conditioned reflex. A reflex action which has been modified by experience. The classic example is that studied by Pavlov. When shown food, a dog will salivate (an unconditioned reflex). If a bell is rung each time the food is shown, the dog will eventually salivate when the bell rings, even if it is not accompanied by food.

conduction. *See* heat transfer.

cone. (*a*) (Bot.) (Strobilus) A compact group of reproductive organs found in conifers, cycads, horsetails, and clubmosses. (*b*) A type of nerve cell present in the retina of most vertebrates which is concerned with discrimination of fine detail and colour, but is not sensitive to very dim light. *See* rod. (*c*) (Indust.) In a blast furnace the steel cone suspended at the top and counterweighted so as to sink when the charge is tipped in and rise again afterwards, preventing the escape of gas.

cone of depression. A conical depression in the water-table immediately surrounding a well.

cone-sheet. A funnel-shaped dyke, generally surrounding an igneous intrusion.

confluence. The flowing together of streamlines so as to increase the windspeed rather than so as to cause an increase in the vertical depth of the stream. *See* convergence.

conformable. Refers to rock strata lying one above the other in parallel order.

congelifluction. Mass-wasting of permanently frozen ground. *See* geolifluction, congeliturbation, congelifraction.

congelifraction. Frost-splitting of rocks.

congeliturbation. Disturbance of material by frost action, frost heaving and differential mass movements.

congeneric species. Species belonging to the same genus.

conglomerate. A sedimentary rock composed of rounded grains predominantly over 2mm in diameter.

Coniacian. *See* Senonian.

conifer. *See* Coniferales.

Coniferales (conifers). An order of cone-bearing plants (*see* cone) which includes nearly all the present day Gymnospermae. Most are tall, evergreen trees with needle-like, linear or scale-like leaves (e.g. pines, firs, cedar). They are characteristic of temperate zones, and are the main forest trees of colder regions. They provide much of our timber, also resins, tars and turpentine.

coning. A plume is coning when it expands to form a roughly conical shape. The phenomenon occurs when the path of the plume is determined mainly by the momentum of the efflux or by buoyancy, or by a combination of both.

connate water. Water trapped with the sediment at the time of deposition. Since most waters in deep subsurface reservoirs are quite different from sea-water, connate water often refers to the water existing in a reservoir rock immediately prior to drilling.

connective tissue. A supporting or packing tissue of vertebrates, containing cells, white fibres of collagen and sometimes more elastic, yellow fibres, lying in a jelly-like, polysaccharide-containing matrix. (*See* polysaccharide). It is widely distributed, found beneath the skin and around muscles and many organs. Fat is often stored in large cells in connective tissue, which is then called adipose tissue.

consequent. Refers to river whose course is determined by the existing slope and pattern of the landscape. *cf.* subsequent.

conservation. The planning and management of resources so as to secure their wise use and continuity of supply while maintaining and enhancing their quality, value and diversity. Resources may be natural or man-made, e.g. historic buildings. Nature conservation is the application of this concept to fauna, flora and physiographical features.

conservation area. An area of special architectural or historic interest, the character or appearance of which it is desirable to preserve or enhance (*Civic Amenities Act* 1967). The designation of a Conservation Area is a public statement of intent by the local authority; it is not in itself a proposal for specific action or a planning technique, but defines an objective and a set of problems for which supporting policies will be required.

conservation of energy. A principle that states that energy can never be created or destroyed, but only changed in form. The First Law of Thermodynamics. *See* thermodynamics, Laws of.

conservative properties of air. A property not significantly changed with time, e.g. content of inert pollution or other constitutent. The most important physical property conserved is the wet bulb potential temperature. A conservative property is altered by mixing.

consociation. A climax community of plants dominated by one particular species (e.g. oak woodland dominated by *Quercus robur*).

consocies. A seral community with but a single dominant, or a development consociation. That subdivision of a formation controlled by a facies.

consocion. A layer in which patches of different dominant species alternate with one another.

consociule. A micro-community (*see* community) of consocial rank. *See* consociation.

constancy. The degree of frequency with which a species occurs in different stands or samples of the same association.

constant. A species which occurs in 95 % or more of samples taken at random within a community.

constant level balloon. A balloon designed, by fixing its volume or by means of some servo-device, to remain at a constant height. This is never accurately achieved because of temperature changes in the balloon in sunshine and occasional large vertical motion of the air. Constant level balloons are used to study the worldwide circulation of air. *See* balloons.

constant pressure map. Although it is convenient to represent the pressure field at sea level by the isobars, it is more convenient to use a map of the height contours of a constant pressure surface, e.g. 700, 500, 300, 100 mb, to study the motion at high altitude.

constant species. Species which occurs in at least half of a stand or sample taken from a community for examination.

consummatory act. An act (e.g. eating) which ends appetitive behaviour.

consumer. *See* food chain.

consumption residues. Wastes resulting from the final consumption of goods or services, rather than from their production or distribution (which are termed production residues).

contact. The boundary surface between one body of rock and another, or between different fluids in a subterranean reservoir. Contacts between different rocks may be the result of sedimentation under varied conditions, or of faulting, but the term is more specially used for the boundary between an igneous intrusion and the country rock. *See* fault.

contact insecticide. An insecticide which kills without being eaten, e.g. by penetrating the cuticle or blocking the insect's spiracles.

contact metamorphism. Metamorphism in the vicinity of an igneous body, forming a metamorphic aureole. The width of the aureole depends on the size of the intrusion, and on the temperature of the intrusion and of the country rock. Without metasomatism the chief effects are the baking of clay, the hardening of shale, the conversion of slate to spotted slate or to hornfels, and the conversion of limestone to marble. Very high pressures along the base of thrusts can also produce a metamorphosed rock flour called mylonite.

contact process. A method for producing sulphuric acid using a platinum catalyst for the oxidation of sulphur dioxide to sulphur trioxide.

container transport. Method for conveying any commodity in a standardised container that can easily be loaded and unloaded by standard equipment for road, rail, or waterborne transport. An increasing proportion of international trade is being containerised.

contamination. The introduction to water of substances containing toxins or live pathogens that constitute a hazard to human health.

contemporary city. An ideal city conceived by Le Corbusier (the Swiss architect Charles Edouard Jeanneret) in which the railway station stood at the city centre, linked to subways and other transport facilities, with a helicopter link to the airport. The population was to be housed in skyscraper blocks with hanging gardens, surrounded by parks in which were located restaurants, shops, theatre, etc. . . . . *See* garden city.

contest competition. Where each successful competitor obtains all it requires for survival or reproduction, its rivals obtain insufficient or nothing.

continental climate. A climate characterised by one or more persistent features of the weather associated with the interior of large continents, e.g. hot summers, cold winters, short autumns and springs, rainy and dry seasons, and a general lack of moderation due to excess of features unhelpful to Man such as long frosts, tornadoes, deserts, often necessitating seasonal migration.

continental drift. The theory that the present continents result from the break-up of a larger continent and have moved independently to their present positions. The evidence for the existence of such a supercontinent comes from paleomagnetism, the distribution of orogenic belts, faunas, sedimentary facies and climatic belts, and the morphological fit of the continents along the edges of the continental shelves.

continental rise. The gently sloping region connecting the bottom of the continental slope with the abyssal plains. The continental rise is only well developed in some areas and has a thick blanket of turbidite fans and reworked sediments swept up by sea-bottom currents.

continental shelf. The comparatively shallow area surrounding the continents and rising from the deep ocean floor like a shelf on which the dry land represents the higher regions. The size and formation of the continental shelves varies, but it is generally taken to be the band between the 183 metre line and the coast, but since most shelves slope down rather than ending abruptly, the definition is arbitrary. Continental shelves vary in width but for the purposes of Exclusive Economic Zones proposed by the *Law of the Sea Conference* bands of 200 miles are accepted.

continental slope. The portion of the continental margin that begins at the outer edge of the continental shelf, i.e. at the 183 metre line, and descends into the ocean deeps.

contingency. (Stat.) The degree of dependence or independence between variables arranged in a complex classification.

continuous spectrum. A frequency analysis having components ranged continuously over the spectrum of wave frequencies.

continuous variations. *See* variation.

contour chart. *See* constant pressure map.

contour farming. Performing cultivations along lines parallel to contours, rather than at right angles to them, so reducing the loss of topsoil by water erosion, increasing the capacity of the soil to hold water (since each furrow is a small reservoir) and reducing the pollution of water by soil.

127

contour strip cropping. The growing of crops on strips that run parallel to contours in order to reduce erosion. *See* contour farming.

contrail. *See* condensation trail.

control. That part of an experimental procedure which is like the treated part in every respect except that it is not subjected to the test conditions. The control is used as a standard of comparison, to check that the outcome of the experiment is a reflection of the test conditions and not of some unknown factor.

controlled tipping. *See* land-fill

conurbation. A large area occupied by urban development, which may contain isolated rural areas, formed by the merging together of expanding towns that formerly were separate.

convection. (*a*) The transfer of heat through a fluid by the movement of the fluid. *See* heat transfer. (*b*) The vertical movement of air, leading to cooling. This is the most efficient atmospheric cooling process. Free convection occurs when air heated by conduction from the warm land surface rises, or when a mass of warm air rises above an adjacent mass of cold air due to the lateral movement of either air mass. Forced convection occurs when a moving air mass encounters mountains and is lifted regardless of its initial temperature. Forced convection can produce high rainfall. Orographic lifting occurs under very stable conditions when a slow-moving warm air mass moves gently upward over wedges of cooler air, producing typical winter rains which are light, steady and extensive. Flash floods in the southern US are produced by the combined effects of horizontal convergence of air masses, upglide over cooler air and orographic lifting of unstable, moist maritime air. See Cellular convection. Penetrative convection occurs when the ground or sea surface is warmer than the air immediately above it, as when cold air moves across a warm surface (a common situation in low latitudes). This produces conditional instability leading to the formation of cumulus and cumulonimbus clouds.

convection streets. The same phenomenon as cloud streets but without visible cloud. Sometimes visible in haze or haze tops, or discovered by soaring pilots who find upcurrents in parallel lines.

convective equilibrium. A state in which a vertically displaced parcel of air experiences no buoyancy force.

convective instability. A state in which a vertically displaced parcel of air experiences a buoyancy force in the direction of the displacement, usually produced by heating of the ground by sunshine. *See* wet adiabatic and potential instability.

convergence. (*a*) *See* convergent evolution. (b) (Atmos.) In three dimensions, convergence means the compression of a fluid into a smaller volume (opposite of divergence), but in meteorology it is used to refer to horizontal convergence, which necessitates the vertical extension of an air parcel. (Horizontal) convergence at the ground implies upward motion above. Not to be confused with confluence which is the consequence of acceleration when streamlines are drawn closer together by the flow without vertical motion. Convergence intensifies the vertical component of vorticity and is therefore usually accompanied by falling pressure.

convergent evolution. (**convergence**). The evolution of two different groups of organisms so that they come to resemble one another more closely (e.g. the marsupial and placental moles, which are both adapted for burrowing).

convivium. A population differentiated within the species and isolated geographically. Usually subspecies and/or ecotypes.

cooling pond. (*a*) (Nuclear) A large water tank in which irradiated fuel elements from nuclear reactors are stored while their short-lived fission products decay. (*b*) An artificial lake or pond in which industrial cooling water is allowed to cool naturally.

cooling tower. A device for cooling water by use of a flow of air at ambient temperature.

cool-season plant. A plant which grows mainly during the cooler part of the year usually the spring, but sometimes the winter.

coombe rock . Combe rock.

Copepoda. A group of small, freshwater and marine Crustacea, some of which form a major part of the plankton (e.g. *Calanus* on which herrings feed). Some are parasitic. Many species have enlarged first antennae, used for swimming. *Cyclops* is common in ponds.

copper (Cu). Element. Red, malleable, ductile metal, unaffected by water or steam and the best conductor of electricity after silver. Occurs as the element and as chalcopyrite, cuprite (Cu_2O), copper glance (Cu_2S). It is used in boilers, plumbing, electrical wire, and in many alloys (e.g. bronze, brass, gun metal, etc.). AW 63.54; At. No. 29; SG 8.95; mp 1084°C.

copper compound process. Japanese process for removing sulphur dioxide from flue gases by dry absorption, to produce sulphur dioxide which is recovered.

copper pyrites. Chalcopyrite.

coppice. A crop of coppice shoots. These are produced by felling trees or shrubs near to the ground, causing several shoots to arise from adventitious buds on the stump. Coppicing was formerly a common practice in woodland management (*see* coppice with standards).

coppice forest. *See* forest.

coppice with standards. An ancient method of management in which woodland is selectively thinned, leaving large, timber-producing trees (e.g. oak) at regular intervals, the spaces being occupied by coppiced underwood (e.g. hazel) which is cut near to ground level every 10 to 15 years. The standards draw up the newly-coppiced underwood which produces numerous long, straight growths for stakes, hurdles, etc. *See* coppice.

coprolite. A fossilised faecal pellet. Coprolites are generally phosphatic.

coquina. A soft, porous limestone composed of shells and other bioclastic remains.

corals. Colonial coelenterates which secrete a calcareous skeleton. The larger reef-building corals (Madreporaria) and the alcyonarian corals (e.g. Organ Pipe Coral, *Tubifora*; Precious Coral, *Corallium*;) belong to the Actinozoa. Hydro-corallinae belong to the Hydrozoa, and unlike the corals produce a free-swimming medusa stage.

Cordaitales. Palaeozoic Gymnospermae particularly abundant during the Carboniferous. They were trees, sometimes 30m tall, bearing a crown of branches with large, simple, parallel-veined leaves and small cones.

Core. The core of the Earth is that part lying below the mantle. The core has a radius of approximately 3500 km of which the innermost part only is a solid, with a radius of about 1400 km. The core is thought to be composed predominantly of nickel and iron, the movement of which produces the Earth's magnetic field.

core area. *See* home range.

core city. Planning concept in which the buildings are packed very densely. In some plans the total floor area provided is equal to the total land area of the city, in a continuous body of intense density and activity. Theoretically, a population of 20 million could be accommodated in a 10 mile radius. In most plans, the floor space amounts to about one-tenth of the total area, so that the core city amounts to a radical departure from conventional planning.

Coriolis Force. The force at right angles to the velocity relative to the Earth experienced by a moving body when referred to coordinates fixed to the Earth. It is only important in horizontal motion and is equal to twice the product of the speed with the vertical component of the Earth's rotation at the place in question. It is to the right in the Northern and left in the Southern Hemispheres. It is named after a French engineer, Gaspard Gustave de Coriolis (1792–1843).

Coriolis Parameter. Denoted by f or by 1, and equal to $2v \sin \phi$, where v is the horizontal speed and ϕ is the latitude.

cork (phellem). A protective, impermeable layer of dead cells produced by cell division in the phellogen (cork cambium) a secondary meristem which arises in the cortex of woody plants. Cork replaces the epidermis of the young stem or root. In the Cork Oak (*Quercus suber*) the cork layer is exceptionally thick.

cork cambium. *See* meristem, cork.

corm. The swollen, food-storing base of a stem, lying underground and bearing buds and shrivelled leaf bases (e.g. crocus). A corm is an organ of perennation and vegetative propagation which lasts only one year. Next year's corm arises on top of the old one. *See* tuber, rhizome.

corn. (*a*) European usage: any cereal crop. (*b*) US usage: maize (*Zea mays*). *See* Graminae.

cornucopian premises. Term coined by Preston Cloud to describe an attitude often encountered to problems of resource depletion, population growth, etc., based on five tenets: (1) Nuclear fuels will provide an inexhaustible supply of abundant energy; (2) New technology will, with (1) enable much lower grades of resources to be extracted, so making reserves virtually inexhaustible; (3) The development of synthetics and substitutes will permit the replacement of resources that become exhausted; (4) Economics is the sole factor governing technology and the use of resources; (5) Populations will stabilise of their own accord before crises occur. *See* technological fix.

corolla. *See* flower.

corona. The coloured rings seen around the sun or moon in tenuous or thin cloud in which the droplet size is fairly uniform. The same phenomenon as iridescence.

Corpus luteum. A temporary hormone-producing area (*see* hormone) on a mammalian ovary, which forms after the ripe egg has been shed. It produces progesterone and persists only if the egg is fertilised. Its formation and activity are promoted by hormones secreted in the pituitary gland.

corrasion. Effects of mechanical erosion on rocks by water, ice and wind when carrying particles that can act as abrasives.

correlation (Stat.) The relationship between two variables according to a specific range of values.

corrie. *See* cirque.

corrosion. The conversion of iron and other metals to oxides and carbonates by the action of air and water. The process commonly takes place at ambient temperatures and is fastest in the presence of seawater or a humid atmosphere. There are several ways of preventing corrosion and some metals, e.g. aluminium, are naturally corrosion-resistant.

cortex. An outer layer, e.g. of gland, kidney, brain, plant stem.

corticotropin. *See* ACTH.

cortisone. Hormone secreted by the cortex of the adrenal gland. One of its effects is to reduce local inflammation.

corundum. A mineral with the formula Al_2O_3 noted for its extreme hardness (9 on Moh's scale). Corundum occurs in some alkaline pegmatites, in some regionally metamorphosed rocks, and in placer deposits. Gem corundum is known as sapphire or ruby, but most corundum is used as an abrasive. Emery is an impure form of corundum.

Corvidae. A family of large, omniverous Passeriformes, including the crow, magpie, jay, jackdaw, rook, raven and chough.

Corylus avellana **(Hazel, Cob Nut) (Corylaceae).** A deciduous shrub or small tree, the common shrub-layer dominant of lowland oakwoods and sometimes of ashwoods in Britain. The plant is wind-pollinated, with pendulous male catkins and bud-like female inflorescences with projecting red stigmas. For centuries it has been cultivated for its nuts and coppiced in woodlands to yield small timber. *See* coppice with standards.

corymb. A broad, flattish inflorescence in which the outer flowers open first.

cosere. A series of unit successions in the same place.

cosmic rays. Complex system of radiation reaching earth from outer space.

cost-benefit analysis. Economic technique for estimating the desirability of a proposed course of action that requires the advantages and disadvantages to be listed, expressed in monetary terms, and the totals compared.

cost-benefit ratio. The ratio of gross costs to gross benefits. Both costs and benefits are discounted over the life of the project at an annual rate of interest. The difference between them is the present value of the net benefit (or cost) and their ratio is the gross cost-benefit ratio.

cotyledon. A seed leaf, part of a plant embryo. Monocotyledonous plants have one, dicotyledonous plants two. In some plants (e.g. beans) the cotyledons are large and fleshy and store food. In others (e.g. grasses) they absorb food from the endosperm and pass it to the growing embryo. Some cotyledons (e.g. Sycamore) grow out of the ground, develop chlorophyll and carry out photosynthesis.

cotype. Syntype.

coulomb (C). The derived SI unit of electric charge, being the quantity of electricity transferred by one ampere in one second. Named after Charles Augustin Coulomb (1736–1806).

coumarin. *See* warfarin. A white, crystalline, aromatic substance ($C_9H_6O_2$) found in Sweet Vernal Grass and other plants, which gives hay its characteristic smell. Used as a vanilla flavouring and in perfume.

Council for Nature. Body representing the voluntary natural history movement in Britain, founded in 1958, with local natural history societies among its members.

Council on Environmental Quality (CEQ). US Federal agency formed under the terms of the *National Environmental Policy Act* (1969) (NEPA) for the enforcement of environmental protection measures. The aim of the Act is to 'encourage productive and enjoyable harmony between Man and his environment; to promote efforts which will prevent or eliminate damage to the environment and biosphere and stimulate the health and welfare of Man; to enrich the understanding of the ecological systems and natural resources important to the Nation; and to establish a Council on Environmental Quality.' The Act also calls for environmental impact statements.

counter. *See* Geiger-Müller counter.

country park. Area of land, or water and land, normally not less than 25 acres (10 ha) in extent, designed to offer to the public, with or without charge, opportunity for recreation pursuits in the countryside.

country rock. The body of rock which encloses an igneous intrusion or mineral vein.

Countryside Commission. British official body, formed under the terms of the *Countryside Act* (1968) to assume the functions of the earlier National Parks

133

Commission. The Countryside Commission is charged with keeping under review all matters relating to the provision and improvement of facilities for the enjoyment of the countryside in England and Wales. Its members are appointed by the Secretary of State for the Environment and the Secretary of State for Wales. The Commission's headquarters are in Cheltenham, Gloucestershire.

County Trusts for Nature Conservation. *See* naturalists' trusts.

coupled modes (acoustic). Modes of vibration which influence one another.

COV. Cross-over value. *See* crossing over.

covellite. A copper sulphide mineral, CuS, and a major ore of copper, usually found in the zone of secondary enrichment.

cow month. The amount of feed or grazing required to maintain a mature cow in good condition for 30 days.

Coypu. A large, aquatic rodent introduced into England from South America for fur farming. It escaped in the 1930's and is now established in eastern England, where it causes damage to crops and river banks.

Cr. Chromium.

cracking. Thermal decomposition process. e.g. hydrocarbons such as CH_4 to C and H_2 or NH_3 to N_2 and H_2. *See also* hydrocarbon cracking.

crag-and-tail. A hill with a steep mass of highly resistant rock at one end and a streamlined tail of a more gently sloping body of softer rock or moraine. Crag-and-tails are a product of glacial erosion.

Craniata. Vertebrata.

Crataegus (Hawthorn). Deciduous trees or shrubs (Rosaceae), usually thorny, numerous in northern temperate regions. The two species native to Britain are *Crataegus monogyna*, the commonest scrub dominant on most soils apart from wet peat and poor acid sand, and *C. laevigata*, Woodland or Midland Hawthorn, which is much less common but more tolerant of shade. This has broader leaf lobes than *C. monogyna*, and usually two styles instead of one. *C. monogyna* is the shrub most commonly planted for field hedges in Britain.

craton. (Geol.) A continental block which has been stable over a relatively long period of Earth history, and which has only undergone faulting or gentle warping.

creep. The plastic flow of solids under constant stress. *See* soil creep.

crenate. Having a notched, or toothed, edge. Applied to leaves whose edges are notched. If the notches are similar to those of a crenate leaf, but very small, they are termed crenulate. *Cf.* serrulate.

crenulate. *See* crenate.

creode. The pattern of development of a single organism from fertilised egg to its death and dissolution.

crepuscular rays. Literally 'evening' rays, namely those seen on account of shadows cast by cloud on haze. Usually referring to rays seen diverging from the sun which is obscured by cloud, but the rays may sometimes be seen converging to the antisolar point at about sunset.

Cretaceous. Refers to the youngest Period of the Mesozoic Era, usually dated as beginning 135–136 million years before present and lasting about 70 million years. Cretaceous also refers to the rocks deposited during the Cretaceous Period: these are called the Cretaceous System, which is divided into Lower and Upper Cretaceous. The Lower Cretaceous is divided into the Neocomian, Aptian and Albian Stages. The Upper Cretaceous is divided into the Cenomanian, Turonian, Senonian, Maastrichtian and, usually, the Danian, although the latter is considered by some to be Palaeocene. The name Cretaceous is derived from the Latin *Creta* (chalk) which characterises the rocks of the Period.

crickets *See* Orthoptera.

Crinoidea (sea lilies). A class of Echinodermata abundant as fossils and present since the Cambrian. Most adults are sedentary and stalked, with long, branched, feathery arms bearing cilia which set up feeding currents. The Feather Star (*Antedon*), the only surviving British genus, breaks from its stalk and swims by waving its arms.

critical frequency. *See* coincidence.

critical links. Organisms in a food chain which are responsible for energy capture and flow and which play a critical role in the assimilation of nutrients and their subsequent release in forms available at higher trophic levels.

critical mass. That mass of fissile material that will sustain a chain reaction. At this mass, escaping neutrons have sufficiently good a chance of colliding with more atoms, rather than escaping, for the process to be self-sustaining. The critical mass for Uranium-235 is a sphere about 80mm in diameter (weighing about 10 kg).

critical organ. An organ within the human body that has a special capacity for concentrating within itself a specific radioisotope, or that is particularly sensitive to radiation, and that is studied with particular care when assessments are made of the effects of radiation on the body.

critical reaction. Very violent form of fighting behaviour which occurs when an animal is motivated strongly by fear, but cannot flee from the animal which threatens it because of lack of space or because of overriding ties, e.g. those which forbid it to desert its young. In this 'cornered' situation an animal will fight desperately.

crocidolite. *See* asbestos, chrysotile. A commercially important form of asbestos.

Crocodilia. Large carniverous reptiles with powerful jaws, and elongated hard palate, numerous conical teeth, webbed feet, closable nostrils placed at the end of the snout, and heavy bony plates as well as scales covering the body. The alligator, crocodile and caiman are present-day representatives.

Cro-magnon man. *See Homo.*

cropmilk (pigeon's milk). A secretion of the crop lining in male and female pigeons, on which the young are fed. Its production is stimulated by the lactogenic hormone secreted by the pituitary gland.

crop rotation. An important method of maintaining soil fertility in cultivated areas. Various plants need different nutrients and ground water at different levels, so growing crops in rotation avoids exhausting the soil of any one nutrient, or of ground water at one level. A leguminous crop is included because of the nitrogen-fixing bacteria in the root nodules. *See* nitrogen fixation.

cross-bedding (cross lamination, false bedding). The arrangement of bedding, or lamination within a bed or at an angle, to the main planes of stratification of the strata concerned. The laminae are the non-eroded parts of the layers of dunes and ripple-marks formed in arenaceous sediments such as sands, silts, and ooliths (see Oolite). The nature and orientation of cross-bedding is used in determining the way-up of the strata and the conditions (e.g. sediment supply, current speed and direction, and if an aqueous deposit, water depth) at the time of deposition.

crosscut. In mining, a horizontal tunnel running through the country rock or ore at a large angle to the strike of the ore-body.

cross-fertilisation. The fusion of gametes from two different individuals (not ramets from a clone) of the same species. *See* self-fertilisation.

crossing over. An equal exchange of material between the chromatids of homologous chromosomes which occurs during the pairing of the chromosomes at meiosis. This causes a reassortment of alleles and increases the diversity of gametes. The frequency of crossing over between two genes at different loci on the same chromosome is the cross-over value (COV). This is proportional to the distance apart of the genes, and is used to map their relative positions.

Crossopterygii. A group of primitive bony fishes (Choanichthyes) now usually considered distinct from the lungfishes (Dipnoi), from which they differ, for instance in having conical teeth. The Crossopterygii were believed to have been long extinct until the coelacanth *Latimeria* was discovered in 1939 off the coast of East Africa.

cross pollination. The transference of pollen from the stamen of one plant to the stigma of another plant of the same species. *See* self-pollination.

Crown of Thorns Starfish (*Acanthaster planci*). A starfish (*See* Asteroidea) that reaches a diameter of 60 cm and that eats hard coral. It is associated with coral reefs in tropical waters, and an increase in its numbers since the early 1960's has led to serious damage to some reefs and concern has been expressed about the effect these animals might have on the Great Barrier Reef.

Cruciferae (Brassicaceae). A family of dicotyledonous, mostly herbaceous plants, chiefly of northern temperate regions. The flowers have four sepals alternating with four petals which are usually white or yellow. Many are weeds of cultivation, e.g. Charlock (*Sinapis arvensis*) and Shepherd's Purse (*Capsella bursa-pastoris*). Others are food plants, e.g. *Brassica*, White Mustard (*Sinapis alba*), Garden Cress (*Lepidium sativum*) and watercress (*Rorippa nasturtium-aquaticum*).

crude oil. Liquid petroleum. Crude oils range from colourless liquids as thin as petrol to viscous black asphalts. Very dense crudes, (i.e. those with low values on the API scale) may exceptionally underlie the water in an oil pool. Crude oil containing less than 0.5% sulphur is termed low-sulphur crude, and is more valuable than the corresponding high-sulphur crude.

crust. The crust of the Earth is the outer shell of the Earth, defined by its composition and the properties of some seismic waves. The crust rests on the mantle, with the intervening boundary being called the Mohorovičić Discontinuity. The crust varies in thickness from an average of about 6 km in ocean areas to 35–70 km in continental regions. In oceanic areas the crust is apparently young (Jurassic or younger) and of basaltic composition, the top layer (Layer 1) being sediment (e.g. turbidites, oozes, red clay), the next (Layer 2) being basaltic pillow lavas, and the lowest (Layer 3) being an intrusion complex of parallel dolerite dykes and gabbro. This igneous material is broadly known as sima. In continental regions the sima underlies a less dense sial, broadly composed of granodiorite and its metamorphic equivalents, overlain by sedimentary and metasedimentary rocks. Continental crust underlies the continental shelves (e.g. the Barents Sea), and also forms some submerged microcontinents, such as Rockall, which appears to be fragments broken off during Continental Drift.

Crustacea. A class of mainly aquatic Arthropoda including crabs, shrimps, woodlice, water fleas, barnacles, etc. *See* Branchiopoda, Ostracoda, Cirripedia, Malacostraca, Branchiura.

Cryptogamia. A large group of plants comprising the Thallophyta, Bryophyta and Pteridophyta. In these the reproductive organs are not prominent, as they are in the Phanerogamia. Pteridophyta are vascular cryptogams.

Cryptozoic. (*a*) (Zool.) Term applied to animals which live in crevices. (*b*) (Geol.) Refers to geological time before the Cambrian Period. Cryptozoic is synonomous with the more usual term Precambrian, as is compared to the Phanerozoic.

cryogenic system. System in which a local temperature lower than the surrounding temperature is produced, e.g. refrigerated transport.

cryopedology. The study of frozen ground.

cryophyte. A plant which grows on snow or ice. Cryophytes include algae, fungi, and mosses, and may be sufficiently abundant to colour the snow.

cryptic coloration. A colour pattern which protects an animal by making it closely resemble its surroundings. Many moths (e.g. Peppered Moth, *Biston betularia*) and ground-nesting birds are camouflaged in this way.

cryptocrystalline. Composed of crystals too small to be resolved by a microscope.

Cryptophyta. *See* algae.

cryptophyte. *See* Raunkiaer's Life Forms.

crystal. A homogenous solid possessing long-range three dimensional, internal order.

crystalline rock. A term usually used to refer to igneous and metamorphic rocks.

Cs. Caesium.

CS gas. o-chlorobenzylidene malonitrile. A fine powder used in riot control for its property of inducing acute irritation to the eyes, to any skin abrasion, and to the tissues of the respiratory passages causing coughing and nausea. The powder is absorbed readily by soils and may find its way into surface waters, where its toxicity and that of its two main hydrolysis products (o-chlorobenzaldehyde and malonitrile) to fish is low.

CSM. 'Corn, Soya, Milk'. A mixture of 70 % processed corn (maize), 25 % soya protein concentrate and 5 % milk solids developed as a protein-enriched food for developing countries. *See* novel protein foods.

Ctenophora. Comb jellies, sea gooseberries, etc. A small group of pelagic, marine animals, sometimes included in the Coelenterata, sometimes considered to constitute a different phylum, as they have no thread cells and show a measure of bilateral symmetry. Their bodies are jelly-like and rounded, and they swim by means of comb-like rows of fused cilia.

Cu. Copper.

cuckoos. *See* Culicidae.

Cuculidae (cuckoos). Long-tailed, slender-winged birds with two toes pointing forwards and two backwards. Some are brood parasites. The European Cuckoo (*Cuculus canorus*) lays eggs in the nests of small passerines (e.g. warblers, also the Meadow Pipit, Pied Wagtail, Robin, Linnet). Individual birds usually parasitise only one species, and their eggs resemble those of the host species. The family Cuculidae belongs to the order Cuculiformes.

Cuculiformes. Cuckoos. Cuculidae.

Cucurbitaceae. A family composed mainly of tropical and sub-tropical tendril-climbing plants, including the melons, cucumbers, marrows and squashes. Some species yield gourds and calabashes, and the fibrous fruit wall of *Luffa aegyptiaca* forms the loofah. *Bryonia dilica* (White or Red Bryony), a common hedgerow plant, is the only species native to Britain.

cuesta. An asymmetric ridge with one slope steep and the other long and gentle and roughly parallel to the underlying beds.

Culicidae. Mosquitoes and certain gnats. A large family of delicate flies (Diptera, Nematocera) usually with scaly wings and a piercing, blood-sucking proboscis in the female. The larvae and pupae are aquatic. Species of Anopheles transmit human malaria; *Adoes (Stegmyia) aegypti* transmit yellow fever, *Culex fatigans* transmits the worm causing elephantiasis. *See* filaria.

cullet. Broken scrap glass that can be collected and re-used.

culm. The flowering stem of a grass. *See* Graminae.

culm and gob banks. Hills, or banks, made from inferior fuels and wastes discharged from plants processing coal. Banks of culm (anthracite) and gob (bituminous waste) are often unsightly and present a fire hazard.

cultch. Materials pumped into the sea to provide sites for the growth of the larval stages of shellfish intended for eventual commercial exploitation. Cultch is often composed of shells and pebbles.

cultivar. A plant variety which is found only under cultivation. It may be maintained by asexual propagation or controlled breeding.

cumulonimbus. A cumulus producing rain. A shower cloud. Often used with special reference to raining cumulus with an ice anvil.

cumulus. A 'heap' cloud, produced by upward buoyant convection, and having growing protuberances on top with a cauliflower-like appearance.

cupola. (*a*) Vertical shaft furnace used for melting metals, as distinct from a blast furnace in which the ore is melted. Source of grit, dust and metallurgical fumes, especially from the 'hot blast' type of cupola. (*b*) (Geol.) A small dome-like upward projection from a batholith. Some stocks and bosses may well be cupolas.

curare. *See* strychnos.

curie. The unit of measurement for radioactivity. One curie is equal to 3.7×10^{10} disintegrations per gram per second. This is the intensity of the radioactivity of radium and thus the curie effectively compares the radioactivity of substances with that of radium. Named after Marie Curie (1867–1934) the Polish-born, but French by marriage, physicist.

curvature. The rate of change of direction of the tangent to a curve as a function of the distance along it. Inverse of the radius of curvature.

Cuscuta (dodder) (Cuscutacea). Leafless, rootless, totally parasitic plants, lacking chlorophyll but with thread-like twining stems bearing suckers which penetrate the host plant and absorb food. *Cuscuta epithymum*, the common British species, usually parasitises gorse and ling, and the more local *C. europea* is found mainly on Stinging Nettle

cuticle. (*a*) A non-cellular covering to a plant or animal, made by the epidermis

140

and often in terrestrial forms serving to prevent excessive loss of water as well as to protect the organism. In higher plants the cuticle is made of waterproof cutin, a mixture of compounds derived from fatty acids. In invertebrates it is made of chitin or a substance similar to collagen. It forms the exoskeleton in Arthropoda, and in insects is covered with a waterproof layer of wax. (*b*) The outer protective layer of dead cells in the skin of terrestrial vertebrates is sometimes called the cuticle.

cutin. *See* cuticle.

cut-off low. A low (cyclone) cut off from the main series of cyclones associated with the polar front, its front having long since been occluded so that no warm air mass is identifiable at the Earth's surface.

cutting oils. Petroleum products containing additives to enable them to resist very high pressures as lubricants; used to cool and lubricate machine tool cutters.

cwm. *See* cirque.

cyanides. Highly toxic class of compounds whose inhalation or ingestion leads to nausea and eventual death. Used in metal plating industry, for the heat treatment of metals, in some pesticides, and disinfectants. Cyanides often survive in industrial wastes to cause disposal problems, although they do not persist in the environment.

cyanocobalamin. Cobalamine.

Cyanophyceae. *See* Cyanophyta.

Cyanophyta (Myxophyta, Blue-green algae, Cyanophyceae, Myxophyceae). A primitive group (Division) of unicellular, colonial or filamentous algae whose cells have no nuclear membrane, mitochondria or chloroplasts. They contain chlorophyll and various other pigments, so exhibit a variety of colours. They are widespread in marine and fresh water and in the soil, and occur even in hot springs. Some are symbiotic (*see* symbiosis) (e.g. in some lichens) and others help to maintain soil fertility by carrying out nitrogen fixation.

cybernetics. The theory of communication and control mechanisms in systems. Now largely embraced by the study of general systems.

cyborg. A fusion of man and machine within a single physical entity.

cycadales (cycads). An order of largely fossil Gymnospermae, with living representatives in tropical and sub-tropical areas. The tall, unbranched stem bears a crown of fern-like leaves. Cycads produce cones and are the most primitive present-day seed-producing plants.

Cycadofilicales (Pteridospermae). Extinct Palaeozoic Gymnospermae abundant during the Carboniferous. They exhibited fern-like characters (e.g. in leaf form) but produced seeds on structures similar to the vegetative fronds.

cycads. Cycadales.

cycles per second. *See* frequency.

cyclodiene insecticides. A subgroup of the organochlorine insecticides including chlordane, heptachlor, aldrin, dieldrin and endrin.

cyclogenesis. Sometimes called development, the intensification of vorticity (rotation) of air by horizontal convergence accompanying upward motion and cloud formation. Common at fronts where the warm air mass is lifted over the cold air.

cyclone. (*a*) A centre of low pressure around which the air circulates cyclonically, i.e. in the same direction as the earth. (*b*) *See* tropical cyclone.

cyclone collector. Structure without moving parts in which the velocity of an inlet gas stream is transformed into a confirmed vortex from which suspended particles are driven to the wall of the cyclone body and collected. Used widely as a pollution control for dusts.

cyclops. *See* Copepoda.

cyclonic curvature. Curvature in the direction of flow around a cyclone, curving to the left in the Northern and to the right in the Southern Hemispheres.

cyclonic vorticity. Circulation in a cyclonic direction, i.e. in the same sense as the circulation of the rotating Earth. A parallel straight airstream possesses cyclonic vorticity if the speed decreases to the left across the stream, and a stream with anticyclonic curvature may possess cyclonic vorticity.

Cyclorrhapha. A large group of Diptera including bluebottles, houseflies, hoverflies and fruit flies. Other members are frit flies, whose larvae bore into crops and are serious pests; warble and bot flies, the larvae of which are endoparasites of stock; and the tsetse fly, which transmits the Trypanosomidae causing sleeping sickness in Man and cattle in Africa. The Tachinidae are important parasites of insects, especially Lepidoptera larvae. The larvae of Cyclorrhapha have much reduced heads and the pupae are enclosed in barrel-shaped cases.

Cyclostomata. The lampreys and hagfish, living members of the Agnatha. They are eel-like, scaleless, jawless and have no paired fins. They attach themselves to fish and suck their blood.

cyclostrophic force. The centrifugal force evident in the wind due to curvature of the paths of the particles. Taken into account only when the curvature is large, in calculation of the gradient wind.

cyclothem. A vertical sequence of sedimentary rock which occurs many times in a succession, and which denotes a repetition of physiographic and sedimentary events in a more or less constant order.

cyme. An inflorescence that branches repeatedly, with the oldest flower at the end of each branch.

Cyperaceae. The sedge family, monocotyledonous grass-like herbs of wet places, including the true sedges (*Carex*), bulrushes, clubrushes, cotton-grasses, etc.

Cyprinidae. A large family of freshwater fishes with scales on their bodies but not on their heads, including bream, carp, chub, dace, roach, rudd, tench, barbel, bleak, gudgeon and minnow.

Cypselurus (flying fishes). Fishes belonging mainly to the family Exocoetidae (order Beloniformes or Synentognathi) and to a limited number of unrelated fishes that are able to sustain a gliding flight by spreading their enlarged, stiff pectoral fins and aiding their forward motion by beating rapidly with the tail fin, which is lifted when sufficient speed has been attained, the smaller pelvic fins are spread to make a second pair 'wings' and the fish glides. Some fish have been known to cover one-fourth of a mile, taking about 43 seconds. The Californian flying fish is *Cypselurus californicus*.

cysteine. Amino acid with the formula $SH.CH_2CH.(NH_2).COOH$. Molecular weight 121.1.

cysticercus. *See* bladderworm.

cytochromes. A group of iron-containing proteins which play an important part in aerobic respiration, acting mainly as coenzymes.

cytogenetics. The science linking the study of heredity with that of the physical appearance of the chromosomes.

cytokinins (phytokinins). Plant hormones which stimulate cell divison. Their effects include the promotion of bud formation and germination, and are thought to be associated with increased nucleic acid and protein metabolism. *See* auxins, gibberellins, zeatin.

cytology. The study of the structure and functions of living cells.

cytolysis. Cell destruction, especially when brought about by the disintegration of the cell membrane.

cytoplasm. All the protoplasm, except for the nucleus, of a cell. *See* cell.

cytoplasmic inheritance. *See* plasmagene.

cytotaxonomy. A method of classifying plants based on observation of the number, shape and size of chromosomes.

cytotype. An individual member of a population which is composed of indivduals of a similar karyotype differing from individuals in other populations.

D

d. Deci.

da. Deca-.

DAC. Development Assistance Committee.

dacite. A fine-grained, calc-alkaline igneous rock with a silica content between that of rhyolite and andesite. Dacites contain quartz phenocysts and occur with andesites in large volumes in orogenic belts.

dalapon. A herbicide used for the control of grasses and to prevent watercourses from becoming clogged by the growth of reeds and other monocotyledonous plants. *See* translocation.

DALR. Dry adiabatic lapse rate.

damage risk criterion. That noise level, as a function of frequency, waveform (pure tone or random noise), intermittency, etc., above which more than a specified degree of permanent hearing loss is likely to be sustained by a person exposed to it.

damping. The removal of energy from an oscillating system or particle by means of friction or viscous forces, the energy removed being converted into heat.

Danian. A sub-division (usually ranking as a stage) of geological time, which is considered either uppermost Cretaceous or lowermost Palaeocene.

daphnia. *See* Branchiopoda.

dark minerals. In petrology, usually refers to the ferromagnesian minerals present in igneous rocks.

Darrieus generator (rotor). Vertical axis aerogenerator consisting of thin aerodynamic strips, designed to produce mechanical work or electricity.

Darwin, Charles Robert (1809–1882). The author of *The Origin of Species by Means of Natural Selection*, which revolutionised concepts of evolution, and *The Descent of Man*, which put forward evidence for the evolution of Man from subhuman forms. Darwin collected evidence for his theories for many years, notably during his voyage as naturalist on *HMS Beagle*, when he studied the unique fauna of the Galápagos Islands. He and Wallace jointly published the first work advancing the theory of evolution by natural selection in 1858, and this was elaborated in 1859 in his *Origin of Species*. The theory may be summarised as follows: Organisms produce large numbers of offspring, but the overall numbers of a particular species remain relatively constant. Therefore a struggle for existence occurs. Individuals of a species exhibit variation. These differences may confer advantages on certain individuals, increasing their chances of survival and reproduction. This results in survival of the fittest, and implies the adaptation of the organism to its environment. The possession of advantageous variation is handed down to the offspring. (Darwin did not know how random variations were inherited, as his work predated that of Mendel). Thus, when conditions change or organisms spread to new areas, new forms will arise, each adapted to its own environment.

dawn. Sunrise.

day-degrees. The sum of degrees of temperature above a threshold (e.g. a daily mean of 4°C) over a certain period (e.g. the growing season of a particular crop).

dB. Decibel.

DCMU. Diuron.

DDT (dichlorodiphenyltrichloroethane). Persistent organochlorine insecticide introduced in the 1940s and used widely because of its low toxicity to mammals and cheapness of manufacture. It is dispersed all over the world, and with other organochlorines, has had a disruptive effect on species high on food chains, esp. on the breeding success of certain predatory birds. It is very stable, relatively insoluble in water but very soluble in fats. Health effects on humans are not clear, but it is less toxic than related compounds such as lindane, aldrin and dieldrin. It is poisonous to vertebrates, esp. fish, and stored in fatty tissue of animals as sub-lethal amounts of the less toxic DDE. Because of its side effects usage in many countries including the UK is now limited. It is also known as Dicophane and Chlorophenotone.

Deadly Nightshade. *See* Solanaceae.

deamination. The removal of the amino (NH_2) group from molecules. In mammals, amino acids are deaminated by enzyme action in the liver and kidney, leaving carbon compounds which may be used in respiration. The waste product of deamination, ammonia, is converted into the less harmful urea in the liver.

debris (Mining). Tailings.

Deca- (da). Prefix used in conjunction with SI units to denote a quantity equal to the unit × 10.

decapod. (*a*) An order of Crustacea. (*See* Malaconstraca). (*b*) A sub-order of Mollusca (see Cephalopoda) whose members possess ten tentacles, including cuttlefish, squids and ancestral Mesozoic forms (Belemnoidea). *See* Octopoda.

deci- (d). Prefix used in conjunction with SI units to denote the unit × 10^{-1}.

decibel (dB). A unit used to measure the intensity of sound, on a logarithmic scale based on measurements of sound intensity in watts per square metre and related to a reference, 10^{-12} watts/m², which is the intensity of the quietest sound perceptible to the human ear. Because the scale is logarithmic ($\log_{10}$), each doubling of intensity increases the decibel value by 3. The scale is then weighted according to three further systems, A, B and C (of which A is that most commonly used to give the dBA unit) to reduce the response of sound meters to very high and very low sound frequencies and to emphasise those within the range that is audible to the human ear. Some meters have a further (D) weighting, to measure 'perceived noise' (PNdB) often used in assessing aircraft noise. On the dBA scale, the rustle of leaves is 10 dBA, a quiet office 40, an alarm clock at one metre 80, a Saturn moon rocket lifting off at 300m 200. One decibel is equal to ten bels (named after Alexander Graham Bell).

deciduous. (*a*) (Bot.) Plants which shed all their leaves annually at a particular season, usually the autumn. *See* evergreen, tropophyte. (*b*) Deciduous teeth, the first set of the two sets possessed by most mammals. There are fewer grinding teeth in the first set, but otherwise the two sets are similar. *See* dental formula. (*c*) Deciduous scales. Fish scales that are shed readily, esp. on contact with a solid object.

Declaration on the Human Environment. The declaration agreed at the 1972 UN Conference on the Human Environment accepting a common attitude by nations to environmental issues. *See* UNEP.

decomposition. The separation of organic matter into simpler compounds.

decomposer. Reducer.

decumbent. (Bot.) Lying on the ground with the ends curving upward.

decurrent. (Bot.) Projecting downwards, below the point of attachment.

deep sea. An imprecise term usually restricted to that part of the ocean beyond the continental shelf. *See* abyssal.

deficiency disease. Disease caused by the deficiency of an essential food substance (e.g. scurvy, caused by lack of vitamin C).

definitive host. *See* host.

deflation. The picking up and removal of loose material by the wind.

defoliant. Herbicide designed to remove leaves from trees and shrubs or to kill plants. Such defoliants as 2, 4, D and 2, 4, 5-T disturb hormonal balances in plants and so induce metabolic disorders.

deforestation. The permanent removal of forest and undergrowth.

deformation. Any change in the original form or volume of a rock-body produced by tectonic forces. Deformation can be a contraction or an extension, and can be formed by folding, faulting or solid flow.

degenerative diseases. Diseases caused by the degeneration of organs or tissue, rather than by infection. *See* diseases of civilisation, Saccharine disease.

degreasing. Process of removing grease, oils and dirt from machine parts by dipping them into a tank containing a degreasing agent, commonly an organic solvent (e.g. trichloroethylene, known as 'tri' or 'trike'). Exposure to 'trike' followed by the consumption of alcohol can cause a skin inflammation called 'degreaser's flush'. Trichloroethylene is also a suspected carcinogen.

dehiscent. Term applied to fruits which open to release seeds (e.g. gorse, poppy). *See* capsule, follicle, legume.

dehumidifier. Device incorporated into many air conditioning systems to dry incoming air by passing it across a bed of a hygroscopic substance or through a spray of very cold water. *Cf.* humidifier.

dehydrogenase. An enzyme which catalyses the removal of hydrogen from a substance. *See* oxidase, respiration.

delayed density dependent. Applied to a situation in which the mortality among a host population is dependent on the population density of the host in successive generations, this affecting the size of the parasite population.

Delphinoidea. Odontoceti.

delta. An accumulation of sediment at a river mouth. Conditions for delta building occur when the rate of sediment deposition into sea or lake exceeds the rate at which it can be removed. As sediment blocks river channels, new channels must be found for the water and its load. Three sets of bedding are usually observable in delta structure. Bottomset beds of fine sediment are deposited farthest into the sea or lake. Foreset beds of coarser sediment are deposited at the mouth of the river and advance the delta into the sea or lake as they accumulate. Topset beds overlie the foreset beds and build up to sea level. They are composed largely of alluvial sediment as in the flood plain. Deltas are typically triangular, with the apex upstream (the Nile Delta being the prototype) but other forms occur, notably the Mississippi Delta, which has a digital or bird's foot form projecting into the sea.

deme. Suffix used in experimental taxonomy to indicate a group of plants which exhibits clearly definable characteristics (e.g. a gamodeme is a group of individuals capable of interbreeding). *Cf,* ecotype.

demersal. A term sometimes used as a synonym for 'benthic'; living in the lowest layer of a sea or lake.

demetron-S-methyl. Systemic organophosphorus insecticide and acaricide used to control aphids and red spider mites on agricultural and horticultural crops. It is poisonous to vertebrates.

demographic transition. A transition in the pattern of growth of a human population from one characterised by high birth and low death rates to one characterised by low birth and low death rates, probably caused by increasing prosperity, better education, the greater economic activity of women and reduced infant mortality.

demography. The study of population dynamics (e.g. the age and sex structure, distribution, rate of change of size, etc.) applied to human populations.

denaturing. The addition of a noxious substance to render a product unfit for human consumption. Denaturing has been used to prevent wheat from being sold for human consumption when economic policies required it to be used for feeding livestock. Fish caught surplus to market requirements is often denatured and dumped at sea. Denaturing substances are often dyes. Methyl alcohol pyridine is added to industrial ethanol to denature it, producing 'methylated spirits', the drinking of which leads to extreme intoxication due to the impurity and the high concentration of ethanol.

dendrites. Nerve cell processes, usually branched, which conduct impulses into the nerve cell body. *See* axon.

dendritic crystals. Ice crystals commonly found in snow, characteristically formed when ice particles fall through supersaturated air and grow by sublimation. They are hexagonal and have many branches, complex but symmetric, producing patterns that inspire much Christmas card art.

dendritic drainage. River systems resembling the branching of a tree.

dendrochronology. The science of dating and investigating historical climates through the study of differences between successive annual rings of trees. Such differences may result from the correlation between ring growth and climate, esp. in certain species. Using bristle-cone pines (*Pinus aristata*) the oldest living example of which is 4600 years old, and by correlating rings in living and dead wood in the same area, dendrochronologically dated wood up to 8200 years old can then be dated by carbon-14 dating (*see* radiometric dating) and the latter checked. This has shown that radiocarbon dates before about 1000 BC were much too young (a calibration curve published in 1970 showed that 6000 BP on radiocarbon should be nearly 7000 BP) and consequently radiocarbon dates, and inferences from dates, published before 1970, must be regarded with suspicion.

denitrification. The breakdown of nitrates by soil bacteria (e.g. *Bacterium denitrificans*) resulting in the release of free nitrogen. This process takes place under anaerobic conditions such as those found in waterlogged soil, and reduces soil fertility. *See* nitrogen cycle.

density. The mass of unit volume of a substance.

density-dependent factor. A limiting factor in the growth of a population which is dependent upon the existing population density (e.g. disease, reproductive rate, access to food).

density independent. Applied to a situation in which the percentage mortality or survival of a species varies independently of population density.

dental formula. A formula used to indicate the number and type of teeth present in a mammalian species. The human formula is $i\frac{2}{2}. \ c\frac{1}{1}. \ p\frac{2}{2}. \ m\frac{3}{3}. \ i$ = incisors, c = canines, p = premolars, m = molars, the top and bottom teeth on one side only being indicated. The molars are not represented in the deciduous set.

denticles. *See* dentine.

dentine. The bone-like substance which makes up the bulk of a tooth and lies inside the enamel. Ivory is composed of dentine. It also occurs in the tooth-like scales (denticles) of present-day cartilaginous fishes and some fossil fishes.

denudation. The combined effects of weathering and erosion in wearing away the surface of the land.

deodoriser (US usage). Equipment for the removal of noxious gases and odours. It may consist of a combustion, absorption or adsorption unit.

deoxyribonucleic acid. DNA.

Department of the Environment. (DoE). The British government department responsible for a wide range of matters relating to the environment, headed by a Secretary of State, and comprising the Ministries of Housing and Construction, and Local Government and Development.

depression. (Meteor.) A low pressure region or cyclone.

derelict land. Land damaged by extractive or other industrial processes, and/or by serious neglect, which in its existing state is unsightly and incapable of reasonably beneficial use unless treated.

Dermaptera (earwigs). A small order of cryptozoic insects (Exopterygota) which bear a pair of forceps at the posterior end of the abdomen (used in a few species for seizing prey), and usually having well developed hind wings folded under the short, leathery forewings. Earwigs eat plants and insects, and are sometimes a nuisance in gardens.

dermis. The thick inner layer of the skin in vertebrates. In mammals it is composed of connective tissue in which lie blood and lymph vessels, sense organs, nerves, fat cells, sweat glands and hair follicles (invaginations of the epidermis) with their erector muscles.

Dermoptera (flying lemurs). A small order of insectivorous placental mammals, which glide by means of the skin stretched between the limbs and tail.

derris (rotenone). Insecticide and acaricide used for the control of aphids, thrips, red spider mites, etc. It is harmful to fish, but not to mammals or birds, and it breaks down very rapidly after application. It is extracted from the root of (*Derris elliptica* and other leguminous plants. Rotenone was the active ingredient of AL63, an insecticide used against lice in the 1939–45 war, before the introduction of DDT.

DES. Diethylstilbestrol.

desalination. The extraction of fresh water from sea or other salt water by the removal of salts, usually by distillation.

desert. An area where evaporation exceeds precipitation, for whatever cause, with consequent lack of vegetation. Evaporation rates will vary according to temperature, but less than 25 cm of rain annually will produce a desert in almost any temperature range. A semi-arid area has a ratio of precipitation to evaporation that is less than one, i.e. a deficiency of rainfall for the year as a whole. A true desert has one half the precipitation that would separate semi-arid climates from humid climates at that temperature range. *See* arid zone, drought.

desert crust. Desert varnish.

desert pavement. A single layer of closely spaced stones collecting on the surface of silt and sand that grades down to gravel in arid and semi-arid areas.

desert rose. A coarsely-crystalline mass of tabular gypsum or baryte crystals found buried in deserts and bearing a vague resemblance morphologically to a rose. The crystals usually contain sand grains.

desert soils. Soils of arid regions where the annual rainfall is generally less than 25.5 cm, though temperatures may vary from cool to hot. The vegetation is sparse and/or sporadic due to the net deficiency of rainfall, rather than an inherent lack of nutrients, and consequently the organic layer is thin or even discontinuous. A pebble layer may accumulate at the surface. The leached layer is usually less than 15 cm thick and characteristically there is a carbonate layer within 30 cm of the surface. *See* soil classification.

desert varnish (desert crust, patina). A hard coating of iron and manganese oxides on the surface of rocks in the desert. Desert varnish is probably formed by deposition from evaporating mineral-charged water drawn to the surface by capillary action. The coating is usually black.

design rule. Rules that require manufacturers to design products so that they conform to environmental or other standards.

Desmidiaceae. A large group of unicellular freshwater Chlorophyta often forming a film on mud and aquatic plants. A desmid cell is usually composed of two symmetrical halves with sculptured or spinous walls.

desorption. The reverse of absorption or adsorption.

desoxyribonucleic acid. DNA.

dessicator. A laboratory glass vessel with a close-fitting lid and a chamber in the base for moisture-absorbent material, in which substances can be placed for the gradual extraction of water at ambient temperatures.

destructive distillation. The distillation of solids accompanied by their decomposition (e.g. of coal to produce coke and coal tar). Solid domestic refuse is heated in a retort without air being added to temperatures between 500°C and 1000°C, its weight being reduced by 90%. Combustible gases, volatile fluids, tar and charcoal are produced and the useless residue can be disposed of at land-fill sites.

desulphurisation. (*a*) In petroleum refining, the removing of sulphur or sulphur compounds from an oil. (*b*) In coal processing, the removing of sulphur by elutriation, froth flotation and magnetic separation. (*c*) The removing of sulphur from iron, non-ferrous metal, or an ore.

detention period. The average length of time that a unit volume of a fluid is retained in a tank during a flow process.

detergent. A surface-active agent used to remove dirt and grease from a surface. Early synthetic detergents, containing alkylbenzene sulphonate (*see* alkyl sulphonates) proved resistant to bacterial decomposition, so causing foaming in rivers and difficulties in sewage treatment plants. These 'hard' detergents were replaced in the mid 1960s, in Europe, North America and Australia, by 'soft' biodegradable detergents, containing straight alkyl chains. Problems remain, arising from the use of phosphate compounds (mainly sodium tripolyphosphate) which can cause eutrophication and for which no satisfactory substitute has been found.

determinant. (*a*) Genetic. A hereditary factor. *See* gene, plasmagene. (*b*) Antigenic. The part of an antigen molecule which combines with the corresponding antibody molecule.

detrital sediments. Sediments formed from fragments of pre-existing minerals and rocks, and from the alteration products of rocks (e.g. clay minerals) that have been transported to the site of deposition and then compacted into sedimentary rocks. *Cf.* clastic ('broken') sediments formed from fragments of rock, mineral or organic remains, but not from alteration products of pre-existing rocks.

deuteric. A term applied to the alteration of igneous rocks by the action of volatiles derived from the magma during the later stages of consolidation. Kaolinisation, tourmalinisation, greisening, and serpentinisation are examples of deuteric processes.

deuterium (D). Element. The isotope of hydrogen with mass number 2 and atomic mass 2.0147. 0.0156 % of natural hydrogen is deuterium and in water about one part in 5000 has hydrogen displaced by deuterium, giving deuterium oxide (D_2O), 'heavy water', used as a moderator or coolant in some nuclear reactors. Heavy water has a specific gravity of 1.1, freezes at 3.82°C and boils at 101.42°C.

developed countries. *See* economic development.

developing countries. *See* economic development.

development. (Meteor.) The generation of motion by buoyancy forces in the atmosphere, involving the ascent of warm air, and a direct circulation. An agent of cyclogenesis.

developmental zero. The temperature, T_0, at which a given developmental process would cease if the rate of the process were proportional to $(T - T_0)$.

development assistance committee. (DAC). Committee of OECD concerned with development and aid to developing countries.

development, economic. Economic development.

deviating force. Coriolis force.

Devonian. Refers to the fourth oldest period of the Palaeozoic Era, usually taken as beginning some time between 405 and 425 million years before the present. Devonian also refers to the rocks formed during the Devonian Period: these are called the Devonian System, which in Europe is divided into 3 series, Lower, Middle and Upper. These series are subdivided into stages: the Lower into Gedinnian, Siegenian and Emsian, the Middle into Eifelian and Givetian, and the Upper into Frasnian and Fammenian. The Devonian Period lasted about 45 to 50 million years. Some geologists place the Downtonian series into the Devonian rather than the Silurian. The non-marine facies of the Devonian in northern Europe is commonly called the Old Red Sandstone (commonly abbreviated to ORS).

dew. Water vapour that condenses on to solid objects when the dew point is reached, usually in the evening when surfaces cool more rapidly than the surrounding air. If surfaces cool to below freezing point, hoar frost will form by sublimation.

dew bow. A rainbow phenomenon seen in dew drops on the ground, rarely noticeable because of the weak intensity.

dew point. The temperature at which air becomes saturated with water vapour on being cooled. If the cooling is due to adiabatic expansion, cloud droplets are formed by further cooling; if caused by cooling at a cold surface, condensation occurs on the surface as dew. If cooling is due to mixing with colder (but unsaturated) air the mixture may be colder than its dew point, and then cloud is formed: this happens in fogs and condensation trails.

dextral fault. A wrench fault, or a fault with a considerable component of strike-slip motion, in which the distant block shows relative displacement to the right when viewed from across the fault-plane.

dextrose. *See* carbohydrate.

dia-. Prefix denoting 'through'.

diabase. (*a*) Dolerite (N. American usage). (*b*) Dolerite older than the Tertiary, which is so altered that few, if any, of the original minerals survive (obsolescent British usage derived from Continental usage). *See* microgabbro.

diachronous. A term used to describe a rock unit which is apparently continuous but which represents the development of the same facies at different places at different times. The bed immediately above an unconformity is usually diachronous.

diagenesis. The changes undergone by a sediment after depositon. These include changes caused by organisms, compaction and the resulting decrease in porosity, and changes related to solution and deposition of minerals by connate and circulating waters. Diagenesis grades into metamorphism.

diageotropism. *See* geotropism.

diallate (di-allate). Soil-acting herbicide of the thiocarbamate group used to control wild oat and blackgrass in brassica and beet crops. It can be irritating to the skin and is harmful to fish.

dialysis. The separation of smaller molecules from larger ones in a solution by means of a semi-permeable membrane which allows the passage of the smaller molecules but not the larger.

diamond. A mineral composed of the high-pressure form of carbon. Diamonds are found in ultrabasic pipes of kimberlite and in placer deposits. Diamond is the hardest mineral known (10 on Moh's scale) and is chiefly used in abrasives. Gem-quality stones are cut for jewellery.

diapause. A dormant stage in the life cycle of some invertebrates, during which metabolic rate is much decreased (e.g. hibernation in insects).

diapir. An intrusion which domes up the overlying layers, having cut through lower layers. Salt-domes are examples of diapirs.

diaspore. (*a*) (Geol.) One of the major ore minerals of aluminium, diaspore, with the formula AlO(OH), is one of the main constitutents of bauxite and laterite (*b*) (Bot.) A disseminule. Any part of a plant (e.g. spore, seed, turion), which is dispersed and can produce a new individual.

diastases. *See* Amylases.

diastem. A minor break in a sedimentary sequence of rocks.

diastrophism. The processes of large scale deformation of the Earth's crust, producing continents, land masses, seas and ocean basins, and mountain ranges. Diastrophism is usually divided into orogenesis (mountain building) and epeirogenesis (vertical movements without major crustal shortening).

diatom. *See* Bacillariophyta.

diatomaceous earth (kieselguhr). A friable, siliceous deposit composed of the skeletal remains of microscopic, unicellular plants called diatoms (*See* Bacillariophyta). It is used as an abrasive, a filtering medium, a filler, a physical insecticide, and as a thermal and acoustic insulator.

dicaryon. dikaryon.

2, 4 D (2, 4–dichlorophenoxyacetic acid). Translocated hormone weed killer used as a herbicide to control many broad-leaved weeds in cereals and grass. *See* translocated herbicides.

dichlorprop. Translocated hormone weed killer used to control many broad-leaved weeds in cereals. *See* translocation.

dichlorvos. An organophosphorus insecticide and acaricide of short persistence, used domestically, in glasshouses, and outdoors on fruit and vegetables for a rapid kill close to harvest. Resistant strains of aphids and red spider mites have appeared in some areas.

dichogamy. The maturation at different times of the male and female parts of a flower. This prevents self-pollination. *See* momogamy, protandrous, protogynous.

dichotomous. (Bot.) Equally forked.

dicophane. DDT.

dicotyledon. *See* cotyledon, Dicotyledoneae.

Dicotyledoneae. The larger of the two classes of flowering plants (Angiospermae). The embryo has two (rarely more) cotyledons. The leaves are mostly net veined, and the vascular bundles in the stem usually contain cambium and are arranged in a ring. The flower parts are commonly in fours or fives, or multiples of these numbers. The class includes many trees as well as herbaceous plants. *See* Monocotyledoneae, hormone weed killers.

Dictyoptera. An order of insects (Exopterygota) which includes the swift-running, omnivorous cockroaches and the slow-moving mantids which have forelegs enlarged to grasp insect prey. Both groups have somewhat flattened bodies, long legs and usually leathery forewings covering membranous hind wings. Formerly included in the Orthoptera.

Didelphia. Marsupialia.

dieldrin. Organochlorine insecticide used widely in the 1950s and 1960s that is highly persistent and fat-soluble, thus becoming concentrated along food chains to produce adverse effects on species high on food chains (e.g. (in the UK) the peregrine falcon whose numbers decreased). UK uses are now very restricted. *See* Cyclodiene insecticides.

diethylstilbestrol (DES). Synthetic oestrogenic hormone used in livestock husbandry, that has been implicated in cancer of the vagina in humans. *See* oestrogen.

differential cooling. *See* differential heating.

differential heating. Refers specifically to heating by sunshine. Surfaces undergo different rates of rise of temperature in the same sunshine because of different thermal capacities and conductivities (e.g. sand becomes hotter than solid rock), absorption over different depths (e.g. land becomes warmer than water), different albedo (bare earth or dark surfaces become warmer than a snow covered surface), or cooling by evaporation (a wet surface or one covered by vegetation warms more slowly than a dry or bare surface). Breezes, most notably sea breezes and anabatic winds, are generated by differential heating. The complementary phenomenon of differential cooling is less spectacular because albedo and condensation effects are small, most bodies being equivalent black bodies when radiating. Furthermore, heating at the bottom of the air produces motion more intense and over a greater depth than cooling at the bottom.

differentiation. The development during the growth of an organism, or the regeneration of one of its parts, of different cells and organs from unspecialised cells. Differences in structure enable the cells or organs to perform different functions (e.g. the development of xylem and phloem elements from cambium in higher plants.)

diffluence. The flowing apart of air particles with the consequent separation of streamlines accompanied by deceleration, the motion being horizontal.

diffraction. The passage of waves around sharp edges that are not large compared with the wavelength. Diffraction of light around particles in the air produces the separation of wavelengths (e.g. in the corona and glory). The diffraction of sound waves can cause a sound shadow behind an acoustic screen.

diffraction analysis. Application of a diffraction technique of X-rays, electrons, neutrons, to study the structure of matter, esp. solids. Involves the diffraction of electromagnetic radiation or particle beams.

diffuse field (Acoustics). A sound field in which sound pressure is equal at every point and sound waves are likely to be travelling in all directions.

diffusion. Process by which molecules intermingle as a result of their random thermal motion. Spreading or scattering of a gaseous or liquid material. *See* molecular diffusion, eddy diffusion.

digestion. The breaking down of complex food substances into simpler compounds which can be used in metabolism. Digestion is brought about by enzymes (e.g. lactase which is secreted by the small intestine of mammals and splits lactose into glucose and galactose).

digital computer. Device, originally mechanical but now generally electronic, used in solving problems and in the construction of models. The digital computer works with data in the form of digits, rather than with physical quantities (*See* analogue computer) and can handle large amounts of information at high speed.

digitalis. *See* Scrophulariaceae.

digitigrade. Applied to animals which walk on the ventral surface of the digits only, not on the whole foot (e.g. cat, dog). *See* plantigrade, unguligrade.

dikaryon (dicaryon). A fungus hypha or mycelium made up of cells each containing two identical or different haploid nuclei. *Cf.* monokaryon.

dike. Dyke.

dilute sulphuric acid process. Japanese process for removing sulphur dioxide from flue gases.

dilution. The disposal of effluent by discharging it into a much larger receiving volume of air or water.

diluvium. Term applied to glacial deposits which were at one time attributed to the Great Flood of Noah. The word persisted in literature as a synonym for glacial drift, but it has fallen increasingly into disuse. Its literal meaning of 'flood deposit' has not been greatly used in the true sense. *Cf.* alluvium.

dimethoate. Systemic organophosphorus insecticide and acaricide used to control red spider mites and insects such as aphids on agricultural and horticultural crops. Strains of mites and aphids resistant to this chemical have arisen in some areas. It is poisonous to vertebrates.

dimorphism. (*a*) The existence within a single species of two different forms (e.g. male and female individuals). (*b*) The capability of possessing two forms, as when a substance crystallises in two forms, or, e.g., hydranth and medusa in some coelenterates.

Dinantian. Lower Carboniferous. The Dinantian usually ranks as a series and comprises the Tournaisian, Visean and Namurian Stages.

dinitro-cresol. DNOC.

dinitro insecticides and herbicides. Compounds used as contact herbicides, fungicides and insecticides (e.g. Dinoseb and DNOC). Although very poisonous to plants and animals, they are broken down rapidly after application, so delayed environmental contamination does not occur. *See* dinocap.

dinocap (DNOPC). Fungicide and acaricide of the dinitro group used to control powdery mildew on horticultural crops and to suppress red spider mites. Can be irritating to the eyes and skin, and is harmful to fish.

Dinosaurs. Mesozoic reptiles belonging to the orders Saurichia and Ornithischia, numerous for ten million years at the end of the Triassic. Throughout most of the Jurassic and Cretaceous the large, bipedal Saurischia (e.g. *Allosaurus, Tyrannosaurus*) were the world's dominant carnivores. Jurassic Saurischia also included the largest terrestrial vertebrates, the herbivorous, quadripedal *Brontosaurus* and *Diplodocus*. The Ornithischia were smaller herbivores which appeared in the Jurassic and were most abundant in the Cretaceous. They included the bipedal *Iguanodon*, the heavily armoured, quadripedal *Stegosaurus* and *Ankylosaurus*, and *Triceratops*, a quadruped with a large head bearing horns and a bony frill.

dinoseb (DNBP, DNSBP, DNOSBP). Contact dinitro herbicide used to control many broad-leaved weeds in leguminous crops, cereals, etc. Very poisonous.

dioecious. Applied to organisms in which male and female reproductive organs are borne on different individuals. *Cf.* monoecious, hermaphrodite, unisexual.

diorite. A coarse-grained intermediate igneous rock consisting essentially of a plagioclase feldspar which is more calcium-rich than that in granodiorite, together with one or more ferromagnesian minerals (e.g. biotite, hornblende, augite). Diorites are the plutonic equivalent of andesite.

dioxin. TCDD.

dip. The angle between the line of greatest slope in a rock surface and the horizontal. The direction of dip is at right angles to the strike. Dip is the complement of hade.

diphyodont. Applied to an animal which has two sets of teeth, deciduous and permanent. This is a characteristic feature of mammals. *Cf.* polyphyodont, monophyodont.

Diploblastica. Metazoan animals whose bodies are made up of only two layers of cells (ectoderm and endoderm) separated by a layer of jelly (mesogloea). Coelenterata are the only diploblastic animals. In many species there is considerable invasion of the mesogloea by cells of the endoderm and ectoderm. *Cf.* triploblastic.

diploid. *See* chromosomes.

diplont. The diploid stage of an organism's life history. In almost all animals this is the whole life cycle apart from the gametes. The sporophytes of Bryophyta, ferns, seed-bearing plants and some algae are diplonts. In other algae and many fungi only the zygote is diploid. *Cf.* haplont.

Diplopoda. *See* Myriopoda.

dipneusti. Dipnoi.

Dipnoi (Dipneusti, lung-fish). A sub-class of the Osteichthyes which first appeared in the Devonian. The three living genera are air-breathing and inhabit tropical rivers which dry up or become very stagnant. *Neoceratodus* is found in two Queensland rivers. The South American species *Lepidosiren* and the African *Protopterus* can lie dormant in mud for at least six months. Dipnoi have characteristic broad tooth-plates for crushing food, which in modern forms consists of decaying vegetable matter and small invertebrates.

dip plating. Method for producing a thin coating of a different metal on a metal object by immersion in a solution of a salt or salts of the metal to be deposited. It is often a source of metallic pollution of waste water streams.

Dipsacaceae (Dipsaceae). A family of dicotyledonous, mainly herbaceous plants which bear dense heads of flowers. *Dipsacus fullonem* (Fuller's Teazel) has hooked bracts on the fruit heads which are used for raising the nap on cloth.

Dipsaceae. Dipsacaceae.

dip-slope. An inclined land surface which dips in the same direction as the underlying rocks and at approximately the same angle. The term is often applied to the back-slope of a cuesta.

Diptera. Two-winged flies, a large order of insects (Endopterygota) which have hind wings reduced to stumps and larvae which are usually legless. *See* the three sub-orders Cyclorrhapha, Nematocera, Brachycera.

diquat. Contact herbicide used to control broad-leaved weeds in still and slow-flowing water, and to dry up foliage to facilitate harvesting of potatoes, cloverseed, etc. Harmful to mammals.

direct circulation. The sinking of cold air and the ascent of warm air with consequent development (cyclogenesis). The direct circulation at the entrance to a jetstream provides the energy to accelerate the air. It is through the agency of the coriolis force that the circulation is converted into wind energy at right angles to the plane of the circulation.

directive evolution. Orthogenesis.

dirty. (Geol.) Refers to an arenaceous or rudaceous rock with a matrix of clay minerals.

disaccharide. *See* carbohydrate.

disc floret. A flower at the centre of a flower head (e.g. in the Daisy family, *Bellis* spp.).

discharge. The volume of water flowing through a given point in a stream channel in a given period of time.

disclimax. A sub-climax that endures for a long time and is prevented from reaching a full climax by human or animal interference. A modification or replacement of a true climax by human or (domestic) animal disturbance.

disconformity. An unconformity with no angular divergence between the old and younger strata.

discontinuous distribution. A pattern of distribution in which similar species are found in widely separated parts of the world. This is usually taken to indicate that the group is ancient and was once more generally distributed, but has become extinct over most of the original range. *See* Dipnoi.

discontinuous variations. *See* variation.

diseases of civilisation. Those diseases, mainly degenerative, that occur much more commonly in industrial ('civilised') societies than in non-industrial societies and are probably associated with environmental factors such as diet, industrial pollution, and general lifestyle, although their apparent incidence may be exaggerated by the greater frequency of infectious and deficiency diseases among poor, non-industrial communities, affecting younger age-groups. *See* saccharine disease.

dishpan experiments. The name given to model experiments to study the motion of fluid in a rotating vessel which was heated below or at an outside or inside vertical wall. Most notably the experiments showed that a meandering jetstream was formed with a small number of waves, and a circumpolar vortex relative to the vessel.

disinfection. The destruction of pathogens usually by applying chlorine.

disintegration (Nuclear). The disruption of the nucleus of an atom, with the release of an alpha or beta particle.

dispersion, atmospheric. Atmospheric dispersion.

displacement activity. An apparently irrelevant action which is performed by an animal when it is stimulated to carry out two incompatible behaviour patterns (e.g. grooming activity performed by a rat when it is presented simultaneously with stimuli which normally elicit approach and flight).

display. A method of communication between animals during courtship, mating, territory holding, etc. It involves showing off conspicuous features (e.g. peacock's tail, newt's crest, red breast of stickleback), ritual performance of actions (e.g. elaborate courtship dance of Great Crested Grebe) or producing sounds (e.g. birdsong, 'drumming' of Snipe).

disruptive coloration. Apatetic coloration.

disseminule. Diaspore.

dissipation trail. Distrail.

dissolved load. The weathered rock constituents carried in chemical solution by moving water.

dissolved oxygen. Oxygen molecules that are dissolved in water, usually expressed in parts per million (ppm). The presence of dissolved oxygen is vital to aerobic organisms in water. Normal saturation at $0°C$ is about 10 ppm, but it falls as temperature rises, to about 6.5 ppm at $20°C$ and 5.5 ppm at $30°C$, and the saturation point also depends on the atmospheric pressure and on the chemical content of the water. In still water, oxygen dissolved from the atmosphere diffuses slowly through the lower levels of water, but in moving water the constant exposure to the air of unsaturated water usually leads to a higher dissolved oxygen content.

distilled water. Water of great purity, prepared by repeated distilling and used in electrical conductivity measurements, etc.

distilling. Process of heating a mixture to separate its components by condensation of the volatile components driven off by the heating.

distrail (dissipation trail). The downwash behind an aircraft accompanying the trailing vortices may force a line of clear air into a thin layer of cloud, causing a clear lane in the layer called a dissipation trail, or d. It often takes the form of a series of holes, which correspond with the blobs of a condensation trail. Sometimes the shadow of a condensation trail on a cloud is imagined by observers to be a distrail.

distribution. Arrangement, or pattern. Statistically, it is the way in which variate values are apportioned.

district heating. A system that uses hot water from a single source (e.g. cooling water from a power station) to heat buildings nearby. Block heating is the heating of one or two apartment blocks or a shopping precinct from a central source. Group heating is the heating of a small group of houses from a central source.

disturbance. (Acoustic) Excitation.

disulfoton. Systemic organophosphorus insecticide used to control carrot fly and aphids in many crops. Poisonous to vertebrates.

ditch. *See* soil drainage.

dithiocarbamate. *See* carbamate.

ditokous. Producing young two at a time. *Cf.* monotocous, polytocous.

162

diurnal. Daily, usually applied to events or cycles that repeat at daily intervals (*See* Circadian rhythm). The diurnal cycle of air pollution is of interest to those concerned with pollution control.

divergence. (Meteor.) In three dimensions, divergence is the rate of increase of unit volume of a fluid. In the atmosphere this is approximately proportional to the vertical velocity, on account of the adiabatic expansion in the hydrostatic pressure field. In regard to horizontal motion envisaged in meteorological charts, divergence is considered in two dimensions: horizontal divergence at the ground is accompanied by descending motion above and the dissipation of clouds. Horizontal divergence decreases vertical vorticity and cyclonic rotation and is therefore accompanied by rising pressure.

divers. Gaviidae.

diversity. (*a*) The degree of uncertainty attached to the specific identity of any individual selected at random. (*b*) A collection is said to have a high degree of diversity if it contains many species of fairly equal abundance; diversity is low when species are few or their abundance uneven.

diuresis. An increase in the volume of urine produced by the kidneys, usually caused by an increase in the amount of liquid drunk.

diurnal rhythm. Circadian rhythm.

diuron (DCMU, DMU). Soil-acting herbicide of the urea group used for total weed control on land not intended for cropping, and for selective control of annual weeds in fruit crops and tree nurseries. Its effects can last for 12 months after application. It can be irritating to the eyes and skin, and is harmful to fish.

division. *See* classification.

dizygotic twins (fraternal twins, non-identical twins). Twins produced as a result of the simultaneous fertilisation of two ova. The twins are not genetically identical, and may be of different sexes. *Cf.* monozygotic twins.

DMU. Diuron.

DNA (Deoxyribonucleic acid, desoxyribonucleic acid, thymonucleic acid). The principal material of inheritance, found in chromosomes. DNA molecules are long, unbranched chains made up of many nucleotides. Each nucleotide is a combination of phosphoric acid, and monosaccharide deoxyribose, and one of four nitrogenous bases: thymine, cytosine, adenine, or guanine. Usually two DNA strands are linked together in parallel by specific base pairing, and helically coiled. Adenine will link (by hydrogen bonding) only to thymine, and guanine only to cytosine. The number of possible arrangements of nucleotides along the DNA chain is immense. Replication of DNA molecules is probably accomplished by separation of the two strands, followed by the building of matching strands by means of base pairing, using the halves as templates. By a mechanism involving RNA, the structure of DNA is translated into the structure of proteins, during their synthesis from amino acids. *See* Genes.

DNBP. Dinoseb.

DNC. DNOC.

DNOC (DNC, dinitro-cresol). Contact dinitro herbicide, insecticide and acaricide used to control broad-leaved weeds and the overwintering stages of many insect and mite pests. Very poisonous.

DNOPC. Dinocap.

DNOSBP. Dinoseb.

DNSBP. Dinoseb.

dodder. *See* cuscuta.

DoE. Department of the Environment.

dogger. (Geol.) (*a*) Strata deposited during Middle Jurassic times. (*b*) A large (boulder size) calcareous concretion.

Dog's Mercury. *See* Euphorbiaceae.

doldrums. The region of small pressure gradients between latitudes 5°N and 5°S, where winds are light because the coriolis force is negligible. It is a region of widespread showers over the ocean.

dolerite. A medium-grained basic hypabyssal rock, equivalent to gabbro and basalt. Dolerites are termed diabases in North America.

doline. A feature of karst landscape consisting of flue, funnel or dish-shaped hollows, with varied outlines and dimensions (ranging from about 2 to 300 m in depth). The floors are filled with fallen rocks.

dolomite. (*a*) A carbonate mineral with the formula $CaMg(CO_3)_2$. Dolomite occurs in evaporite deposits, as a replacement in limestones, as a cement, as a gangue mineral in hydrothermal deposits, and in carbonatites. (*b*) A rock consisting of a high percentage (usually 50%) of the mineral dolomite. To avoid confusion, such rocks are sometimes called dolomite-rock or dolostone.

dolomitisation. The alteration of original calcite limestones by percolating magnesium-carbonate solutions.

dolostone. Dolomite-rock, i.e. a rock consisting predominantly of the mineral dolomite.

domestic wastes. Water-borne wastes from households, including sewage and sullage water.

dominance frequency. The number of sampling units in which a particular species is most numerous.

dominant. (*a*) (Ecol.) The characteristic and often the tallest species in a particular plant community. The dominant species is the one which exerts the greatest influence on the character of the community (e.g. oak in oakwood, reed in reed swamp). (*b*) (Genetic) A dominant character is the one of a pair of contrasted characters which is fully developed whether the individual is heterozygous or homozygous. *See* recessive. (*c*) (Social). The leader in a group of animals (e.g. the most aggressive male in a herd of red deer during the rutting season).

Donora smog incident. An air pollution incident that occurred in Donora, Penna., in October 1948, when fog accumulated in very stable atmospheric conditions for a total of 7 days, before it was washed down by rain. The population of this industrial town was 14000. 42% suffered illness, 10% were seriously ill, and 18 persons died. The principal pollutants were believed to be sulphur dioxide and particulate matter.

Doppler effect (shift). The apparent change in the frequency of sound or electromagnetic waves caused by the relative motion of the source and the observer. Thus an approaching source will emit waves each of which begins from a point closer to the observer than the previous one, producing an apparent increase in frequency and consequent rise in the pitch of a sound or increase in blue light. A departing source will emit waves whose frequency appears to decrease, producing a lowering of sound pitch or increase in red light.

dormancy. A resting condition in which the growth of an organism is halted and metabolic rate is slowed down. Dormancy may involve the whole organism or only its reproductive bodies. It may be caused by unfavourable conditions or it can be part of a rhythmic cycle (e.g. winter dormancy in deciduous trees, regulated in some by photoperiod; summer dormancy of daffodil bulbs). *Cf.* hibernation, aestivation, diapause.

dormin

dormin. Abscisin.

dorsal. (*a*) (Zool.) The part of an animal or organ at, or nearest to, the back, which in most species is directed upwards. In vertebrates the dorsal surface is the one adjacent to the vertical column. In bipedal vertebrates it is directed backwards, and in bony flat fishes the upper side is anatomically lateral, not dorsal. A few invertebrates move about with the dorsal surface downwards (e.g. back-swimming water boatmen, *Notonecta*) or lie on their sides (e.g. freshwater shrimps, *Gammarus*). (*b*) (Bot.) Abaxial. The surface of a leaf facing away from the stem. *Cf.* ventral.

dorsiventral. A term applied to dicotyledonous leaves which lie more or less horizontally, and whose upper and lower sides show differences in structure. *See* mesophyll.

double recessive. An individual homozygous in respect of a particular recessive gene.

doubling time. *See* exponential growth.

downdraft. A descending current of air of interest in the region behind buildings in which pollution from chimneys may be brought to ground level, and under rainstorms, in which the rain cools air by evaporation into it and causes it to descend and spread out at the ground. *See* squall.

downland vegetation. The vegetation characteristic of the chalk downs of southern England is short turf rich in flowering herbs. Until recently the natural succession was held at this stage because of grazing by sheep and rabbits. Since myxomatosis drastically reduced the rabbit population, scrub has spread, to the detriment of many attractive herbs. *See* grassland.

Downtonian. The youngest series of the Silurian System in the UK, and equivalent to the Pridolian of Europe. The Downtonian is placed by some geologists in the Devonian System.

dowsing. Detection of underground water or other substances or objects by feeling the motion of a split stick (often hazel) held in the hands. Used by experienced persons the technique has a long history of successful application, though the principles are not understood.

drag. *See* aerodynamic drag.

dragonflies. *See* Odonata.

dragon reactor. Experimental nuclear reactor, funded by ten nations and built at

Winfrith Dorset, UK, to investigate the principles of high temperature gas-cooled reactors.

drainage basin. *See* catchment.

drainage morphometry. The study of drainage patterns.

dreikanter. A pebble abraded by wind-blown sand, and with three facets.

drift. (*a*) The 'drift' edition of a UK geological survey map has the superficial deposits (e.g. glacial and fluvioglacial deposits) coloured in, in contrast to the 'solid' edition which has the bedrock (underlying rocks) coloured. (*b*) Drive. (*c*) An oceanic current. (*d*) Any superficial deposit caused by a current of air or water, esp. a deposit of wind-blown sand in the lee of a gap between two obstacles. (*e*) Glacial drift. *See* boulder clay. (*f*) continental drift.

drive. (*a*) In animal behaviour, a state of activity and responsiveness to stimuli which normally leads to the satisfaction of a need (e.g. sex drive, hunger drive). (*b*) In mining, a horizontal tunnel or opening, lying in or close to the ore-body, and parallel to the strike of the ore-body.

drizzle. Falling drops of water which are carried significantly by the air motion, having diameters less than about 0.5 mm. A form of rain of soft character, usually produced in clouds less deep than those generating larger raindrops. Sometimes produced close to the ground in dense fogs, particularly away from towns where condensation nuceli are less numerous and the moisture distributed among fewer larger droplets.

drop-sonde. A radiosonde device released at high altitude from a balloon or aircraft and descending on a parachute. Used to obtain a sounding of the air in the same way as a radio-sonde balloon.

Drosera. *See* Droseraceae.

Droseraceae. A family of dicotyledonous herbs, all insectivorous, and mostly found in acid bogs. *Drosera* (Sundew) catches insects by means of sticky tentacles on the rosettes of its leaves. Once trapped, the insect is pressed down on to the leaf blade by bending of the tentacles, which secrete a protein-digesting enzyme. *Drosera* is able to live in very poor soil because of its insectivorous habit.

Drosophila. A genus of fruit fly, small, yellowish or brownish Diptera (Cyclorrhapha), much used in genetical research because of its short life cycle and the large chromosomes in the salivary glands of the larvae. *See* salivary gland chromosome.

drought. A long period of unusually low rainfall resulting in parched ground and abnormal withering of vegetation. A drought is defined arbitrarily to suit the region. Thus a period of four months without rain will be regarded as a drought only if people were unprepared for it and it was not normal. In desert regions a drought is a succession of unusually dry years or an unusually long succession of drier than average years. A drought may be recognised as such only if alternative (e.g. underground) water is not available.

drumlin. A smooth, oval hill of glacial drift (usually boulder clay), characteristically with one end blunter in plan and with a steeper slope. A drumlin-field, with many drumlins, is sometimes called 'basket-of-eggs' topography.

drupe. A single-seeded fruit in which the pericarp consists of a skin (epicarp), a thick, usually fleshy middle layer (mesocarp) and a stony endocarp enclosing the seed (e.g. plum, cherry). A coconut is a drupe with a fibrous mesocarp. A blackberry fruit is a collection of small drupes (drupelets). *Cf.* berry.

drupelet. *See* drupe.

dry adiabatic lapse rate. The same as the adiabatic lapse rate but distinguished from the wet adiabatic lapse rate.

dry ice. Solid carbon dioxide. It sublimes at temperatures above $-72°C$ and is therefore a convenient portable refrigerant, used by itinerant ice-cream vendors, etc.

dry impingement. A process that pushes particulate matter carried by a gas stream against a retaining surface, which may be coated with adhesive.

drying agents. Substances that remove water, including calcium oxide and silica gel.

dry-weather flow. The rate of flow of material through a sewer, or of water through a river, in dry weather.

dry weight rank method. DWR.

dual-purpose sewer. Combined sewer.

duck decoy pond. A star-shaped pond with curving arms (pipes) ending in traps, into which ducks are enticed, usually by intermittent sight of a dog specially trained to move around reed screens placed along the pipes. Decoy ponds were formerly much used for the commercial trapping of ducks, but few are now in existence. One, in the Cambridgeshire Fens, UK, is used exclusively to trap ducks for ringing. *See* bird ringing.

ducks. *See* Anatidae.

ductility. Capacity of metals for cold flow, which is accompanied by progressively increasing resistance to such flow, known as 'work hardening'. This
property makes possible the drawing of wire, cold-pressing and similar
operations.

ductless gland. Endocrine organ.

dugong. *See* Sirenia.

dumping. *See* sanitary landfill.

dun. *See* Ephemeroptera.

dunite. An ultrabasic rock consisting entirely, or almost entirely, of olivine.

Duplicidentata. Lagomorpha.

duramen. Heartwood.

durilignosa. Broadleaf sclerophyll (evergreen) forest and bush.

duripan. *See* hardpan.

durum. *Triticum durum.* Variety of wheat used for making pasta and grown in
the Mediterranean region, the USSR, Asia, and North and South America,
especially in arid areas. *see* Graminae.

dust. Solid particles, 1 to 100 microns in size, in the atmosphere. Dust particles
settle under the influence of gravity. They are produced in many industrial
processes, and naturally, and are generally injurious to health, esp. of the lungs
and respiratory system.

dust bowl. An agricultural region from which the soil is carried away by wind,
more specifically a large region of the central USA which experienced low rainfall
in the 1930s, at a time when the soil was ploughed or bare for other reasons.
Before the introduction of agriculture, the soil was retained by prairie grasses.
Some areas lost as much as 60 to 90 cm of soil.

dust burden. The weight of dust suspended in a unit volume of a medium (e.g. flue
gas), expressed in grams per cubic metre at standard temperature and pressure.

dust collectors. Equipment to remove and collect dust from process exhaust
gases. May employ the principles of sedimentation, inertial separation (cyclones,

impacters, impingers), precipitation (thermal and electrostatic) and filtration.

dust devil. A rotating convection current made visible by particles carried off the ground by the whirlwind into the air, ascending in the vortex.

dustlice. *See* Psocoptera.

dust storm. A storm of dust blown up from the ground when wind speed exceeds a critical value (commonly between 24 and 48 km per hr) that depends on the specific gravity, size, shape and dampness of the surface particles, and when the ground surface provides suitable particles. Dust storms are of two main types, into one of which most storms fall: (*a*) Haboob, which is local, associated with a thunderstorm or cumulonimbus cloud from which rain has begun to fall. The rain evaporates before reaching the ground and the dust is blown into the air with the appearance of smoke, having a bulge at the leading edge and a slope at the upper surface. Dust is carried to heights of 1500 to 1800 metres or more. (*b*) Khamsin (an Egyptian word for which the Libyan equivalent is gibleh). This is an extensive storm covering a large area and associated with an area of low atmospheric pressure. In air which is thermally unstable, hot dust and sand particles rising rapidly may increase the instability.

Dutch elm disease. Disease of Elms, endemic in Britain at least since 1927, caused by the fungus *Ceratostomella ulmi* and spread by the Elm Bark Beetle, *Scolytus scolytus*. *See* Ulmus.

Dwale. Deadly nightshade. *See* Solanaceae.

DWR (dry weight rank method). A method for estimating the percentage contribution each species makes to the total yield of a pasture.

dyke (dike). (*a*) A tabular igneous intrusion which is discordant (i e it cuts across the bedding-planes in sedimentary rocks or across the foliation in metamorphic rocks). *See* metamorphism. (*b*) Ditch used for drainage. (*c*) Low wall or bank used to prevent water from invading low-lying land.

dynamic soaring. A soaring technique used by sea birds whereby they maintain or increase air speeds by flying across wind gradients, either behind wave crests or in the general increase of wind in the friction layer. It is not relied upon significantly by birds over land, but is undoubtedly important to the larger sea birds, particularly the Wandering Albatross (*Diomedea exulans*) which remains airborne for many hours, and allegedly for many days, on end.

dynamic stability. In the atmosphere, in the presence of turbulence inducing factors (notably wind shear), a condition in which small perturbations of the flow do not tend to grow. *Cf.* static stability.

dynamometer. An instrument for measuring power (e.g. the power of an engine). It is also used to assess the rate of emission of motor vehicle exhausts under test conditions.

dysphotic zone. The zone of water in a sea or lake lying between the euphotic and aphotic zones. The dysphotic zone is subject to dim light, and usually spans depths of approximately 100 to 600 m.

dystrophic. Applied to freshwater bodies which are deficient in calcium, very poor in dissolved plant nutrients, and therefore unproductive. These waters are typical of acid peat areas, and have bottoms covered with undecomposed plant remains harbouring a poor fauna. The water is usually stained brown with peat. *Cf.* eutrophic, mesotrophic, oligotrophic.

E

ear. A sense organ of vertebrates which is both a sound receptor and an organ by means of which the animal is made aware of its movements and its position in relation to gravity. The sense of hearing may be absent in some fishes and reptiles. Different kinds of receptor cells in the inner ear are stimulated by vibrations initiated by sound waves, movements of liquid, caused by angular acceleration, and movements of otoliths (granules of carbonate) in response to gravity. The auditory ossicles (malleus, incus and stapes in mammals) are small bones which transmit vibrations of the ear drum to the inner ear.

Early Stone Age. Palaeolithic.

earthquake. A series of shock waves generated by a transient distubance within the Earth's crust or mantle. The point of origin of the earthquake is called the focus, while the point on the Earth's surface above this is called the epicentre. The shock waves are classified into body waves (P and S waves) which travel within the Earth and the surface waves (L and R waves). The damage, caused by the surface waves, is classified on the Modified Mercalli Scale, which is based on local structural damage which is itself dependent on the nature of the underlying soil and bedrock. Many of the deaths associated with earthquakes occur after the shock and are due to tsunamis, landslides, fires, epidemics caused by polluted water, exposure, etc. Earthquake magnitudes are given using the Richter Scale which is based on the amplitude of the largest trace recorded by a standard seismograph 100 km from the epicentre. The Richter Scale is always given in Arabic numerals, with 7 being considered a major earthquake. An earthquake of magnitude 8 probably releases about 30 times more energy than one of magnitude 7. The intensity on the Modified Mercalli Scale is given using Roman numerals. Earthquake prediction is a growing science based on such phenomena as changes in magnetic fields, pressure in wells, underground electrical currents, seismic velocities, radon gas underground and the build up of stress in the Earth. Many animals seem able to sense imminent earthquakes and leave the area.

earth resources technology satellite (ERTS). An unmanned, earthorbiting satellite equipped to scan the surface of the Earth to obtain information relating to natural resources and the environment. ERTS-1 was launched in 1972 from the USA. It was developed by NASA and made 14 revolutions of the Earth each day. Its sensors recorded information from overlapping 160 km wide strips and covered the whole Earth every 18 days, producing images for different wave bands from blue to infrared, which can be combined. The chief uses for these images relates to resource mapping (e.g. land use, water resources, potential sites for geothermal energy installations and geological mapping), for mapping inaccessible areas (e.g. polar ice cover) and for mapping ephemeral phenomena (e.g. crop diseases, the movement of polluted water, etc.) The LANDSAT programme replaced the ERTS programme with the launch of the LANDSAT satellite in January, 1975.

earthscan. A news agency, funded by UNEP that commissions original articles on environmental matters and sells them as features to newspapers and magazines, esp. in developing countries. Its headquarters are in London and its first director is Mr Jon Tinker. Earthscan was founded in 1976.

earth's shadow. The darkness that can be seen rising up the eastern sky just after sunset in suitable conditions of haze.

earthwatch programme. A worldwide programme to monitor trends in the environment, established under the terms of the Declaration on the Human Environment in 1972. It is based on a series of monitoring stations and its activities are coordinated by UNEP.

earthworm. *See* Oligochaeta.

earthy. *See* lustre.

earwig. *See* Dermaptera.

easterly wave. Sinuosities in the easterly winds (Trades) of the tropics associated with increased rain, and thought to be connected with the birth of hurricanes.

ecad. A form of plant modified by its habitat. It exhibits non-heritable characteristics.

ecdysone. *See* ecdysis.

ecdysis. Moulting. The periodic shedding of the outer covering of the body, esp. in Arthropoda, Amphibia and Reptilia. In vertebrates the sloughing of the outer epidermis is under the control of pituitary and thyroid hormones. In insects the shedding of the hard exoskeleton, which usually occurs only in the immature stages, is under the control of the hormone ecdysone, produced by the glands in the first thoracic segment. Increase in size in Arthropoda can only occur before the new cuticle hardens.

ecesis. The establishment of colonising plant species.

Echidna. *See* Monotremata.

Echinodermata. A phylum of marine invertebrates usually exhibiting five-rayed symmetry as adults. The skin bears calcareous plates. The coelom is intricate and large, with extensions into the numerous tube feet which protrude from the body surface. The pelagic larvae have affinities with those of the Hemichordata. *See* Echinoidea (Sea urchins), Asteroidea (Star fishes), Ophiuroidea (Brittle stars), Holothuroidea (Sea cucumbers), and Crinoidea (Sea lilies), the five present-day classes.

Echinoidea (Sea Urchins, Heart Urchins, Cake Urchins, Sand Dollars). A class of spiny, armless, cushion-shaped or discoidal Echinodermata with skeletal plates joined to form a rigid exoskeleton. Sea urchins travel by means of their moveable spines and ten meridional rows of tube feet. They feed largely on seaweed, but also ingest mud and detritus. They are either ciliary feeders (*see* Cilia) or obtain food by means of their tube feet, or, most commonly, use a complicated jaw apparatus (Aristotle's lantern).

Echiuroidea. *See* Annelida.

echo. Reflected sound which reaches the observer after a time interval long enough for the echo to be perceived as a separate sensation. The principle is used in echolocation to determine the direction and distance of an object and in echo sounding (*see* Sonar) used to measure the distance between the instrument on board ship and the sea bottom. Radar operates on a similar principle, but uses electromagnetic radiation transmitted as a beam. 'Radar' is an abbreviation of RAdio Detection And Ranging.

echolocation. (*a*) A method used by some animals (e.g. bats, porpoises) to locate surrounding objects by emitting sounds (usually highpitched) and perceiving their reflections. (*b*) *See* echo.

echo sounding. Sonar. *See* echo.

eclipse plumage. *See* moult.

ecliptic. The plane of the Earth's orbit around the sun.

eclogite. A fairly coarse-grained metamorphic rock (*See* metamorphism) with the chemical composition of basic igneous rock but with essential magnesium-rich garnet and sodium-bearing pyroxene indicative of crystallisation, or recrystallisation, at high pressure and temperature. Eclogites are found as xenoliths in blue ground and in some metamorphic belts.

ecocline. (*a*) A cline associated with an environmental gradient. *See* geocline. (*b*) A gradient of ecosystems associated with an environmental gradient.

ecodeme. *See* ecotype, deme.

ecological balance (balance of nature). The components of a natural community are in equilibrium (ecological balance) if their relative numbers remain more or less constant, thus forming a stable ecosystem. Gradual readjustments to the composition of a balanced community are continually taking place in response to natural ecological succession and to alterations in climatic and other influences. Man upsets this balance by removing or introducing plants or animals, by polluting the environment, by destroying habitats and by rapidly increasing the numbers of his own species.

ecological capacity. Carrying capacity.

ecological factor. Any factor of the environment which influences living organisms.

ecology. The study of the relationships between living organisms and their environment.

economic conservation. The management of natural resources, or the environment, so as to sustain a regular yield of a commodity at the highest level feasible.

economic development. The process whereby a country changes its economic base from agriculture or other primary production to industry. The definitions of 'developed' and 'developing' countries is arbitrary. A developing country is likely to have a per capita annual income of less than $500 and primary industries will provide 50% or more of employment and up to 70% of export earnings. The rate of population growth is likely to be more than 2% per annum. Because exceptions can be found of 'developed' countries exhibiting any one of these characteristics, all of them may need to be present before a country can be considered as 'developing'. In a 'developed' country, incomes are likely to be $1000 or more, and primary production will provide much less employment (probably 20% or less) and a smaller proportion of export earnings. Population will be growing at 1% per annum or less. *See* Third World.

economic efficiency. The efficiency with which stipulated economic ends are achieved, often measured as the cost per unit of output. Provided product quality is not sacrificed, the lower the unit cost the greater the efficiency. The concept has been criticised for under-valuing social and environmental costs. *See* cost-benefit analysis.

economic ends. The objectives of economic activity, both quantitative and qualitative. Economics is not concerned with the nature of the ends, but only with their number and relative importance.

economic entomology. The study of insects with particular reference to pests of agricultural crops and the control of their populations.

economic growth. The annual rate of change in the Gross National Product, usually expressed as the percentage difference from the previous year.

economiser. Apparatus for transferring heat from flue gases to boiler feed water, thus increasing the efficiency of the heating system.

ecoparasite. *See* parasitism.

ecopornography. Self-seeking statements, often advertisements or public relations brochures, that illustrate work being done by the issuing organisation to protect or improve the environment, while failing to mention much greater damage to the environment being done by the same organisation elsewhere. Also applied to statements of environmental improvements made by organisations only because present or anticipated legislation required them.

ecospecies. One or more ecotypes in a single coenospecies.

ecosphere. The biosphere together with all the ecological factors which operate upon organisms.

ecosystem. A community of interdependent organisms together with the environment which they inhabit and with which they interact (e.g. a pond, an oakwood).

ecotone. A transitional zone between two habitats (e.g. a woodland edge bordering grassland). *See* mictium.

ecotype (ecodeme, ecospecies). A sub-specific group (by some regarded as a distinct species) which is genetically adapted to a particular habitat, but can interbreed with other ecotypes (or ecospecies) of the same species (or coenospecies) without loss of fertility.

ectoblast (ectoderm). *See* germ layers.

ectoderm (ectoblast, epiblast). *See* germ layers.

ectoparasite. *See* parasitism.

ectoplasm. (*a*) (Bot.) Ectoplast (Plasmalemma). The external plasma-membrane lying just inside the cell wall. (*b*) (Zool.) Cell cortex. The outer layer of cytoplasm which in many cells (e.g. ova, Protozoa) is semi-solid (gel) and relatively free of granules and organelles. *See* endoplasm.

ectoplast. *See* ectoplasm.

Ectoprocta. *See* Polyzoa.

ectotropic. *See* mycorrhiza.

ecumenopolis. The ultimate city, occupying most of the Earth's land surface and accommodating all of its population, that would result from the continued growth of human population. The concept was enunciated by C. A. Doxiadis, who defined 15 spatial units starting with the human individual, proceeding to the large city, and thence to the metropolis, conurbation, megalopolis and ecumenopolis.

edaphic factors. The chemical, physical and biological characteristics of the soil which affect an ecosystem.

eddy. A current in a fluid that moves in a direction contrary to that of the main stream, often having a rotary motion.

eddy diffusion. The movement of a bulk quantity of one substance through another, to give mixing with local variations of concentration. The most important mixing process in the atmosphere. *See* diffusion, molecular diffusion.

Edentata (edentates). Placental mammals with teeth absent or much reduced, comprising the South American anteaters, sloths and armadillos. The term was formerly used more widely to include the African aardvark and the pangolins (scaly anteaters) of Africa and Asia, but the three groups are only superficially similar. *See* Pholidota, Tubulidentata.

edentates. Edentata.

edominant. Secondary or accessory species that exhibit no dominance.

eelworms. Free-living and plant-parasitic Nematoda (e.g. potato root and sugar beet eelworms).

176

EEZ. Exclusive Economic Zone.

effective. Applied to quantities (e.g. effective sound pressure) to mean root-mean-square value.

effective height of emission. The height above ground at which rising waste gases are estimated to spread horizontally. This will be higher than the top of the chimney, owing to their upward momentum and buoyancy.

effector. An animal organ or cell organelle which carries out movement (muscles, cilia), secretion (glands) or other actions (e.g. chromatophores, thread cells) in response to stimuli.

effluent. General term for a fluid emitted by a source.

efficiency. *See* economic efficiency.

effluent charge. A charge levied against a polluter for each unit of waste discharged into public water. The charge may be general, or variable according to the nature of the waste being discharged and the absorptive capacity of the receiving water, levied at all times or only when deteriorating conditions warrant it.

effluent standard. The maximum amount of specified pollutants permitted in effluents. In the case of gaseous discharges, effluent standards are generally known as emission standards. *See* environmental quality standards.

effusive. Extrusive.

egg membrane. Membranes which surround the ova of animals (e.g. in a bird's egg the fine membrane surrounding the yolk (ovum) is the vitelline membrane, secreted by the ovum itself. The white, shell membranes and shell are also egg membranes, secreted by the oviduct). The chorion of insect eggs is a membrane secreted by the ovary.

egocentric. Term applied to traits which favour the survival of the individual. *cf.* altruistic.

EGR. Exhaust gas recirculation.

Eifelian. The fourth oldest stage of the Devonian System in Europe.

einkorn. A primitive wheat, first domesticated in the Near East and South West Asia, probably about 11000 years BP. Domesticated diploid einkorn (*Triticum monococcum*) was derived from the wild form *T. boeoticum* which occurs in several forms. *See* Graminae.

ekistics. The science dealing with human settlements, and involving research and experience in architecture, engineering, town planning and sociology.

Ekman spiral. The spiral traced out by the velocity vector in the ocean as depth increases, or in the atmosphere as height increases. It is caused by interaction of the drag force between the air and the ocean (or ground), the coriolis force, and the shear stress (*see* wind shear) between layers of the air or water.

elaioplast. A plastid in which oil is stored.

Elasmobranchii. *See* Chondrichthyes.

elastic limit. The maximum stress which can be obtained in a structural material without causing permanent deformation.

elastic pavement. *See* pavement.

elbow of capture. *See* river capture.

electric field. (Meteor.) Typically there is a potential gradient in the air in fine weather of about 200 volts per metre. There are great variations in the neighbourhood of thunderstorms, and it is reduced by convection and rain.

electricity, static. Static electricity.

electrometer. Instrument used to measure the atmospheric electrical field. The first electrometer was probably the one used in 1766 by Horace Benedict de Saussure, who measured the change in potential over time between a conductor raised a measured distance above the surface. Modern instruments also measure the potential in free air, between two balloons at different altitudes, and may accelerate the measurement by using a small amount of radioactive material to ionize the air immediately surrounding the instrument (then called a collector).

electro-refining. Process for removing impurities from metals by making the crude metal the anode in an electrodeposition bath (where the metal is deposited on the electrodes), the required metal being deposited in a purified form on the cathode.

electrostatic field. A region in which a stationary electrically charged particle would be subjected to a force of attraction or repulsion as a result of the presence of another stationary electric charge.

electrostatic filters. Filters where an electrostatic charge is applied to the filter to improve collection efficiency of small particles.

electrostatic precipitators. Devices that separate particles from a gas stream by passing the carrier gas between two electrodes across which a high voltage is applied. The particles pass through the field, become charged, and migrate to the oppositely charged electrode. Electrostatic precipitators are highly efficient collectors for minute particles, widely used in the cement industry, etc.

elements, periodic table of. Periodic table of elements.

Elm bark beetle (*Scolytus scolytus*). The beetle that spreads the fungus *Ceratostomella ulmi*, which causes Dutch Elm Disease. *See* Ulmus.

elutriation. (*a*) Separating lighter particles of a powder from the heavier by means of an upward stream of fluid, often air. (*b*) The separation of the lighter from the heavier material in domestic refuse. (*c*) (Geol.) Mechanical separation and analysis of grain sizes of sediments by passing them through currents of water of differing velocities.

eluviation. The physical transport of insoluble soil particles (usually clay minerals) in water from upper to lower soil horizons. The similar movement of soluble salts and minerals is called leaching.

elvan. A term used by Cornish miners and quarrymen to denote a microgranite. Many elvans have been used as aggregate and as building stone.

elytra. *See* Coleoptera.

emagram. A thermodynamic diagram, having temperature and logarithm of pressure as coordinates, used for plotting and analysing atmospheric soundings of temperature and humidity.

Embioptera. A small order of insects (Exopterygota) with soft, flattened bodies, all the females and some males being wingless. They produce silk from glands in the front legs and make webs and tunnels in crevices in the soil, under bark, etc., where they live, often in small groups.

embryo. An organism in the process of developing from a fertilized or parthenogenetically activated ovum. In animals the embryonic stage terminates with birth or hatching from the embryonic membranes. In seed plants a well developed embryo consists of a bud (plumule), a root (radicle) and one or more cotyledons. The embryonic stage terminates at germination of the seed.

embryonic membranes (extraembryonic membranes). *See* egg membranes, allantois, amnion, chorion, yolk sac.

Embryophyta. Metaphyta.

embryo sac. The megaspore of a flowering plant, contained wthin the ovule. The embryo sac is a large cell, containing several nuclei when mature. One of these is the egg nucleus, another the primary endosperm nucleus. At fertilisation, one male nucleus from the pollen grain fuses with each of them, producing a zygote (which develops into an embryo) and an endosperm nucleus (which gives rise to the endosperm). The contents of the mature embryo sac represent the female gametophyte. *See* alternation of generations.

emerald. Green beryl of gem quality.

emery. A naturally-occuring abrasive composed of a mixture of the minerals corundum, magnetite and spinel. Emery occurs in some regionally metamorphised (*see* metamorphism) rocks. Emery is used in non-skid road surfaces and non-slip paints.

emission standards. *See* effluent standard.

emmer. A primitive wheat, first domesticated in the Near East and South West Asia about 11000 years BP. Domesticated emmer (*Triticum dicoccum*) was probably derived from the wild emmer (*T. dicoccoides*) which in turn was derived from the crossing of wild einkorn, (*T. monococcum*) and a species of goat-grass (*Aegilops* spp.) *See* Graminae.

emphysema. *See* bronchial diseases.

Emsian. The third oldest stage of the Devonian System in Europe.

enamel. The smooth, hard outer layer of the crown of a vertebrate tooth or of the exposed part of the scales (denticles) of present-day cartilagenous fishes and some fossil fishes. Enamel consists largely of crystals of a calcium phosphate-carbonate salt.

enation. An outgrowth on a leaf caused by localised multiplication of cells in response to a virus infection.

endemic. (*a*) General. Confined to a given region and having originated there. (*b*) Of pests or disease-producing species. The normal population level of a species which occurs continuously in a given area.

endergonic reaction. A chemical reaction which requires an input of energy (e.g. photosynthesis). *Cf.* exergonic.

end moraine. *See* moraine.

endoblast (endoderm). *See* germ layers.

endocrine organ (ductless gland). A gland (e.g. adrenal, pituitary) which produces a hormone.

endocytosis. The ingestion of particles of fluid by a cell. *See* phagocytosis, pinocytosis.

endoderm (entoderm, endoblast, entoblast, hypoblast). *See* germ layers.

endodermis. The innermost layer of the cortex of roots, and of the stems of some higher plants. The cells of the endodermis typically each have a complete band of waterproofing suberin and lignin (Casparian strip) laid down in radial and transverse walls.

endogamy (inbreeding). Sexual reproduction between closely related individuals, which have at least some genes in common. If continued intensively this leads to pure lines *Cf.* exogamy.

endogenous. Growing inside a plant or animal.

endometrium. The glandular lining of the uterus of mammals which undergoes cyclical thickening and regression during the oestrous cycle.

endomitosis. Doubling of the chromosome number of a cell without subsequent nuclear division. This may occur repeatedly, giving large multiples of the original chromosome number. Endomitosis occurs esp. in flowering plants and insect tissues.*See* polyploid, allotetraploid, autopolyploid, polyenergid.

endoparasite. *See* parasitism.

endoplasm. (*a*) (Bot.) Endoplast. The cytoplasm without the external plasma-membrane of a plant cell. (*b*) (Zool.) The inner part of the cytoplasm which in many cells (e.g. Protozoa, ova) is more fluid than the outer layer, and contains many granules and organelles. *Cf.* ectoplasm.

endoplasmic reticulum (ergatoplasm). A complex network of membrane-bound channels ramifying through the cytoplasm and probably forming a conducting system within the cell. By means of electron microscopy it has been shown that the endoplasmic reticulum may join up with the nuclear membrane and the Golgi apparatus and that its membranes are often lined with ribosomes.

endoplast. *See* endoplasm.

Endoprocta. *See* Polyzoa.

Endopterygota (Holometabola). A large group of insects (Pterygota) in which the immature stages (larvae) are very different from the adults (e.g. housefly, butterflies, beetles). The transformation of larva into adult (metamorphosis) is a drastic remodelling which occurs within a pupa. *Cf.* Exopterygota.

endoskeleton. A skeleton which lies inside the body. Vertebrates have a bony endoskeleton. Some Arthropoda have an internal skeleton composed of inpushings (apodemes) of the exoskeleton.

endosperm. The food-storing tissue around the embryo in a seed. In endospermic seeds (e.g. wheat, pine, castor oil) some of the endosperm remains until germination. In non-endospermic (exalbuminous) seeds (e.g. pea, bean) the endosperm is absorbed by the embryo before the latter is fully developed.

endosulfan (thiodan). Organochlorine insecticide and acaricide used to control mites and insect pests such as aphids. Because of its side effects, use of this chemical is now restricted in the UK.

endothelium. A single-layered sheet of flattened cells lining the blood and lymph vessels and the heart in vertebrates. Endothelium is derived from the mesoderm *Cf.* epithelium.

endothermic. Term describing a reaction, process or change in which heat is absorbed. *Cf.* exothermic.

endotrophic. *See* mycorrhiza.

endozoochore. Seed, spore, etc. that is disseminated by being carried within the body of an animal.

endrin. Persistent organochlorine insecticide which is extremely toxic to vertebrates, much more so than other organochlorine compounds (e.g. DDT). Endrin is no longer approved for agricultural use in the UK. *See* cyclodiene insecticides.

energy budget. A record of the flow of energy through a system. Applied originally to ecosystems, energy budgeting is now applied to industrial processes as a means of measuring the efficiency with which energy is used and as an alternative to conventional financial budgeting.

energy flow. The passage of energy through the trophic levels of a food chain. Energy, almost all from sunlight, is trapped by the autotrophic organisms of the first trophic level. Because much energy is dissipated during respiration, about 90 % of the available chemical energy is lost each time energy is transferred from one trophic level to the next higher one.

energy of storms. Storms are very inefficient users of energy by comparison with the larger wind systems. In a tropical or sub-tropical cyclone, the energy released as latent heat of the rain is equivalent to between 5 and 100 H-bombs a day. This causes the conversion of gravitational potential energy due to the sinking of cold air under warm air, equal to between 1 and 10 H-bombs a day, into wind systems. Except in fluids confined within walls, such low efficiencies cannot be equalled by energy released by man, so that storm control by energy release is not a practical proposition.

Engel's Law. *See* income elasticity of demand.

enology. The study of the production, history, and use of wines and the cultural associations with wine-drinking.

enteric bacteria. Bacteria which inhabit the human gut, including those which may cause disease. *See* pathogen, coliform bacteria.

enteropneusta. Hemichordata.

enthalpy. The heat content of a body or system, usually given by the formula $H = U + pV$, where H is the heat content, U is the internal energy, p the pressure and V the volume.

entisols. *See* soil classification.

entoblast (entoderm). *See* germ layers.

entoderm (endoderm). *See* germ layers.

entomogenous. Growing parasitically on insects.

entomophily. Insect pollination.

entropy. A measurement of the degree of disorder within a system. The term is derived from thermodynamics: heat passes from warmer to cooler bodies, thus becoming dispersed more generally with time, so that within a closed system eventually all heat will be distributed evenly. Since organisation represents the local concentration of energy, its even distribution represents disorganisation. As order decreases, entropy increases. This is summed up in the second law of thermodynamics. *See* thermodynamics, Laws of.

environment. (*a*) The physical, chemical and biotic conditions surrounding an organism. (*b*) Internal. The intercellular fluid which bathes body cells. In vertebrates esp. the composition of this medium is maintained constant.

environmental forecasting. The technique of predicting the environmental effects of proposed developments, esp. as these may impinge on public health or welfare.

environmental geology. The application of geological data and principles to the solution of problems created by human occupancy or other activity (e.g. the geological assessment of the effects of mineral extraction, the construction of septic tanks, the erosion of land surfaces, etc.).

environmental impact statement. A report, based on detailed studies, that discloses the environmental consequences of a course of action as an aid to decision making. In many countries, industrial organisations planning new projects are required by law to conduct such studies and to produce an environmental impact statement which can then be examined critically, sometimes in public.

environmental protection. That part of resource management that is concerned with the discharge into the environment of substances that might be harmful, or with harmful physical effects (e.g. noise, or the release of radiation), and with safeguarding beneficial uses.

Environmental Protection Agency. Independent federal agency of the US Government, established in 1970, that is responsible for dealing with the pollution of air and water by solid waste, pesticides, radiation, and with nuisances caused by noise.

environmental quality standards. The maximum limits or concentrations of pollutants permitted in specified media (e.g. air or water). US standards are based on: (primary) estimates of maxima which, with an allowance for safety, present no hazard to human health; (secondary) maxima which, with an allowance for safety, present no hazard to public welfare. They may take the form of emission standards, or relate to the content of products (e.g. to additives or pesticide residues in food, or phosphates in detergents).

environmental resistance. The restriction of population growth by the interaction of environmental factors.

Environment, Department of. Department of the Enviornment.

enzyme (zymoprotein). An organic catalyst (a substance which, when present in very small amounts, promotes a chemical reaction without itself being used up). Enzymes are unstable proteins or protein-containing compounds, and are inactivated by heat. They control the many chemical reactions of metabolism, each enzyme being responsible for only one, or for a very limited range of, reactions. An enzyme acts by combining with a specific substance (substrate) and very rapidly activating it, so that it undergoes a chemical change, simultaneously losing its combination with the enzyme. Enzymes each work best at a specific pH and some are dependent for their action upon the presence of co-enzymes. *See* amylase, oxidase, dehydrogenase, proteolytic enzymes, ATPase, respiration, digestion.

Eocene. Refers to a sub-division of the Cenozoic Era. The Eocene usually ranks as an epoch, following the Palaeocene and lasting from about 54 to 38 million years BP. Some authors do not recognise the Palaeocene, and consider the Eocene to follow the Cretaceous directly, and thus last from about 65 to 38 millions years BP. Eocene also refers to rocks deposited during this time: these are called the Eocene series.

eoclimax. (*a*) The climax of a given period of dominance of a specified plant group. (*b*) The climax of the Eocene Epoch.

eolian deposit. Aeolian deposit.

eon (aeon). (*a*) Refers to either all Phanerozoic time or all Cryptozoic time. (*b*) In North American usage, eon also equals one thousand million years (1 US billion years).

eosere. The development of vegetation during an eon or era. A major developmental series within the climatic climax of a geological period.

EPA. Environmental Protection Agency.

epeirogenesis. *See* Diastrophism.

ephemeral. A plant which completes more than one life cycle, from seed to seed, in one year (e.g. groundsel).

Ephemoptera (mayflies). An order of insects (Exopterygota) with long-lived, aquatic nymphs and short-lived, non-feeding adults with membranous wings and long tail filaments. A winged, sub-imago stage (the dun) occurs before the final moult yields the adult. All the stages are eaten in large numbers by freshwater fishes.

epiblast (ectoderm). *See* germ layers.

epicentre. The point on the Earth's surface vertically above the focus of an earthquake.

epideictic display. A behaviour pattern used to mark out the territory of an animal or group of animals.

epidemiology. The study of epidemics and the patterns of incidence of diseases.

epidermal cells. Cells of the epidermis. In Man, the epidermis forms distinct layers. The basal cell layer (*stratum basale*) consists of a single layer of columnar epithelial cells arranged in a regular palisade fashion, resting on the dermis. Between and slightly below the cells of this layer is a network of melanocytes (*see* melanins) which communicate with one another and with the cells of the deeper epidermis to which which they can transfer pigment. Above this layer is the *stratum spinosum*, several cells thick. These two strata form the Malpghian layer. The next layer is the *stratum granulosum*, two to four cells thick and containing keratohyalin, a granular substance, in their cytoplasm. Above this layer is the *stratum lucidum*, which may not be well developed but, where it is, contains a lipid-like substance, eleidin. This layer provides the major barrier to the passage through the skin of water and salts. The uppermost layer, the *stratum corneum*, consists of 10 to 20 layers of flat, dead, dry, keratinised cells which adhere to one another closely except at the surface, where they are shed as minute flakes. The surface of the skin supports an ecosystem of its own, composed of a range of microorganisms.

epidermis. The outer layer of cells of an animal or plant. In many invertebrates and plants the epidermis is one cell thick and secretes a cuticle. In vertebrates the epidermis is many cells thick, and in land-living forms the outer layers are dead and keratinized, protecting the body against excessive water loss. *See* epithelium.

epigamic character. An animal character, other than one associated with the reproductive organs, which is concerned with sexual reproduction (e.g. bird song, distinctive coloration of many male birds and fish).

epigeal. (*a*) (Bot.) Applied to germination in which the cotyledons grow out above the ground (e.g. sycamore). *Cf.* hypogeal. (*b*) (Zool.) Applied to animals which live above ground.

epigene. Applied to processes operating in and on the ground.

epilimnion. The warmer uppermost layer of water lying above the thermocline in the lake. The epilimnion is subject to disturbance by the wind. *Cf.* hypolimnion. *See* thermal stratification.

epinephrine. Adrenaline.

epiorganism (superorganism, supraorganism). An entity such as a bee colony or a stand of vegetation, made up of a group of individual organisms.

epipalaeolithic. Mesolithic.

epipedon. Organic and upper-leached layer of the soil. *See* pedon, soil horizon.

epiphyte. A plant which grows on the outside of another plant, using it for support only and not obtaining food from it (e.g. lichens on trees).

epistatic gene. A gene whose presence prevents the expression of another, non-allelic gene. *See* Allelomorphs. *Cf.* hypostatic gene.

epithelium. (*a*) (Zool.) A sheet of cells which covers any free surface of an animal's body (e.g. the outer surface, the lining of the body cavity, gut, glands). Epithelia may be one or many cells thick, and may be ciliated (e.g. lining of respiratory tubes in vertebrates) or glandular (e.g. lining of stomach in vertebrates). The epidermis of vertebrates is stratified epithelium, the lowermost layer actively dividing and giving rise to more superficial layers, the outer layers being dead, flattened cells. (*b*) (Bot.) A layer of secretory cells surrounding an inter-cellular cavity or canal (e.g. resin canal in coniferous plants).

epithermal. Refers to ore deposits formed from an ascending, essentially aqueous solution within about 1000 m of the surface, and in the approximate temperature range of 50°C to 200°C. Most such ores occur as vein-fillings, stockworks, pipes and irregularly-branching fissures. *Cf.* hypothermal, meso-thermal.

epizoite. An animal which lives attached to another animal, but is not a parasite (e.g. limpet on crab shell).

epoch. A unit of geological time such that two or more epochs constitute a period. Epochs are subdivided into ages.

equation of state. The equation $p = R \rho T$, relating the pressure p, the gas constant R, the density ρ and the absolute temperature T of air (or any gas), which embodies the gas laws of Charles and Boyle.

equilibrium level. The level at which a parcel experiences no buoyancy force in a stably stratified fluid medium.

equilibrium population. Stable population.

equisetales. An order of cone-bearing Pteridophyta numerous in the Carboni-ferous, when tree-like forms (e.g. *Calamites*) were present. The few living forms (*Equisetum* is the only genus) are herbaceous plants with furrowed, photo-synthetic stems bearing whorls of branches and scale-like leaves.

era. The second largest division (after the eon) of geological time. The Palaeozoic, Mesozoic and Cenozoic are eras, and, on some schemes, so is the short Quaternary. Eras are divided into periods.

eradication. The complete and final extinction of a species throughout its range.

erg. (*a*) A desert region of shifting sand. Erg is contrasted with hamada and reg. (*b*) Unit of work or energy. The work done by a force of 1 dyne through 1 centimetre. 1 erg $= 10^{-7}$ joules in SI units.

ergatoplasm. Endoplasmic reticulum.

ergosome. *See* ribosomes.

ergosterol. A precursor of vitamin D, which is present in the skin of some animals, including Man, and which is converted to vitamin D by ultraviolet radiation. *See* vitamins.

ergot. *Claviceps* a genus of fungus (Ascomycetes) which infests the ovaries of cereals and grasses, gradually replacing the grains with black, dense, banana-shaped masses of interwoven hyphae. These drop off in the autumn and remain dormant over the winter in the soil. The fungus is very poisonous and was responsible for the disease 'Holy Fire' in medieval times. Ergotamine is extracted from it for use as a drug which constricts blood vessels.

ergotamine. *See* ergot.

Ericaceae. A cosmopolitan family of woody dicotyledonous plants, mainly shrubs, which form ecologically important communities on peaty soil and in swamps. The family includes *Erica* (Heaths), *Calluna* (Ling), Rhododendron, Vaccinium (Cowberry, Bilberry, Cranberry, etc.) and *Arbutus* (Strawberry Tree). All the British species apart from *Arbutus* are calcifuge. The roots of many species (e.g. *Erica*) have endotrophic mycorrhizas.

ericfruticeta. Ericilignosa.

ericilignosa (ericfruticeta) .Vegetation dominated by heaths.

erosion. The breakdown of solid rock into smaller particles and its removal by wind, water, or ice. *See* wind erosion, weathering.

erosion of thermals. The mechanism whereby the diluted exterior of a thermal, rising in a stably stratified air mass, is removed to find its own equilibrium level while the interior warmer air continues to rise, itself to be eroded at a higher level.

erratic. (Geol.) A glacially-deposited piece of rock which is different in composition to the bedrock beneath it. Erratics are used to reconstruct the movements of ice sheets.

error. (Stat.) The difference between observed and expected values, usually caused by chance.

ERTS. Earth Resources Technology Satellite.

erythrism. A colour variation in animals (esp. birds) in which chestnut red replaces black or brown.

erythrocytes. *See* blood corpuscles.

escape. An organism that has escaped from cultivation or captivity and has established itself in the wild (e.g. the coypu in Britain).

esker. A long, narrow, usually sinuous ridge of sand and gravel. Eskers were deposited by melt-water within a glacier or ice sheet.

essential minerals. The mineral constituents of a rock which are necessary to its nomenclature. An essential mineral need not be a major constituent. The opposite of accessory mineral.

esters. Compounds derived by replacing the hydrogen in an organic acid by an organic radical (group). e.g. acetic acid (CH_3COOH) becomes the ethyl ester, ethyl acetate ($CH_3COOC_2H_5$). Many esters are liquids with a pleasant smell and are used in flavourings. Many vegetable and animal fats and oils are esters.

estrogens. Oestrogens.

estrous cycle. Oestrous cycle.

etesian winds. *See* mistral.

Ethiopian region. *See* zoographical regions.

ethnobotany. The study of the uses to which plants and plant products are put by peoples of different cultures.

ethnocentrism. The evaluation of cultural traits in terms of a particular culture that is assumed, often unconsciously, to be superior (e.g. the view that a culture cannot be considered important unless it produces artifacts similar to those produced by one's own culture).

ethnozoology. The study of the uses to which animals and animal products are put by peoples of different cultures.

ethogram. A complete catalogue of the behaviour of an animal under conditions as natural as possible and of the contexts in which it occurs. *See* motor pattern.

ethology. The study of the behaviour of animals in their natural environment.

ethyl acrylate. A foul-smelling gas produced in the manufacture of acrylates used in the plastics industry (e.g. in 'Perspex'). They can be removed from waste gases by carbon filters, or by containing them in storage ponds by maintaining a thick layer of foam on the surface of the liquid, but should they escape even small amounts can be smelt for considerable distances downwind.

etiolation. A condition produced by growing a green plant in the dark. The plant fails to develop chlorophyll, so is yellow. It has weak, elongated stems and its leaves are smaller than normal.

-etum. Suffix used in ecology to indicate a plant community dominated by a particular species (e.g. Fagetum, woodland dominated by *Fagus sylvatica*, the Beech).

Eucalyptus (Myrtaceae). The characteristic trees of Australia (gums, stringy bark, iron bark, etc.), now grown widely in warm climates. The leaves contain oil glands. Some species attain great size. They grow rapidly and many species yield valuable timber, oils and gum.

eucaryotic (eukaryotic). A term applied to cells or organisms in which there is a nuclear membrane separating the cytoplasm from the nucleus. The cytoplasm contains the membrane-bounded organelles (e.g. mitochondria, plastids) and the chromosomes consists of DNA and protein. The cells of all organisms except bacteria and blue-green algae are eucaryotic. *Cf.* procaryotic.

eudominant. A dominant more or less peculiar to a particular climax (e.g. Beech or Chestnut in their respective communities).

eugenics. The study of genetics based on the view that the evolutionary future of man may be susceptible to human guidance by the selective breeding of persons possessing characteristics that are considered desirable. The study of eugenics can be traced back to ancient times and Malthus and Darwin were influenced by it, but it was formulated by Francis Galton, a cousin of Darwin, in 1869. The English Eugenics Society was founded by Galton in 1907, and the American Eugenics Society in 1905, by Madison Grant, Henry H. Laughlin, Irving Fisher, Fairfield Osborn and Henry Crampton. In the 1920s and 1930s eugenic laws were passed in several European countries and in 27 US states, permitting the sterilisation of 'defective' persons. These laws were later repealed as doubts were expressed regarding the relative contribution to characteristics, esp. behavioural characteristics, of inheritance and environment (the 'nature-nurture' controversy), and regarding definitions of 'desirable' traits.

eugeosyncline. A geosyncline with a thick sequence including volcanic rock, developed off shore of a craton and its miogeosyncline.

Euglenophyta. A division of unicellular, mostly freshwater algae, whose members move by means of flagella and lack a rigid cell wall. Some forms possess chloroplasts, others are colourless. Even the photosynthetic forms require a supply of vitamin B 12. The group is also regarded as belonging to the Protozoa (Euglenoidina, class Flagellata), i.e. as being animal rather than plant. *Euglenia viridis* is often abundant in stagnant water containing much nitrogenous organic matter.

eukaryotic. Eucaryotic.

Euphausiacea. *See* Malacostraca.

Euphorbiaceae. A family of mainly tropical dicotyledonous plants with unisexual flowers. Many are xeromorphic, often cactus-like, and most are poisonous. They include cassava (Manihot), rubber trees (*Hevea* and *Sapium*), and the castor oil plant (*Ricinus*). Most Euphorbiaceae are trees and shrubs, but the British species are all herbs. These include Dog's Mercury (*Mercurialis perennis*), frequently the dominant plant of the field layer in woodland, and the spurges (*Euphorbia* spp.).

euphotic zone. The upper zone of a sea or lake into which sufficient light can penetrate for active photosynthesis to take place. This zone is usually 80 to 100 m deep, and is the upper part of the photic zone.

euploid. Applied to an organism whose body cells contain an exact multiple of the haploid number of chromosomes. Each chromosome is represented the same number of times as the rest, so the organism is genetically balanced. *Cf.* polyploid, aneuploid.

euroclydon. *See* mistral.

European ash. *Fraxinus excelsior.*

euryhaline. Able to tolerate a wide range of saline conditions (i.e. a wide variation of osmotic pressure) in the environment. *Cf.* stenohaline.

eurythermous. Able to tolerate a wide range of temperature in the environment. *Cf.* stenothermous.

eurytopic. Applied to organisms with a widespread distribution. *Cf.* stenotopic.

eustatic. Refers to worldwide and simultaneous changes in sea-level.

eutheria. Placentalia.

eutrophic. Applied to freshwater bodies which are rich in plant nutrients and are therefore highly productive. The large number of organisms present may render their waters cloudy. The hypolimnion of eutrophic lakes becomes depleted of oxygen in the summer. *Cf.* oligotrophic, mesotrophic, dystrophic. *See* eutrophication.

eutrophication. Enrichment of a water body, e.g. by the input of organic material or of surface runoff containing nitrates and phosphates. Eutrophication leads to an increase in the growth of aquatic plants, and often to a deficiency of oxygen. In extreme cases this results in the death of most of the aquatic animals and macrophytes. *See* eutrophic, oligotrophic, algal bloom, Saprobic classification.

evaporation (Meteor.) Because of the large latent heat of evaporation, vegetation can avoid becoming very hot in sunshine by evapotranspiration whereby moisture taken in by roots is evaporated at a leaf surface. The lapse rate above cloud base is very stable between clouds because of the large difference between the wet and dry adiabatic lapse rate and rain falling out of clouds above cloud base can, by cooling due to evaporation, bring stably stratified air from above base down to the ground.

evaporation pond. A body of seawater enclosed for natural evaporation and used to produce sea salt, traditionally a pure material. Much sea salt is produced in the Mediterranean, where marine pollution has reduced its purity.

evaporimeter. Instrument for measuring the rate of evaporation of water from a surface.

evaporite. A rock composed of minerals which have been precipitated from aqueous solution as a result of the evaporation of the water (e.g. oolitic limestones (*see* oolite), gypsum, anhydrite, sylvite and carnalite).

evapotranspiration. The combined evaporation from the soil surface and transpiration from plants. The opposite of precipitation.

evergreen. A plant which bears leaves throughout the year (e.g. Scots Pine, Yew, Holly, Ivy). Many evergreens have leaves which remain on the plant for longer than a year. The old leaves persist until a new set appears. *Cf.* deciduous.

evolution. The process of cumulative change which occurs in successive generations of living organisms, and which has led to the development from a common ancestor of different species and sub-species. *See* Darwin.

evolution of the atmosphere. The composition of the atmosphere (which is almost completely of biological origin) has been stable for 2 or 3 thousand million years,

192

having almost certainly been completely transformed over a previous thousand million years by life forms from a reducing one containing mainly carbon and hydrogen compounds into an oxidising one containing mainly nitrogen and oxygen produced by vegetation, thus paving the way for animal (breathing) life forms and stabilising mechanisms for its composition and restriction of climatic variations by control of radiation screens (e.g. carbon dioxide) and turbidity (e.g. natural photochemical smog). The dominance of biological control of the atmospheric composition in the long term must be better understood if predictions of Man's effect on the atmosphere are to be meaningful.

exalbuminous. *See* endosperm.

excepted land. In British planning, land included within an area of an access agreement or order but from which the public is excluded. As defined by the Countryside Commission.

excitation. A forced variation in pressure, position or similar quantity, whereby energy is added. The excitation of a particle transfers it from its ground state to a higher energy level, the excitation energy being the difference in energy between the two states.

exclusive economic zone. Concept proposed at the UN Law of the Sea Conference, whereby coastal states assume jurisdiction over the exploration and exploitation of marine resources in their adjacent section of continental shelf taken arbitrarily to be a band extending 200 miles from the shore.

exclusive species. A species confined completely, or almost completely, to one community.

excretion. Getting rid of the waste products of metabolism. Some are stored (e.g. nitrogenous compounds in the fat body of insects and in the hollow wing scales of some butterflies (Pieridae). The elimination of the carbon dioxide and water produced in respiration is excretion, but the term is particularly applied to the removal from the body of nitrogenous products (e.g. urea, ammonia compounds) by the kidneys in vertebrates and by nephridia, Malpighian tubules and other organs in invertebrates.

exergonic. A chemical reaction which gives out energy, (e.g. respiration). *Cf.* endergonic.

exfoliation. The separation of scale-like layers from a mineral, or from an exposed or soil-covered massive rock. The process can be due to a variety of causes including alternating heating and cooling of the rock, alteration of layers of minerals (esp. by water), and pressure-release as a result of erosion. One type of exfoliation is sphaeroidal weathering.

exhaust gas recirculation (EGR). Technique to improve the combustion of pollutants in internal combustion engines, so reducing pollutant emissions.

exodermis. The outer layer of the cortex in a non-woody root. It is composed of cells impregnated with waterproofing suberin, and replaces the piliferous layer where this withers behind the root hair region.

exogamy (outbreeding). Sexual reproduction between individuals not closely related. This maintains heterozygosity. Crossing between two genetically distinct lines may lead to hybrid vigour (heterosis).

exogenous. Growing on the outside of a plant or animal.

Exopterygota (Hemimetabola, Heterometabola). A large group of insects (Pterygota) in which the young stages (nymphs) are similar to the adults apart from being sexually immature and having wings incompletely developed or absent, (e.g. dragonflies, mayflies, grasshoppers, earwigs, cockroaches). There is no pupal state, and the transformation of the nymph into an adult (metamorphosis) is a gradual process, accomplished by a series of moults. *Cf.* Endopterygota.

exoskeleton. A skeleton which covers the outside of an animal's body or lies in the skin. The exoskeleton of Arthropoda is a cuticle containing chitin overlying the epidermis. Some vertebrates (e.g. armadillo, tortoise) have exoskeletons of bony plates covered with keratin and lying on the skin. *Cf.* endoskeleton.

exosphere. The outermost layer of the atmosphere, lying beyond the ionosphere, in which air density is such that a molecule moving directly outward has a 50% chance of escaping, rather than colliding with another molecule.

exothermic. Applied to a chemical reaction in which energy, as heat, is released. *See* exergonic. *Cf.* endothermic.

exotic (Geol.). Autochthonous.

expansion wave. Every heating or cooling of a body of air causes an expansion or contraction of volume which spreads around the rest of the atmosphere with the speed of sound (as from a big bang, but not abruptly). Thus heating effectively produces an instant change of density because the motions produced by buoyancy forces are slow compared with the velocity of sound. The expansion waves which make the adjustments of density produce no observable phenomena.

explantation. Tissue culture.

explosion wave. The wave travelling outwards from an explosion, whereby the extra volume of the explosion is accommodated. Explosion waves travel as shock waves near the source, then as spherical sound waves, but at distances beyond a few tens of kilometres they assume the form of external or sonic gravity waves in the atmosphere.

explosive. A substance which undergoes a rapid chemical change on heating or detonation, with the evolution of great heat and a large volume of gas.

exponential growth (geometric growth, compound interest). The growth of a value over a period by a fixed percentage of the capital, which is the original capital plus interest accumulated in earlier periods, so that while the rate of increase remains constant, the amount of increase is greater in each successive period. Linear growth (simple interest, arithmetic growth) is based on interest calculated on the original capital only, so that the sum added at the end of each period remains constant. The time required to double the original value is called the 'doubling time', and for exponential growth it is approximately 70 divided by the percentage rate of growth (e.g. a growth rate of 10% will double the value in 7 periods).

exponential reserve index. *See* static reserve index.

exposure. The amount of a physical or chemical agent that reaches a target or receptor.

exsic. Exsiccata.

exsiccata (exsic.). Herbarium material preserved by drying.

externalities. An economic concept covering those costs and benefits attributable to an economic activity that are not reflected in the price of the goods or services produced. Thus damage to the environment may not be counted as a cost or benefit in production. It is an aim of the 'polluter pays' principle to require polluters to meet the cost of avoiding pollution or of remedying its effects, so internalizing the externalities.

exteroceptor. A sense organ (e.g. eye) which detects stimuli originating from outside an animal. *Cf.* interoceptor.

extinction of species. *See* Red Book.

extracellular. Outside the cells of an organism (e.g. digestion in the rumen of an animal's gut or around the hyphae of fungi).
extraclinal. *See* topotype.

extraembryonic membranes. *See* embryonic membranes.

extraneous. Closer to the periphery than to the centre.

extrusive. Refers to igneous rocks formed from magmas which flowed out at the surface of the Earth. Extrusive is the opposite of intrusive, used of rocks forming intrusions.

eye. A sense organ which detects light. An ocellus is a simple eye found in many invertebrates. A compound eye, found in Crustacea and insects, is a collection of units (ommatidia) each of which forms an image.

eye of storm. In the centre of a tropical cyclone the pressure is very low, lower than could be produced by the warming of the air from below. Thus the centre is filled with air drawn down from above, and air from the stratosphere descends to the sea surface, with some mixing and moistening in the lowest few kilometres. This descending air is cloud-free except sometimes for a layer of cloud at 1 km or less. It occupies the rain-free centre, or eye of the storm, whose width is 10 to 50 km. The wind there is relatively calm, but it is surrounded by the strongest winds and heaviest rain of the cyclone.

eye spot. Stigma.

F

F. (*a*) Farad (*b*) fluorine.

f. Femto-.

F_1 **(first filial generation).** The first generation of offspring produced in a genetical experiment by crossing the parental generation P_1. Crossing members of the F_1 generation gives the F_2 (second filial) generation.

facies. The sum total of the characteristics of a rock (e.g. the texture, mineral or organic composition, form and structure) from which its environment of deposition and subsequent history can be deduced.

factor. Hereditary factor. *See* gene.

facultative parasite. *See* parasitism.

faecal (fecal) streptococcus. A group of bacteria which are normally abundant in the intestinal tracts of warm blooded animals other than Man, and are indicators of the contamination of water by the faeces of these animals.

faeces (feces). Material voided via the anus from the alimentary canal, consisting mainly of indigestible food residue along with bacteria. (In some animals, excretory products are expelled along with faeces). The droppings of birds consist of faeces covered by whitish, semi-solid, nitrogenous waste expelled from the urinary system.

***Fagus sylvatica* (Beech) (Fagaceae).** A tall tree of northern temperate regions often forming homogenous forests. It is the characteristic dominant of chalk and soft limestone in South East England. Beech bears catkins, and pairs of nuts enclosed by prickly, woody, four-valved cupules. Beechwoods have a peculiar ground flora including White Helleborine (*Cephalanthera damasonium*) and Truffle (*Tuber aestivum*). Beech is useful for hedging and yields a hard timber.

Fairy Shrimps. *See* Branchiopoda.

Falconiformes (Raptores). The birds of prey which hunt by day, having sharp, hooked beaks, strong talons, and powerful flight. Many soar on updraughts. The families represented in the British Isles are the Pandionidae (the fish-eating Osprey), Falconidae (e.g. Peregrine, Hobby, Merlin, Kestrel, all with pointed wings and narrow tails) and Accipitridae. The latter includes eagles (large birds with broad wings and tails), vultures (eagle-like but mostly with naked heads and necks), buzzards (medium sized, broad wings and tails), harriers (long, broad wings, long tails), kites (long, angular wings, long, forked tails) and accipiters or bird hawks (e.g. Goshawk, Sparrow Hawk, with short-rounded wings and long tails).

falling pressure. The most familiar prognostication of increasing cloud and the possibility of rain. The upward motion necessary for cloud formation produces convergence with increased cyclonic rotation at the surface, with decreasing pressure at the centre accompanying the creation of centrifugal forces.

fallout. (*a*) Particles falling out of the atmosphere because their fall speed relative to the air significantly exceeds the characteristic vertical velocities present in the air (e.g. hail, snow, rain, drizzle, graupel (soft hail), sleet). (*b*) Measurement of air contamination based on the mass rate at which solid particles are deposited from the atmosphere. Applied to dust, soot, and radioactive particles resulting from atomic or thermonuclear explosions.

fallout front. The lower boundary of a region of fallout. Thus when the fallout front from a rain shower reaches the ground the rain begins.

fallstreak. A streak of falling cloud particles (*see* Virga) almost always ice particles which are not evaporating.

fallstreak hole. A hole in a cloud layer composed of supercooled water droplets

197

caused by the local freezing of some droplets and their conversion into fallout, often in streak form, by the Bergeson-Findeisen mechanism.

false-bedding. Cross-bedding.

family. *See* classification.

Fammenian. The youngest stage of the Devonian System in Europe.

fanning. Describes the behaviour of a plume when, in stable air, gases reach their equilibrium level quickly and move horizontally with some meandering but very little vertical mixing, to produce a thin but concentrated layer of pollutant that may come into direct contact with hillsides or high buildings.

FAO (The Food and Agriculture Organization of the United Nations). One of the first of the UN agencies to be formed, and one of the largest. Based in Rome, its aim is to increase food production and availability among those sections of the world population where hunger is prevalent. It is closely associated with its 'Green Revolution', conducts many field projects, initiates and cooperates in research, and seeks to reform world trading policies relating to food items in the furtherance of its aims. *See* Indicative World Plan for Agricultural Development, High-yielding varieties of cereals.

Farad (F). The derived SI unit of capacitance, being the capacitance of a capacitor between the plates of which there appears a potential difference of one volt when it is charged with one coulomb of electricity, i.e. ampere seconds per volt. Named after Michael Faraday (1791–1867).

farmyard manure (FYM). Mixture of straw and animal excrement used, either raw or after composting, as a manure in agriculture or horticulture. FYM is produced when cattle are housed in yards or buildings and straw is used as a litter, fresh straw being placed over soiled and trampled straw from time to time. FYM provides plant nutrients (*see* macronutrients, micronutrients) and improves the structure of soils.

fascicle. (Bot.) A compact cluster.

fast breeder reactor. Liquid metal fast breeder reactor.

fat (lipid, lipide, lipoid). A combination of an alcohol with a fatty acid. A neutral, or true fat is glycerol combined with a fatty acid (e.g. oleic acid in olive oil). The term 'fat' may be used more widely for any substance (e.g. sterols, steroids, carotene) which can be extracted from tissue by using a fat solvent such as ether, but which does not necessarily contain a fatty acid. Fats are commonly stored as food in the bodies of animals and in seeds. Fatty substances occur in suberin and cutin and are important constituents of protoplasm (e.g. lecithin, a phospholipid found in unit membranes).

Fata Morgana. A superior image caused by a pressure anomaly, whereby the air pressure increases with height for a very short distance above the surface of cold water. Under perfect conditions, the mirage can transform cliffs or distant houses into great castles partly in the air, part beneath the sea. It is named after the fancied reflection of the submarine palace of Morgan le Fay (Fata Morgana) the fairy sister of King Arthur, in the Strait of Messina.

fault. A fracture in the Earth along which there has been displacement parallel to the fault-plane. Faults are classified by the relative motions of the faulted blocks into (*a*) dip-slip faults (e.g. normal and reverse faults and thrusts) in which the movement is parallel to the dip of the fault-plane; (*b*) strike-slip faults (e.g. wrench, tear and transform faults) in which movement is parallel to the strike of the fault-plane; and (*c*) oblique-slip faults in which movement is at an appreciable angle to both dip and strike. The upper surface above a fault is called the hanging wall, the lower surface the footwall. The vertical displacement of a point perpendicular to the fault-plane is called the throw, the corresponding horizontal displacement the heave, and the horizontal displacement along the fault-plane is called the strike slip.

fault-breccia. A rock composed of broken fragments lying along a fault-plane.

fault-plane. The surface of movement of a fault. Material along an active fault-plane can become broken off (to form fault-breccia) and ground up to a powder (a fault-gouge) which can become fused to a mylonite. The surfaces can become polished and striated, and these striations are called slickensides.

fauna. The animals of a particular region or period of time.

Fe. Iron.

feces. Faeces.

feedback. In systems theory, the informational response to a cause that tends to inhibit further repetition of the cause (e.g. the feeling of satiation which reduces an animal's motivation to continue eating). Normal feedback works in a negative sense, so having a stabilising effect on the system. Positive feedback, which encourages the cause to continue, tends to produce 'vicious spirals' (e.g. wage-price inflation, where higher wages cause prices to rise, so causing wages to rise again).

feeder reservoir. *See* reservoir.

feedlot. A large area of small pens in which beef cattle are fattened for slaughter. Food, mainly grains, is brought to the animals. The concentration of large quantities of sewage from feedlots causes heavy contamination of waste streams. Feedlots are common in the USA, but rare in Britain.

feldspar (felspar). The most abundant group of rock-forming silicate minerals, feldspars are grouped into plagioclase and alkali feldspar. Plagioclase ranges in composition from albite, $NaAlSi_3O_8$, to anorthite, $CaAl_2Si_3O_8$, to orthoclase, $KAlSi_3O_8$. Alkali feldspar is characteristic of the alkali igneous rocks, whereas plagioclase of varying composition is found in igneous rocks ranging from acid to basic. Feldspars are abundant in metamorphic rocks (*see* metamorphism) and in arkoses.

feldspathoids. A group of silicate minerals all containing sodium and/or potassium, which are related to feldspar but contain relatively less silica. Two of the simpler feldspathoids are nepheline, $NaAlSiO_4$, and leucite, $KAlSi_2O_6$. Feldspathoids occur in alkali igneous rocks, though never in rocks containing quartz.

felsic. A term applied to the light-coloured minerals, feldspar, feldspathoids, and quartz (silica) in igneous rocks. Mafic is the corresponding term for the dark-coloured minerals.

felsite. An igneous rock which is fine-grained to cryptocrystalline, of acid to intermediate composition, and which may or may not be porphyritic. Because of this imprecision, felsite is best used as a convenient field term. Felsites are important sources of aggregate and roadstone.

felspar. Feldspar.

femto-(f). Prefix used in conjunction with SI Units to denote the unit $\times 10^{-15}$.

fen. An area of waterlogged peat which, unlike bog, is alkaline or only slightly acid. Typical fen plants include Reed (*Phragmites communis*), Reed Canary Grass (*Molinia caerulea*), Ragged Robin (*Lychnis flos-cuculi*), Meadowsweet (*Filipendula ulmaria*) and Great Spearwort (*Ranunculus lingua*).

fenitrothion. Organophosphorus insecticide used to control insects such as aphids and caterpillars in fruit crops, moths and weevils in peas, and leatherjackets in cereals. Also used to control beetle pests in grain stores. It is poisonous to vertebrates.

fentin. Fungicide used to control diseases such as potato blight. Fentins are organic compounds containing tin. They are poisonous to vertebrates.

feral. Applied to animals once domesticated which have established themselves in the wild (e.g. cat and 'domestic' pigeon (Rock Dove) in the British Isles).

fermentation. The breakdown of organic substances by organisms, with the release of energy. Particularly used for the anaerobic breakdown by yeasts and bacteria of carbohydrates, forming carbon dioxide and alcohol or other organic compounds.

Fernau glaciation (Little Ice Age). A climatic oscillation that led to a cooling of probably about 1°C over much of western Europe from about 1590 to 1850, causing the extension of glaciers. The change in climate amounted to much less than an ice age despite its popular name, and is known more correctly by the name of the Fernau Glacier in Austria. This glacier ends in the Bunte Moor peat bog and its advances and retreats are recorded in layers in the bog by the formation of peat which can be dated radiometrically (*see* radiometric dating) during ice-free periods, and by the deposition of morainic sand during periods of glaciation.

ferns. Filicales.

Ferrel's Law. Named after the American meteorologist W. Ferrel (d. 1891), the law states that the wind is deflected (by the coriolis force) to the right in the Northern Hemisphere. Ferrel deduced this on theoretical grounds some time earlier than Buys Ballot, who later acknowledged Ferrel's prior claim to the discovery.

ferromagnesian mineral. A rock-forming silicate mineral containing essential iron and/or magnesium. Olivine, augite, hornblende and biotite are ferro-magnesian minerals.

Fertile Crescent. The approximately crescent-shaped area in the Near East bounded by the Rivers Tigris and Euphrates, that was the site of the earliest Near Eastern civilisations and one of the areas in which agriculture was first practised by members of what became 'hydrological civilisations' (i.e. civilisations dependent on sophisticated irrigation and drainage systems to sustain their agriculture). At its height, the Fertile Crescent supported a population density equal to that of much of modern Europe.

fertilisation. The fusion of two gametes to form a zygote. This is the essential process of sexual reproduction, and results in the bringing together of an assortment of genes from two haploid nuclei.

fertiliser. Any material applied to land as a source of nutrients for plant growth. It can be natural in origin, (e.g. farmyard manure, crop residues or compost) or artificial (e.g. compounds of ammonium or phosphorus). The main fertiliser constituents are nitrogen (N), phosphorus (P) and potassium (K) and fertiliser values are commonly measured as weights of N, P or K nutrients. Calcium (Ca), although required by plants and to reduce soil acidity, is not usually regarded as a fertiliser. Excessive use of salts that are highly soluble in water (e.g. nitrates) can cause contamination of water (*see* methaemoglobinaemia) or its over-enrichment (*see* eutrophication). Excess nitrates may also lodge in the tissues of certain plants, and excesses of any one nutrient may lead to nutritional imbalances in crops. *See* macro-nutrients.

fibre. (*a*) (Bot.) An elongated cell whose walls are thickened with cellulose or lignin, and which gives mechanical support to the plant. Fibres are extracted from many plants and used in the manufacture of such products as paper, hemp ropes (from *Cannabis sativa*) and linen (from *Linum usitatissimum*). *See* Sclerenchyma. (*b*) (Zool.) Strands of material such as the collagen fibres found in connective tissue, tendon, cartilage of vertebrates. Fibres are made by irregularly-shaped cells, fibroblasts (fibrocytes). *See* nerve cell, muscle.

fibreglass. Manufactured non-flammable fibre made from glass. Used for insulation and bonded with resin to give a strong, light construction material. Fibres penetrate skin easily and affect lungs as a dust. Fibreglass is resistant to most chemicals and solvents.

fibrin. *See* clotting of blood.

fibrinogen. *See* clotting of blood.

fibroblasts. *See* fibre.

fibrocytes. *See* fibre.

fibrous root system. A root system without a tap root. It consists of a tuft of roots of fairly equal diameter bearing numerous smaller lateral roots (e.g. grasses, Groundsel, Strawberry).

Ficus (Moraceae). A genus of trees, shrubs, climbers and epiphytes of warm regions. The flowers, lying inside a globular or pear-shaped receptacle, are pollinated by fig-wasps (Hymenoptera) which lay eggs inside the ovaries. *Ficus carica* is the Fig Tree. *F. elastica*, the Indiarubber Tree, is a stout tree with large buttress roots which usually starts life as an epiphyte. It yields rubber when tapped (*see* Hevea). *F. benghalensis* is the Banyan, which has large aerial roots supporting the branches.

fidelity. The degree to which a system reproduces accurately at its output the characteristics of the signal impressed on its input.

field. (*a*) The study of rocks or living organisms in situ or in their natural habitat is termed 'field work'. (*b*) (Geol.) An area of workable mineral riches (e.g. coalfield, oil field).

field capacity. The greatest amount of water it is possible for a soil to hold in its pore spaces after excess water has drained away.

filament. *See* flower.

filaria. A genus of Nematode worms, some of which cause serious diseases in Man. *Filaria (Wuchereria) bancrofti* lives in the blood and lymph vessels, causing elephantiasis. The worms are transmitted to new hosts by the mosquito *Culex fatigans*, inside which they pass through an essential stage of development. *Filaria (Dracunculus) medinensis*, the Guinea worm, grows to a length of 100 cm. It lives under the skin and discharges its eggs through ulcers into water. The intermediate host is a small aquatic crustacean, *Cyclops*, which may be swallowed in drinking water.

Filicales (ferns). A large group of Pteridophyta whose members produce spores on the underside of the leaves or on special leaf segments. The leaves are typically large and compound. Most ferns have rhizomes, but some (e.g. *Cyathea*, tropical and subtropical tree ferns) have aerial stems several metres high. Most species are terrestrial, a few (e.g. *Azolla*) are aquatic. *See* Bracken.

Fire Algae. Pyrrophyta.

firn. Névé.

first filial generation. F_1.

fish. Pisces.

Fish, Flying. Cypselurus.

Fish, White. Gadidae.

fissile. (*a*) Capable of being split along closely spaced, parallel planes. (*b*) Applied to isotopes of elements that are capable of undergoing fission upon impact with a slow neutron. *See* critical mass, nuclear reactor.

fission. (*a*) Asexual reproduction by splitting into equal parts. Binary fission (splitting into two) occurs in many unicellular organisms (e.g. bacteria, Protozoa) and a few multicellular ones (e.g. corals). Multiple fission occurs in Sporozoa. (*b*) A nuclear reaction in which a heavy atomic nucleus splits into two parts, emitting neutrons and energy. *See* critical mass, nuclear reactor.

fission energy per unit mass of uranium. The energy released from the fission of one gram atom (235 g) of uranium 235, equal to 12×10^{12}J, or 3.3×10^6 kWh. *See* nuclear reactor, uranium.

fissure. A fracture in a rock with displacement perpendicular to the break.

fissure eruption. Extrusion of lava or pyroclastic material from a linear vent.

fitness. The response of a population of organisms to natural selection measured in terms of the number of offspring produced related to the number needed to maintain a constant population size.

fixative. A chemical (e.g. formaldehyde or alcohol) used to preserve cells, ideally with as little distortion as possible.

fixed action patterns. Motor patterns that are extremely stereotyped (e.g. the invariable use of the beak by the Greylag Goose to retrieve eggs that have rolled from the nest). *See* ethogram.

fjord. A long-steep-sided inlet of the sea, in a mountainous region. Fjords are the result of over-deepening of river valleys by glaciers. Characteristically, fjords have a shallower bar at their seaward end.

flag. Flagstone.

Flagellata (Mastigophora). A class of Protozoa whose adult members swim by means of flagella (*See* flagellum). They include plantlike forms (e.g. *Euglenia* (see Euglenophyta) and Dinoflagellata) which usually contain chloroplasts and so are able to photosynthesize, and animal-like forms with holozoic nutrition. Important examples of the latter group are *Trypanosoma* (see Trypanosomidae) a parasite which causes sleeping sickness in Africa, and *Trichonympha*, a wood-digesting symbiont which lives in the gut of termites. *See* Pyrrophyta.

flagellum. A thread of cytoplasm capable of lashing movements, projecting from a cell. Flagella are usually few in number and do not move in unison (*Cf.* cilia). They are responsible for the movement of many unicellular organisms, gametes, spores and bacteria, and occur in some multicellular organisms (e.g. sponges, Coelenterata). Apart from bacterial flagella, each of which consists of a single filament, the structure of a flagellum resembles that of a cilium.

flagstone (flag). A sandstone or sandy limestone which splits along bedding planes into slabs.

flame. A reaction or reaction product, partly or entirely gaseous, yielding heat and more or less light, as the result of a chemical reaction, usually combustion.

flame photometry. An analysis technique for elements. The process of emission spectroscopy is in the visible and ultraviolet regions, using flame sources. Some 70 elements can be determined by this method of excitation in an arc or high-voltage spark.

flash colours. Brightly coloured parts of animals which are exposed suddenly to confuse predators (e.g. patterns on the wings of Lepidoptera).

flat plate solar collector. *See* solar collector.

Flatworms. Platyhelminthes.

Flea. *See* Aphaniptera.

flint. A cryptocrystalline variety of silica with a conchoidal fracture. Flint and chert are often used synonymously, though many geologists reserve the term 'flint' for the irregular, siliceous nodules in the chalk, in which the most abundant organic remains are sponge spicules.

flocculation. (*a*) Process of contact and adhesion whereby the particles of a dispersion form larger clusters. *Cf.* agglomeration, coagulation. (*b*) (Geol.) The process of aggregation of colloidal (*See* colloid) particles into small lumps, which are then capable of settling out. This process occurs when river water carrying electrically charged colloidal clays mixes with sea water, which contains electrically charged particles in solution.

floccus. Fleece-like clouds, which are evaporating castellatus.

flood-basalt. An accumulation of basaltic lava many thousands of square kilometres in extent, with the individual flows themselves being very extensive. A high rate of extrusion of very runny lava from fissures is implied.

flood plain. Relatively level part of a river valley, adjacent to the river channel, formed from sediments deposited by the river during periods of flooding. *See* alluvium.

floods. Unusual accumulations of water above the ground caused by high tide, melting snow or rapid runoff from artificially paved areas or unusual lack of motion of storm clouds causing local accumulation of fallout. Many rivers have natural flood plains, and although floods may be caused by urbanisation leading to immediate runoff, they occur naturally and may be disastrous in human terms when rain, onshore wind and wave, induced by low atmospheric pressure in a cyclone, coincide with natural high tide.

flora. (*a*) The plants of a particular region or period of time. (*b*) A descriptive list of plant species of such a time or place, often with a key to their identification.

floral formula. A formula which summarizes the number and arrangement of the parts of a flower. The floral formula for a wallflower (*Cheiranthus cheiri*) is: $K_{2+2}C_4A_{2+4}G_{(2)}$. This shows that the flower has a calyx (K) composed of four free sepals, an inner and an outer pair; a corolla (*C*) of four free petals; an androecium (*A*) of six stamens, two outer and four inner ones; a gynoecium (*G*) of two fused carpels. () indicates that the carpels are united and that the gynoecium is superior.

floret. *See* flower.

florology. The study of the genesis, life and development of vegetative formations.

florula. The plants of a small, confined area (e.g. a pond).

flowers of tan (*Fuligo septica*). *See* Myxomycophyta.

flower. The reproductive shoot of Angiospermae. In a typical flower the axis (receptacle) bears: (*a*) the calyx, composed of sepals, the outermost organs, usually green and leaf-like, which protect the flower in bud; (*b*) the corolla, composed of petals, usually brightly coloured and attractive to insects; (*c*) the androecium, the male parts of the flower, composed of stamens (microsporophylls) each of which consists of a stalk (filament) bearing an anther which produces pollen; (*d*) the central gynoecium (pistil), the female parts of the flower, composed of one or more carpels (megasporophylls) each of which consists of an ovary containing an ovule or ovules, and bearing a style ending in a stigma. The stigma is the surface on which the pollen grains are received. The carpels may be separate, as in the buttercup, or united as in the wallflower. The calyx and corolla are accessory flower parts, together making up the perianth. A tepal is a unit of the perianth (a term used particularly when the perianth is not distinctly differentiated into sepals and petals as in the tulip). In wind pollinated flowers (e.g. grasses, poplars) and some plants relying on pollen or nectar to attract insects (e.g. willows) the perianth is reduced or absent. Some plants (e.g. Hazel) have male and female organs in separate flowers. A floret is a small individual flower of a compound flower head, such as Dandelion (*See* Compositae).

flow structure. (*a*) A texture of igneous rocks, esp. lavas in which the flow lines are shown by the alignment of prismatic crystals or elongated inclusions and by alternating bands of different minerals or crystal size (also called Flow banding). (*b*) Sedimentary structures resulting from sub-aqueous flow.

flue. Passage for combustion or other gases. *See* chimney, flue gas, flue gas scrubber.

206

flue gas. Waste gas, usually from a combustion process. *See* flue, flue gas scrubber.

flue gas scrubber. Equipment for removing fly ash and other materials from the products of combustion by sprays or wet baffles. This reduces the temperature of the effluent.

fluidised bed. A two-phase system, consisting of a mass of small particles suspended in a fluid, usually a gas, which is flowing upwards and maintaining the stability of the system. In many respects, the bed acts like a fluid.

fluidised bed combustion. A technique for extracting energy by the combustion of fossil fuels in a fluidised bed system. Sometimes this bed consists mainly of limestone or dolomite particles, as a method of pollution control; 90 % or more of the sulphur is retained in the bed, the lower combustion temperatures result in reduced emissions of nitrogen oxides, and higher thermal efficiencies can be achieved.

flukes. Trematoda.

fluorescence. The emission of visible light by materials subjected to electro-magnetic radiation of shorter wavelength (e.g. X-rays, ultraviolet and violet light). The name is derived from 'fluorite' since some varieties of this mineral exhibit the phenomenon in the ultraviolet light present with the visible daylight. Geologically, the property of fluorescence is used in prospecting for scheelite, and in the rough identification of the API of crude oil in drilled samples of rock.

fluoridation. The addition to public water supplies of fluorides in order to prevent dental caries or to delay its onset. Many water supplies contain fluoride naturally and a concentration of 1 ppm is considered optimal. The aim of fluoridation is generally to bring to this level of concentration waters whose fluoride content is lower or from which natural fluoride is absent. The measure has aroused great controversy, although the consensus of medical opinion probably favours it.

fluorides. Compounds containing fluorine, the lightest and most reactive of the halogens. Fluorides occur widely in nature and so may be released into the atmosphere in the course of a number of industrial processes, including the manufacture of cement and bricks, the refining of aluminium, the processing of rock phosphate, the processing of electric furnace slag, and steel manufacture where a fluorspar flux is used. Fluorides may also be emitted by the tapping of geothermal energy sources. As atmospheric pollutants, fluorides are very harmful to plants and to mammals and there have been many cases of serious economic damage to farm crops and livestock downwind of emission sources.

fluorine (F). Element. A pale yellowish-green gas, resembling chlorine but more reactive. It occurs naturally as fluorite and cryolite. Organic fluorine compounds have a number of industrial uses. AW 18.9984; At. No. 9.

fluorite (fluorspar). A mineral with the formula CaF_2. Fluorite is found mainly in sedimentary rock, in acid igneous rocks, and as a gangue mineral. Its main uses are in the chemical industry, particularly in the manufacture of hydrofluoric acid, and in ceramics. A little is still used as a flux in steel making.

fluorocarbon. *See* chlorofluoromethanes.

fluorosis. Disease in ruminants caused by over-consumption of fluorine compounds, often as a result of air pollution which deposits fluorine compounds on vegetation that is then consumed. Fluorosis produces mottling and weakening of teeth and excessive thickening of bones. Fluorine compounds are emitted from a number of industrial processes. *See* fluorides.

fluorspar. Fluorite.

fluvial. Pertaining to the actions of rivers.

fluvioglacial deposits. Outwash deposits.

fluviomarine. Pertaining to the action of both rivers and sea, and used particularly of sediments deposited at the mouths of large rivers.

flux. (*a*) The rate of flow of a fluid or of radiation across an area. In the case of a vector quantity, this is the product of the area and the component of the vector at right angles to the area (e.g. electric intensity, magnetic intensity, etc.). In nuclear physics, the product of the number of particles (including photons) per unit volume and their average velocity. (*b*) A substance added to assist fusion (e.g. in metal working). (*c*) A morbid or excessive discharge of blood, excrement, etc. ('flux' was formerly the name for dysentry).

fly ash. Finely divided particles of ash entrained in flue gases resulting from combustion of fuel or other material. The particles of ash may contain incompletely burned fuel and other pollutants.

flying fish. Cypselurus.

flying lemurs. Dermoptera.

flysch. A rapidly alternating, marine succession of sandstones and shales, of late Cretaceous to early Tertiary age, found on both flanks of the Alps. The sediments were derived from erosion of the rising mountain chain during deformation, and were themselves later deformed. The term 'flysch' has been used for many broadly similar sequences of greywacke and shale. Flysch is contrasted with molasse, which is a facies developed after an orogeny (*see* orogenesis).

foam. A gas in liquid dispersion. Together with the bursting of air bubbles, the fragmentation of thin films in foam is a major contributor of the small droplets entering the air from the sea which evaporate and leave airborne salt particles. *See* spray.

focus. The point of origin of an earthquake.

Foehn wind. A warm wind descending from the mountains, originally in the European Alps, where physiological and psychological discomfort is produced by the big changes in temperature, humidity and wind strength following its onset. It is warm and dry because of the adiabatic descent. This often, usually wrongly, is attributed to the latent heat released when rain falls from the air following ascent on the other side of the mountains. The effect is due mainly to the blocking of a cold air mass by the mountain range so that the air previously near the mountain top descends the lee slope. Very strong winds often occur in the troughs between lee waves and the sky is characteristically filled with wave clouds. Glider pilots have exploited the waves for soaring to great effect. The chinook is mechanically similar.

foetus. A mammalian embryo in the later stages of development, after the main features have become recognisable.

fog. Visible moisture in the atmosphere that reduces horizontal visibility to below 1000 metres. Fog is caused by the cooling of relatively warm, moist air when it encounters a land or sea surface that is colder, so reducing the temperature of the air in immediate contact with the surface to below the dew point. Advection fog is caused by the movement of moist warm air across a colder surface as part of a general weather pattern. Radiation fog is caused by the cooling of the land surface at night by radiation, so that while the land and air temperature were equal during the day, at night the land cools more rapidly than the air. *See* mist, smog, steam fog, frozen fog, pea soup fog, sea fog, fog showers, valley fog.

fog bow. An arc, usually white because of the wide range of drop sizes, analogous to a rainbow, but seen in fog rather than rain droplets. Physically the same as a cloud bow.

fog showers. Phenomenon characteristic of isolated high mountains lying above the cloud base and from time to time engulfed in passing convection clouds. When supercooled, the cloud droplets are captured as rime or glazed frost on wires, twigs and other small objects.

folacin. Folic acid.

fold. A bend in strata or in any planar structure in rocks or minerals. Folds are classified geometrically by the dips of their limbs and attitude of their axial plane (the plane dividing the fold as symmetrically as possible). *See* anticline. syncline monocline, terrace, isocline, recumbent fold. Folds are described as open if the interlimb angle is more than 70°, tight if it is less than 30°. Folds are said to plunge if their axes are not horizontal. Folds can also be classified by their mode of formation.

foliation. Layering in rocks caused by parallel orientation of minerals or bands of minerals. The texture is characteristic of such metamorphic rocks (*see* metamorphism) as slate, phyllite, schist and gneiss and is also found in igneous rocks which flowed during cooling.

folic acid (folacin, pteroylglutamic acid). One of the B group of vitamins, xanthopteryl-methyl-para-aminobenzoyl glutamic acid (folic acid) is one of a group of related substances that are essential in the diet of many animals. Folic acid deficiency in Man may lead to some types of anaemia and because of a relationship between folic acid and cobalamine that is poorly understood, folic acid deficiency may also be implicated in pernicious anaemia.

follicle. (*a*) (Bot.) A dry fruit formed from a single carpel which splits down one side only to liberate its seeds (e.g. Larkspur). (*b*) (Zool.) 1. Hair follicle. An inpushing of the epidermis in mammals which surrounds the hair root and produces the hair. 2. Ovarian follicle. An envelope of cells surrounding the developing oocyte in many animals. In vertebrates it secretes oestrogen. In most mammals the follicle (Graafian follicle) becomes the *corpus luteum* after bursting at ovulation.

follicle-stimulating hormone (FSH). A hormone secreted by the pituitary gland of vertebrates which stimulates the growth of the oocyte and ovarian follicles in the female and the development of spermatozoa in the male. *See* gonadotrophic hormones.

Food and Agriculture Organisation of the United Nations. FAO.

food chain. A number of organisms forming a series through which energy is passed. At the base of the chain (the producer level, first trophic level, T_1) there is always a green plant or other autotroph which traps energy, almost always from light, and produces food substances, thereby making energy available for the other (consumer) levels. At the second trophic level (T_2) is a herbivore (primary consumer). At subsequent trophic levels are smaller, then larger carnivores (secondary consumers). e.g. unicellular algae-daphnia-dragonfly nymph-smooth newt-grass snake. The biomass of a lower trophic level is always, in a balanced community, higher than that succeeding it, because at each level a large amount of energy is dissipated during respiration. About 90 % of the available chemical energy is lost each time energy is transferred from one trophic level to the next higher one. Saprophytes are present at all consumer levels. Any natural community will have many interlinked food chains, making up a food web, or food cycle. *See* energy flow.

fool's gold. Iron pyrites (pyrite).

footwall. The lower side of an inclined fault or vein, or the ore-limit on the lower side of an inclined ore-body.

Foraminifera. *See* Rhizopoda.

forbs. Herbaceous plants, excluding grasses, sedges and other grass-like groups.

forced convection. The transport of air (or other fluid) across the mean stream and consequent mixing induced by the turbulence over a rough surface. This is equivalent to mechanical turbulence as opposed to free convection which is generated by buoyancy, when the surface is heated.

forced oscillation (vibration). An oscillation maintained by the application of a fluctuating energy supply. *See* natural frequency.

foreset beds. Inclined layers of sediment deposited on the advancing edge of a delta, or on the lee slope of an advancing sand dune.

foreshock. A relatively small earthquake which precedes a larger earthquake by a few days or weeks, and originates near or at the focus of the larger earthquake.

foreshore. The shore zone covered only by exceptionally high spring tides. Its vegetation is sparse and specialised, including plants such as Sea Rocket (*Cakile maritima*).

forest. (*a*) An extensive area of woodland, either unmanaged or maintained for the production of timbers, etc. Coppice forest consists of trees derived from coppice shoots (produced from trees cut near the ground) and root suckers. High forest is mature woodland usually composed of tall trees derived from seeds, their tops forming a closed canopy. Rain forest is evergreen forest growing in regions of high rainfall where the dry season is short or absent. Epiphytes and climbers are abundant. The term is often used in the restricted sense, meaning only tropical rain forest. Monsoon forest occurs in regions with a well-marked rainy season. In some, trees are deciduous, losing their leaves for at least part of the dry season. (*b*) In Britain, an area originally unenclosed and not necessarily wooded, preserved for hunting (e.g. New Forest, deer forests of the Scottish *form*

Highlands). (*c*) Mixed forest. Forest composed of two or more species of tree, with at least 20% of the canopy consisting of species other than the dominant one.

form. (*a*) (Bot.) The smallest sub-specific grouping of plants, based on trivial characteristics (e.g. colour of petals). (*b*) (Zool.) 1. A term used loosely to cover various minor groupings of animals. 2. The resting place of a hare, where it lies concealed by vegetation.

formation. (*a*) (Biol.) The largest natural vegetation type (e.g. tropical rain forest). The plants of a land biome (*b*) (Geol.) A bed, or collection of beds of a distinct rock type that can be traced over a considerable area of country. A formation is the basic mapping unit, subdivided into members or beds. Several formations may form a group.

formation-type (-class). A group of geographically widespread communities of similar physiognomy and life form and related to major climatic and other environmental conditions.

fosse. (*a*) A depression separating two terminal moraines or an outwash plain and a terminal moraine. (*b*) A narrow excavation such as a canal, ditch or trench. (*c*) (French usage) An ocean deep.

fossil. (*a*) Remains or traces of an organism that have been preserved in the Earth's crust by natural processes. Fossils may be preserved in a number of ways: 1. the hard parts are preserved without chemical alteration; 2. replacement of the hard parts by another mineral (e.g. pyrites, silica). In addition to shells of animals, wood may be 'petrified' in this manner; 3. internal or external casts of the organism in the rock; 4. as carbon residues after organic decomposition; 5. as impressions of soft parts in fine-grained sediment; 6. traces of the activity of organisms when alive (e.g. tracks and burrowing of animals, coprolites and rooting of plants). Fossils are normally found in sedimentary rocks, but recognisable, albeit distorted, fossils may be found in metamorphic rocks if metamorphism has not been too intense. Occasionally whole animals may be preserved if conditions are right (e.g. Wooly Mammoths in frozen ground in Siberia, insects preserved in amber). (*b*) The term is sometimes extended to include inorganic remains of geological age (e.g. fossil sand-dune, fossil beach).

fossil fuels. The fuels derived from ancient organic remains: peat, coal, crude oil, and natural gas. Tar-sands and oil-shales, though currently not widely exploited, are included in resources of fossil fuels.

fossil turbulence. Inhomogeneities of temperature and humidity remaining in the air after the motion which produced them has subsided and the density, though not humidity and temperature, has become uniform. The inhomogeneities cause scattering of radio waves, and lumpy clouds when the air is made to ascend.

Foucault's Pendulum. A pendulum hung from the ceiling which, swinging in a plane in space, appears to swing in a rotating plane because the Earth is rotating. Named after the French physicist J. B. L. Foucault (1819-1868) and often seen in museums of science.

fracture. (*a*) The nature of the broken surface of a mineral usually described using the words even, uneven, hackly, and conchoidal (*b*) A break in a rock with no displacement. A break with displacement along the break is a joint; one with displacement perpendicular to the break is a fissure.

fracture zone. The zone along which faulting has taken place. The term is more particularly used for the linear zones of ridges and troughs approximately perpendicular to mid-oceanic ridges, which they offset. Such zones are the topographic expression of transform faults.

fragipan. *See* hardpan.

Franklin, Benjamin (1706–1790). American physicist who demonstrated the electrical nature of lightning by flying kites in thunderstorms.

Frasnian. The sixth oldest stage of the Devonian System in Europe.

fraternal twins. Dizygotic twins.

Fraxinus excelsior **(European Ash) (Oleaceae).** A deciduous tree with pinnate leaves, black buds and winged fruit (keys), esp. common on calcareous soils, where it may form woods. It is frequently found in oak woods. The tree yields valuable timber.

free acceleration test. A test for measuring exhaust emissions from vehicles. The engine is accelerated rapidly, in neutral gear, and the exhaust gases are fed through a smoke meter.

free convection. The motion and mixing of a fluid induced by buoyancy forces.

free field (Acoustics). A region in which no significant reflections of sound occur.

free-martin. A sterile and partly hermaphrodite female hoofed animal, whose peculiarities come about because its placental circulation fuses with that of its twin brother.

free progressive wave. A theoretical wave propagated in an infinite medium.

freestone. A sandstone or limestone which does not to split in one direction but can be cut and dressed equally well in any direction.

freezing level. The height of the 0°C isotherm in the atmosphere. Although cloud droplets in rising air may remain supercooled liquid above this height, descending ice particles begin to melt at it, although large hail may reach the ground unmelted.

freezing nuclei (ice nuclei). Atmospheric particles on to which water will freeze. Normally water vapour condenses into liquid, even at temperatures well below freezing point, unless a freezing, or ice nucleus is present. Ice nuclei are of mineral origin and possess a crystal structure similar to that of ice so that within a supercooled droplet they will initiate the formation of an ice crystal with an efficiency that increases the lower the temperature. Silver and lead iodide crystals initiate freezing at about −5°C and −7° respectively, but naturally occurring nuclei begin to function at below −10°C. At −40°C and below spontaneous freezing occurs without the presence of freezing nuclei. The most efficient freezing nuclei of all are ice crystals themselves, which may grow by accretion as they fall through air containing super-cooled water droplets.

freons. *See* chlorofluoromethanes.

frequency. (*a*) The number of times a vibrating system or particle completes a repetitive cycle of movement in a period of time. The derived SI unit of frequency is the Hertz (after Heinrich Hertz, 1857–94), equal to one cycle per second, symbol Hz. 1 kilo-hertz (kHz) is 1000 Hz, 1 megahertz (MHz) is one million Hz. (*b*) The average number of statistical units found in an area. In practice this may be expressed as the percentage of total samples or quadrats in which a species occurs.

friction layer. The boundary layer, within which friction occurs, between two air layers with different wind speeds. At the ground surface this friction is manifested as shear stress and at higher levels it causes turbulence in bands a few hundred metres or more thick.

friction velocity. If ρ is the air density and T the shear stress at the surface, the quantity (T/ρ) is called the friction velocity, usually denoted by u_*.

Frog. *See* Anura.

front. A narrow transitional zone between air masses. The front is named after the incoming air mass, so that if cold air is replacing warm air, the front is cold, and if warm air is replacing cold air, the front is warm. Where a warm front and a cold front meet and begin to merge, the front is 'occluded', or an 'occlusion'.

frontal analysis. The analysis of weather charts by marking in the positions of fronts between different air masses.

frontal slope. The inclination of a frontal surface to the horizontal denoted by α is given in terms of gravity (g), the coriolis parameter (f), the temperature (T), and its discontinuity at the front (ΔT), and the discontinuity in geostrophic wind (ΔV_G) by:

$$\Delta V_G = \frac{\Delta T}{T} \frac{g}{f} \ \sin \alpha \text{ to a good degree of approximation.}$$

frontal structure. In nature, a front is not always a sharp discontinuity but is more in the nature of a zone of transition from one air mass to the other, sometimes with more than one fairly sharp transition. Cold fronts are more often sharply defined than warm ones, and on passage the transition may occur in a few minutes.

frontal wave. The wave-like perturbation at the frontal surface (usually the 'polar front' in temperate latitudes) of a warm air mass as pressure falls and cyclonic rotation of the air commences. As the pressure continues to fall, the wave-like deformation increases and near the centre the front becomes narrower as the cold air mass moves faster than the warm air. The wave then moves rapidly along the front with the centre of low pressure at the crest until the cold air overtakes the warm air to produce an occluded front.

frontal zone. The broad band of weather etc., associated with a front or the zone of transition from one air mass to another when the front is not sharply defined.

frost. *See* dew.

frost hollow. A relatively small low-lying area subject to frequent and severe frosts because of the accumulation of cold air at night. Typically severe where hills shade the ground from afternoon sunshine.

frost point. The temperature to which air must be cooled for frost to begin to form on solid surfaces. It is best measured by the temperature to which a solid surface must be raised in order that ice (frost) on it shall evaporate. The frost point is higher than the dew point by an amount which increases from zero at $0°C$ to about $3.5°C$ at $-40°C$.

frost wedging. The process by which freezing water in rock pores or fissures shatters the rock.

froth. *See* foam.

froth flotation. The separation of a mixture of finely divided minerals by mixing them in a froth of oil and water, so that some float and others sink.

215

frozen fog. A fog of cloud composed of ice crystals. When supercooled fog begins to freeze, fallout is generated by the Bergeson-Findeisen mechanism so that frozen fogs usually clear soon by fallout from the almost motionless air. Supercooled fogs have been dispersed by seeding to cause freezing.

fruit. The ripened ovary of a flower containing the seed or seeds. *See* achene, berry, capsule, drupe, follicle, nut, legume, pome, samara, schizocarpic, caryopsis, siliqua.

Fruit Fly. Drosophila.

frutescent (fruticose). Shrubby.

fruticeta. Shrub forest.

fruticose (frutescent). Shrubby.

FSH. Follicle-stimulating hormone.

fuel efficiency. The proportion of the potential heat of a fuel that is converted into useful energy.

fuel element. A unit in the core of a nuclear reactor containing fissile material.

fulgurite. A tube of glassy rock resulting from the fusing by lightning of grains in a loose sand or more compact rock.

Fulham-Simon-Carves process. A process for removing sulphur dioxide from flue gases, using ammonia liquor as the washing medium. Scrubbing produces ammonium salts which can be converted into ammonium sulphate and sulphur, so yielding a usable end product. The process was designed in 1939 and pilot plants were working in 1957 to 1960, but probably it will not be applied on a full scale.

Fuller's Earth. (*a*) An argillaceous rock or clay with strong powers of adsorption for water, colouring matter, grease and some oils. This property is due to the presence of montmorillonite clay. Fuller's earth was formerly used for degreasing (fulling) fleeces. (*b*) A stratigraphic name for a Middle Jurassic rock unit in Britain with substantial beds of Fuller's earth.

fumarole. A volcanic vent emitting only gases at a temperature above that of the atmosphere. The most abundant product is usually steam, and from this minerals are deposited. Sulphur and chloride minerals, esp. ammonium chloride, are the most common, but numerous metallic minerals have been found.

fume. Solids in air generated by the condensation of vapours. Fumes can also arise from sublimation, condensation and chemical reactions. Particles are less than 1 micron in diameter and are often metals or metallic oxides which may be toxic.

fumigation. (*a*) A rapid increase in air pollution close to ground level, sometimes leading to very high concentrations of pollutants for an hour or more. The phenomenon occurs when a nocturnal temperature inversion has caused pollutants to accumulate aloft. In the morning the warming of the ground initiates mixing upcurrents of air and these bring down the pollutants held by the inversion, which prevents their escape upwards. The turbulence gradually draws clean air from above the inversion into the lower layers, so diluting the pollutants. (*b*) A technique used to apply pesticides (esp. fungicides) in an enclosed space (e.g. a glass-house) as a fume, the area then being sealed for a period of time before workers are allowed to re-enter.

function. (*a*) The rate of biological energy flow through an ecosystem, i.e. the rates of production and respiration of the populations in the community. (*b*) The rate at which materials or nutrients are cycled, i.e. the rate at which the biogeochemical cycles proceed. (*c*) The biological or ecological regulation of species by the environment (e.g. photoperiodism) and the regulation of the environment by organisms (e.g. nitrogen fixation). (*d*) A quantity which varies as a result of variations in another quantity.

fundamental frequency. The frequency with which a periodic function reproduces itself. The first harmonic.

fundamental particle. Any particle that cannot be demonstrated to contain simpler units.

fungicide. Chemical used to control fungal diseases. *See* Bordeaux mixture, carbamate, dinitro groups of pesticides, organomercury and organotin fungicides, Streptomycin.

Fungi Imperfecti. Fungi in which a sexually reproducing stage is unknown. This makes them difficult to classify, but most are thought to belong to the Ascomycetes.

fungus. Mycophyta.

funnel cloud. A cloud which appears in the core of a tornado or water-spout on account of the low pressure. It resembles the air core of a bath plug vortex in shape, and its outline is approximately the isobar of the condensation level.

furnace. General term for a container in which a material (often metal or metal ore) is heated to a high temperature.

Furze (Gorse, Whin)

Furze (Gorse, Whin). Ulex. *See* Leguminosae.

fusion reactor. A nuclear reactor that derives energy from the fusion of two atoms (deuterium, tritium or lithium, or some combination of them) to form one helium atom, with the release of energy. Still at an early experimental stage. The operating temperature is about 100 million degrees C and the fuel is contained, as a plasma, in a magnetic field. Although in theory, deuterium is plentifully available it requires much higher operating temperatures than the much rarer lithium; consequently the deuterium reactor is probably a more distant prospect.

FYM. Farmyard manure.

G

g. Giga-

gabbro. A coarsely-crystalline basic igneous rock consisting essentially of calcium-rich plagioclase feldspar and pyroxene with or without olivine. Dolerite is the medium-grained and basalt the fine-grained equivalent of gabbro.

Gadidae. A large and economically important family of marine fish including the cod, whiting and haddock.

gallactose. *See* carbohydrate.

galena. The mineral lead sulphide, PbS. Galena accounts for most of the world's production of lead, used principally in storage batteries and in the antiknock additive lead tetraethyl, in petrol. Galena usually occurs with sphalerite in hydrothermal deposits, in sedimentary stratiform deposits, and in deposits formed by the replacement of limestone by metasomatism. Many of the sedimentary ores are of Precambrian age and have undergone metamorphism. Galena is also an important source of silver, which it contains as an impurity.

gall (plant). An abnormal growth of plant tissue produced in response to the invasion of insects, mites, eelworms or fungi. Gall wasps (Hymenoptera) cause Robin's pincushion on wild roses, also oak-apples, spangle galls, currant galls and marble galls on oak; big bud of blackcurrant is due to a mite; witches' broom on birch is caused by a fungus.

Galliformes (Game birds). The order of birds which includes grouse, ptarmigan, capercaillie, partridges, pheasants, quails, turkeys and peacock. These are mainly grain-eating, heavy-bodied, ground-nesting birds, capable only of short, rapid flights. The cocks are usually more colourful than the hens.

galvanise. To coat steel or iron with zinc, either by immersion in a bath of molten zinc or by deposition from a solution of zinc sulphate, to give protection against corrosion.

Game birds. Galliformes.

gamete (germ cell). (*a*) A haploid cell or nucleus which fuses with another gamete during fertilisation to produce a zygote which developes into a new plant or animal. Usually the two gametes are different, the female one (ovum) non-motile with a large amount of cytoplasm, the male one (spermatozoon in animals; spermatozoid (antherozoid) in many plants; male nucleus in seed plants) motile, usually by means of flagella (*see* flagellum), and small. In some algae, fungi and Protozoa, similar gametes (isogametes) are produced. (*b*) A cell similar to a female gamete, but usually diploid, which develops by partheno-genesis into a new individual.

gametocyte. A cell which gives rise to a gamete or gametes by meiosis. *See* oocyte, spermatocyte.

gametophyte. *See* alternation of generations, embryo sac.

gamma. γ

gamma BHC. Lindane.

gamma rays. One of the radiations produced by atomic transformations, as in nuclear reactors and bombs. They are extremely penetrative and cause radiation sickness.

ganglion. A discrete mass of tissue containing nerve cell bodies.

gangue. The waste minerals in an ore. The term is essentially economic as material which is gangue in one mine may, in higher concentration or under different economic conditions be a valuable component. Common gangue minerals in hydrothermal veins are quartz, tourmaline, chlorite, fluorite, haematite, pyrite, baryte, chalcedony, dolomite and calcite.

gannister. A fine-grained arenaceous siliceous rock which underlies some coal seams.

garbage. Organic wastes resulting from the handling, preparation and cooking of foods.

garden city. A concept of town planning devised by Sir Ebenezer Howard (1850–1928) that would extend to all town dwellers the advantages of suburban conditions. The garden city was to be surrounded by countryside and expansion was to be achieved by developing new garden cities on the other side of the green belt, leading to clusters of cities grouped around a central city. Several garden cities (e.g. Welwyn) were built in Britain. Howard was influenced by the city of Adelaide, which retained park lands on its northern side and expanded beyond them to form North Adelaide.

garnet. A group of minerals with the general formula $A_3B_2Si_3O_{12}$, where A can be iron, magnesium, manganese or calcium, and B can be iron, aluminium or chromium. Garnets have no cleavage and are hard, so are used as abrasives. Different garnets are found in a wide range of rocks.

garnierite. A nickeliferous serpentine mineral, $H_4Ni_3Si_2O_9$, occurring in laterites developed from ultrabasic igneous rocks. Garnierite (also called nouméite in New Caledonia) is one of the major ores of nickel, which is used in alloys and in stainless steel. *See* pentlandite.

gas-cap. An accumulation of natural gas above an oil pool.

gas chromatography. An analytical technique for separating mixtures of volatile substances. The sample is placed on a separating column and is washed through with an inert gas. The column selectively retards, and thus separates, the substances. Packing with absorbent material coated with relatively non-volatile material gives gas-liquid chromatography; without liquid coating, gas-solid chromatography. The technique gives quantitative results on small samples and it is used widely in pollen analysis.

gas field. *See* oil field.

gas/oil ratio. GOR.

gas pool. *See* pool.

gas thermometer. A thermometer in which the expanding and contracting material is gaseous.

Gastropoda. A large class of Mollusca, most of which have a single shell, usually spirally coiled. The Prosobranchia (Streptoneura) are mainly marine species (e.g. winkles, whelks, sea limpets). The Opisthobranchia (e.g. sea slugs) are marine, and many have the shell reduced or absent. The Pulmonata (e.g. land and pond snails, slugs) have lungs and are mainly terrestrial and freshwater.

gastrula. An animal embryo at the stage of development after the blastula. In the gastrula the cells move about, forming the germ layers.

220

GATT. General Agreement on Tariffs and Trade.

gauging station. Station established for the measurement of water flowing through a stream channel. The water surface level, channel shape, stream velocity and dissolved or suspended sedimentary matter are recorded. Data is provided for calculating water resource, potential flood damage, stream pollution, and for projects connected with damming and irrigation schemes.

Gaussian distribution (normal distribution). A distribution that shows the maximum number of occurrences at or near to a centre or mean point, a progressive decrease in occurrences with increasing distance from the centre, and symmetrical distribution of occurrences on all sides of the centre.

Gaviidae (divers). A family of diving birds, order Gaviiformes, including the Red-throated, Black-throated, and Great Northern Divers. They inhabit open waters, are clumsy on land, and come ashore only to breed. They have webbed feet and feed mainly on fish.

GDP. Gross Domestic Product.

Gedinnian. The oldest stage of the Devonian System in Europe.

geese. Anseriformes. *See* Anatidae.

Geiger-Müller counter. Often called simply a Geiger, it is an apparatus for counting charged particles by means of the ionisation they produce. It is a basic tool of the nuclear scientist and those searching for radioactive minerals.

Geiger threshold. The lowest voltage which, when applied to a Geiger counter, will produce pulses which in each case are of about the same size, irrespective of the number of primary ions produced.

gel. Material formed when a colloidal (*see* colloid) solution is allowed to stand, often having a jelly-like appearance. Gels may contain as little as 0.5% solid matter, but their properties resemble those of solids more closely than those of liquids.

gelatin. Protein extracted from collagen and used in medicine, biology, paper-making, textile processing, food processing and in the production of films and adhesives.

gemmation. A form of asexual reproduction in plants and animals in which new individuals or members of a colony develop from groups of cells arising on the parent's body. In mosses and liverworts, small groups of cells (gemmae) become detached from the parent and then develop into new plants. In some coelenterata (e.g. *Hydra*) the new individuals develop while still attached to the parent. (Gemmation in animals is usually called budding.)

genecology. The study of the genetics of plant populations in relation to their environments.

gene exchange. Sexual reproduction within an ecotype, species or coenspecies, which results in recombination of the parental genes.

gene flow. The movement of genes as a result of sexual reproduction between populations.

gene frequency. The frequency with which a certain gene occurs in a population, compared with the frequency of all its allelomorphs.

General Agreement on Tariffs and Trade (GATT). Intergovernmental agency concerned with facilitating trade by reducing tariff barriers.

general circulation. The average, world-wide system of winds. Air movement is caused by differential heating of the Earth's surface and atmosphere and by the Earth's rotation, with topographic differences causing local variations. The distribution of alternating belts of high and low pressure between the Equator and the Poles causes a general flow of air from high to low pressure areas, which the Earth's rotation swings to the right in the northern hemisphere and to the left in the southern. Buys Ballot's Law states that winds flow around high and low pressure areas and this produces the characteristic northern hemisphere north-east tradewinds, westerlies and north-easterlies, and the southern hemisphere south-east trades, westerlies and south-easterlies, moving from lower to higher latitudes in each case.

generation curve. The population density of a given developmental stage plotted against generation number for a sequence of generations.

genes. Physical units of inheritance transmitted from one generation to another and responsible for controlling the development of characters in the new individual. A gene is a short length of chromosome. Structural genes determine the sequence of amino acids in the synthesis of proteins (e.g. enzymes). Regulative genes control the activities of structural genes. *See* cistron, plasma-gene, allelomorphs, mutation.

genetic code. *See* DNA, RNA.

genetic drift. A change in the genetic composition of a population which occurs by chance and not as a result of natural selction.

222

genom. Genome.

genome (genom). The set of chromosomes characteristic of a particular species.

genotype. (*a*) The genetic constitution of an organism (*See* phenotype); (*b*) A group of organisms with identical genetic constitution; (*c*) The type species of a genus.

genus. *See* classification.

geobenthos. *See* benthos.

geobotanical anomaly. The indication of enrichment or depletion of elements in the soil according to the presence or absence of particular plant species or gross physical changes in plants (e.g. the 'Copper Flower' growing in soils containing 100 to 5000 ppm of copper). *See* biogeochemical anomaly, indicator species.

geobotany. The geography and ecology of plants.

geochemical anomaly. The local enrichment or depletion of an element in soil or rock. *See* geochemical dispersion.

geochemical dispersion. (*a*) Primary: The dispersion of elements at depth within the Earth. (*b*) Secondary: The dispersion and redistribution of elements at or near the Earth's surface.

geocline. A cline associated with a geographical gradient. *Cf.* Ecocline.

geode. A hollow, roughly globular body in a rock, lined with crystals projecting inwards.

geolifluction. Mass-wasting associated with permafrost. *Cf.* congelifluction.

geological time. The history of the Earth is divided into Eras, subdivided into Periods, and these subdivided again into Epochs. The earliest Eras belong to what is generally known now as the Precambrian, which embraces all Earth history up to the base of the Cambrian Period, and this is thought to have had a duration of 4000 million years. The Precambrian has also been called the Cryptozoic (meaning 'hidden life') although in fact Precambrian life forms have now been discovered and primitive cell-like aggregates must have existed for at least 2500 million years before multicellular animals arose. Phanerozoic ('evident life') Eras account for time from the base of the Cambrian to the present day. The Precambrian Eras are followed by the Palaeozoic Era, in which large numbers of life forms appeared, divided into the Cambrian (from about 570 million years ago), Ordovician (from about 500 million years ago), Silurian (from about 430 million years ago), Devonian (from about 395 million years ago), Carboniferous (from about 345 million years ago) and Permian (from about 280 million years ago). The Mesozoic Era began about 225 million years ago with the Triassic Period, followed by the Jurassic (from about 190 to 195 million years ago) and the Cretaceous (from about 136 million years ago). The Cenozoic Era, which includes the present, is divided into the Tertiary and Quaternary Periods. The Tertiary Period is sub-divided into Palaeocene (from about 65 million years ago), Eocene (about 54 million years ago), Oligocene (about 38 million years ago, Miocene (about 26 million years ago), and Pliocene (about 7 million years ago) Epochs. The Quaternary Period covers the Pleistocene (from about 2 to 2.5 million years ago), and the Holocene, present, Epochs. The Holocene began about 10000 years ago. There are some who believe this latter division to be unreal, holding that at present we live in an interglacial period, still in the Pleistocene Epoch. *See* period.

geomagnetic induction. The induction of the Earth's magnetic field postulating an imaginary dipole at the centre of the Earth and measuring the Earth's magnetic induction with reference to three axes at right angles: geographic north, geographic east and vertically downwards to the Earth's core. This dipole field is overlain by an irregular non-dipole field, which is continually changing.

geometric growth. Exponential growth.

geomorphology. The study of the form and development of the Earth, especially its surface and physical features, and the relationship between these features and the geological structures beneath.

geophone. A microphone buried in soil or towed by a vessel to record shock waves in a seismic survey.

geophyte. *See* Raunkiaer's Life Forms.

geosere. Series of climax formations developed through geological time. The total plant succession of the geological past.

geosphere. The mineral, non-living portion of the Earth. The Earth, excluding the atmosphere, hydrosphere and biosphere.

geostrophic wind. Horizontal wind that blows parallel to the isobars indicating a balance between the horizontal pressure-gradient force and the horizontal components of the Coriolis force.

geosyncline. An elongated basin which has been filled with a great thickness of sediment, usually with intercalated volcanic rocks. The strata of most geosynclines have been affected by orogeny, often with the intrusion of batholiths.

Geosynclines have been classified into at least ten named types, of which only the terms miogeosyncline and eugeosyncline are in common use. Miogeosynclines have a relatively thin sequence of sediments without volcanic rocks, and develop adjacent to a craton whereas eugeosynclines develop, with thicker sediments and abundant volcanics, further from the craton.

geotaxis. *See* taxis.

geothermal energy. Energy, as heat, derived from anomalies in the temperature gradient in the Earth's crust. Normally, temperature in the crust increases with depth at a constant rate. Locally, however, water or rock may be much hotter than the surrounding rocks. The hot water may be tapped and its heat used. Hot rock may be exploited by pumping cold water into a borehole and extracting it from a second borehole after it has been in contact with the rock, which may have been crushed by explosives or pressurized water to make it permeable. Geothermal energy is delivered as hot water and can be used only close to its source. It is not inexhaustible, since the extraction of heat from the localised anomaly causes cooling, so that eventually the source is cooled to the temperature of the surrounding material. Nor is geothermal energy necessarily non-polluting, since the heated water may bring to the surface dissolved substances of many kinds, including halogens, some of which are highly corrosive and toxic. *See* fluorides, hot brine, hot rock, hydrofracturing.

geothermal gradient. The change of temperature in the Earth with depth, usually expressed in degrees per unit depth. The geothermal gradient is usually lowest in shield areas and highest in mid-oceanic ridge zones.

geotropism. (*a*) (Zool.) Once a synonym for geotaxis (*See* taxis). (*b*) A growth response of plants in which the stimulus is gravity. Main roots are positively geotropic (i.e. they grow downwards). When placed horizontally, elongation of the cells on the upper side of the growing region is increased, causing a downward curvature. Main stems are negatively geotropic. An organ (e.g. root, or stem branch) growing at right angles to the direction of gravitational force is plagiogeotropic. Diageotropic organs (e.g. some rhizomes) grow at right angles to the direction of gravitational force. (Diageotropism is often considered to be a type of plagiogeotropism). Growth curvatures are under the control of auxins.

Gephyrea. A small, miscellaneous group of marine worms, formerly regarded as a phylum, which includes the Siphunculoidea and Echiuroidea. *See* Annelida.

germ cell. Gamete.

germ layers. The layers which can be distinguished during the gastrula stage of an animal embryo. In Diploblastica these are the endoderm and ectoderm. In Triploblastica there is a third, the mesoderm. Each layer gives rise to different organs. Epidermis, nervous tissue and nephridia develop from ectoderm. The gut lining and associated glands develop from endoderm. The intervening tissues (e.g. muscle, blood system, kidneys, connective tissue) develop from the mesoderm.

germ plasm. *See* Weismannism.

germule. An agent of migration.

gibberellins. A group of auxins which control growth and development in plants. They cause a marked increase in stem elongation in some plants, and control processes such as flower and fruit formation and dormancy. *See* cytokinins.

gibbsite. One of the major ore minerals of aluminium, gibbsite, with the formula $Al (OH)_3$, is one of the main constituents of bauxite and laterite.

gibleh (khamsin). *See* dust storm.

Giga- (g). Prefix used in conjunction with SI units to denote the unit $\times$ 10^9. (Pronounced 'jigga').

gill fungi. Agarics.

gill pouches. *See* gills.

gills. (*a*) (Bot.) Lamellae on which spores are formed in agaric fungi. (*b*) (Zool.) Respiratory organs of aquatic animals. In vertebrates (fish, Amphibian tadpoles) the gills are associated with gill slits. These develop as outpushing (gill pouches) of the pharynx which break through to the exterior, meeting inpushings of the epidermis. Gill slits (or gill pouches and corresponding epidermal grooves) develop in all Chordate embryos, even those of reptiles, birds and mammals, none of which subsequently develop gills. In invertebrates (e.g. many Mollusca, Crustacea, Annelida, aquatic insect larvae) gills are borne on various parts of the body.

gill slit. *See* gills.

Ginkgoales. An order of Gymnospermae abundant during the Mesozoic, and with a single representative in the present day, the Maidenhair Tree (*Ginkgo biloba*) found growing in eastern China in 1758. This is a deciduous tree with fan-shaped leaves, catkin-like male cones and female cones containing a pair of ovules in each.

Givetian. The fifth oldest stage of the Devonian System in Europe.

glabrous. (Bot.) Smooth, not hairy.

glacial drift. The sediments deposited directly by glaciers or indirectly in meltwater streams, lakes or the sea. *See* boulder clay.

glacial striation. Scratches made on ice-smoothed rock by rocks or grit frozen into an ice-mass. Striations are used to reconstruct local directions of movement of an ice-mass.

glacier. A body of ice originating on land by the compaction and re-crystallisation of snow, and showing evidence of present or past movement. Glaciers occur where winter snowfall exceeds summer melting. Altogether they occupy about 10% of the Earth's land surface and contain about 98% of the planet's fresh water. There are several types of glacier: (*a*) Ice sheets; (*b*) Valley glaciers, which are ice streams flowing down mountain valleys; (*c*) Piedmont glaciers, intermediate between valley glaciers and ice sheets, and comparatively rare, are valley glaciers that spread out across lowland at the foot of a mountain range.

gland. An organ or cell which makes and pours out one or more specific substances (secretions). Examples are nectaries in plants and sweat, digestive and mammary glands in animals. Some excretory organs are also called glands (e.g. hydathodes). Endocrine glands secrete hormones directly into the blood stream.

glass. A hard, amorphous, mixture, often transparent, made by fusing oxides of silicon, boron or phosphorus, followed by rapid cooling. This is a general term including many types of mixture having the same typical composition and physical characteristics.

glass fibre. Fibreglass.

glass fibre reinforced cement. *See* cement.

glazed frost. Clear ice, like glass, deposited on objects by impact of supercooled water droplets in a cloud or fog. The water is spread over the surface of the object before the droplet freezes. This happens when the rate of deposition is high, so that the latent heat of freezing raises the temperature of the capturing body to near 0°C.

glazing. Process giving a smooth, lustrous surface to pottery by means of a colourless glass fired on to the surface.

gley. Sticky, organic-rich soil layer that develops on ground that is frequently or continuously saturated with water. *See* bog boil.

globigerina ooze. *See* Rhizopoda.

globulins. A group of widely distributed proteins including antibodies and many plant seed proteins.

Gloger's Rule. As the mean temperature of the environment decreases (e.g. with increase in latitude) the pigmentation in warm blooded animal species tends to decrease.

glory. A system of coloured, rainbow-like rings around a shadow cast on a cloud surface. On mountain peaks, when the sun is low, an apparently enormously magnified shadow of the observer may be cast on to thin cloud, the apparent magnification being caused by the observer's assumption that the image is at a much greater distance than it is because of other objects that can be seen dimly through the mist. The shadow of an aircraft on cloud is often surrounded by coloured rings, as is the shadow cast from the ground.

Glossina (Tsetse Fly). *See* Cyclorrhapha, Trypanosomidae.

glucagon. A hormone secreted by the pancreas of vertebrates which promotes the breakdown of glycogen to glucose. *See* insulin.

glucose. *See* carbohydrate.

glume. (Bot.) Basal bracts in grass spikelets.

glutamic. Amino acid with the formula $COOH . (CH_2)_2 CH . (NH_2) . COOH$. Molecular weight 147.1.

glutamine. Amino acid with the formula $NH_2 CH . (CH_2)_2 (CO . NH_2) . COOH$. Molecular weight 146.1.

glycine. Amino acid with the formula $CH_2 (NH_2) . COOH$. Molecular weight 75.1.

glycocholate. *See* bile.

glycogen. *See* carbohydrate.

glycolysis. The anaerobic first stage in the liberation of energy from food during respiration. Glucose is broken down to lactic or pyruvic acid by a series of enzyme-controlled reactions, and a small amount of ATP is built up. Glycolysis occurs in all types of organism. It is an important source of energy during short bursts of intense muscular activity which outrun the available oxygen supply.

glycosides. Complex carbohydrate substances which, in the presence of amino acids or enzymes and on hydrolysis, produce one or more simple sugars and a non-sugar product, aglycon, which is toxic. Glycosides are water-soluble, bitter, often produce an odour, and may be coloured or colourless. They fall into three groups: (*a*) cyanogenetic glycosides, found in species of *Sorghum*, *Prunus* and *Linum*, in which the poison is hydrocyanic ('prussic') acid; (*b*) saponin glycosides, found in species of *Agrostemma, Digitalis* and *Actinea*; (*c*) solanin glycosides, found in some species of Solanaceae.

Gnathostomata. Vertebrates which possess jaws. *Cf.* Agnatha.

gneiss. A coarse-grained banded metamorphic rock with alternating layers of dissimilar minerals.

Gnetales. An order of Gymnospermae showing some resemblance to Angiospermae (e.g. in the possession of vessels in the wood). Gnetales comprise only three genera, *Ephedra*, shrubs of warm temperate regions with scale-like leaves; *Gnetum*, tropical, evergreen, mostly climbing shrubs, *Welwitschia*, a long-lived woody plant of South West African deserts, with two oblong leaves which grow throughout the plant's life.

GNP. Gross National Product.

goats. *See* Bovidae.

gob. *See* Culm and Gob banks.

goethite. A hydrated iron-oxide mineral, with the formula FeO. OH, goethite is a weathering product of iron-bearing minerals and is a major component of limonite.

gold (AU). Metallic element. AW 196.967; At. No.79; mp 1063°C; SG 19.3. A bright yellow, rather soft metal that is not corroded by air or water, is not attacked by most acids, but which does dissolve in aqua regia (a mixture of one part of nitric to four parts of hydrochloric acid by volume). The metal is found native in hydrothermal veins and in placer deposits. Gold is also concentrated in some gossans.

Golden Triangle, the. (*a*) North West European Megalopolis (*b*) The roughly triangular island that accommodates the business centre of Pittsburgh, Penna.

Golgi Apparatus (Golgi Body). A structure present in the cytoplasm of plant and animal cells, thought to be concerned with secretion. Electron microscopy has shown that the Golgi apparatus consists of a group of flattened, membrane-bounded sacs and associated vessicles, often continuous with the endoplasmic reticulum.

gonad. An organ (ovary, testis) which produces gametes in animals.

gonad hormone. A hormone produced by a gonad. *See* androgen, oestrogen, progesterone, testosterone.

gonadotropic hormones (gonadotrophic hormones, gonadotropins). Hormones of vertebrates secreted by the pituitary gland which control the activity of the gonads, including their production of hormones. *See* FSH, lactogenic hormone, luteinizing hormone.

gonadotropins. Gonadotropic hormones.

GOR (gas/oil ratio). The ratio of oil to gas in a produced crude oil. This is expressed as standard cubic feet of gas per barrel of oil, and may vary from under 100 to several thousand.

Godiacea. Nematomorpha.

Gorse (Furze, Whin). Ulex. *See* Leguminosae.

gossan. A cellular mass of hydrated iron oxides (essentially limonite), often with quartz and other gangue minerals, from which sulphide minerals have been oxidised and leached out by downward percolating waters. The presence of a gossan at the surface is usually indicative of primary sulphides, as well as secondary enrichment of the ore vein, at depth.

Gossypium. Genus of tropical and subtropical plants (Malvaceae) whose seeds yield cotton, long cellulose fibres which cover the seed coat. The seeds also yield oil when crushed, and the residue (oil cake) is used as cattle food.

Graafian follicle. *See* follicle.

graben. A downthrown block between two normal faults.

grab sampling. Obtaining a sample of an atmosphere in a very short time, such that the sampling time is insignificant compared with the process or period being sampled.

grade. (*a*) Refers to a system in equilibrium: 1. With reference to slopes, a graded slope is one that is dynamically stable and will maintain itself in the most efficient configuration. 2. With reference to rivers, a graded river is one where slope and channel have evolved to provide the exact velocity needed to transport the sediment load. Grade is first established downstream and gradually extends upstream. Theoretically the profile is smooth and hyperbolically curved steep at the source and becoming more nearly horizontal at base level. (*b*) Has been used to indicate gradient. (*c*) Applied to things having the same quality or value.

graded aggregates. *See* sorting.

graded bedding. A sedimentary structure in which the coarsest material is concentrated at the bottom of a bed, and the average grain-size decreases to the top of the bed. Waning currents produce graded bedding, which is a feature of turbiditis.

gradient, geothermal. Geothermal gradient.

gradient wind. A generalisation of the geostrophic wind which disregards friction and assumes the wind to flow parallel to the isobars. This gives a truer representation of the actual wind, esp. at high windspeeds and along very curved trajectories. Whereas the geostrophic wind can be calculated from the distribution of pressure along a level surface or pressure surface, the gradient wind must also take account of the curvature of the trajectory.

Graminae (grasses). A very large and widespread family of monocotyledonous plants with over 10000 species. Most are herbaceous, a few are woody (e.g. Bambusaceae). The stems are jointed, the long, narrow leaves originating singly at the nodes. The flowers are inconspicuous, with a much reduced perianth, and are wind pollinated or cleistogamous. The fruit is single-seeded, usually a caryopsis. Grasses are very important as food for animals and Man. They provide cereal crops such as rice (*Oryza sativa*), maize (*Zea mays*), millet (*Sorghum vulgare, Setaria italica, Pennisetum typhoideum* and *Panicum* spp.), wheat (*Triticum* spp.), oat (*Avena sativa*), barley (*Hordeum vulgare*) and rye (*Secale cereale*). The sugar cane (*Saccharum officinarum*) yields sugar from the soft central stem tissues. Valuable pasture grasses in Britain include Cock's Foot (*Dactylis glomerate*), fescues (Festuca spp.), rye grasses (*Lolium* spp.) and Timothy (*Phleum pratense*). Grasses also yield fibres, paper, adhesives, plastics and thatching and building materials. *See* einkorn, emmer.

gram-reaction. A bacteriological staining technique used to differentiate between gram-positive bacteria (e.g. *Streptococcus, Staphylococcus*) which retain the stain, and gram-negative ones (e.g. *Gonococcus*, which causes gonorrhoea) which do not. This contrast reflects marked differences in the biochemistry of the bacteria.

grana. *See* chloroplast.

Grandpa's Knob generator. An experimental wind-powered generator designed to produce 1250 kW, installed at Grandpa's Knob, Vermont, USA, in the 1940s. The generator rotor failed in a high wind, but provided sufficient information to encourage further investigation of the possibilities of generating large amounts of electrical power by the use of windmills.

granite. A coarsely-crystalline acid igneous rock with quartz (at least 10 %) and alkali feldspar as the essential minerals. Mica is commonly present, as is a sodium-rich plagioclase feldspar. Granite is chiefly used as aggregate and as polished facing for buildings.

granodiorite. A coarsely-crystalline acid igneous rock with quartz and both plagioclase and orthoclase feldspar (the plagioclase predominates). Mica is commonly present. Granodiorite is probably the most voluminous of the plutonic igneous rocks and predominates in most of the batholiths in the world.

granophyre. Microgranite with a graphic texture of intergrown quartz and feldspar.

granophyric. *See* graphic, microgranite.

granule. *See* rudite.

granulite. A high grade metamorphic (*see* metamorphism) rock with a granular texture.

granulocytes. *See* blood corpuscles.

graphic. (Geol.) A term used to describe a texture consisting of inter-grown quartz and alkali feldspar crystals which is seen in certain pegmatites, granites and microgranites.

graphite. A crystalline form of carbon, the atoms being joined together by strong C-C bonds in two dimensions, but only by weak Van der Waal forces in the third dimension. It is found in nature, but is also manufactured from coke and pitch to form electrodes or blocks of nuclear reactor moderator graphite. It is also a component of the 'lead' in pencils, being soft owing to the characteristic structure of the crystal. It can be machined, but health precautions are taken in graphite machine shops.

graptolites. A class of Palaeozoic hemichordata whose fossils are used to date rocks from the Lower Palaeozoic Era. True graptolites appeared first in the Ordovician Period and graptolites became extinct in Britain in the Silurian and elsewhere in Europe in the early Devonian. They consisted of one or more branches along which cup-like *thecae* were arranged, cylindrical in the Lower Ordovician but showing a variety of shapes in the Upper Ordovician. The name is derived from the resemblance to writing of the carbonaceous film left by their remains (Greek *grapho*, write, and *lithos*, stone).

grasses. Graminae.

grasshoppers. *See* Orthoptera.

grassland. Well over half of the British Isles is grassland. Grassland above the tree limit on mountains or subject to winds and spray on coastlines is natural in origin, but most of it has been created over the last 2000 years by clearing forests and grazing, both by domestic animals and rabbits. Cessation of grassland management (e.g. grazing, cutting firing) leads to the formation of scrub and the re-establishment of woodland (see Downland). The vegetation of permanent hay meadows includes tall perennial herbs such as Dog Daisy (*Chrysanthemum leucanthemum*), Meadow Buttercup (*Ranunculus acris*) and Yarrow (*Achillea millefolium*). Permanent pasture is distinguished by the abundance of rosette plants such as Daisy (*Bellis perennis*), Dandelion (*Taraxacum officinale*) and Ribwort (*Plantago lanceolata*). Calcareous grassland is rich in attractive herbs such as orchids, Sheep's Scabious (*Scabiosa acaulon*), Rockrose (*Helianthemum chamaecistus*), Felworts (*Gentianella*) and Horseshoe Vetch (*Hippocrepis comosa*). Ley, temporary grassland, sown as a crop and containing only a few species, is replacing permanent grassland in many areas. Extensive grasslands in other parts of the world are the steppes of Asia and Europe, the savannas of Africa, the prairies of North America, and the pampas of South America.

graupel. Soft hail.

grauwacke. Greywacke.

gravel. *See* Rudite.

gravimeter. Instrument for measuring the variation in the Earth's gravitational field.

gravitational water. Vadose water.

gravity anomaly. *See* isostasy.

grayling zone. *See* river zone.

grazing. *See* grassland, Graminae.

greasy. *See* lustre.

Great Barrier Reef. An extensive area of continental shelf extending from

233

Gladstone on the coast of Queensland, Australia, north to the Torres Strait, a total length of almost 2400 km, with a width varying from 30 to 300 km. The Reef is composed of calcareous skeletal remains of coral, molluscs and other marine organisms and has a large area of living coral with its associated fauna. Parts of the Reef are exposed at low tide, but most lies beneath a shallow sea. *See* Crown of Thorns Starfish.

Grebes. *See* Podicipedidae.

green bans. Work bans, often imposed officially by trades unions, on projects that are likely to cause environmental damage. A number of green bans have been very effective in Australia, where they were first attempted.

green belt. An area of land, not necessarily continuous, which is near to and sometimes surrounds a large built-up area, and which is kept open by permanent and severe restriction on building. *See* Garden city.

greenhouse effect. Heating of the air caused by permitting incoming solar radiation but inhibiting outgoing radiation. Incoming, short-wave radiation, including visible light and heat, is absorbed by materials which then behave as black bodies re-radiating at longer wavelengths. Certain substances (e.g. carbon dioxide) absorb long-wave radiation, are heated by it, then begin to radiate it, still as long-wave radiation, in all directions, some of it downwards. Because of this phenomenon it has been suggested that the generation of carbon dioxide by the combustion of fossil fuels might increase the carbon dioxide content of the atmosphere, so leading to a general warming and consequent climatic changes. Despite its name, the actual heating in a greenhouse is caused mainly by the physical obstruction of the glass, which prevents warm air from leaving and cooler air from entering.

green manuring. The agricultural and horticultural practice of growing a plant crop specifically in order to plough or dig it into the soil, where it improves the structure and releases nutrients as it decomposes.

Green Revolution. *See* Indicative World Plan for Agricultural Development.

greenstone. An imprecise term for basic igneous rock.

gregale. *See* mistral.

grey brown podzolic (podsolic) soils. Acidic soils, but less leached than a podzol. A forest soil with surface litter layer and a thin organic layer over a greyish-brown leached layer (the A horizon). Below this is a darker brown depositional layer (the B horizon). *See* soil classification, soil horizons, brown podzolic soil.

grey matter. The part of the vertebrate nervous system which contains the nerve cell bodies. It lies mainly inside the white matter but forms a superficial layer in the cerebral hemispheres and cerebellum of higher vertebrates. *See* central nervous system, cerebral hemispheres.

greywacke (grauwacke). A poorly sorted (*see* sorting) sandstone charactersitic of geosynclinal sequences. Greywackes are mainly composed of angular or subangular rock fragments in an argillaceous matrix.

greywethers. A popular name for sarsens from their fancied resemblance to grazing sheep.

grid pattern. Urban planning design based on streets spaced at regular intervals and intersecting at right angles. Grid patterns were adopted in many North American city centres, but the concept is very ancient.

grignard reagents. Organo-metallic reagents of considerable importance in synthesis, consisting of an ether-soluble organo-magnesium halide, with the general formula $RMgX$, where R is an alkyl group (i.e. C_nH_{2n+1}) and X is a halide. Named after François A.V. Grignard (1871–1935).

grike. An enlarged fissure in the surface of limestone caused by chemical weathering.

Grillo-Ags Process. A German process for removing sulphur dioxide from flue gases with up to 90% efficiency by scrubbing with a suspension of manganese and magnesium oxides at 60°C to 80°C. The reagent and sulphuric acid are recovered.

grit. (*a*) Solid particles larger than about 76 microns in size, released into the atmosphere, usually as a result of industrial activity. Particles smaller than grit are dust. Particles larger than 5 microns cannot penetrate the alveoli and so grit presents no great hazard to human health, although it may be a nuisance. (*b*) (Geol.) An imprecise term for an arenaceous rock which feels gritty. Limestones, sandstones, arkoses and greywackes can be grits. Often the cause is angular to subangular shaped grains.

gross domestic product (GDP). The value of all goods and services produced within a nation in a period of time (usually one year) charged at market prices and including taxes on expenditure, with subsidies treated as negative taxes. Essentially it is a measure of national income. Divided by the number of the population, it yields per caput GDP, or national average income.

gross national product (GNP). The total monetary value of all goods and services produced within a country during a period of time (usually one year) making no

235

allowance for depreciation of stock or other consumption of capital, but including investment. GNP is a convenient indicator of the level of economic activity and of changes from one year to another, but beyond that its high degree of aggregation of data makes it a crude tool.

gross primary production. *See* production.

gross production rate. *See* production.

ground layer. *See* layer.

ground moraine. *See* moraine.

grounds. Solids deposited from suspension on standing, esp. in brewing.

ground water. Water that occupies pores and crevices in rock and soil, below the surface and above a layer of impermeable material, as opposed to surface water, which remains at or close to the land surface. The upper limit of the ground water is the water table, whose level varies according to the quantity of water entering the ground water, compared to the quantity lost (e.g. through abstraction). *See* aquifer, meteoric water.

group heating. *See* district heating.

growing point. *See* meristem.

growth. Permanent increase in mass and volume as a result of an organism taking in chemical substances unlike itself and converting them into its own substance. In higher plants growth is localised (see meristem) and more or less continuous, and involves division, enlargement and differentiation of cells. In higher animals there are no specialised growing regions, and growth is confined to an early phase in the life cycle.

growth hormone (somatotrophic hormone, STH). A hormone which causes growth of the entire body and is produced by the pituitary gland of vertebrates.

growth inhibitor. Antimetabolite.

growth rings. Annual rings.

groyne. A barrier, made from wood or other material, built on a beach at right angles to the water's edge and entering the sea. Groynes are often built as a series, and they serve to hold back the longshore currents that tend to carry sediments along the beach, so reducing erosion.

Gruiformes. An order of birds including diverse forms such as cranes, bustards, crakes, rails, coots and the moorhen. Many are waders living in marshy country.

grumusols. Soils composed mainly of clay that swell in wet weather and crack in dry weather. *See* soil classification.

guano. Deposits of bird excrement used as a fertiliser. Found almost exclusively on islands or near coasts, esp. on the west coast of South America.

guard cells. *See* stoma.

guillemots. *See* Alcidae.

gulls. *See* Laridae.

gully reclamation. The levelling and stabilization of gullies to arrest erosion. The soil may be stabilized by planting shrubs or trees and if the gully is too wide for it to be ploughed and made level, dams may be constructed within it to collect silt until the gully fills.

gumbo. Clay-rich ground, forming wet and sticky mud.

guncotton. A nitrocellulose made by nitrating cotton. It is a very safe and convenient explosive much used in military applications.

gunmetal. An alloy of copper, zinc and tin, sometimes with the addition of lead and nickel. It has good resistance to corrosion and wear.

gunpowder. An explosive consisting of a mixture of potassium or sodium nitrate, charcoal and sulphur, in varying proportions. It is said to be the first man-made explosive.

Gunung. Mount Agung.

gustiness. Turbulence close to the ground caused by buildings or other obstacles that prevent the direct flow of air.

guttation. The excretion of drops of excess water from glands (Hydathodes) on the leaves of many plants (e.g. at the tips of grass leaves).

guyot. A flat-topped sea-mount, thought to have been truncated by wave action.

Gymnophiona. Apoda.

Gymnospermae. Primitive woody seed plants (Spermatophyta) whose ovules are

237

not protected by ovaries and are usually borne in cones (*see* Angiospermae). The orders of Gymnospermae are the Coniferales, Taxales, Gnetales, Cycadales, Ginkgoales, Bennettitales, Cordaitales and Cycadofilicales.

gynandromorphism. An abnormality found in birds, mammals and insects in which one part of the body is male and the rest female. *See* intersex.

gynoecium. *See* flower.

gypsum. An evaporite mineral, hydrated calcium sulphate ($CaSO_4$, $2H_2O$). Used extensively in making wall-board and plaster, in paint and paper fillers and in the production of sulphuric acid, etc. Different forms of gypsum are known as alabaster, selenite, satin spar and desert rose.

gyroscope. A heavy symmetrical disc free to rotate about an axis which itself is confined in a framework so that it is free to take on any orientation in space. Used as a stabiliser and in instruments in aircraft and ships which show changes in the orientation of the vehicle in which they are contained.

H

H. (*a*) Henry (*b*) Hydrogen.

h. Hecto.

ha. Hectare.

Haber Process. Industrial process for synthesising ammonia (NII_3) from atmospheric nitrogen and hydrogen from a hydrocarbon source by passing the gases at high temperature and pressure through a catalyst bed (osmium). The gases combine according to the equation $N_2 + 3H_2 \rightarrow 2NH_3$. Named after Fritz Haber (1868–1934), the process was developed in Germany under the pressure of an acute ammonia shortage during the 1914-18 war.

habit. The general appearance of a plant (e.g. creeping, erect).

habitat. Dwelling place of a species or community, providing a particular set of environmental conditions (e.g. forest floor, sea shore).

habitat analysis. The evaluation of the non-living factors in a habitat.

habitat diversification. Beta-diversity.

habitat type. Group of communities resembling one another through similarities in their habitats.

habituation. A diminishing response to repeated stimulation.

haboob. *See* dust storm.

hackle. A metal comb, made from teeth mounted vertically on a wooden base, used in the preparation of certain vegetable fibres (e.g. flax) to remove the partially decomposed outer sheath of the plant stem, leaving the fibres. *See* retting.

hackly. Refers to a fracture consisting of a surface covered with sharp, jagged projections.

hade. The angle between a fault-plane, or the plane of a mineral vein, and the vertical. Hade is the complement of dip.

Haeckel, Ernst Heinrich (1834–1919). German zoologist credited with having first defined and used the word 'ecology', to mean 'the study of the economy, of the household, of animal organisms'. Haeckel supported enthusiastically the views of Darwin, who believed that Haeckel was instrumental in advancing the theory of evolution in Germany. Later, Haeckel attempted to expand Darwinian evolutionary theory into a philosophical and religious system, eventually denying the existence of a personal God, the freedom of the will and the immortality of the soul, based on his belief that higher forms of life evolved from simpler forms and that every cell has psychic properties, so that psychic and psychological processes are merely extensions of the physical.

haematite (hematite). An iron oxide mineral, with the formula Fe_2O_3, and the most important ore of iron. It occurs as a widespread accessory mineral in igneous rocks, in hydrothermal veins, and as ooliths and as a replacement mineral and a cement in sedimentary rocks. The major exploited occurrence is in the Precambrian banded ironstone formations. Haematite has a cherry-red streak and is used as a pigment, and in anti-corrosion paints. Varieties of haematite include kidney ore, specularite and micaceous haematite.

haemocoel. *See* coelom.

haemocyanin. A blue-green, copper-containing respiratory pigment found in the blood of some Mollusca and Arthropoda.

haemoglobin (hemoglobin). An iron-containing respiratory pigment found in the blood of vertebrates and a few invertebrates. The oxygenated form is scarlet and the deoxygenated form bluish-red. Substances that are able to form stable compounds with haemoglobin (e.g. carbon monoxide) reduce the oxygen supply reaching tissues and can cause anoxia. *See* methaemoglobinaemia.

Hagfish. *See* Agnatha.

hair follicle. *See* follicle.

halarch succession. Halosere.

half-life. The time required for the decay or disappearance of half of a substance that decays in a regular, exponential way. The term is used most often to express one half-life of the atoms of radioactive substances.

half value layer. The thickness of a given material that will reduce the intensity of a beam of radiation to one half of its original value.

halinokinesis. The tendency of a halite to flow under the pressure of overburden.

halite (rock salt). An evaporite mineral with the formula $NaCl$. The tendency of halite to flow under the pressure of overburden (called halinokinesis) is important in creating oil traps on the flanks and over the crests of the resulting salt-domes.

halo. A misty circle or series of circles around the sun or moon, preceding the advance of an area of rain.

halocarbons. *See* chlorofluoromethanes.

halocline. The boundary between two masses of water whose salinity differs.

halogenation. The incorporation of one of the halogen elements, usually chlorine or bromine, into a chemical compound.

halomorphic soils. Soils containing excess salt or alkali.

halophyte. A plant which grows in soil containing a high concentration of salt (e.g. samphire (*Salicornia*) which grows in salt marshes).

halosere (halarch succession). The stages in a plant succession beginning under saline conditions.

hamada (hammada). A desert region with a surface of bedrock. Hamada is contrasted with erg and reg. *See* desert pavement.

hammer mill. A crushing machine in which swinging hammers are pivoted to a revolving element and crush material against a grid of steel bars. Hammer mills are used in the preliminary treatment of ores and waste materials for disposal.

hamra. Red, sandy soil, also containing clay.

hanging valley. A valley formed by a tributary glacier flowing into a main glacier. Although the ice surface may have had a common level, the trough formed by the tributary glacier is often less deep than that formed by the main glacier. Thus when the ice melted the main valley was connected with tributary valleys whose floors stand considerably higher than their own. *See* truncated spur.

hanging wall. The higher side of an inclined fault or vein, or the ore-limit on the upper side of an inclined ore-body.

haploid. *See* chromosomes.

haplont. The haploid (*see* chromosomes) stage of an organism's life history. In animals this is the gamete. The haplont is the dominant stage (gametophyte) in Bryophyta and many algae and fungi. In ferns and seed-bearing plants it is reduced. *See* diplont, alternation of generations.

haptotropism (thigmotropism, stereotropism). A growth response of plants in which the stimulus is localized contact (e.g. a tendril coiling round a support).

hardness (*a*) Of water. A measure of the amount of mineral salts dissolved in water, so that the more salts there are, the harder the water is said to be. Hardness increases the amount of soap or detergent required in washing, and it causes scale in boilers and pipes, and furring in kettles. (*b*) Resistance to deformation. The hardness of metals is usually determined by pressing a hardened steel ball or a diamond pyramid into the metals for a given time under a known load. The hardness of minerals is normally measured using a scratch test invented by Friedrich Mohs in 1812, who arranged ten minerals in order of hardness so that each will scratch those lower in the order. Diamond, at 10 on Mohs' Scale, is the hardest, and talc, at 1, the softest.

hardpan. A strongly compacted sub-surface soil layer. When cemented with silica this is called a duripan or silcrete, with calcium carbonate a caliche or petrocalcic horizon, with clay a clay pan, with ferric oxide a laterite or plinthite. A thin crust of ferric oxide may be called an iron pan. A fragipan is an acid cemented horizon with platey structure occurring between the depositional horizon of the soil and the parent material.

hardwoods. Broadleaved trees (Dicotyledoneae) or their timber. Hardwood species possess vessels in their wood. *Cf.* softwoods.

harem. *See* polygyny.

Hares. *See* Lagomorpha.

harmattan. A hot, dry wind that blows from the north-east over the southern Sahara, usually in winter. It is accompanied by blowing sand and dust which can be seen from afar and which reduces visibility considerably inside the affected area. The Harmattan is known locally as 'the doctor' for the relief it brings from the damp heat that precedes it.

harmonic. A pure tone, a sinusoidal component in a complex periodic wave, of a frequency that is an integral multiple of the fundamental frequency of the wave. If a component (an overtone) in a sound has a frequency twice that of the fundamental, it is called the second harmonic.

Harvestmen. *See* Phalangida.

haustorium. An organ of parasitic plants (e.g. fungi, dodder) which withdraws food material from the tissues of the host plant.

Hauterivian. *See* Neocomian.

Hawaiian Goose (Nene). *Branta sandivicensis. See* Anseriformes.

Hawks. *See* Accipitrinae.

Hawthorn. *See* Crataegus.

hazardous wastes. Wastes that contain any substance harmful to life. These may be toxic (e.g. pesticides, compounds of arsenic, cyanides), flammable (e.g. hydrocarbons), corrosive (strong acids or alkalis) or oxidising (e.g. nitrates or chromates) and some may be hazardous on more than one count.

Hazel. *Corylus avellana.*

HCH. Lindane.

He. Helium.

head. Combe rock.

headstream (highland brook). *See* river zones.

hearing loss. The amount, in decibels for a specified ear and frequency by which the threshold of audibility for that ear exceeds the normal threshold.

242

Heart Urchins. Echinoidea.

heartwood (duramen, truewood). The central part of the wood, which, in the living tree, no longer contains live cells. It functions only for support and not for water conduction. The heartwood is generally darker than the sapwood and is more resistant to decay.

heat exchanger. Apparatus for transferring heat from one fluid or body to another (e.g. a coiled pipe carrying hot water through a tank of cooler water in order to warm one, cool the other, or both).

heath. (*a*) A shrub of the genus *Erica* (*see* Ericaceae). (*b*) An area with poor, acid soil, typically dominated by ling (*Calluna*) or heaths (*Erica*). Although heathland is usually associated with sandy or gravelly soils, it may develop in chalk or limestone areas where the topsoil has become acid through leaching.

heat of formation. The heat evolved when one gram-molecule of a compound is formed from its constituent elements.

heat of fusion. The heat necessary to convert a given weight of solid to its liquid form.

heat of solution. The amount of heat taken in or given out when a substance is dissolved in a large amount of solvent.

heat pump. Normally a device that provides space heating or cooling (air conditioning) by exploiting the temperature gradient between the space to be heated and a colder, adjacent space. A chemical substance whose boiling point is close to ambient temperature is circulated by a pump between the two bodies or air masses so that it condenses in the cooler and evaporates in the warmer, so gaining heat as it liquefies and losing it as it vaporises and using the cooler body to heat the warmer, the energy transferred being the latent heat of condensation and evaporation. A refrigerator uses the same system but with the aim of cooling a small space while heating a larger one. The chemicals used are commonly chlorofluoromethanes.

heat recovery. The transfer of heat from a substance which has been heated during a process to the substance about to be treated, in order to avoid a waste of energy (e.g. the recovery of heat from waste gases from a furnace or incinerator). The term also refers to the use of waste heat for another purpose (e.g. the use of hot cooling water from power generation for district heating.)

heat transfer. Heat is transferred by three different processes: (*a*) conduction, where the heat diffuses through solid materials or stagnant fluids; (*b*) convection, where heat is carried from one point to another by actual movement of the hot material; (*c*) radiation, where heat is transferred by means of electromagnetic waves.

heave. The horizontal displacement between the upthrown and downthrown sides of a fault.

heavenly cross. A sun pillar crossed by a horizontal bar.

heavy hydrogen. Deuterium (D).

heavy liquid. In geology, any of a group of dense liquids used to separate out dense accessory minerals (the heavy minerals). Commonly used heavy liquids are bromoform (SG 2.87), methylene iodide solution (SG 3.2) and Clerici's solution (SG 4.25): the SG of quartz is 2.65, of calcite 2.71, and of feldspar 2.55 to 2.76.

heavy minerals. The accessory detrital minerals of a sediment or sedimentary rock which are of high specific gravity (usually set arbitrarily at over 2.87, the specific gravity of bromoform). Heavy minerals are concentrated in placer deposits, such as black sands.

heavy spar. *See* barytes.

heavy water. Deuterium oxide (D_2O), or water in which the hydrogen has been replaced by deuterium. *See* nuclear reactor.

Hebb-Williams maze. A standard device used to measure intelligent behaviour in experimental animals. It consists of an enclosure with start and goal boxes at opposite ends. The floor is marked out into 36 five-inch squares and barriers are arranged along the boundaries of the squares so that there are many blind alleys but only one correct path to the goal. The measure is then the number of times a hungry rat must be released from the start box in order to learn the correct path, and the total number of errors it makes in learning its way through the maze.

hectare (ha). Metric unit of area, equal to 10000 square metres, or 2.471 acres.

hecto- (h). Prefix used in conjunction with SI units to denote the unit $\times$ 10^2.

hekistotherm. A plant which thrives with very little heat (e.g. Arctic mosses and lichens; alpine plants).

heliophobe (skiophyte, shade plant, sciophyte). A plant which grows best in shady places (e.g. Dog's Mercury, *Mercurialis perennis*). *Cf.* heliophyte.

heliophyll. A plant with isobilateral (equal sided) leaves. *Cf.* skiophyll.

heliophyte (sun plant). A plant which grows best in full sunlight (e.g. cactus). *Cf.* heliophobe.

heliosis. Discoloration on leaves caused by high intensities of sunlight. *See* solarisation.

heliotaxis. Phototaxis.

heliotropism. Phototropism.

helium (He). Element. An inert gas which occurs in some natural gases in the USA, occluded in some radioactive ores (monazite, pitchblende) and in the atmosphere at about one part in 200000. It is used to fill balloons for its properties as a lifting agent, in some metal working processes, as a tracer in determining the migration of oil and gas in geological structures, and it has medical, commercial and scientific uses. AW 4.0026; At.No. 2.

helminth. A parasitic worm, esp. a member of the Platyhelminthes or Nematoda. *See* Anthelminthics.

helophyte. *See* Raunkiaer's Life Forms.

helm wind. A steady, strong wind that blows down the westward (lee) slopes and for some distance across the lowlands in the northern Pennines of England, when the prevailing wind over the wider area is easterly. It is caused by wavelike disturbances in the air stream as the wind flows over the ridges of low hills, which accelerate the wind down the lee slope.

hematite. Haematite.

Hemichordata (Enteropneusta). A small sub-phylum of the Chordata marine animals very different from other chordates. Either worm-like and burrowing (e.g. *Balanoglossus*) or sedentary and colonial.

hemicryptophyte. *See* Raunkiaer's Life Forms.

Hemimetabola. Exopterygota.

hemimetabolous. *See* metamorphosis.

Hemiptera (Rhychota). The Bugs. A large order of insects (Exopterygota) with piercing and sucking mouth-parts, most of which feed on plant juices, a few on animals. The sub-order Homoptera, whose members have forewings of uniform consistency, includes aphids, cicads, fog-hoppers, mealy-bugs and scale insects. These plant bugs often do a great deal of damage to crops by sucking the sap, discharging honeydew which blocks the stomata, and transmitting diseases. The sub-order Heteroptera, whose members typically have half of the forewings horny, includes many other plant bugs, the water bugs (e.g. Water Boatmen) and the wingless bed bug (*Cimex lectularius*).

hemoglobin. Haemoglobin.

hemp. *See* Cannabiaceae.

Henry (H). The derived SI unit of self and mutual induction, being the inductance in a closed circuit such that a rate of change of current of one ampere per second produces an induced electromotive force (EMF) of one volt. Named after Joseph Henry (1707–1878).

heparin (antithrombin). *See* clotting of blood.

Hepaticae (liverworts). A class of Bryophyta whose members live in damp places or in water. A simple liverwort has a small, flat, green, repeatedly forked, ribbon-like body, lying close to the ground, to which it is attached by unicellular hair-like rhizoids. Some liverworts resemble mosses in possessing leaf-like expansions, but unlike mosses they have no strands of conducting tissue.

heptachlor. An organochlorine insecticide which breaks down in the soil to heptachlor-epoxide, a stable and more poisonous substance than heptachlor. Heptachlor is no longer approved for agricultural use in the UK. *See* cyclodiene insecticides.

herb. (herbaceous plant). A non-woody vascular plant having no parts which persist above the ground.

herbicide. A chemical used to kill weeds. *See* carbamate, dinitro group of pesticides, hormone weedkillers, soil-acting herbicides, translocated herbicides, urea group of herbicides.

herbivore. A plant-eating animal. A primary consumer in a food chain.

herb layer. *See* layer.

herbosa. Vegetation which is made up of non-woody plants.

hercynian. Refers to the processes and products of the orogeny which affected large parts of Europe in the Upper Palaeozoic, (Devonian to Permian). Hercynian massifs are exposed in SW England and Ireland, Brittany, central France, the Iberian Peninsula, Bohemia, and in the Ardennes-Eifel-Rhenish

Schiefergebirge belt. Fragments of hercynian massifs occur within the massifs of the Alpine orogeny. *See* Variscan.

hereditary factor. *See* genes.

heritability. The estimate of the extent to which a characteristic is inherited, as the percentage variation in the characteristic in a population that can be accounted for by genetic variation within the population (e.g. in Man, height has a heritability of 90 to 95%). This estimate contributes nothing to the nature-nurture controversy (one cannot say that 90% of height is determined genetically and 10% by environmental factors) and use of the heritability factor in descriptions of behaviour has been criticised.

hermaphrodite (bisexual). Applied to an individual which contains both male and female functional reproductive organs (e.g. earthworm, buttercup). *Cf.* unisexual, monoecious, dioecious.

heroin. Diacetylmorphine, obtained by the alkylation of morphine. It is a narcotic alkaloid with an effect resembling that of morphine and is classed as a dangerously addictive drug.

herptile. A member of the Reptilia or Amphibia.

Hertz. *See* frequency.

heterocaryon. Heterokaryon.

heterochrosis. Abnormal coloration (e.g. pale plumage in birds).

heterocyclic compounds. Cyclic carbon compounds in which other atoms, typically nitrogen, oxygen or sulphur, form part of the ring structure.

heterodont. Possessing different types of teeth (e.g. incisors, canines and molars in a mammal). *Cf.* homodont.

heteroecious (heteroxenous) parasite. *See* parasitism.

heterogametic sex. The sex in which the sex chromosomes are dissimilar (X and Y) or in which there is only an unpaired X chromosome. In Lepidoptera and many vertebrates (birds, reptiles, some Amphibia, some fish) the female is heterogametic. In most other organisms the male is heterogametic. *Cf.* homogametic sex.

heterogamy (*a*) Anisogamy. The production of unlike gametes which differ in size and/or form. *See* oogamy. (*b*) The production of unlike gametes which differ as to the chromosomes they contain. *See* Homogamy (*b*). (*c*). (Bot.) The production of flowers of more than one sexual type in the same inflorescence. *See* homogamy. (*d*) (Zool.) The alternation during the life cycle of two types of reproduction involving gametes (e.g. parthenogenesis and syngamy in aphids).

heterokaryon (heterocaryon). A cell containing two or more nuclei of differing genetic constitutions. *Cf.* dikaryon, homokaryon.

heterokontae. Xanthophyta.

heterometabola. Exopterygota.

heterophyte. A parasite (*see* parasitism) or saprophyte.

heterosis (hybrid vigour). Increased vigour (e.g. in terms of fertility or growth) resulting from the crossing of two genetically different lines.

heterospory. The production, as in seed-producing plants and some clubmosses, of two kinds of spores. Microspores give rise to male gametophytes and megaspores grow into female gametophytes. *See* pollen, embryo sac. *Cf.* homospory.

heterostyly. Variation in the length of the style in different flowers of the same species (e.g. thrum eye and pin eye in primrose). This ensures cross pollination by insects, as the anthers of one type of flower are at the same height as the stigmas of the other.

heterothallism. The condition in which the thallus of an alga or fungus is not self-fertile. *Cf.* homothallism.

heterotrophic. Applied to organisms which need ready-made organic food material from which to produce most of their own constituents and (except in the case of a few organisms such as phototrophic Protozoa) to obtain all their energy. Animals, fungi, most bacteria and a few flowering plants are heterotrophs. Whether they are herbivores, carnivores, parasites or saprophytes, they are dependent on autotrophic (*see* autotroph) organisms for their supply of food.

heteroxenous (heteroecious) parasite. *See* parasitism.

heterozygote. An organism with dissimilar allelomorphs in respect of a particular character. *Cf.* homozygote.

Hettangian. A stage of the Jurassic System.

Hevea (Euphorbiaceae). *Hevea brasiliensis* is the source of the best natural rubber. The tree was introduced from Brazil into the Far East in the late 19th century. *See* Ficus, Manihot, Moraceae.

hexacanth. Six-hooked.

hexapoda. *See* insects.

hexose. *See* carbohydrate.

Hg. Mercury.

hibernation. Dormancy of animals in winter, during which metabolic rate is much reduced and the body temperature of homoiothermic species drops to that of the surroundings. In Britain, reptiles, Amphibia, some mammals and many invertebrates hibernate. Some (e.g. dormouse) do not stir during the hibernation period, while others (e.g. bats and hedgehogs) become active and feed during warm spells. *See* diapause, aestivation.

hiemilignosa. Monsoon forest and bush. The woody plants have small leaves which are shed during the hot, dry season.

high-. Prefix used in metallurgy and other fields to denote a large content of the substance associated with it (e.g. high-chromium steel: steel with a high content of chromium; high-alumina cement; cement with a high content of alumina).

high alumina cement. Aluminous cement.

high forest. *See* forest.

highland brook (headstream). *See* river zones.

high level inversion. An atmospheric temperature inversion formed high (300 m or more) above the surface by the descent of air in anticyclonic (*see* anticyclone) conditions and its warming as it is compressed by the higher pressure at a lower level. The inversion may continue provided wind speeds below it are sufficient.

high pressure area. *See* anticyclone.

high-temperature gas-cooled reactor (HTGR). Nuclear reactor using enriched uranium as a fuel, made into particles coated with carbon and silicon dioxide. Uses helium as a coolant and graphite as a moderator. One HTGR has been built in Britain (the Dragon reactor) and one is on-stream at Fort St. Vrain, Colorado, owned by Gulf General Atomic. Operating temperatures are much higher (1000°C) than in the Magnox or AGR and only ceramic materials can be used. All orders for HTGRs have been cancelled.

249

high-yielding varieties (HYVs). Varieties of wheat, maize and rice developed in Mexico at the Rockefeller Institute, at the International Rice Research Institute (IRRI) in the Philippines and elsewhere. HYVs respond well to fertiliser provided the water supply is adequate and they mature quickly. They are short-strawed, which makes them less prone to lodging and they form the basis of the 'Green Revolution'. *See* Indicative World Plan for Agricultural Development.

Hirudinea (leeches). A class of mainly aquatic Annelida whose members have a sucker at each end of the body. Many leeches prey on invertebrates, but some are able to pierce vertebrate skin and suck blood.

histic. Applied to soil surface layers that are high in organic carbon and seasonally saturated with water. *See* soil classification.

histidine. Amino acid with the formula $C_3H_3N_2.CH_2CH.(NH_2).COOH$. Molecular weight 155.2.

histogens. Zones of tissue primordia present in apical meristems and elsewhere in plants.

histolysis. The breakdown of tissue.

histosols. *See* soil classification.

hoar frost. A white crystalline deposit of ice on the surface of solid objects caused by the direct sublimation of water vapour from the air. *See* dew.

holarctic (arctogea). *See* zoographical regions.

holism. The belief that systems may be understood only when viewed in their entirety. *Cf.* reductionism, vitalism.

holocephali. *See* chondrichthyes, chimaera.

Holocene. Refers to the younger subdivision of the Quaternary in which we are living, the previous subdivision being the Pleistocene. The Holocene, or Recent, is approximately the time since the last glaciation, or about the last 10 000 years. Holocene also refers to rocks deposited during this time. However, some climatologists propose that the present is merely another interglacial of the Pleistocene and that the use of the term Holocene is premature.

holocoen. The whole environment, comprising the biocoen and abiocoen.

Holometabola. Endopterygota.

holometabolous. *See* metamorphosis.

holophytic nutrition. The characteristic mode of nutrition of a green plant, i.e. the synthesis of organic compounds from carbon dioxide, water and mineral salts, by means of light absorbed by chlorophyll (photosynthesis). *Cf.* autotroph, holozoic.

Holothuria. Holothuroidea.

Holothuroidea (Holothuria, sea cucumbers). A class of Echinodermata whose members have sausage-shaped bodies and no arms, but tube feet around the mouth enlarged to form tentacles. In some species food is extracted from mud and shovelled into the mouth by the tentacles, in others the tentacles are sticky and are used to entangle small organisms.

holotype. Type specimen.

holozonic nutrition. The characteristic mode of nutrition of animals i.e. the eating of plants, animals or their solid products. *Cf.* heterotrophic, holophytic.

homeostasis (homoiostasis). The maintenance of constancy within a biological system, either in terms of interaction between the organisms of a community or as regards the internal environment of an individual.

homeothermous. Applied to organisms whose body temperatures remain constant through the operation of physiological mechanisms.

home range. The total area occupied over the years by a group of animals or an animal. That part of the home range in which it spends most of its time is the core area.

Hominoidea. The superfamily of Primates which includes the family Hominidae (modern and ancestral Man) and the apes.

Homo. The genus which includes modern Man (*Homo sapiens sapiens*). *Homo habilis*, whose remains were found in the Olduvai Gorge, Tanzania, was a very early species, alive more than 1.5 million years ago. He was able to make stone tools and to stand erect. The cranial capacity was small. Some authorities dispute the inclusion of these remains in the genus *Homo*. *Home erectus* includes Java Man (formerly *Pithecanthropus erectus*) who lived about half a million years ago, walked erect and had prominent ape-like brow ridges. Fossil remains from other parts of the world are now attributed to this species, including China Man (formerly *Sinanthropus pekinensis*), who used fire and made tools. *Homo heidelbergensis* was a robust species with powerful jaws, thought by some to belong to *Homo erectus*. *Homo sapiens neanderthalensis* had a large cranial capacity, well-developed brow ridges and no chin prominence. Neanderthal Man probably became extinct about 50,000 years ago. It is associated particularly with the Mousterian Culture in which the dead were buried ritually. *Homo sapiens rhodesiensis* (Rhodesia Man), found in Broken Hill, Zambia, had massive brow ridges combined with some modern features. *Homo sapiens fossilis* (Cro-Magnon Man) had a high cranium and a broad face without prominent brow ridges. His remains, found throughout Europe, are associated with sculpture and paintings. Some Cro-Magnon people were contemporaneous with the last Neanderthalers. The successor of Cro-Magnon Man was *Homo sapiens sapiens*. *See* Palaeolithic, Mesolithic, Neolithic.

homocaryon. Homokaryon.

homocyclic compounds. Cyclic compounds which contain a ring consisting wholly of atoms of the same element.

homodont. Possessing teeth which are all similar, as in most reptiles, Amphibia and fish. *Cf.* heterodont.

homogametic sex. The sex in which each nucleus contains a pair of X chromosomes (*see* sex chromosomes). In Lepidoptera and many vertebrates (birds, reptiles, some Amphibia, some fish) the male is homogametic. In most other organisms the female is homogametic. *Cf.* heterogametic sex.

homogamy. (*a*) Inbreeding brought about by isolation. (*b*) The production of gametes all alike as regards the chromosomes they contain. *Cf.* heterogamy (*b*). (*c*) The production of flowers all of the same sexual type (male, female or hermaphrodite). *Cf.* heterogamy (*c*). (*d*) The condition in which the anthers and stigmas of a flower mature simultaneously. *Cf.* dichogamy.

homogeneity. The even distribution of species or characteristics, so that a sample taken at any one point resembles one taken at any other. An association is homogenous if the individuals of its component species are distributed as regularly as water and alcohol in a mixture of the two liquids.

homoiosmotic. Applied to animals (e.g. vertebrates) which maintain their body fluids at a constant osmotic pressure. *Cf.* poikilosmotic.

homoiostasis. Homeostasis.

homoiothermy. 'Warm-bloodedness'. The maintenance of a constant body temperature irrespective of variations in the external temperature. This occurs in birds and animals. *Cf.* heterokaryon.

homologous. (*a*) An organ of one species is homologous to that of another species when the two organs have a fundamentally similar structure, development or origin. This is clear during embryonic development, but later on organs may become much modified. Homologous organs may have very different functions (e.g. the wings of a bat, the paddles of a whale and the forelegs of a horse are homologous; the ear ossicles of a mammal are homologous with certain bones involved in the articulation of the jaw in fishes). *Cf.* analogous. (*b*) Homologous chromosomes. *See* chromosomes.

homologous chromosomes. *See* chromosomes, meiosis.

homospory. The production, as in many Pteridophyta of only one kind of spore. This gives rise to a gametophyte which generally produces both male and female reproductive organs. Occasionally environmental conditions may lead to male and female organs being borne on different gametophytes. *Cf.* heterospory.

homothallism. The condition in which the thallus of a fungus or alga is self-fertile. *Cf.* heterothallism.

homozygote. An organism with identical allelomorphs in respect of a particular character. *Cf.* heterozygote.

honeydew. Sweet, sticky substance produced by certain sap-eating insects. In hot, dry weather some species of aphids and scale insects which eat sap excrete large amounts of modified sap, sometimes in such large quantities that it coats the leaves, esp. of maples, lindens and roses, and may fall from the tree as a fine mist. Several insect species, including some ants, feed on honeydew.

hoodoo. Unusually shaped erosion remnant of rock.

Hookworms. Parasitic Nematoda responsible for much debilitation in Man throughout tropical and subtropical regions. Heavy infestation causes anaemia and retardation of mental and physical development. *Ancylostoma duodenale* and *Necator americanus* are common in Man; other species parasitize other mammals. Adult hookworms feed on blood and tissue from the wall of the intestine. Eggs pass out in faeces, and the larvae subsequently enter a new host by burrowing through the skin. Wearing shoes and practising the sanitary disposal of human faeces prevents infection.

Hop. *See* Cannabiaceae.

horizon. (Geol.) *See* soil horizon.

hormone. A chemical produced by one part of an organism and transported to another part, where a minute quantity exerts control over specific metabolic functions. *See* auxins, endocrine organ, adrenal gland, corpus luteum, parathyroid, thyroid gland, pituitary gland, androgen, insulin, oestrogen, intermedin.

hormone weedkillers. (auxin-type growth regulators) These herbicides (e.g. 2,4 D and MCPA) are synthetically produced organic compounds which have effects similar to the natural growth regulating substances of plants (auxins). They are absorbed by roots or leaves, translocated to the growing points, and inhibit growth or cause deformed growth which results in the death of the plant. Some are toxic to dicotyledonous plants and non-toxic to monocotyledonous plants and have great importance in the selective control of weeds in cereals and grass. *See* translocated herbicides, 2,4,5-T, mecoprop, dichlorprop.

horn. A pyramidal mountain-peak formed between three or more cirques.

Hornbeam. *Carpinus betulus.*

hornblende. A calcium magnesium aluminosilicate mineral which is one of the amphibole group.

hornfels. A hard, splintery rock formed by the thermal metamorphism of argillaceous rock adjacent to a large igneous intrusion.

Hornwort. *See* Bryophyta.

Horsehair worms. Nematomorpha.

Horseshoe crab. Xiphosura.

horsetails. *Equisetum. See* Equisetales.

horst. An uplifted block between two normal faults.

host. (*a*) An organism which harbours a parasite (*See* parasitism). A definitive (primary) host harbours the mature stage of a parasite, and an intermediate (secondary) host only immature stages (e.g. the sheep is the definitive host for the liver fluke *Fasciola hepatica*, and the snail *Lymnea truncatula* is its intermediate host). (*b*) In commensalism, the partner which receives no benefit from the relationship. (*c*) An organism into which an experimental graft is transplanted.

hot blast. Technique, devised in 1828 by James Neilson, for reducing fuel consumption in iron foundries. A blast of air is required in a furnace to increase the temperature at which the fuel burns and to assist the purification of the metal (see Bessemer process). The use of hot air was found to reduce fuel consumption by about 40 %. In modern furnaces the hot gas leaving the furnace is used to pre-heat the incoming blast. *See* blast furnace, cupola.

hot brine. Brackish subterranean water whose temperature is markedly higher than that dictated by the normal geothermal gradient and that may therefore be used as a source of geothermal energy. In some areas (e.g. New Zealand) hot brines have a low salinity, but elsewhere the mineral content may be high, making the water highly corrosive and so difficult to handle and to dispose of once its heat has been removed.

hot rock. Subterranean rock whose temperature is higher than would be dictated by the normal geothermal gradient and which may be used as a source of geothermal energy. *See* hydrofracturing.

Howard, Sir Ebenezer. *See* Garden city.

Howden-ICI Process. British process for removing sulphur dioxide from flue gases by washing the gases in a solution of calcium hydroxide and removing calcium sulphate by precipitation. The process was invented in 1935 and a pilot plant was working in 1938, but it has never been applied on a full scale.

HTGR. High temperature gas-cooled reactor.

humate. Salt or ester of a humic acid derived from humus during the decomposition of organic matter in the soil.

humidification. Process for increasing the water content of air or other gases. It is often used in central heating systems to reduce the build-up of static electricity.

humidifier. Device for increasing the water content of air, usually incorporated into an air conditioning system. Humidifiers bring air at a desired temperature (which may be achieved by heating or cooling it) into contact with water, the operation of the device often being controlled automatically by a sensor device that monitors atmospheric humidity (a humidistat). The water used must be at the same temperature as the air. If it is very cold it can be used to cool and dehumidify the air simultaneously (using a dehumidifier). Air may also be dried by passing it over a bed of hygroscopic crystals (e.g. lithium chloride).

humidity. *See* absolute humidity.

humidity mixing rate. *See* absolute humidity.

humification. The microbial breakdown of organic matter in the soil to form humus.

humus. The more or less decomposed organic matter of the soil. Besides being the source of most of the mineral salts needed by plants, humus improves the texture of the soil and holds water, so preventing the nutrients from being leached out. Mild humus (mull) is produced in soil containing abundant earthworms, where decay is rapid. *See* mor, mull.

hurricane. *See* bath plug vortex, tropical cyclone.

Huygens' construction. Each point of a wave front may be regarded as a new source of secondary wavelets. From this construction, if the position of the wave front at any given time is known, its position at any subsequent time can be determined. Named after Christian Huygens (1629–1695).

hybrid. (*a*) An organism produced by crossing parents of different taxa (e.g. species, subspecies or varieties). Hybrids are often sterile. *See* allotetraploid. (*b*) An organism produced by crossing parents of different genotypes.

hybrid swarm. A large, often very varied population resulting from hybridization with subsequent crossing and back-crossing.

hybrid vigour. Heterosis.

hydathodes. *See* guttation.

hydragyrum. *See* mercury.

hydrarch succession. Hydrosere.

hydrates. Salts which contain water of crystallisation, i.e. water retained on crystallisation from aqueous solution.

hydration. The chemical addition of water to a compound.

hydraulic geometry. The study of the shape of stream channels, based on measurements of the width (the shortest distance from bank to bank), depth at a particular point, vertical cross section area at right angles to the direction of flow, and slope or gradient.

hydrazine. Reactive chemical intermediate in the production of explosives, photographic chemicals and antioxidants. It is also used as a rocket fuel.

hydrocarbon. A compound composed of the two elements hydrogen and carbon only.

hydrocarbon cracking. Decomposition by heat, with or without catalysis, of petroleum or heavy petroleum fractions, to give lower-boiling materials used as motor fuels, domestic fuel oil and other products. It is generally performed in oil refinery complexes. *Cf.* cracking.

Hydrocharitaceae. A family of tropical and temperate freshwater and marine monocotyledonous plants. Some (e.g. Frogbit, *Hydrocharis morsus-ranae*) have floating leaves, but most have ribbon-like, submerged leaves. The female plant of *Elodea canadensis* (Canadian Pondweed) was introduced into Europe in the mid 19th century, and has spread rapidly through inland waters. *Stratiotes aloïdes* (Water Soldier) floats at the surface in summer and sinks in the autumn.

hydrochore. A plant whose seeds, turiens or other reproductive structures are dispersed by water (e.g. alder, *Alnus*; coconut, *Cocos*).

hydrochory. Dissemination by water.

hydrofluoric acid. One of the strongest and most corrosive acids known, reacting with a wide variety of materials. It is a solution of hydrogen fluoride in water and is used for etching glass, in metal processing, and for cleaning stonework.

hydrofoil. An arrangement of flat struts beneath the hull of a high powered boat that pushes the boat out of the water at speed due to the upward pressure from the struts. This reduces drag, as less of the hull is in contact with the water, and so permits greater speeds. It is used mainly in lakes and rivers where waters are calm.

hydroforming. Catalytic process for the dehydrogenation of paraffins and their conversion to cyclic and aromatic hydrocarbons.

hydrofracturing. Technique for making large cracks in subterranean rocks by injecting water under high pressure. The technique has been used experimentally for the extraction of geothermal energy from hot rock.

hydrogen (H). Gaseous element, giving the lightest gas. Inflammable. It has a wide range of uses in synthesis (e.g. of ammonia (NH_3), hydrochloric acid (HCl), methanol (CH_3OH), and in the hydrogenation of coal. It is also used widely in laboratories esp. as a carrier gas in gas chromatography. It is converted to helium, with the release of large amounts of heat and other radiation, in the hydrogen bomb, which uses heavy hydrogen (Deuterium). AW 1.00797; At.No. 1.

hydrogen cyanide. Intensely poisonous gas implicated in many industrial accidents, used as a fumigant and as a starting material for nylon and other polymers, pharmaceuticals and dyestuffs.

hydrogen ion concentration. pH.

hydrogen peroxide. Somewhat unstable, highly reactive, colourless liquid,

257

H_2O_2, often used in aqueous solution as a bleaching agent, antiseptic, oxidising agent, and as an oxidant for rocket fuels (e.g. hydrazine).

hydrogen sulphide (H_2S). A poisonous, evil smelling gas, inhalation of small amounts of which can cause headaches. It is used in the chemical industries, rayon manufacture, and in analytical laboratories. It is also a product of the anaerobic decomposition of sulphur-containing materials. In high concentrations it can cause death.

hydrograph. A graph indicating the level of water in watercourses or aquifers or the rate of flow of water during a period of time.

hydroid. *See* Coelenterata.

hydrological civilisation. A civilisation that depends on sophisticated water management to sustain its agriculture. *See* Fertile Crescent, Indus Valley, Chinampas.

hydrological cycle. The constant cycling of water by evaporation and precipitation. One of the major global cycles. 97 % of the Earth's water is in the oceans. Of the remaining 3 %, 98 % is frozen in the polar icecaps. Each day about 875.3 cu km of water evaporates from the seas and about 160.5 cu km is lost from the land surface by evaporation and transpiration. About 775.3 cu km falls on the seas as precipitation and 100 cu km is carried in the air from the sea over the land, which receives about 260.5 cu km of precipitation (equal to the 100 cu km carried from the sea plus the 160.5 cu km lost from the land surface). The cycle is completed by the return to the sea of about 100 cu km through rivers. Small absolute losses to the cycle are made good by juvenile water.

hydrologic sequence. Sections of soil from uniform parent material showing increasing lack of drainage downslope. *See* catena.

hydrology. The study of the chemistry and physics of water and water movement upon and beneath the ground.

hydrolysis. (*a*) The formation of an acid and a base from a salt, by the ionic dissociation of water. (*b*) The decomposition of organic compounds by interaction with water (e.g. the formation of alcohols and acids from esters).

hydrophilous. Hygrophilous.

hydrophyte. *See* Raunkiaer's Life Forms.

hydromorphic soils. Soils containing excess water.

hydroponics. The growing of plants without soil by suspending them with their roots immersed in water enriched with essential nutrients or by rooting them in an inert material (e.g. quartz sand) and supplying them with a nutrient solution. The technique can produce large yields in a small space, but the best nutrient mixtures are very complex.

hydrosere (hydrarch succession). The stages in a plant succession beginning in water or a wet habitat, and progressing towards drier conditions. *Cf.* xerosere.

hydrosphere. That part of the Earth which is composed of water: oceans, seas, the icecaps, lakes, rivers, etc.

hydrostatic pressure. Pressure exerted by water at rest equally at any point within the water body.

hydrothermal. Refers to any geological process involving heated or superheated water. Such process may be divided into: (*a*) Alteration (e.g. of feldspar to kaolin, and of olivine and pyroxene to serpentine), and (*b*) Deposition. Many metalliferous ore deposits are thought to be of hydrothermal nature, originating as a concentration of volatiles in a magma. The major subdivisions of hydrothermal deposits are hypothermal, mesothermal and epithermal.

hydrotropism. A growth response (tropism) in which the stimulus is water.

Hydrozoa. A large class of Coelenterata whose members usually have an alternation of sessile polyp colonies and free-swimming medusae. The group includes hydroids, some of which (e.g. *Hydra*) are solitary and have no medusa stage, the stinging corals, and pelagic polyp colonies such as the Portugese Man-of-War. *See* Actinozoa, Scyphozoa.

hygrometer. Instrument used to measure atmospheric moisture.

hygropetric. Inhabiting wet rock surfaces.

hygrophilous (hydrophilous). Inhabiting wet places.

hygrophyte. Plant that is found only in a moist habitat and that is very sensitive to dry conditions.

hygroscopic. Applied to substances which absorb water readily from the atmosphere.

hygroscopic moisture. Water present in soil but held by surface tension forces too great for its use by plants.

259

Hymenoptera. A large order of insects (Endopterygota) whose winged members have two pairs of membranous wings coupled together in flight by small hooks. The Symphyta (e.g. saw-flies and wood-wasps) are unwaisted and have saw-like or drill-like ovipositors which enable them to insert eggs into leaves, stems or wood. The Apocrita, in which there is a narrow waist between the thorax and abdomen, include bees, wasps, ants and ichneumons. *See* Aculeata, Parasitica.

hypabyssal. Refers to a rock, or to an igneous intrusion, which crystallized near the surface. In general, such a rock is medium-grained, and the intrusion is typically a sill or dyke. Hypabyssal is intermediate between plutonic and volcanic.

hyperkinesis. Excessive motility of an organism or of a muscle.

hyperparasite. *See* parasitism.

hyperphagia. *See* aphagia.

hyperplasia (hyperplasy). An abnormal increase in the number of cells in part of an organism, as in some plant galls or animal tumours.

hyperplasy. Hyperplasia.

hypertonic solution. *See* osmosis.

hypertrophy. An increase in size caused by hyperplasia or by enlargement of cells or fibres, as in well exercised muscles.

hypha. One of the mass of filaments making up a fungal mycelium. Hyphae are tubular and branched, and may have cross walls.

hypoblast (endoderm). *See* germ layers.

hypocentre. The point on the Earth's surface directly below the centre of a nuclear bomb explosion.

hypogeal. Applied to germination in which the cotyledons remain underground (e.g. pea). *Cf.* epigeal.

hypogene. Refers to mineral deposits formed by ascending aqueous solutions. Hypogene is contrasted with supergene.

hypolimnion. The colder, non-circulating layer of water in a lake, lying beneath the thermocline. *See* thermal stratification. *Cf.* epilimnion.

hypophysis cerebri. Pituitary gland.

hypostatic gene. A gene whose expression is prevented by the presence of another, non-allelic (*see* allelomorphs) gene. *Cf.* epistatic gene.

hypothalamus. Part of the brain floor of vertebrates. In mammals it controls body temperature and produces hormones which influence the pituitary body lying below it.

hypothermal. Refers to deposits formed from an ascending, essentially aqueous solution at high temperatures and at great depths. The general temperature range is 300°C to 500°C. *Cf.* epithermal, mesothermal.

hypotonic solution. *See* osmosis.

hypoxia. Anoxia.

hypsodont teeth. High-crowned teeth typical of herbivorous mammals (e.g. ungulates). The crowns must be high because the teeth are continually worn away by grazing. Some hypsodont teeth grow continually from open roots (e.g. in rodents).

Hyracoidea (Hyraxes). An order of small, herbivorous placental mammals, resembling short-tailed squirrels, with four toes on the front feet and three on the hind feet. They show similarities to both rodents and elephants.

Hyraxes. Hyracoidea.

HYVs. High-yielding varieties.

Hz. Hertz. *See* frequency.

I

I. Iodine.

IAA (indole-3-acetic acid). A growth-regulating hormone (auxin) produced by many plants.

IAEA. International Atomic Energy Agency.

IBP. International Biological Programme.

Ice Age. Period of prolonged cold climatic conditions characterised by snow and ice sheets that persist throughout the year in regions free from summer ice and snow in non-glacial periods, and by major extensions of glaciers. The cause of ice ages is uncertain, but M. Milankovich (Yugoslav geophysicist) proposed in the 1950s a correlation between climatic change and variations in the Earth's orbit and the inclination of its axis. Alternative hypotheses are based on changes in solar activity and on changes in the atmospheric carbon dioxide content (the "greenhouse effect"). There have been many ice ages in Earth history, dating back as far as the Precambrian in which at least 15 major groups of ice ages (periods of glaciation that experience remissions) occurred in the Laurasian and Gondwanaland continents. Further groups occurred during the Carboniferous, with glaciation confined to the southern hemisphere. Throughout history, central and southern Africa have experienced glaciation more frequently than any other land area. The third main group of ice ages occurred during the Quaternary, the most recent being the Pleistocene ice age that receded about 10000 years BP.

ice anvil. The anvil-shaped top of a cumulonimbus cloud, formed from minute ice crystals.

ice cap. Ice sheet.

ice evaporation level. That level in the atmosphere at which the temperature is sufficiently low for water to change between the gaseous and solid state with, at most, a very transitory liquid phase. The formation of ice directly from saturated air occurs at any temperature below $-40°C$ and if ice crystals in air at or below this temperature encounter very dry air they will change to the gaseous state by sublimation.

ice nuclei. Freezing nuclei.

ice sheet (ice cap). The largest form of glacier, ice sheets cover extensive areas and are often thick enough to bury all but the highest peaks of entire mountain ranges. Almost all of Antarctica is covered by ice that locally is 2500 m thick and the Greenland ice sheet is more than 3000 m thick. Smaller ice sheets occur in Iceland, Spitsbergen and other Arctic islands and still smaller ones in the highlands of Norway.

Ichneumon. *See* Parasitica.

Ichthyosauria. An order of large, marine, Mesozoic reptiles. They were fish-like, with four paddle-shaped limbs, a vertical tail-fin and elongated jaws bearing many teeth.

ICSU. International Council of Scientific Unions.

IDA. International Development Association. *See* International Bank for Reconstruction and Development.

identical twins. Monozygotic twins.

idiobiology. The study of individual organisms.

IFC. International Finance Corporation. *See* International Bank for Reconstruction and Development.

igneous. Refers to rocks formed as a result of solidification of magma. Igneous is contrasted with sedimentary and metamorphic (*see* metamorphism), the other two fundamental groups of rocks. Igneous rocks are classified on their silica content (into acid, intermediate and basic), the minerals present (into ultrabasic, alkaline and calc-alkaline), the grain-size of the groundmass (i.e. ignoring phenocrysts), and the rock texture. There is a plethora of names: some of the commoner ones are granite, microgranite, rhyolite, granodiorite diorite, andesite, syenite, gabbro, dolerite, basalt and peridotite.

ignimbrite. A pyroclastic rock composed of unsorted pumice and other material, characteristically with glass shards, formed by the explosive disintegration of pumice, which has been flattened. Ignimbrites often show flow-structures and range from loose granular deposits to completely glassy, obsidian-like rocks: many rhyolites have been reinterpreted as ignimbrites. No eruption producing an ignimbrite has been recorded, but it is believed that it would be related to one producing nuées ardentes.

illuviation. The deposition and precipitation of material leached and /or eluviated from the A soil horizon into the B horizon. *See* soil horizon.

ilmenite. The main ore mineral of titanium, with the formula $FeTiO_3$. Ilmenite is widely distributed in rocks but the major ores result from magmatic differentiation in anorthosites and in placer deposits. Nearly all the titanium produced is used as the oxide, a white pigment, in paints. Titanium alloys are used in the aerospace industry.

imaginetic centre. Term coined by Alvin Toffler in *Future Shock* (published in 1970) to describe places where people noted for their creative ability might be provided with technological assistance and encouraged to examine present or anticipated crises and to speculate freely about the future.

imago. The adult, sexually mature, stage of an insect.

imbricate. Refers to a structure in which tabular mass (e.g. flat pebbles, thrust sheets) overlap one another.

IMCO. Intergovernmental Maritime Consultative Organisation.

index fossil. The fossil by whose name a particular unit of rock strata (called a zone), often containing a characteristic assemblage of fossils (known as a biostratigraphic zone) is known. A zone represents a period of time during which these organisms had reached a particular evolutionary stage and zone fossils are used for correlating chronologically rocks of different facies. A good zone fossil should be common and distinctive, have had a wide geographical range and tolerated varying environments and have had a relatively rapid rate of evolution. Important examples include Trilobita (Cambrian System), Graptolite (Ordovician and Silurian Systems), Ammonites (Jurassic and Cretaceous Systems).

index mineral. A mineral whose first appearance in passing from lower to higher grades of metamorphism indicates that the zone in question has been reached.

IMF (International Monetary Fund). *See* International Bank for Reconstruction and Development.

Imhoff tank. Tank in which sedimentation treatment for sewage is combined with anaerobic biological treatment, sewage entering into an upper chamber, solids settling through slots into the lower digestion chamber, and sludge removal being automatic. Imhoff tanks are used extensively in Australia.

immission. The reception of pollution from a remote source of emission.

immunity. The ability of an organism to combat infection by parasites. *See* phagocyte, antibody, interferon, phytoalexin, lysozyme.

impedance. (*a*) The quantity which determines the amplitude of a current of given voltage in an alternating current circuit, expressed as:

$$Z = \left(R^2 + \left[L\omega - \frac{1}{C\omega} \right]^2 \right)^{\frac{1}{2}}$$

where Z = impedance, R = resistance, L = self-inductance, C = capacitance, ω = angular frequency (a constant equal to $2f$, where f = frequency of the alternating current). (*b*) (Acoustics) A measure of the complex ratio of force or pressure to velocity. *See* characteristic impedance.

Impennae (Penguins). Aquatic birds of the southern hemisphere which have lost the power of flight and swim mainly by means of the fore-limbs which are modified as flippers. Many make no nest, and some incubate their eggs by carrying them on their webbed feet. *See* Sphenisciformes.

impermeable. *See* permeability.

impingement. Bringing of matter into contact. Most often applied to dust impinging on a collector, such as a cyclone collector. *See* dust collector.

implantation (nidation). The initial attachment of the mammalian embryo to the uterus of the mother.

impounding reservoir. *See* reservoir.

imprinting. Rapid and stable learning occurring in a young animal, which results in reaction to a particular object (e.g. a duckling becomes imprinted on its mother and subsequently follows her).

improductive forest. Forest that is incapable of yielding products other than fuel because of adverse conditions. Forest that grows slowly or whose trees are dwarfed or stunted.

impulse turbine. One of the two principal types of turbine, in which the whole available head is transformed into kinetic energy before reaching the wheel.

inbreeding. Endogamy.

incaparina. A new food made by adding oilseed (maize and cottonseed meal enriched with vitamins A and B) protein concentrates to a staple cereal food, developed by the Institution of Nutrition for Central America and Panama (INCAP) for use in developing countries.

inceptisols. *See* soil classification.

incinerator. Equipment in which solid, semi-solid, liquid or gaseous combustible materials are burned as a method of disposal. The technique has been applied to many types of industrial and domestic wastes. If the waste will not support combustion, auxiliary fuel is added. There are many incinerator types to deal with different applications.

incipient lethal level. LD_{50}. *See* application factor.

incised meander. A meander where the lateral curve in the river channel remains the same, while the channel itself cuts deeper vertically. *Cf.* ingrown meander.

income elasticity of demand. The mathematical expression of the relationship between changes in income and demand for particular commodities. Starting with surveys of actual consumer behaviour, this can be expressed by the formula: $\dfrac{dE}{d Y}\dfrac{Y}{E}$ where E is the expenditure on the commodity, Y is income, dY is the change in income and dE is the change in expenditure. If the result is less than unity, increases in income will not result in proportional increases in expenditure. The technique is based on the first study of family budgets, published in 1857, by the German civil servant and statistician Ernst Engel (1821–96). Engel's Law states that the poorer a family is, the greater the proportion of its income that will be spent on food.

incompetent bed. A layer of rock which accommodates itself to the shape of interbedded competent beds during folding.

indehiscent. Applied to fruits which do not open spontaneously to release seeds. *See* achene, berry, caryopsis, drupe, nut.

independent assortment (independent segregation). The Second Law of Mendel which asserts the chance distribution of alleles (*See* allelomorphs) to gametes. Independent assortment does not apply to genes lying on the same chromosome. *See* linkage.

independent segregation. Independent assortment.

index species. An organism adapted only to a narrow range of environmental conditions, which is used to characterise those conditions.

Indian rice. Wild rice.

Indicative World Plan for Agricultural Development (IWP). A comprehensive scheme drawn up in the 1960s and published in 1970 by the Food and Agriculture Organisation of the UN (FAO), whose aim is to increase the production and availability of food in developing countries. It provides the strategy for the introduction of new high-yielding varieties of cereals and the technology to grow them, which has been nicknamed the 'Green Revolution'.

indicator species. Species which indicate the presence of certain environmental conditions. See Biotic index, *Clematis vitalba*, Coliform bacteria, Faecal streptococcus, Saprobic classification.

indifferent species. Species that has no pronounced affinities for any community.

indigenous. Native or original to an area.

indole-3-acetic acid. IAA.

indore. Method of making compost (*See* composting) devised by Sir Albert Howard while he was working at an agricultural research station at Indore, India. Most garden compost is made today according to his method or some variant of it.

induration. The process of hardening of sediments through the action of cementation, pressure, or heat.

industrialisation. Economic process whereby an increasing proportion of national or regional income is derived from industrial manufacturing associated with increasing investment in manufacturing industry. As industrialisation proceeds, economic activity becomes more concerned with processing, and therefore adding value to, primary products, and less with primary production (mining, agriculture, etc.).

industrial melanism. The occurrence of dark (melanistic) forms of animals in industrial areas. The dark form increases through natural selection at the expense of the normal pale variety because the former is more effectively camouflaged in blackened surroundings. Industrial melanism has occurred in the Peppered Moth (*Biston /Pachys betularia*) and in spiders in northern England.

industrial waste. Solid materials discarded from trading, commercial and industrial premises and requiring disposal. It can be divided roughly into five categories: (*a*) General factory rubbish uncontaminated by factory process waste; (*b*) Relatively inert process wastes; (*c*) Flammable process wastes; (*d*) Acid or caustic wastes; (*e*) Indisputably toxic wastes.

Indus Valley. The site of an early advanced civilisation whose agriculture depended on sophisticated irrigation and drainage systems. *Cf.* Fertile Crescent.

infiltration. Process by which water seeps into the soil. *See* infiltration capacity.

infiltration capacity. The rate at which water penetrates the soil, governed by the texture of the soil, vegetation cover and the slope of the ground. *See* field capacity.

inflorescence. A flowering shoot, comprising stems, stalks and bracts, as well as flowers.

infraneustronic. *See* neustron.

infrared photography. Photography based on exposure to radiation in the infrared waveband, outside the visible light spectrum. It is used increasingly for environmental surveying from the air and reveals details of heat distribution, waste discharges and atmospheric conditions not revealed by conventional photography.

infra-sonic. Applied to sounds whose frequency is below the range of audio frequency.

infusoria. Formerly a collective term for all the minute organisms found in infusions of organic substances such as hay. Now usually synonymous with Ciliophora.

ingrown meander. The curve in a stream channel that is enlarged laterally at the same point at which the channel cuts downward. *Cf.* incised meander.

initially-complex model. See model.

initially-simple model. *See* model.

inlier. An outcrop of older rock surrounded by younger rock. Inliers may be the result of faulting, folding, deep erosion, or a combination of these processes.

inositol. A fat-soluble substance that may be a necessary constituent of the diet of some animals, possibly including Man. Inositol is necessary for the growth of isolated human tissue, but it is possible that in most animals (including Man) it is normally synthesised in the body. *See* vitamins.

inquiline. An animal which lives in the home of another animal of a different species, sharing its food. *See* commensalism.

Insecta (Hexapoda). The largest group of animals, a class of about 7000000 species of Arthropoda. Most are terrestrial, at least as adults. Typically they have three pairs of legs, two pairs of wings, one pair of antennae, compound eyes and the body divided into head, thorax and abdomen. *See* Anoplura, Aphaniptera, Apterygota, Coleoptera, Dermaptera, Dictyoptera, Diptera, Ephemeroptera, Hemiptera, Hymenoptera, Isoptera, Lepidoptera, Mallophaga, Mecoptera, Neuoptera, Odonata, Orthoptera, Phasmida, Plecoptera, Psocoptera, Strepsiptera, Thysanoptera, Trichoptera.

insecticide. Chemical used to kill insects. *See* carbamate, dinitro groups of pesticides, derris, nicotine, organochlorine and organo-phosphorus groups of insecticides, pyrethrum, quassia, systemic insecticides.

Insectivora. *See* Insectivore.

Insectivore. (*a*) An insect-eating (insectivorous) animal or plant (e.g. sundew); (*b*) A member of the order Insectivora, a primitive group of small placental mammals, including shrews, mole and hedgehog.

inselberg. An isolated hill of steep-sided form caused by exfoliation processes.

in situ. Term used to describe a fossil, mineral or rock found in its place of deposition, growth or formation.

insolation. The reception of solar radiation at a surface.

instar. Any stage of development between moults in an insect larva or nymph.

instinct. Innate (unlearned) behaviour, consisting of an elaborate system of reflexes which, when activated in response to internal or external stimuli, produces a fixed pattern of action.

insulin. A sulphur-containing hormone produced by the Islets of Langerhans in the pancreas of vertebrates. Insulin stimulates the conversion of glucose to glycogen (*see* carbohydrate) and fat. Deficiency results in excess blood-sugar (diabetes mellitus). *See* adrenaline.

intensity. (*a*) The strength of an energy field. (*b*) The rate of flow per unit area. Sound intensity, expressed in watts per square metre, for plane or spherical free progressive waves is equal to $\dfrac{p^2}{\rho c}$, where p = sound pressure, ρc = the characteristic impedance of the system.

interbedded. Laid down in sequence between one layer and another. The term is also used to describe a repetitive sequence of two or more rock types.

intercalary meristem. *See* meristem.

interception. Rainwater caught and held on the leaves of trees and vegetation before reaching the ground and returned to the atmosphere by evaporation.

interferon. A protein made by animal cells when they are invaded by viruses. Interferon is non-specific and inhibits reproduction of viruses.

interfluve. The area of land between two adjacent streams.

interglacial period. Period of retreat during an ice age.

Intergovernmental Maritime Consultative Organisation (IMCO). The international body that regulates many aspects of the operation of ships on the high seas, including the pollution of the sea. In 1972, for example, IMCO drew up the International Convention for the Prevention of Pollution from Ships (the London Convention) forbidding the discharge of certain substances at sea. The earlier Oslo Convention deals with the discharge of substances resulting from sea-bed operations. *See* Barcelona Convention.

intermediate. A term applied to igneous rocks containing intermediate amounts (commonly set at 50 to 60 %) of silica (SiO_2) in their chemical composition. Most of the silica is in the form of silicate minerals, such as feldspars, micas, amphiboles, pyroxenes, and feldspathoids, and there is less than 10% quartz. Syenite, andesite, diorite and trachyte are intermediate rocks. In petrology, intermediate is contrasted with acidic, basic and ultrabasic.

intermediate host. *See* host.

intermedin. A hormone secreted by the pituitary gland. It stimulates the expansion of pigment cells, causing darkening of the skin in some fish, Amphibia and reptiles.

intermittent sampling. Sampling successively for limited periods of time throughout an operation or for a predetermined period of time.

internal combustion engine. Engine powered by the combustion of a fuel in an enclosed space, the power being used to produce mechanical motion. The high temperatures generated lead to the production of such pollutants as nitrogen oxides, carbon monoxide and hydrocarbon residue.

International Atomic Energy Agency (IAEA). A United Nations agency concerned with all aspects of atomic energy and commercial and scientific uses of radioisotopes. It is a partner with FAO in the Division of Atomic Energy in Food and Agriculture. IAEA has its headquarters in Vienna, Austria.

International Bank for Reconstruction and Development (World Bank). A bank formed as a result of the 1945 United Nations Monetary and Financial Conference at Bretton Woods, New Hampshire, USA, to facilitate trade and development. The Bank now has 117 members, and works in association with other agencies, including development banks for Africa, Asia and Latin America, and International Finance Corporation (IFC) and the International Development Association (IDA). The International Monetary Fund (IMF) was also founded at Bretton Woods, with the aim of reforming and stabilising currencies.

International Biological Programme (IBP). A world study of biological productivity and human adaptability initiated by the International Council of Scientific Unions, to run from 1964 to 1974. The study was divided into 7 sections: productivity of terrestrial communities, production processes on land and in water, conservation of terrestrial communities, productivity of freshwater communities, productivity of marine communities, human adaptability, and the use and management of biological resources. More than 40 countries participated, many of which undertook special studies to complement and extend the general investigation.

International Council of Scientific Unions (ICSU). A non-governmental organisation, based in Paris, which encourages the exchange of scientific information,

270

initiates programmes requiring international scientific cooperation (e.g. the International Biological Programme) and studies and reports on matters relating to the social and political responsibility and treatment of scientists.

International Development Association (IDA). *See* International Bank for Reconstruction and Development.

International Finance Corporation (IFC). *See* International Bank for Reconstruction and Development.

IGY. International Geophysical Year.

International Geophysical Year (IGY). An international programme of cooperative research, involving more than 70 nations, and conducted during the period (of 30 months) from July, 1957 to December, 1959 after the 18 month programme planned originally had been extended by a further 12 month programme called International Geophysical Cooperation. The programme was organised by the International Council of Scientific Unions and was directed toward a systematic study of the Earth and its environment, covering 11 fields: aurora and airglow, cosmic rays, geomagnetism, glaciology, gravity, ionospheric physics, longitude and latitude determinations, meteorology, oceanography, seismology and solar activity. The IGY is recognised as the basis for the Antarctic Treaty, dedicating the Antarctic to peaceful uses and signed by 12 nations active in that region and, through the ICSU, IGY generated new international organisations to continue the work it had begun.

International Hydrological Decade. A ten-year programme of international scientific cooperation in research on water problems, planned and directed by UNESCO, which ran from January, 1965. National committees were formed in 96 countries and among the Decade's scientific achievements are a preliminary survey of sediment transport to the oceans providing data concerning more than 120 river basins, a study of the water budget, energy balance and circulation of one of the Great Lakes, and a study of the hydrology of the Chad Basin, in Africa. Postgraduate training courses in hydrology were started in a number of countries as part of a general acceleration in the training of hydrologists generated by the programme.

International Monetary Fund (IMF). *See* International Bank for Reconstruction and Development.

International Planned Parenthood Federation (IPPF). International non-governmental organisation with headquarters in London, which aims to study and publicise matters related to family planning and population policies. National Family Planning Associations are affiliated to the IPPF, which is recognised officially by the UN.

International Referral System. A programme for the exchange of information on environmental problems, started as part of the Action Plan agreed at the 1972 UN Conference on the Human Environment.

International Union for Conservation of Nature and Natural Resources (IUCN). An independent international body, founded in 1948, with headquarters at Morges, Switzerland. It promotes and initiates scientifically-based conservation measures and cooperates with United Nations and other inter-governmental agencies, and with its sister organisation the World Wildlife Fund, which exists primarily to raise and allocate funds, mainly by national appeals.

International Whaling Commission (IWC). The international body that regulates commercial whaling by setting annual quotas for each species.

International Years of the Quiet Sun (IQSY). An international cooperative programme for the study of the Earth's environment based on observations made during 1964 and 1965, when solar activity (the number of sunspots) was at a minimum. IQSY (the initial letters were transposed because it was believed that in this form the acronym would be easier to pronounce) covered meteorology, geomagnetism, aurora, airglow, ionosphere, solar activity, cosmic rays, aeronomy and space research, with emphasis on space measurements. The IQSY programme was planned under the auspices of a body formed by the International Council of Scientific Unions.

IQSY. International Years of the Quiet Sun.

interoceptor. A sense organ (e.g. a receptor in the wall of the gut) which detects stimuli originating inside an animal's body or resulting from substances introduced into the body. *Cf.* proprioceptor.

intersex. An animal which is intermediate between male and female because of abnormality in the sex chromosomes or hormones (*See* freemartin). Its cells are all genetically identical, unlike those of a gynandromorph (*see* gynandromorphism).

interstadial period. A relatively minor period of ice retreat during a major advance of the ice during an ice age.

intraclinal. *See* topotype.

intrinsic rate of increase (r). $r = b - d$, where b is the instantaneous birth rate and d is the instantaneous death rate in a population.

introgression (introgressive hybridisation). The infiltration of genes from one species or population into another, closely related one. This is accomplished by hybridisation and subsequent back-crossing.

introgressive hybridisation. Introgression.

intrusion. (*a*) A body of igneous rock which invades re-existing rock. (*b*) The process of formation of an intrusive body. Intrusive bodies are classified according to their size, shape and relationship to the country rock (the rocks they invade). Batholith, stock, boss, plug, laccolith, lopolith, sill and dyke are the most commonly used terms for different varieties of intrusions.

inverse density dependence. A proportionate decrease in mortality, or increase in fecundity, as population density increases.

inversion. A condition in which air temperature does not decrease with height. Normally, air temperature decreases as height increases (*see* adiabatic lapse rate). Where, locally and invariably in still air, cold air lies beneath warmer air, or the temperature of the air remains constant with increasing height, rising warm air from the surface will be trapped when it encounters air at the same temperature and loses its buoyancy. This will tend to reinforce the inversion, so that inversions tend to be very stable. The upper limit of the inversion, at which rising warm air is held, is the inversion layer.

inversion layer. *See* inversion.

Invertebrata. All animals not belonging to the Vertebrata.

inverted relief. A landscape in which synclines form high ground and anticlines low ground.

invisible hand. A concept proposed by Adam Smith (1723–1790) in *The Wealth of Nations* in which he advocated free trade. If every individual seeks to promote his own economic advantage, then he will unconsciously promote the public good, supporting domestic rather than foreign industry for his own security, and producing as much value as he can. Thus he is 'led by an invisible hand to promote an end which was no part of his intention'. *Cf.* Tragedy of the Commons.

involucre. A number of free or united bracts.

involuntary muscle. *See* muscle.

involution (*a*) A decrease in size of part of an organism (*Cf.* Hypertrophy). (*b*) The development of abnormal forms in an old culture of micro-organisms. (*c*) (Bot.) Rolling in at the edges in organs such as leaves.

iodine (I). Element. A blackish-grey, crystalline solid that is very volatile and gives off a violet vapour. It is slightly soluble in water and readily soluble in alcohol (to give 'tincture of iodine'). Compounds are found in seaweeds and sodium iodate ($NaIO_3$) occurs in Chile saltpetre. It is used in medicine, in chemical analysis, and photography, and the radioisotope $^{131}_{53}I$ (with a half-life of 8.6 days) is used to diagnose and treat disorders of the thyroid gland. AW 126.90; AT. No. 53; SG 4.95; mp 114°C; bp 184°C.

ion. An atom or molecule which has gained or lost one or more electrons and thus has a negative or positive electrical charge.

ionisation. The process whereby atoms acquire an electrical charge through the gain or loss of electrons.

ionosphere. Layer of the upper atmosphere extending upwards from about 80 km above the surface, in which atoms tend to be ionized by incoming solar radiation.

IPPF. International Planned Parenthood Federation.

iridescence. *See* corona.

iron. (*a*) A meteorite composed mainly of nickel-iron alloys (*b*) Element Fe (Ferrum). AW 55.847; At.No. 26; SG 7.86; mp 1535°C. The magnetic properties of iron are modified in natural ores. *See* lodestone, haematite, steel, smelting. Iron is an essential micronutrient.

iron ore. *See* haematite.

iron pan. *See* hardpan.

iron pyrites. Pyrite.

irradiation. Exposure to rays, esp. to ionizing radiation (*see* ionisation).

irritability. The characteristic ability of organisms to respond to stimuli.

isallobaric winds. *See* wind classification.

island arc. Arcuate belts of andesitic and basaltic (*see* andesite, basalt) volcanic islands. Island arcs are paralleled by oceanic trenches, and are associated with the deep-focus earthquakes of a Benioff Zone and with intense magnetic and gravitational anomalies.

isobar. A line drawn on a map connecting points of equal surface atmospheric pressure.

isobilateral. Applied to leaves in which the blade lies more or less vertically and has the same structure on both sides. *Cf.* heterogamy (anisogamy).

isocline. A fold in which the adjacent limbs are parallel.

isogamete. *See* gamete.

isogamy. The production of gametes which are all alike. This occurs uncommonly in Algae, Fungi and Protozoa. *Cf.* heterogamy (anisogamy).

isogeneic (isogenic, syngeneic). Having the same genes. *Cf.* allogeneic.

isogenic. Isogeneic.

isohaline. A line drawn on a map connecting points of equal salinity.

isohyets. A line drawn on a map connecting points that receive equal rainfall.

Isokontae. *See* Chlorophyta.

isolating mechanisms. Mechanisms (e.g. geographical isolation, sterility of hybrids) which tend to prevent gene exchange between related populations, so facilitating evolutionary divergence.

iso-leucine. Amino acid with the formula (CH_3). $CH_2CH(CH_3)CH.(NH_2)$. COOH. Molecular weight 131.2.

isoline. Isopleth.

isomer. (*a*) One of two or more chemical compounds with the same chemical formula but possessing different properties due to the different arrangement of atoms within the molecule. (*b*) In nuclear physics, one of two or more nuclei with the same atomic number and mass number, but different energy states.

isophytochrones. Lines drawn on a map to connect places with growing seasons of the same length.

isopleth (isoline). A line drawn on a map through points of equal value for the phenomenon plotted.

Isopoda. *See* Malacostraca.

Isoptera (Termites, white ants). An order of primitive, mostly tropical insects (Exopterygota) which live in colonies with highly developed caste systems. Termites inhabit tunnels in the ground, inside wood, or in large mounds which they construct out of wood or earth. They do serious economic damage because they feed on cellulose and may destroy wooden buildings, living trees, paper, etc.

isoseismal line. An imaginary line joining points of equal intensity of earthquake shock, and normally forming a closed curve round the epicentre. Isoseismic lines can be deduced from questioning witnesses or from a field survey, or from closely-spaced seismographs. Earthquake intensities are measured on the 12-point Modified Mercalli Scale, which goes from instrumental (1) to catastrophic (12).

isostasy. Equal standing. A concept that an overall constancy of mass (m) of the Earth's crust is maintained above a theoretical level of compensation within the Earth, so that in principle any column of identical cross-sectional area above that level will have the same mass. Thus, as mass is the product of volume and density ($m = vd$), variations in volume (v) on the surface, such as a mountain range, are compensated for by variations in density (d) under these surface features. The mountain range will be itself less dense and have a less dense root than the thinner, denser crust underlying the ocean floor. This relates to the uniformity of the Earth's gravitational attraction (g) over the Earth's surface, except in localised areas of gravity anomaly, where geological processes have artificially lowered or raised the crust out of isostatic balance. For example, areas depressed by ice sheets in Scandinavia acquired a negative gravity anomaly and, over the last 10 000 years, have been and still are rising to regain isostatic equilibrium. Land may also rise to compensate for loss of load when material is eroded from the tops of mountains and transported to the sea. This process is known as isostatic readjustment. *See* raised beach platform.

isostatic readjustment. *See* isostasy.

isotonic solution. *See osmosis.*

isotropic. Exhibiting uniform properties throughout, in all directions.

Itai-Itai. 'Ouch ouch'. A disease that leads to bone deterioration caused by cadmium poisoning that has occurred in Japan and so is known by its Japanese nickname.

IUCN. International Union for Conservation of Nature and Natural Resources.

IWC. International Whaling Commission.

IWP. Indicative World Plan for Agricultural Development.

J

J. Joule.

jasper. A form of chalcedony.

Java Man. *See* Homo.

Jeanneret, Charles Edouard (1887–1965) 'Le Corbusier'. One of the most influential architects and urban planners of modern times. Swiss by birth, but French by training, he studied and later worked in Paris, after a short stay in Berlin. He regarded the city as a machine and produced an ideal design for the 'contemporary city' to accommodate three million persons. He advocated dwellings with flat roofs and roof gardens. His concepts influenced the design of many buildings and cities and he produced plans for Algiers, Saõ Paulo, Rio de Janeiro, Buenos Aires, Barcelona, Geneva, Stockholm, Antwerp and Moscow.

JET. Joint European Torus.

jet propulsion. Propulsion by the discharge of a fluid through an orifice, the reaction to this producing a thrust in the direction opposite to that of the discharge. The technique is best known for its application in aircraft, where the fluid is the heated product of the combustion of air and kerosene.

jetstreams. Cores of fast-moving air, with speeds of 100 to 200 knots, which occur near the tropopause in temperate latitudes. They are a few thousand metres in depth and some tens of kilometres wide and move horizontally, but often in an irregularly wavy pattern, from west to east in both hemispheres. A sharp difference in temperature occurs across them, the warmer air lying to the right of the jet stream in the northern hemisphere and to the left in the southern.

JOIDES (Joint Oceanographic Institutions for Deep Earth Sampling). The JOIDES Deep-Sea Drilling Project consists of coring the Earth's crust, using the drilling ship *Glomar Challenger*. The expeditions, involving successive multinational teams of scientists, have yielded immensely important results (e.g. information on sea-floor spreading rates, deep sea sedimentation, the history of the Mediterranean, and underwater resources). The JOIDES project was one outcome of the abandoned Project Mohole, whose aim was to drill through the crust to the Moho and beyond.

joint. A fracture in rock, often across bedding planes, along which little or no movement has taken place. *See* columnar joints.

Joint European Torus (JET). An EEC project to develop a fusion reactor.

Joint Oceanographic Institutions for Deep Earth Sampling. JOIDES.

Joule (J). The derived SI unit of energy, being the work done when a force of 1 newton displaces a point 1 metre, and the work done per second by a current of 1 ampere flowing through a resistance of 1 ohm. Named after James Prescott Joule (1818–89).

Juncaceae. Monocotyledonous herbaceous plants which live mainly in wet or cold situations. Rushes (*Juncus* spp.) the chief members of the family, form valuable fodder for sheep in hill pastures poor in grasses, and are used to make baskets, mats, etc. The flowers of the Juncaceae are more complete than those of the grasses, having a perianth of six brownish or green segments.

Jurassic. Refers to the middle period of the Mesozoic Era, usually dated as beginning between 190 and 205 million years ago and lasting about 60 million years. Jurassic also refers to the rocks deposited during the Jurassic Period: these are called the Jurassic System, which is divided into Lower (or Lias), Middle (or Dogger), and Upper (or Malm). The Lower Jurassic is divided into four stages (Hettangian, Sinemurian, Pliensbachian, and Toarcian), the Middle into two (Bajocian and Bathonian) and the Upper into five (Callovian, Oxfordian, Kimmeridgian, Portlandian or Volgian, and Purbeckian).

juvenile water. Water arising from underground magmatic sources.

K

K. (*a*) Kelvin (*b*) Potassium.

k. Kilo-

Kainozoic. Cenozoic.

kame. The term kame has been used to describe a wide range of depositional forms of glacial drift, including eskers, moraines, steep-sided hummocks, alluvial cones, valley-side terraces, and crevasse-fillings. Most, but not all kames are composed of stratified sand and gravel. Kame is also a Scottish word for a long, steep-sided ridge.

kame terrace. A terrace-like deposit of fluvioglacial sediment along the side of a valley, formed by streams flowing between a glacier and the sides of its trough.

kaolin (china clay). A decomposition product of feldspar, essentially hydrated aluminium silicate (kaolinite is $Al_2Si_2O_5(OH)_4$). Feldspar itself occurs in granite, from which matrix the clay is extracted. Deposits are very limited globally. China clay mining is an open-cast process, often producing deep workings, since the decomposition producing the clay proceeds from the lowest levels upwards, so that the quality of the clay often improves at greater depths. The white spoil heaps, made from sand and micaceous wastes, have created spectacular landscapes of low-lying white 'plains', deep blue 'lakes' and very conical white 'mountains'. Modern techniques lead to spoil heaps built in lifts of about 20 m, giving a stepped effect, and in South West England the heaps are now seeded with grasses as soon as work on them ends. The wastes can be used for construction and they have been shown experimentally to support plant growth. Some other metals, including tin, are extracted as by-products of china clay working. The principal uses of china clay are in paper making, as a filler in rubber and plastics, in medicines, and in the manufacture of china.

karst. Limestone topography, where solution channels have formed deep caverns and depressions occur where the roofs of these caves have collapsed. A rough, uneven terrain.

karyogamy. The fusion of two nuclei in syngamy.

karyokinesis. Mitosis.

karyotype. The chromosomal characteristics (number, shape, size) of the body cells of an individual or species.

katabatic. Wind caused by cold air flowing downhill, which occurs most commonly when a hillside cools at night. The air in contact with the surface is cooled and flows down into the valley. Katabatic is the opposite of anabatic. *See* frost hollow.

katabolism. Catabolism.

katadromy. Catadromy.

kata-front. A front at which cold air is descending significantly. *See* cold front.

keel. (Bot.) A ridged lower petal, or fused petals, resembling the keel of a boat (e.g. as in the pea family). *See* Leguminosae.

kelvin (K). The SI unit of thermodynamic temperature. That temperature, pressure and volume of water at which the solid, liquid and gaseous states are in balance (the triple point) contains 273.16 kelvins. A temperature expressed in K is equal to the temperature in degrees C minus 273.15°C, the interval between kelvins and degrees Celsius (Centigrade) being identical. The name 'degree kelvin' and the symbol '°K' have been discontinued, the correct usage now being simply 'kelvin' or 'K'.

Kenaf (Ambari Hemp). *Hibiscus cannabinus.* A tropical plant which will also grow in temperate climates and is cultivated for its fibre. In some countries (e.g. New Zealand, the USSR and the USA) it is being considered as a crop for making pulp for the paper industry. Like a number of other fibre plants it is commonly called hemp, although it is unrelated to *Cannabis sativa*, true hemp.

keratin. A sulphur-containing protein constituting epidermal products such as nails, hooves, claws, hair, beaks, feathers and the outermost layer of skin and horn in vertebrates. Keratin also occurs in the skeletons of some sponges and Coelenterata.

keratinised. Applied to cells that are dead and have been transformed into scales consisting largely of keratin.

kerogen. Insoluble organic material found in sedimentary rocks. Microscopically, kerogen appears to be macerated plant remains, and chemically it differs from crude oil in its high content of oxygen and nitrogen.

kerosene (kerosine, paraffin oil). A mixture of hydrocarbons obtained in the distillation of petroleum. Used as a fuel for lighting, heating, and in some engines.

ketones. A series of organic compounds with the general formula RR'C:O, where R and R' are univalent hydrocarbon radicals (groups). Acetone is $(CH_3)_2CO$, ethyl methyl ketone is $CH_2(CH_3)_2CO$.

ketose. A monosaccharide (*see* carbohydrate) which contains a ketone group.

kettle-hole. A depression in glacial drift caused by the melting of ice which once formed part of the deposit.

kettle lake. Lake filling a kettle-hole.

kg. kilogram.

khamsin (gibleh). (*a*) *See* dust storm. (*b*) *See* sirocco.

kidney iron-ore. A variety of haematite occurring in a kidney-shaped form.

kieselguhr. Diatomaceous earth.

killas. A miners' term for the slates and phyllites in South West England.

kiln. Furnace in which the heating operations do not involve fusion. Most frequently used for calcining and also for the firing of bricks, pottery and refractory products.

kilo- (k). Prefix used in conjunction with SI units to denote the unit x 10^3.

kilocalorie. 1000 calories, or 1 Calorie. A unit of heat, now replaced by the joule. 1 calorie was the heat required to raise the temperature of 1 gram of water from 14.5°C to 15.5°C, equal to 4.1868 joules. *See* calorie, calorific requirement.

kilogram (kg). The SI unit of weight, 1000 grams. Defined in terms of the international prototype in the custody of the Bureau International des Poids et Mesures at Sèvres, near Paris.

kimberlite. An altered peridotite containing mica and pyroxenes found in blue ground. Some contain diamonds.

Kimmeridgian. A stage of the Jurassic System.

kinaesthetic sense. The ability to detect movement by means of proprioceptors.

kinase. An enzyme which activates another enzyme (e.g. thrombokinase. *See* clotting of blood).

kinesis. A random locomotory movement of an organism in response to a stimulus, the direction of movement bearing no relation to the position of the stimulus. *See* taxis.

kinetin. One of the cytokinin group of plant hormones.

kinetochore. Spindle attachment.

kinetoplast. *See* centriole, cilia, flagellum.

kinetosome (kinetoplast, blepharoplast, basal body). *See* centriole, cilia, flagellum.

King Crab. Xiphosura.

kingdom. *See* classification.

kingfishers. Alcedinidae.

kinomere. Spindle attachment.

klendusity. The ability of an organism to avoid disease because of the way it grows (e.g. when the susceptible stage in the life history does not coincide with the seasonal occurrence of a parasite).

klepto-parasite. *See* parasitism.

knot. One nautical mile per hour.

kopje. South African term for butte. Many kopjes are capped by dolerite sills.

Krakatau (Krakatoa). Volcanic islet in the Sunda Strait, Indonesia, between Java and Sumatra, which erupted very violently on May 20th, 1883, emitting very large amounts of particulate matter and gases into the atmosphere. *See* natural pollutant, sulphate, sulphur dioxide, Mount Agung.

Krebs' Cycle. Citric acid cycle.

Krebs, Sir Hans Adolf (1900–   ). German-born British biochemist who won the 1953 Nobel Prize for medicine and pharmacology jointly with Fritz Lipmann for his contributions to biochemistry and in particular for his discovery of the citric acid cycle in the metabolisation of carbohydrates.

kwashiorkor. Deficiency disease caused by protein deprivation. The name means 'the sickness the child develops when another is born' and the disease is caused most frequently in very poor communities by premature weaning of infants on to an inadequate diet when the birth of a new infant creates a more imperative claim for the mother's milk.

L

Labiatae. A cosmopolitan family of dicotyledonous plants, mostly aromatic, square-stemmed herbs (e.g. dead-nettles) or small shrubs. Many (e.g. thyme, sage, mint, origanum) are used as culinary herbs, others (e.g. lavender, rosemary) yield oils and perfume.

Labyrinthodontia. A large group of extinct Amphibia with teeth in which the dentine and enamel were much folded.

laccolith. An igneous body with a dome-shaped upper surface and a flat base, which is concordant with the strata into which it is intruded.

Lacertilia (Sauria). *See* Squamata.

lactic acid. An organic acid produced during the metabolism of many types of cell (e.g. vertebrate muscle fibre, bacteria causing the souring of milk). *See* glycolysis.

lactoflavin. Riboflavin.

lactogen. Lactogenic hormone.

lactogenic hormone (lactogen, prolactin, luteotropic hormone, LTH). A hormone produced by the pituitary gland of vertebrates. It stimulates the secretion of progesterone from the corpus luteum and initiates milk production in mammals. In pigeons it stimulates the production of cropmilk.

lactose. *See* carbohydrate.

lacustrine. Pertaining to a lake.

ladybirds (ladybugs). Coccinellidae. *See* Coleoptera.

Lagomorpha (Duplicidentata). Rabbits and hares. An order of very successful herbivorous animals, with chisel-shaped, continually-growing incisor teeth used for gnawing. (In contrast to Rodentia the upper pair are accompanied by a smaller, second pair.) Lagomorphs are serious pests in many parts of the world, including Australia, where rabbits were introduced in the 18th century. *See* myxomatosis.

lagoon. Pond or pit for the disposal of toxic and useless wastes from industry or intensive livestock husbandry.

lahar. (*a*) A mudflow of water-saturated volcanic ash (*b*) The deposit formed by a mudflow of water-saturated volcanic ash.

laissez-faire. An economic doctrine of non-interference by government or other institutions with the free play of market forces. The underlying theory holds that human affairs, like processes in nature, are subject to general laws which operate for the good of the system as a whole. If each individual is permitted to work for his or her economic advantage, then this will tend toward the general good and harmony will be achieved. The concept was first enunciated, and the phrase coined ('*laissez faire, laissez passer*') by the Physiocrats in the 18th century and it came to be supported by many leading economists, including Adam Smith (1723–1790), David Ricardo (1772–1823), Thomas Robert Malthus (1776–1834), John Stuart Mill (1806–1873) and John Locke (1632–1704).

Lake Pedder. An isolated lake in south west Tasmania, whose character is unique. The lake became the centre of controversy in the late 1960s, when the Hydro-Electric Commission of Tasmania produced plans to build a hydro-electric scheme in the banks of the Gordon River, which would have destroyed Lake Pedder by flooding it to make a much larger lake.

Lake Superior-type iron ore. An alternative name for banded ironstones from the region in which they were first studied.

Lamarck, Jean Baptiste de (1744–1829). A French biologist who advanced the first clear theory of evolution, based on the inheritance of acquired characters and more especially acquired habits. He maintained that if, in response to the effects of the environment, part of an animal's body was subjected to an unusual degree of use or disuse, this organ would become modified and the modification would be handed on to the progeny. He postulated, for example, that the long neck of the giraffe has evolved because the shorter-necked ancestor stretched upward for its food, thus lengthening the neck, a development that was inherited by its offspring. Similarly, the poorly developed wings of flightless birds had evolved beccause they ceased to be used for flight. Lamarck's theory of evolution was superseded by that of Charles Darwin *Cf.* Lysenko.

Lamellibranchiata (Pelecypoda). A class of molluscs, including oysters, cockles and mussels, which have a bivalve shell. They are aquatic, feeding and breathing by means of currents of water set up by cilia. *See* Teredo.

lamiation. A layer society in a climax (e.g. a forest with a canopy of medium density may have five or six lamiations above ground level). *See* layers.

lamies. A layer socies of an associes.

laminar motion. One of the two types of motion that occur in fluids (the other being turbulence) in which individual particles of the fluid follow smooth, well-defined paths, the majority of them moving in the same direction.

Lamprey. *See* Agnatha.

lamp shells. Brachiopoda.

lancelets. Amphioxus. *See also* Acrania.

lanceolate. (Bot.) Shaped like a lance, i.e. narrow and tapering.

land breeze. *See* breeze.

land-fill. The disposal of refuse by tipping it on land. Often the refuse is used to fill in old mine workings or low-lying land, to reclaim land from water, or to create a feature on flat land. If the refuse is deposited in prepared trenches or holes, over which earth can be heaped at the end of each day, this is called controlled tipping in Britain and sanitary land-fill in the USA.

landing and take off cycle (lto cycle). All the operations performed by an aircraft between the time it is at a height of 915 m on its landing approach and the time it reaches the same height after its next departure. The Cycle is used in the measurement of and the setting of limits to the rate of emission and duration of pollutants released into the air, leading to standards that relate to specified pollutants and to particular types and sizes of aircraft.

landnam. A short period of time during which a forest was cleared and the clearing occupied by swidden farmers. After a few years the clearing was abandoned and the forest regenerated.

land reclamation. The treatment of any unusable land (e.g. slag heaps, quarries, gravel pits, etc.) usually by filling with refuse or levelling, until the land can be brought into productive use.

landsat. *See* Earth Resources Technology Satellite.

land use. The deployment of land for any use. Competition for limited areas of land requires the establishment of priorities among claims, which is the object of land use planning.

lanugo. The fine covering of hair on a foetus which is shed before birth.

lapilli. Pyroclastic fragments ejected by a volcano and from 4 mm to 32 mm in diameter.

lapse rate. The rate at which temperature decreases in the atmosphere as height above the surface increases. Within the troposphere the average lapse rate is 6°C to 8°C per km. *See* dry adiabatic lapse rate, adiabatic lapse rate, wet adiabatic lapse rate.

Laridae. Gulls and terns. A family of colonial seabirds with webbed feet and powerful flight. Most gulls alight on the water to seize food, whereas terns dive from the air.

larva. An immature stage of an animal, usually very different in form from the adult, able to feed itself, but very rarely capable of sexual reproduction. *See* nymph, cysticercus, trochophore, leptocephalus, metamorphosis, paedogenesis, nauplius, pluteus, veliger.

laser. Light Amplification by Stimulated Emission of Radiation. A device that produces a powerful, highly directional, monochromatic, coherent (i.e. its waves are in phase) beam of light. Lasers consist of a transparent cylinder with a reflecting surface at one end and a partially-reflecting surface at the other. Light waves are reflected back and forth, some of them emerging at the partially-reflecting end. The light source may be a ruby, whose chromium atoms are excited by a flash lamp so that they emit pulses of highly coherent light, or a mixture of inert gases which produce a continuous beam, or a cube of treated gallium arsenide which emits infrared radiation when an electric current passes through it.

lasion. A more or less dense growth of interdependent organisms attached to surfaces submerged in water, above the bottom. *See* periphyton.

latent heat of freezing. About 334.7 joules per gram, released when water freezes. It causes large hail to become wet when falling through a cloud of supercooled droplets many degrees colder than 0°C because the freezing of some droplets on to the hailstones soon raises their temperature to 0°C.

laterite. Earthy, granular or concretionary mass, chiefly of iron and aluminium oxides and hydroxides, occurring as a layer or as scattered nodules in tropical soils. *See* soil classification.

lateritic soil. Soil containing laterite. Such soil is often unsuitable for agriculture, being weathered until the laterite forms a hard, impermeable layer. *See* laterisation, soil classification.

laterisation. The process of forming laterite, resulting from the removal of silica, alkalis and alkaline earths and the consequent enrichment in the iron and aluminium compounds that form laterite. The process may be accelerated by the removal of surface vegetation, which serves to bind together soil particles and to maintain the cycling of nutrients. Removal leaves the soil exposed to rain. *See* lateritic soil, soil horizons, leaching, soil classification, eluviation.

Late Stone Age. Neolithic.

latimeria. *See* Crossopterygii.

laurilignosa. Laurel forest and laurel bush. Subtropical rainforest.

lava. A molten liquid extruded on to the surface of the Earth, or the solidified product subsequently formed. Acidic lavas are more viscous than basic lavas and give rise to more violent eruptions.

law of minimum. Plants have minimum requirements of certain mineral salts (*see* macronutrients, micronutrients). If a soil does not supply this minimum, the plants that require those particular salts in amounts exceeding what the soil does provide cannot grow, regardless of the abundance of other nutrients.

lawn sand. A mixture of mercurous chloride (calomel) and ferrous sulphate used to control moss in turf. Harmful to fish.

layers (*a*) (Ecol.). Strata. Four horizontal vegetation layers are recognised in plant ecology. These are the tree layer, shrub (bush) layer, field (herb) layer, and ground (moss) layer. (*b*) (Bot.) 1. Tissue layers (e.g. annual rings). 2. A branch which strikes root and can grow independently after becoming detached from the parent plant. (*c*) (Zool.) Tissue layers, e.g. germ layers.

LD$_{50}$. That concentration of a substance which causes the death of half of a population exposed to it within a given period of time. *See* application factor.

leachate. *See* leaching.

leaching. The removal of the soluble constituents of a rock, soil or ore by the action of percolating waters. Leaching is a major process in the development of porosity in limestones, in the secondary enrichment of ores and in the formation of soils.

leaching field. A system of open pipes in covered trenches that permits effluent from a septic tank to enter surrounding soil.

lead (Pb). Element. A metal, compounds of which are used in many industries, and may accumulate in biological systems. In humans, small doses produce behavioural changes, larger doses paralysis, blindness and eventual death. Its additon to petrol has caused its widespread distribution, esp. in the vicinity of major roads. It may also be a serious pollutant near heavy metal smelters. Pregnant women are especially vulnerable. AW 207.19; At. No. 82; SG 11.34; mp 327.4°C.

leaf insect. Phasmida.

learning. Behaviour which enables an animal to modify its actions with changing situations as a result of experience. *See* instinct, conditioned reflex.

Le Chatelier Principle. For general physical systems, if the system is subjected to a constraint, whereby the equilibrium is modified, a change takes place which, if possible, partially annuls the constraint.

lecithin. A complex fatty substance containing phosphorus (phospholipid). It is a constituent of unit membranes of plant and animal cells, and is present in the yolk of most eggs. *See* fat, lipin.

Le Corbusier. Jeanneret, Charles Edouard.

lectotype. A specimen chosen from original material as a substitute for a missing type specimen. *See* paratype.

lee. The side away from wind, ice or current. *See* stoss.

leeches. Hirudinea. *See* Annelida.

lee waves. Mountain waves.

legume. (*a*) The fruit of a member of the Leguminosae. A pod which releases its seeds by splitting lengthwise down both sides, often explosively. (*b*) Crops of plants belonging to the Leguminosae (e.g. clover, peas, beans).

Leguminosae. A very large, cosmopolitan family of dicotyledonous plants, including herbs, trees, climbers, water plants, etc. They are of great value as soil-enriching crops because most bear root nodules containing symbiotic, nitrogen-fixing bacteria (*Rhizobium* spp.). Many species are important sources of human food and animal fodder (e.g. peas, beans, lupins, vetches, clover). *Acacia* and other trees yield timber and resin. *Crotalaria* is cultivated for its fibre (Bombay or Madras hemp). Ulex (Gorse, Furze, Whin) is a spiny evergreen shrub (a xerophyte) which covers wide areas of Britain. *See* root nodules.

lek. A social arrangement among animals in which a group of males assembles on a traditional breeding ground. An order of dominance is established among them and the dominant male then undertakes most of the breeding with visiting females. Lek breeding has been observed in such birds as the sage grouse, and in animals such as the Uganda kob. *See* polygyny.

lemurs, flying. Dermoptera.

lenticel. A pore, through which gases can diffuse, in the corky outside layer of a woody stem.

lentic water. Standing water (i.e. in lakes, ponds, marshes, etc.) *See* lotic water.

Lepidoptera (butterflies and moths). A large order of insects (Endopterygota) whose wings and bodies are covered with minute scales. The larvae (caterpillars) are largely herbivorous and almost all the adults feed on nectar, by means of a long, coiled, tubular proboscis. The order includes serious pests such as the Cabbage White Butterflies (*Pieris*), Clothes Moth (*Tinea*), Flour Moth (*Ephes-*

tia) and Cotton Bollworm (*Platyhedra gossypiella*). Lepidoptera are important agents of pollination, and some control weeds. The moth *Cactoblastis cactorum* was successfully introduced into Australia to control the Prickly Pear. *Bombyx mori* is the commercial silk worm moth. Butterflies (Papilionoidea) can be distinguished from moths by their clubbed antennae and the absence of a mechanism for hooking fore and hind wings together. The division of Lepidoptera into butterflies and moths is artificial, and these features may be found separately in some moths.

Lepidosauria. Squamata.

leptocephalus. The transparent, pelagic larva of the eel, which migrates across the Atlantic from the breeding grounds off the West Indies (Sargasso Sea) to the European rivers where it matures.

leste. Sirocco.

lethal gene. A gene which kills the individual bearing it by causing some disorganisation of metabolism (e.g. the production of a toxin). If the lethal gene is recessive it kills only individuals homozygous for it.

leucine. Amino acid with the formula $(CH_3)_2CH.CH_2CH.(NH_2).COOH$. Molecular weight 131.2

leucism. A type of albinism involving the absence of pigment from fur or feathers but not from other parts of the body.

leucocratic. Consisting mainly of light-coloured (i.e. felsic) minerals. *Cf.* melanocratic.

leucocytes. *See* blood corpuscles.

leucoplast. *See* plastids.

leukaemia. A disease of the blood, often fatal, in which white blood cells (leucocytes) proliferate.

levanter. *See* mistral.

leveche. Sirocco.

levee. A ridge of alluvium at the side of a river where the coarse, suspended load carried by flooding water has been dumped as the current is checked at the bank.

level. (Geol.) The horizon at which an ore-body is being worked, and often used to cover all the working at one horizon.

ley. *See* grassland, Graminae.

LH. Luteinising hormone.

Li. Lithium.

lias. (*a*) Strata deposited during Lower Jurassic times. (*b*) An interbedded sequence of shale and limestones.

Lice. *See* Anoplura.

lichen. A dual organism formed by a green or blue-green alga and a fungus living together symbiotically (*see* symbiosis). Lichens live attached to rocks, tree trunks, etc., and may be encrusting, upright and branched, or leaf-like and flat. They are important primary colonisers of bare surfaces, and are dominant in mountainous or arctic situations where conditions are too harsh for other plants to grow. Reindeer Moss is the staple diet of reindeer in Lapland. Most lichens are very sensitive to air pollution.

LIDAR (Light Detection And Ranging). A technique for detecting and tracking chimney plumes, often after they have ceased to be visible to the naked eye. The equipment consists of a laser transmitter, which emits brief, high intensity, pulses of coherent light, and a receiver which measures the energy of that light when backscattered by atmospheric aerosols and interprets the range.

life form. (*a*) The characterisitic form of a plant or animal species at maturity (e.g. herb, worm). (*b*) *See* Raunkiaer's Life Forms.

life system concept. That part of an ecosystem which determines the existence and abundance of a particular population, based on the study of a subject population and its environment as a single system.

life table. A description of the age-specific survival of cohorts of individuals in relation to their age or stage of development.

Light Detection And Ranging. Lidar.

light minerals. In petrology, usually refers to the felsic minerals (e.g. quartz, feldspar and feldspathoids) present in a rock.

light-water reactor. (LWR). Nuclear reactor that uses light water (as opposed to heavy water, or deuterium) as a coolant and moderator. There are two types, the boiling-water reactor and the pressurised-water reactor.

lignin. A complex, non-carbohydrate substance deposited in the walls of many plant cells, esp. those of woody tissue, imparting strength and rigidity.

lignite. *See* coal.

lignosa. Vegetation which is made up of woody plants.

Lilac. *Syringia vulgaris. See* Oleaceae.

Liliaceae. A large, cosmopolitan family of monocotyledonous plants, mostly herbs with rhizomes, bulbs or corms and flower parts arranged in whorls of three. They include ornamental plants (e.g. lilies, croci and daffodils) and food plants (e.g. onions and asparagus).

limb. (Geol.) The part of a fold between adjacent hinges.

lime. Calcium hydroxide used in agriculture to raise the soil pH and to provide calcium. (*See* macronutrients). Also used in cements, as quicklime (calcium oxide) and in mortars.

lime blow-in process. Japanese process for removing sulphur dioxide from flue gases, using dry limestone scrubbing, producing no recoverable sulphur compound, or reagent.

lime gypsum process. Japanese process for removing sulphur dioxide from flue gases, using limestone as a reagent in a wet scrubber to produce gypsum.

limestone. A sedimentary rock composed largely of the carbonate minerals calcite or, less frequently, aragonite. Limestones can be divided into: (*a*) those formed by the evaporation of aqueous solutions (e.g. tufa, travertine and oolitic limestone (*see* oolite), (*b*) those formed of calcite produced biochemically by living organisms (e.g. chalk, coquina, coral limestone and forameniferal (*see* rhizopoda) limestone, and (*c*) those formed by the mechanical accumulation of fragments of pre-existing limestone. Many limestones have been formed by a combination of two or more of these methods, usually followed by a complex diagenesis. Diagenesis or metasomatism may convert the limestone into a dolomite-rock. Limestones are important as aquifers, as reservoir-rocks for hydrocarbons, as building stone and aggregate and, with clay, for making cement.

limestone-injection process. German process for removing sulphur dioxide from flue gases with 20 to 35 % efficiency, by injecting hydrated lime, and ground lime, into the boiler close to the flame, at temperatures between 900°C and 1500°C.

limestone scrubbing. British process for removing sulphur dioxide from flue gases, by reacting sulphur dioxide with an aqueous slurry of calcium hydroxide in a scrubber, to produce calcium sulphate and sulphite.

limiting factor. An environmental factor (e.g. temperature) which restricts the distribution or activity of an organism or population.

limestone rock. *See* cement.

Limits to Growth. *See* Club of Rome.

limnetic. (*a*) Living in the open water of a lake or pond. (*b*) Inhabiting marshes, lakes or ponds.

limnic. Refers to sediments deposited in freshwater lakes. Limnic also refers to the environment of deposition and, in geology, is contrasted with paralic.

limnology. The study of the physical, chemical and biological components of fresh water.

limonite. A mixture of amorphous and cryptocrystalline hydrated iron oxides and hydroxides formed as a weathering product of iron-bearing minerals. Important occurrences of limonite are as bog-iron-ore, laterite, ochre and in gossan.

lindane. (HCH, benzene hexachloride, BHC, gamma BHC). Organochlorine insecticide used to control insect soil pests, aphids, grain weevils, mites, etc. Harmful to bees, fish, livestock, etc. and persistent, though less so than DDT. Somewhat phytotoxic.

linear growth. Exponential growth.

line squall. A belt of severe thunderstorms accompanying a cold front.

linkage. The association of non-allelomorphic genes (*see* allelomorph) so that they are inherited together and do not show independent assortment. Linkage occurs between two genes when they are situated on the same chromosome. *See* crossing over.

linnaeite. One of the major ore minerals of cobalt, with the formula Co_3S_4, linnaeite is found in hydrothermal veins, typically with nickel and/or copper sulphides. Cobalt is used in various high-temperature alloys.

Linnaeon. Linneon.

Linnaeus, Carolus (1707–1778). Swedish biologist, the author of *Systema*

Naturae, in which he evolved the system of binomial nomenclature still used today. Each organism was designated by a generic name, referring to the group to which it belonged, and a specific name for itself (e.g. *Homo sapiens*). Linnaeus did not believe in evolution and his classification was an artificial one, based on obvious external characteristics and not on relationships between organisms.

Linneon (Linnaeon). A species according to Linnaeus, often a superspecies.

linoleic acid. Constituent of vitamin F. *See* vitamin.

linolenic acid. Constituent of vitamin F. *See* vitamin.

linseed. *See Linum.*

Linum (Linaeceae). A genus of dicotyledonous plants, including flax (*Linum usitatissimum*) which yields linen fibre, obtained by retting away the softer tissues in water. Linseed oil and 'cattle cake' are obtained from the seeds.

Linum usitatissimum. Flax. *See Linum.*

linuron. Translocated and soil-acting herbicide of the urea group used to control corn marigold in cereals and many weeds in root crops. It can be irritating to the skin and eyes, and is harmful to fish.

lipase. An enzyme which breaks down fat into its constituent fatty acid and alcohol.

lipid. Fat.

lipide. Fat.

lipin (lipine). A complex fatty substance (e.g. lecithin, a phospholipid found in unit membranes).

lipoid. Fat.

liquefied petroleum gas. LPG.

liquid–metal fast breeder reactor (LMFBR). A nuclear reactor that uses plutonium dioxide or uranium dioxide as a fuel and molten sodium and potassium as a coolant. Being a breeder reactor, it uses no moderator.

lithification. The conversion of sediment into consolidated sedimentary rock. Many processes of diagenesis result in lithification but diagenesis may continue after lithification has taken place.

lithite. *See* statocyst.

lithium (Li). Element. A light, silvery-white alkali metal, the lightest solid known. Chemically it resembles sodium, but is less active. It is used in alloys and may be used as a fuel in fusion reactors AW 6.939; At. No. 3; mp 179°C; SG 0.534.

lithocyst. Statocyst.

lithosere. The stages in a plant succession beginning on an exposed rock surface.

lithosols. Surface deposits with no soil horizons developed. See Soil classification.

lithosphere. (*a*) The solid portion of the Earth i.e. the crust and mantle. (*b*) The crust of the Earth. (*c*) The sialic (*see* sial) upper crust. (*d*) The portion of the Earth above the Low Velocity Zone. Since the advent of the theory of plate tectonics, 'lithosphere' has generally been used in this last sense, as the rigid plates are thought to move around on the rather more fluid material of the Low Velocity Zone.

litoral. Littoral.

Little Ice Age. Fernau glaciation.

littoral currents. Currents moving parallel to the shore, produced within the surf zone by waves breaking obliquely on the shore. *See* littoral drift, longshore currents, longshore drift, beach drift, shore zonation.

littoral drift. The movement of material in littoral currents i.e. within the surf zone. *See* longshore currents, longshore drift, beach drift.

littoral (litoral) zone. Usually regarded as: (*a*) The part of a lake extending from the shore down to the limit for rooted vegetation. (*b*) The intertidal zone of a sea. *See* shore zonation. *Cf.* sublittoral zone.

Liverwort. *See* Bryophyta, gemmation. Hepaticae.

Llandeilian. The third oldest series of the Ordovician System in the UK.

Llandoverian. The oldest series of the Silurian System in Europe.

Llanvirnian. The second oldest series of the Ordovician System in the U.K.

lm. Lumen.

LMFBR. Liquid-metal fast breeder reactor.

LNG. Liquefied natural gas. *See* natural gas.

load on top. A method of carrying oil in tankers, where the cargo floats on top of water in one of the tanks (the 'slop tank'). The slop tank water is used first to wash out the other tanks and residues from ballast water are added to it. The oil carried on top of this water can be refined. The system is designed as an alternative to the highly polluting process of discharging ballast residues and water that has been used for washing tanks into the sea.

local allocation tax. A system of block grants made by central to local government (e.g. in Japan) calculated as a percentage of revenue from certain national taxes (in Japan, 52 % of national income, corporate business and liquor taxes). *See* local transfer tax.

local forecast. A weather forecast for a small enough area for the weather to be uniform over it, and usually involving reference to features of weather which are not shared with neighbouring localities.

local nature reserve. *See* nature reserve.

local transfer tax. A system of block grants made by central to local governments (e.g. in Japan) calculated as a percentage of national revenue from certain taxes (in Japan, local road tax, petrol tax, tonnage tax on sea ports). *See* local allocation tax.

lociation. A local variant of an association, which differs in the composition of its important subdominants and influents.

locus. The position on a chromosome occupied by a particular gene of its allele. *See* allelomorph.

Locust. *See* Acrididae, Orthoptera.

lode. A vein, or system of veins, usually containing a metalliferous mineral, which, with any intervening country rock, may be mined as one unit.

lodestone. A variety of magnetite which acts as a compass needle when it is free to swing.

lodging. The mechanical collapse of a cereal crop. This can occur in heavy rain or hail, and the crop may be rendered more vulnerable if the plants are very tall and unable to give adequate mechanical support to the weight of the ears. Many traditional varieties of cereals responded to additional applications of fertiliser by growing to a greater height as well as by producing a heavier ear, thus making them liable to lodging, and this was remedied by the breeding of short-stemmed varieties. A lodged crop is difficult to harvest, much of the grain falls from the ear and is lost, and if the grain becomes moist it may begin to germinate, so making it unsuitable for most uses.

loess. A very fine, unconsolidated and unstratified, permeable material, generally grey to buff in colour. Loess is composed mainly of angular particles of such minerals as quartz, feldspar, and calcite, in a clay matrix. Most loess seems to be wind-blown glacial rock-flour although some appears to be of desert origin. Loess makes a fertile soil.

logarithmic reproduction curve. A reproduction curve with logarithmic axes.

logarithmic wind profile. A wind profile of a type frequently observed in the lower atmosphere, which is dominated by roughness effects, taking the form: $\dfrac{u}{u_*} = \dfrac{1}{k} \log_e \left(\dfrac{z}{z_0}\right)$ where u is the mean speed at height z, u_* is the friction velocity, k is Von Karman's constant (about 0.4) and z_0 is the roughness length, a parameter approximately equal to one-thirtieth of the average height of the surface obstacles.

logistic curve. An S-shaped curve on a graph, often characteristic of population growth and stabilisation. The curve rises slowly at first, then more and more steeply, finally flattening at a new level.

London Convention. *See* Intergovernmental Maritime Consultative Organisation.

London smog incidents. Two episodes, in 1952 and 1962, when descending cold air, trapped under very stable inversions caused the Thames Valley to become a reservoir of cool, still, moist air. Fog, with chemical and physical pollutants added, remained stationary for four days in 1952 and five days in 1962. In 1952, 4000 people died and in 1962 about 700 died, though most were elderly and suffering from chronic respiratory complaints.

long range forecast. A weather forecast of a speculative nature for a period beyond that for which physical and mechanical laws can be the basis of forecasting because of the complexity or uncertainty of starting conditions. Thus a forecast for a month or season is 'long range'.

longshore currents. Currents moving parallel to the coast, generated by waves, winds and tides. *See* littoral currents, littoral drift, longshore drift, beach drift.

longshore drift. Movement of material in longshore currents along the coast, generally not only in the surf zone. *See* littoral currents, beach drift.

looping. The behaviour of a chimney plume, when large-scale thermal eddies bring puffs of concentrated pollutants to the ground for a few seconds before carrying them aloft again.

lophodont teeth. Cheek teeth which have the cusps on the crown joined up to form ridges. This type of tooth is typical of many hoofed mammals (e.g. rhinoceros, which has a simple lophodont pattern; horse, with a complex pattern) and rodents. *Cf.* bunodont, selenodont.

lopolith. A saucer-shaped igneous intrusion.

Los Angeles smog. *See* smog.

lotic water. Flowing water (rivers, streams). *Cf.* lentic water.

loudness. The intensity of a sound as perceived by the human ear, dependent on sound pressure and frequency. *See* decibel, sound level, perceived noise level.

lower shore. *See* shore zonation.

lowland reach. *See* river zones.

low pressure area. *See* cyclone.

low velocity zone (LVZ). A zone of the Earth where earthquake shock-waves travel at much reduced speeds. The top of the zone varies between 70 and 150 km in depth, and the bottom between 200 and 360 km. The LVZ (or asthenosphere as it is sometimes called) is possibly the zone over which lithospheric plates move.

LPG (liquefied petroleum gas). LPG is composed mainly of butane (C_4H_{10}), propane (C_3H_8), and pentane (C_5H_{12}) and is produced from the gas associated with crude oil. LPG is sold in pressurized containers as 'bottled gas'.

LSD. Lysergic acid diethylamide.

LTH. Lactogenic hormone.

LTO. Landing and take-off cycle.

Ludlovian. The third oldest series of the Silurian System in Europe.

lumen (lm). The derived SI unit of luminous flux, being the amount of light emitted per second in unit solid angle of one steradian by a uniform point source of one candela intensity.

luminous flux. *See* flux.

lung. A highly vascular organ for breathing air. In land vertebrates lungs are derived from branches of the gut. In lungfish (Dipnois) the air bladder is modified as a lung. In terrestrial molluscs the lung is a fold of epidermis enclosing an air space. Spiders and scorpions have lung 'books' in depressions in the body wall.

lungfish. Dipnoi.

Lusitanian flora and fauna. Plants and animals which occur in south west Europe and which, in the British Isles, are confined to the west and south west, mainly in coastal areas (e.g. Pale Butterwort, *Pinguicula lusitanica*).

lustre. The appearance of a mineral in reflected light, usually described by such terms as metallic, submetallic, adamantine (like diamond), vitreous, resinous, greasy, silky, pearly and earthy.

luteinising hormone [LH]. A hormone produced by the pituitary gland of vertebrates. It initiates ovulation, the production of oestrogen by the ovaries, and the development of the corpus luteum in mammals. In males it initiates the production of androgen by the testes.

luteotropic hormone. Lactogenic hormone.

lutite. Argillite.

lux (lx). The derived SI unit of illumination, being the illumination provided by one lumen per square metre.

LVZ. Low Velocity Zone.

LWR. Light-water reactor.

L_x noise levels. Noise levels that exceed the usual, or permitted level for proportion of a measured period. Thus an L_{50} noise level is one that exceeds the usual or permitted level for 50% of the time. *See* decibel.

lx. Lux.

Lycopodiales (Lycopsida, Clubmosses). A group of largely extinct Pteridophyta, abundant during the Carboniferous, when tree-like species (e.g. *Lepidodendron*, common in coal measures) existed. Present-day species are small plants, most superficially resembling mosses. *Lycopodium* and *Selanginella* grow in boggy ground and *Isoëtes* (Quillwort) is an aquatic plant.

Lycopsida. Lycopodiales.

lymphocytes. *See* blood corpuscles.

Lysenko, Trofim Denisovich (1898–1976). Russian geneticist, the chief proponent of neo-Lamarckism, a revival of the theory of the inheritance of acquired characters. Hence, 'Lysenkoism'.

lysergic acid diethylamide (LSD). Hallucinogenic drug ('acid') manufactured from lysergic acid, $C_{15}H_{15}N_2COOH$, a crystalline substance obtained from ergot.

lysine. Amino acid with the formula $NH_2(CH_2)_4CH.(NH_2).COOH$. Molecular weight 146.2.

lysogenic bacterium. A bacterium carrying a non-virulent form of virus (bacteriophage). The genetic material of the virus, attached to that of the bacterium and reproduced when the host cell divides, is known as prophage.

lysosomes. Membrane-bounded particles present in cytoplasm which liberate enzymes responsible for the breakdown of dead or damaged cells (autolysis). Lysosomes are probably involved in dissolution of tissue during metamorphosis.

lysozyme. An enzyme which kills bacteria by destroying their cell walls. Lysozymes occur in mammalian body fluids and are produced by the skin.

M

M. Mega.

m. (*a*) metre. (*b*) milli-.

Ma. Millions of years.

Maastrichtian. A stage of the Cretaceous System.

machair. Calcareous grassland on shell sand along the western coast of Scotland, valuable for grazing.

Mach number. The ratio of the speed of an object to the speed of sound in the undisturbed medium through which the object is travelling. Thus an aircraft travelling supersonically has a Mach number greater than one.

299

macrogamete (megagamete). The larger of the two types of gamete produced during anisogamy. The female gamete. *See* oogamy.

macronutrient. An element or compound needed in relatively large quantities by an organism. Crop plants require amounts ranging from a few kg to a few hundred kg per hectare of carbon, hydrogen and oxygen (supplied from the air) and calcium, nitrogen, potassium, phosphorus, magnesium and sulphur (supplied from the soil). *Cf.* micronutrient.

macrophages. Large, irregularly-shaped cells found in connective tissue and body fluids of vertebrates. Macrophages remove dead cells and foreign particles by engulfing them.

macrophagous. Applied to animals which feed at intervals on pieces of food which are large in relation to their body size. *See* microphagous.

macrophytes. Large aquatic plants (e.g. crowfoot, water lily), as opposed to phytoplankton and other small algae.

macrosporangium. Megasporangium. *See* sporophyll.

macrospore. Megaspore. *See* heterospory.

Mad Hatter's disease. Mental derangement caused by the absorption of mercury in small doses over a long period. The disease affected hatters who used mercury in the manufacture of felt.

Madison Process. Industrial process for fractionating in a dry state the components of mixed urban wastes on site, producing material suitable as a feedstock for a bioplex.

maestrale. Mistral.

maestro. Mistral.

mafic. Refers to the ferromagnesian minerals (i.e. the opposite of felsic) or else to igneous rocks relatively rich in such minerals (i.e. synonomous with basic and melanocratic.

magma. A molten material which is composed of silicates and volatiles (water and gases) in complex solution, and which originates within the lower crust or mantle. Magmas may contain appreciable amounts of solids, and usually have temperatures in the range 700 to 1100°C. Within the molten mass various processes, broadly called magmatic differentiation, operate to produce fractions of different compositions: these fractions can then solidify adjacently, producing mixed magmas and layered intrusions, or one or more fractions may migrate independently. The volatiles become concentrated at the top of the magma and can then migrate through the country rock and/or the Earth's surface producing metasomatism, hydrothermal deposits, fumaroles, etc. Magma which reaches the surface and loses most of its volatile substances is known as lava, and the solidified products of magma are called igneous rocks.

magmatic differentiation. Any process which tends to produce two or more separate fractions in a magma. Processes which have been suggested include the separation of early formed crystals from a melt by gravity settling (suggested for some chromite deposits), and the development of immiscible liquids (suggested for some magnetite and some sulphide deposits).

magnesium (Mg). Element. A light, silvery-white metal, which tarnishes easily in air and burns with an intense white flame to magnesium oxide (MgO). It occurs as magnesite, dolomite, carnallite, and in many other compounds. It is used in alloys, in photography, signalling and in incendiary bombs, and it has medical uses. It is an essential nutrient (*see* macronutrient). AW 24.312; At. No. 12; SG 1.74; mp 651°C.

magnesium oxide process. Japanese process for removing sulphur dioxide from flue gases using a wet scrubber to produce recoverable sulphur dioxide.

magnesium oxide scrubbing chemico. US process for removing sulphur dioxide from flue gases, using a wet scrubber in which the gases react with a slurry of magnesium oxide and magnesium sulphite, to form more magnesium sulphite. The fly ash is removed by settling. The magnesium sulphite is recovered by centrifuging, and calcination with coke or coal makes it possible to recover the magnesium oxide for recycling and sulphur dioxide, which is used to manufacture sulphuric acid.

magnetic anomaly. Fluctuations from the normal value for Earth's magnetic field due to local concentrations or deficiencies of magnetic minerals and changes in the internal magnetic field which is recorded in rocks at the time of their formation. *See* geomagnetic induction, paleomagnetism. The symmetry of magnetic anomalies shown by rocks on either side of oceanic ridges (e.g. mid-Atlantic Ridge) is regarded as substantial evidence for the theory of Sea Floor Spreading.

magnetic flux. *See* flux.

magnetic induction. *See* geomagnetic induction.

magnetite. An iron oxide mineral, Fe_3O_4, of the spinel group and an important source of iron. Magnetite occurs as a widespread accessory mineral of many rocks, and is found in economic deposits in basic igneous rocks as a product of high-temperature magmatic differentiation in metasomatised limestones and in placer deposits. Magnetite also occurs with haematite in the feebly recrystallised banded ironstones. Magnetite is also an important source of the metal vanadium, which is used in iron and steel.

magnetohydrodynamics (MHD). Technique for generating electricity by passing an ionised gas through a magnetic field. Three systems are being considered experimentally: open cycle, which is the most advanced; closed cycle plasma; and closed cycle liquid metal. Theoretical efficiencies of generation could be in the region of 50 to 60%.

magnetosphere. The space surrounding the Earth, or other celestial body, in which there is a magnetic field associated with that body.

magnox. British built nuclear reactor, the first in the world to operate commercially. They use carbon dioxide as a coolant, graphite as a moderator, and derive their name from a magnesium alloy used to clad the uranium fuel rods. The fuel is natural uranium. Eight Magnox stations were built, the first in 1956 at Calder Hall, Cumbria. The Magnox operates at 300°C.

mahogany. *See* Meliaceae.

main sewer. A sewer that collects sewage from a large area.

maize (corn). *Zea mays. See* Graminae, Teosinte.

major quadrat. *See* quadrat, major.

make up water. Water that is used to replenish a system that loses water through leakage, evaporation, etc.

malachite. A copper carbonate mineral, with the formula $CuCO_3Cu(OH)_2$, found in the enriched oxidised zone of a copper sulphide vein which has undergone secondary enrichment and also, rarely, as a cement in a sandstone. Malachite is an important ore of copper.

Malacostraca. A large group of Crustacea of great diversity, with compound eyes and, typically, a tail fan, and a carapace covering the thorax. The most important groups are: (1) Isopoda (e.g. woodlice, water and shore slaters, some fish lice), a large group which includes aquatic, terrestrial and parasitic forms and in which the body is flattened dorso-ventrally; (2) Amphipoda (e.g. sand hoppers, fresh-water shrimps) in which the body is laterally compressed; (3) Decapoda (e.g. crabs, lobster, crayfish, prawns, shrimps) a very diverse group whose members have five pairs of thoracic walking or swimming legs, often with large pincers on the first pair; (4) Euphausiacea (krill) marine, pelagic, mostly luminescent animals which are often abundant and form an important part of the food of whales.

malathion. Organophosphorus insecticide and acaricide used to control aphids, leafhoppers, codling moth, mites, etc. Resistant strains of aphids and mites have appeared. Harmful to fish and bees, but it is one of the less toxic organophosphorus compounds. Toxicity to mammals may be increased by prior exposure to parathion or other organophosphorus compounds which inhibit detoxifying mechanisms.

maleic hydrazide. Growth regulator used to control grass and other weeds on verges and in amenity areas. Enters the plant mainly through the foliage, and inhibits growth by preventing cell division. Has low mammalian toxicity.

malignant pustule. Anthrax.

mallee scrub. A scrub consisting mainly of low eucalyptus bushes, characteristic of the dry sun-tropical regions of the south-east and south-west of Australia.

Mallophaga. Biting lice. Small, wingless, flat-bodied insects (Exopterygota), mostly associated with birds, a few with mammals. Some are important pests of poultry, sheep and cattle, but none affects Man. They do not pierce the skin like the sucking lice (Anoplura) but scrape the skin and chew feathers.

malpais. Bad ground (Spanish).

Malpighian layer. *See* epidermal cells.

Malpighian tubules. The main excretory organs of insects. They are blind vessels varying in number from 2 to 100 or more, lying freely in the body cavity and discharging into the gut a mixture of urine and food residues which forms faeces as it passes along the hind gut. Named after the Italian physiologist, Marcello Malpighi (1628–1694) who is regarded as the founder of the microscopic study of anatomy.

Malthus, Rev. Thomas Robert (1766–1834). English sociologist, who predicted in his *Essay on the Principle of Population* that the human population must ultimately be kept in check by famine, warfare or disease because population increases more rapidly than the availability of food to sustain it. Darwin and Wallace were influenced by his ideas. Neo-Malthusians maintain that the adverse effects of population pressure may be mitigated by the voluntary limitation of population, e.g. by the regulation of fertility.

maltose. *See* Carbohydrate.

mamma. Breast-like cloud formations in the base of anvil cloud, caused by fallout or instability at the cloud base.

Mammalia (mammals). A class of 'warm-blooded' air-breathing vertebrates with many distinctive features, including hair, mammary glands for suckling young, three auditory ossicles and heterodont teeth. *See* Placentalia, Marsupalia, Monotremata.

Mammoth Tree. *See* Taxodiaceae.

manatee. *See* Sirenia.

maneb. Fungicide of the dithiocarbamate group used to control diseases such as potato blight and tomato leaf mould. Can be irritating to the eyes and skin.

manganese. (Mn). Element. A reddish-white, hard, brittle metal. Occurs as pyrolusite and is used in many alloys. It is an essential micronutrient for plants. AW 54.938; At.No. 25; SG 7.20; mp 1244°C.

manganese nodules. Hydrated manganese oxide (with some iron) concretions, up to depths of a few centimetres, covering ocean floor sediments in three major oceans over millions of square kilometres. These were first discovered by the *Challenger* Expedition of 1873–76 and their origin and evolution are still not fully understood. The nodules also contain significant concentrations of metals such as copper, cobalt and nickel.

mangold-wurzel. *See Beta vulgaris.*

mangrove. Trees and shrubs of the genus *Rhizophora*, *Bruguiera* and Avicennia or, more generally, communities dominated by them. They occur along tidal estuaries, in salt marshes, and on muddy coasts in tropical America and Asia, where they form dense thickets. The American mangrove formations consist mainly of the Common, or Red Mangrove (*R. mangle*) of the Rhizophoraceae, and the Black Mangrove (*A. nitida* or *marina*) of the Verbenaceae. Asian formations also include members of other families, such as *Sonneratia* (Lythraceae) and the aculescent Nipa palm. Mangroves produce adventitious aerial roots (or pneumatophores), which descend in an arch, strike at some distance from the parent stem, and send up new trunks, so that the forest spreads in a very dense fashion. The seeds produce a long embryonic root which grows downward while the fruit is still attached to the tree and may take root before the fruit falls, so that the new plant produces shoots almost at once. The roots beneath the mud are aerated by aerial roots which project above the mud and which have minute openings (lenticels) into which air diffuses and passes down to the main roots beneath. Wood from some species is hard and durable, the bark yields a substance used in tanning, the fruit of the Common Mangrove is sweet and wholesome. *See* Rhizophoraceae.

Manidae. Pholidota.

Manihot. (Euphorbiaceae). A genus of herbs and shrubs native to the American continent. M. *esculentus* or *utilissima* is cassava (manioc, mandioca), widely cultivated for its starchy, tuberous roots, from which tapioca is prepared. The juice of some types is poisonous, and must be squeezed out. When evaporated it forms an antiseptic syrup used for preserving meat. (Cassava is very low in sulphur proteins, so subsistence on this food can lead to the deficiency disease kwashiorkor). *M. glaziovii* and other species yield rubber when tapped.

Manila Hemp. *See* Musa.

Manioc. *See* Manihot.

Mantle. The part of the interior of the Earth between the crust and the core. The upper surface of the mantle is the Mohorovičić Discontinuity at a depth of about 5–70 km below the Earth's surface and the lower surface is the Gutenberg Discontinuity at a depth of about 2900 km. The upper mantle is probably composed of ultrabasic rocks, which are the source of most basic and ultrabasic magmas.

mar. Mor.

marasmus. Deficiency disease found most commonly in babies less than one year old and caused by general undernutrition, aggravated by premature weaning. *See* kwashiorkor.

marble. A metamorphosed limestone. Marbles are formed under thermal or regional metamorphism, and the recrystallisation to a sucrosic texture destroys all fossils. Nonetheless, many fossiliferous limestones are sold as 'marbles'.

marine park. A permanent reservation on the sea bed for the conservation of species. Marine parks exist in the USA, Japan, the Philippines, Kenya, Israel and Australia, where two parks have been declared on the Great Barrier Reef.

marl. A calcareous mudstone.

Marram grass. *Ammophila*.

marsh A term often restricted to waterlogged ground with a largely mineral basis, in contrast to the peat of bog and fen.

marsh gas. *See* methane.

Marsupialia (Metatheria, Didelphia). A subclass of mammals with living representatives native only to Australasia and America. In marsupials the placenta either does not develop or is not as efficient as that of the Placentalia and the young are born in a very underdeveloped state. Most female marsupials have a pouch (marsupium) inside which the young are suckled and develop further. The American oppossums are arboreal, with prehensile tails, and usually no pouch. Australian marsupials, lacking competition from placental mammals, show considerable diversity and include herbivores (e.g. koala bear), carnivores (e.g. Tasmanian wolf, devil and tiger-cat), burrowing forms (e.g. wombat, marsupial mole), jumping forms (e.g. kangaroo), gliding forms (flying phalangers), etc.

marsupials. *See* Marsupialia.

marsupium. *See* Marsupialia.

Mastigophora. *See* Flagellata.

mass production. The making of large numbers of identical products in a factory often using production line techniques.

mass spectrometer. Apparatus for separating electrically charged particles in accordance with their masses. Commonly used in chemical analysis, especially of organic compounds.

mass wasting. The downslope movement of weathered rock material by gravity alone. The process may be aided or hastened by the presence of ice and/or water. *See* soil flow, solifluction, creep.

maximum allowable concentration (MAC). The concentration of a pollutant considered (in regulations) harmless to healthy adults during their working hours, assuming that they breathe uncontaminated air the rest of the time.

maximum permissible body burden. The concentration of a radioisotope which will deliver not more than the maximum permissible dose to a critical body organ when breathed or consumed at a normal rate.

maximum permissible dose. The dose of ionising radiation accumulated in a specified time, of such magnitude that no injury may be expected to result during the lifetime of the individual exposed, and no intolerable burden is likely to accrue to society through genetic damage of his or her descendants.

maximum permissible level. A general term covering the greatest degree of contamination that is permitted from any source, but especially from radioactive substances.

mayflies. *See* Ephemeroptera.

maze. *See* Hebb-Williams maze.

MB. *See* Bar.

MCPA (MCP, 4K, 2M). Translocated hormone weed killer used as a herbicide to control many broad-leaved weeds in cereal and grass crops. Not very persistent, breaking down in the soil within a few weeks of application. *See* translocated herbicides.

MCPP. *See* mecoprop.

meadow. A piece of permanent grassland especially one cut for hay. Water meadows are grass fields regularly flooded by river water.

mean, arithmetic. (Statistical) The sum of a set of variables divided by the number of variables in the set. *See* variable.

meander. Freely developed curve in a river. *See* meander belt, meander scroll, ox-bow lake, incised meander, ingrown meander, point bar.

meander belt. The total area covered by a meandering river, usually 15–18 times the width of the river channel when full.

meander scroll. Arc-shaped depressions of former meanders, now partially infilled. *Cf.* ox-bow lake.

mean free path. (*a*) The average distance sound travels in a room between successive reflections. (*b*) The average distance travelled by a molecule in a fluid between collisions with other molecules.

mecoprop (MCPP). Translocated hormone weed killer used to control many broad-leaved weeds in cereals, turf and orchards. *See* translocation.

Mecoptera (scorpion flies). A small, ancient order of mainly carnivorous flies (Endopterygota) some of which have changed little since the Permian. The males often have the abdomen upturned.

median. (Statistical). The value of the variate which divides the total frequency into two halves.

medulla. (*a*) The central part of an organ (e.g. the pith of a stem; the inner part of a kidney). (*b*) Medulla oblongata. The part of the brain in vertebrates which merges into the spinal cord. Its main functions are coordination of impulses from sense organs and regulation of involuntary actions such as heart-beating and breathing.

medullary rays. Thin, vertical plates of living cells running radially through the vascular tissue of stems and roots. Food materials are stored in and conducted radially through medullary rays.

medusa. A disc- or bell-shaped, free-swimming generation in coelenterates (Coelenterata) which reproduces sexually, usually giving rise to the polyp generation. The medusa stage is absent in the Actinozoa and is the dominant or only generation in the Scyphozoa.

medusoid. *See* Coelenterata.

Mega- (M). Prefix used in conjunction with SI units to denote the unit x 10^6.

megachromosomes. Giant chromosomes occurring in the salivary glands and some other tissues of many Diptera, including the fruit-fly *Drosophila*. Each chromosome is greatly thickened, consisting of many identical, parallel chromatids, and each exhibits a pattern of transverse bands marking the arrangement of the genes along its length. These chromosomes have been intensively studied in *Drosophila*, and by associating changes in the banding pattern with genetic differences, many genes have been located on specific chromosomes.

megagamete. Macrogamete.

megalecithal. Telolecithal.

megalopolis. A large urban area produced by the expansion of neighbouring conurbations (e.g. areas of the eastern seaboard of the USA, and the 'North West European Megalopolis' (a term used by the EEC which extends from the Ruhr to the Manchester–Liverpool conurbation.)

megalopolitan corridor. The region between Washington D.C. and Boston, Mass., U.S.A.

megasporangium. *See* sporophyll.

megaspore (Macrospore). *See* heterospory, spore.

megasporophyll. *See* flower, sporophyll.

megatherm. A tropical plant, needing continuously high temperature.

meiosis (reduction division). Cell division during which the chromosome number is halved. It occurs in all organisms which reproduce sexually; in animals during gamete formation and in many plants during spore formation. The process usually consists of two successive divisions. During the first, the chromosomes arrange themselves in homologous pairs on the equatorial plane of the spindle, then the partners move apart to opposite poles of the cell (see Crossing over). The second division resembles mitosis with half the normal number of chromosomes. The final result is the formation of four haploid daughter cells from one diploid cell.

melanins. Dark pigments present in many animals, often in special cells (melanophores, melanocytes). *See* industrial melanism, chromatophores.

melanism. *See* melanins, industrial melanism.

melanocratic. Consisting mainly of dark, i.e. ferromagnesian, minerals. *See* leucocratic.

melanocyte. *See* melanins.

melanophore. *See* melanins.

Meliaceae. A family of dicotyledonous trees and shrubs of warm regions, including many species which yield valuable timber. *Khaya* is African mahogany and *Swietenia* is the mahogany of tropical America.

melting point (mp). The constant temperature at which the solid and liquid phases of a substance are in equilibrium at a given pressure. Melting points are usually quoted for standard atmospheric pressure (760 mm of mercury).

membrane. *See* plasma membrane, unit membrane.

Mendel, Gregor (1822–1884). An Austrian monk who conducted precise experiments in heredity (using the garden pea) and formulated the basic laws of genetics. *See* segregation (Mendel's first law), independent assortment (Mendel's second law).

mendelian population. An interbreeding population.

Mercalli scale. *See* modified Mercalli scale.

mercaptans. Organic compounds with the general formula R-SH, meaning that the thiol group -SH, is attached to a radical, e.g. CH_3 or C_2H_5. The simpler mercaptans have strong, repulsive odours but these become less pronounced with increasing molecular weight and higher boiling points. Mercaptans may be produced in oil refinery feed preparation units as a result of incipient cracking, where the offensive gases are burnt in plant heaters. Mercaptans that arise in cracking units are removed by scrubbing with caustic soda, removed from the caustic soda by stream stripping, then burnt.

Mercurialis perennis (**Dog's Mercury**). *See* Euphorbiaceae.

mercury (Hg) (quicksilver hydragyrum). Liquid metal which damages the nervous system on inhalation or ingestion, especially of its organic compounds. Used extensively in chemical and plastics industries. A.W. 200.59; At.No. 80; SG 13.6; mp -39°C; bp 357°C.

Merioneth series. The third oldest Series of the Cambrian System.

meristem. A region of active cell division in a plant. Many plants (e.g. seaweeds, mosses, ferns) have unicellular meristems, but in flowering plants meristems consists of groups of cells. Cells formed by meristems become differentiated into various tissues depending on the location of the meristem. Apical meristems (growing points) are at the tips of stems, roots and their branches; Cambium lies between xylem and phloem and produces new xylem and phloem during secondary thickening; Phellogen (cork cambium) is a secondary meristem which arises in the cortex of woody plants and produces cells which become corky; Intercalary meristems occur in grasses at the bases of internodes.

merogony. The fertilisation of an egg fragment or of an egg that has no nucleus. In certain marine organisms, an egg that is separated into two parts can develop into two larvae, the larva from the egg fragment that contained no nucleus possessing only the paternal set of chromosomes. An egg fragment treated with a parthenogenetic agent will also develop to the blastula stage.

meromixis. Permanent stratification of water masses in lakes. Sometimes dissolved substances create a gradient of density differences with depth, such that complete mixing and circulation of water masses is prevented.

mesa. A steep-sided plateau of horizontally bedded rock, topped with resistant caprock.

mesarch. All the stages in a plant succession beginning in a moderately damp habitat.

mesh size. Particle size (e.g. of activated carbon granules) as determined by the US Standard Sieve series.

mesoblast (mesoderm). *See* germ layers.

mesoclimate. A local climate effect, over an area several kilometres wide and one or two hundred metres in height, where climate differs from the regional climate.

mesoderm (mesoblast). *See* germ layers.

Mesogloea. *See* Diploblastica.

Mesolithic (Middle Stone Age, Protoneolithic, Epipalaeolithic). The transitional period in the development of human societies between the Palaeolithic and Neolithic. The period occupies different spans of time in different places, and the concept is a developmental rather than chronological one.

mesophilic micro-organism. A micro-organism whose optimum temperature for growth lies between 20 and 45°C. e.g. pathogenic bacteria which infect mammals and birds. *See* psychrophilic, thermophilic.

mesophyll. The tissue inside a leaf blade. In dorsiventral leaves the upper (palisade) mesophyll consists of elongated, tightly packed cells, containing numerous chloroplasts. The lower (spongy) mesophyll consists of cells which contain fewer chloroplasts, and are loosely packed, the air spaces between them communicating with the stomata.

mesophyte. A plant which grows in conditions which are neither very wet nor very dry. *See* hydrophyte, xerophyte.

mesosaprobic. Applied to a body of water in which organic matter is decomposing fast and in which the oxygen level is considerably reduced. *See* polysaprobic, oligosaprobic, catarobic, saprobic classification.

mesosphere. (*a*) That part of the atmosphere extending from the ionosphere to the exosphere, i.e. from about 400 to 1000 km above the surface of the Earth, and sometimes considered to be part of the exosphere. (*b*) That part of the atmosphere between the stratosphere and the thermosphere, i.e. from about 40 to 80 km above the surface of the Earth.

mesothelioma. Major type of lung cancer caused by asbestos.

mesotherm. A plant of warm temperate regions (e.g. maize).

mesothermal. Refers to ore deposits formed from an ascending, essentially aqueous solution, at fairly high pressures in the temperature range 200 to 300°C. Mesothermal deposits are hydrothermal deposits formed in conditions intermediate between those which produced hypothermal and epithermal deposits.

mesotrophic. Applied to freshwater bodies which contain moderate amounts of plant nutrients and are therefore moderately productive.

Mesozoic. One of the Eras of geological time, occurring between the Palaeozoic and the Cenozoic, and comprising the Triassic, Jurassic and Cretaceous Periods. Mesozoic also refers to the rocks formed during this time.

messenger RNA. *See* RNA.

meta-. (*a*) (Geol.) Abbreviation for metamorphosed (*see* metamorphism). e.g. a metabasalt is a metamorphosed basalt. (*b*) Changed or altered. (*c*) (Chem.) Indicates that an organic compound contains a benzene ring substituted in the 1.3 positions. (*d*) Indicates that an acid or salt is the least hydrated of several compounds of the same name.

metabiosis. Beneficial exchange of growth factors among species.

metabolism. All the chemical reactions which take place in a living organism, both anabolism and catabolism. Basal metabolism is the energy exchange of an animal at rest.

metabolite. A substance involved in metabolism. An essential nutrient. *See* antimetabolite.

metagea (arctogea). *See* zoographical regions.

metagenesis. Alternation of sexual and asexual generations in the life cycle of an animal (e.g. many Coelenterata.) *See* alternation of generations.

metal. Substance possessing certain qualities: metallic lustre, malleability, ductility, high specific gravity, good conductivity of heat and electricity. Metals are generally electropositive, and combine with oxygen to give bases. Substances possessing certain of these qualities but other non-metallic properties (e.g. arsenic) are metalloids.

metaldehyde. Molluscicide used to kill slugs and snails.

metalimnion. Thermocline.

metallic. (*a*) *See* lustre. (*b*) pertaining to metal.

metalloid. Substance possessing some of the properties of a metal and some non-metallic properties.

metameric segmentation. Metamerism. *See* segmentation.

metamerism. Metameric segmentation. (*see* segmentation).

metamorphic aureole. The zone of country rock around an igneous intrusion in which new textures and minerals have developed in response to the heat dissipated by the intrusion. The size of the aureole depends on the nature of the country rocks and the size, temperature and volatile content of the intrusion.

metamorphism. The alteration of the texture and/or composition of a rock by the action of heat and/or pressure, and in some cases by the addition of extra material (a process called metasomatism.) Metamorphism is considered to take place in the solid state. It can involve heat alone (producing a metamorphic aureole around an igneous rock), pressure (e.g. along a thrust producing a mylonite), or heat and pressure producing metamorphic rocks over a large area. The last is called regional metamorphism. Increased heat and pressure changes the texture of a rock, and so, for example, a mudstone can change to a slate, phyllite, schist, or possibly a gneiss with increasing pressure and temperature. With increased pressure and temperature the original minerals become unstable and new ones form. From experimental work the stability fields of many minerals are known and so the condition of formation of metamorphic rocks within a metamorphic zone (characterised by an index mineral) can be postulated.

metamorphosis. The transformation of a larval into an adult stage, occurring in Amphibia, some fishes (*see* Leptocephalus) and many invertebrate groups. In insects, incomplete (hemimetabolous) metamorphosis (e.g. in cockroaches) is a gradual process. Complete (holometabolous) metamorphosis (e.g. in housefly) is a rapid and drastic change, involving extensive breakdown of larval tissue and rebuilding of adult organs, which takes place in the pupal stage. *See* lysosomes.

Metaphyta (Embryophyta). A sub-kingdom of plants comprising all groups with multicellular sex organs and an embryo (Bryophyta, Pteridophyta, Spermatophyta).

metaquartzite. A metamorphic (*see* metamorphism) rock consisting of quartz sand grains which have been welded together by pressure. In thin section the quartz can be seen to have been stained, and the contacts between grains to be sutured. Metaquartzite is sometimes called quartzite, but this term also covers orthoquartzite.

metaplasm. Non-living constituents of protoplasm.

metasediment. Metamorphosed sedimentary rock. *See* metamorphism.

metasomatism. The process in which a pre-existing rock is partly or wholly altered by the addition of new material. Metasomatism accompanied by raised temperatures is sometimes termed contact metamorphism.

metatheria. Marsupialia.

metaxenia. The influence of pollen from different parents upon the development of the fruit, causing, for instance, variation in maturation time.

Metazoa. Animals whose bodies consist of many cells, but usually excluding the sponges (Parazoa) *See* Protozoa.

meteoric water. Water that has fallen as rain and percolated down through the vadose zone to the water table. Chemical reactions within the vadose zone introduce sulphate, carbonate, and bicarbonate ions into the water.

meteorite. A solid extra-terrestrial body which has fallen to the surface of the Earth. Meteorites are classified on their chemical compositions into 3 major groups: irons, stony irons, and stones. Irons are alloys of nickel and iron, whereas stony irons also contain silicate minerals. Stones consist predominantly of pyroxenes, with variable amounts of olivine, nickel-iron alloys and other minerals. Stones are divided into chondrites, usually containing small, spherical mineral bodies called chondrules, and achondrites without chondrules. Some chondrites are carbonaceous and contain hydrocarbons, fatty acids and amino-acids. The importance of meteorites is (*a*) their bearing on the composition of the Earth's mantle and core and (*b*) their ages which lie between 4000 and 5000 million years.

meteorology. The study of the atmosphere, its structure, composition and phenomena.

methaemoglobinaemia. Illness caused by the presence in the blood of methaemoglobin, a compound of haemoglobin and oxygen that is more stable than oxyhaemoglobin and so does not yield up its oxygen. In extreme cases methaemoglobinaemia leads to asphyxiation (in infants, the 'blue baby' condition). It can occur in infants, and ruminant animals when nitrates ingested from food or water are reduced to nitrites by bacteria in the digestive system. Nitrite is able to pass through the gut wall into the bloodstream, together with hydroxylamine, another product of the bacterial reduction of nitrate to nitrite, which plays a principal role in the destruction of vitamin A as well as having a haemolytic action that can give rise to anaemia.

metham-sodium (metam-sodium). Soil fumigant of the dithiocarbamate group used to control soil fungi, to kill earthworms and other soil fauna, and to check the growth of weed seedlings. Used to sterilise potting composts and greenhouse soils. Metham-sodium and its decomposition products are irritating to the skin and eyes.

methane. The simplest hydrocarbon (CH_4), product of natural or artificial

314

anaerobic decomposition. In its natural form also known as marsh gas. Methane can be used as a fuel.

methionine. Amino acid with the formula $CH_3.S. (CH_2)_2CH. (NH_2).COOH$. Molecular weight 149.2.

methylene-blue stability test. A test to measure the ability of an effluent to remain in an oxidised condition when incubated anaerobically. The test continues for 5 days at 20°C and if the blue colour formed at the start remains at the end of the period, the effluent is considered stable.

metoecious (metoxenous) parasite. *See* parasitism.

metoxenous (metoecious) parasite. *See* parasitism.

metre (m). The SI unit of length, defined in 1960 as the length equal to 1650763.73 wavelengths in vacuo of the radiation corresponding to the transitions between the levels $2p_{10}$ and $5d_5$ of the isotope $^{86}_{36}$ Kr. 1 metre = 39.3701 inches.

metropolis. The chief city, often the capital, of a country or region.

Meuse Valley incident. An air pollution incident which occurred in Meuse Valley, France, in December, 1930, when cold weather and katabatic winds, together with a temperature inversion, caused fog, contaminated with industrial pollutants, to persist for several days. Several hundred people became ill, about 60 died, and many cattle had to be slaughtered.

Mg. Magnesium.

MHD. Magnetohydrodynamics.

mica. (*a*) A group of silicate minerals with a single perfect cleavage producing characteristic flexible cleavage flakes. Micas are composed of rings of silica tetrahedra forming layers joined by cations such as K, Mg, Fe, Li, Al and Ca. Biotite, a potassium–iron mica, and muscovite, $K_2Al (Si_6Al_2). O_{20} (OH, F)_4$, are the commonest micas. Micas occur in igneous rocks, and muscovite is also common in sediments (*b*) Thermal and electrical insulator, inhalation of fragments of which can cause silicosis.

micaceous haematite. *See* haematite.

micro- (μ). Prefix used in conjunction with SI units to denote the unit $\times 10^{-6}$.

micotrophy. The nutritional relationship between certain fungi and vascular plants and bryophytes. *See* mycorrhiza.

microbial metallurgy. The use of bacteria in metallurgical processes. Bacterial action may possibly be useful in separating metals from some ores, and in producing sulphur from pyrite and gypsum, and sulphuric acid from sulphur.

microclimate. The climate of a very local area.

microcrystalline. Composed of crystals which can be resolved with the aid of a microscope. *Cf.* cryptocrystalline.

microgabbro. A medium grain-size rock of gabbroic composition, more commonly called dolerite in the UK or diabase in North America.

microgamete. The smaller of the two types of gamete produced during anisogamy. The male gamete. *See* oogamy.

microgranite. A medium grain-size acidic igneous rock, having a similar mineralogical and chemical composition to granite. Porphyritic microgranite is usually termed quartz-porphyry, and microgranite showing a small-scale graphic texture, (called micrographic or granophyric) is termed granophyte. Microgranites grade with decreasing grain size into rhyolites. Microgranites form sills, dykes and plugs.

micrographic. *See* graphic, microgranite.

micro-habitat. A small habitat such as a decaying tree stump.

microlecithal. Applied to eggs (e.g. human egg) which contain little yolk. *See* telolecithal.

micrometre. Micron

micron (micrometre, m). One-thousandth of a millimetre, a unit used for measuring microscopic objects. One nanometre (millimicron, nm), one-thousandth of a micrometre, is equivalent to 10 ångströms.

micronutrient. An element or compound required in very small amounts by an organism. Plants require a few grams to a few hundred grams per hectare of iron, manganese, zinc, boron, copper, molybdenum and cobalt. *See* vitamin, trace element. *Cf.* macronutrient.

micro-organism. Any plant or animal too small to be seen by the naked eye.

microphagous. Applied to animals (e.g. Balaenoid whales) which feed, usually continuously, on particles (e.g. plankton) which are minute in relation to their body size. *See* macrophagous.

Micropodidae. Apodidae.

Micropodiformes (Apodiformes). *See* Apodidae.

microrelief. Small scale topography, measured in centimetres.

microsere. All the stages in a plant succession that occur within a microhabitat.

microspecies. A small species or sub-species which shows very little variability (e.g. minor variants in dandelions, roses or blackberries).

microsporangium. *See* sporophyll.

microspore. *See* heterospory, spore.

microsporophyll. *See* flower, sporophyll.

microtherm. A plant which grows in cool temperate regions (e.g. oats)

mictium. A mixture of species such as that occurring in a transition zone between two distinct habitats. *See* ecotone.

middle shore. *See* shore zonation.

Middle Stone Age. Mesolithic.

Midland Hawthorn. *See* Crataegus.

mid-oceanic ridge. *See* ridge.

mie scattering. The scattering of sunlight predominantly in a forward (down-beam) direction by atmospheric particles of more than about 0.1 micron radius. *See* backscatter, Rayleigh scattering, aerosol.

migmatisation. The process of alteration of a high grade metamorphic rock to granite. Migmatisation involves alteration of the composition of the metamorphic rock, chiefly by increasing the sodium and potassium content. Migmatites are rocks consisting of thin alternating layers or lenses of granite and gneiss in which 'ghosts' of the pre-existing banding can often be seen in the granitic parts.

migration. (*a*) (Bot.) The culmination of dissemination in ecesis when the organism becomes established in a new area. (*b*) (Zool.) Movements of animals carried out regularly, often between breeding places and winter feeding grounds. *See* leptocephalus, anadromy, catadromy.

migrule. The unit, or agent, of migration. A diaspore.

mildew. (*a*) A type of fungus causing disease in plants. Powdery mildews (Ascomycetes) grow on the surface of host plants. Downy mildews (Phycomycetes) penetrate deeply. (*b*) A synonym for mould, definition 'a'.

millet. *Sorghum vulgare, Setaria italica, Pennisetum typhoideum* and *Panicum* spp. *See* Graminae.

milli- (m). Prefix used in conjunction with SI units to denote the unit x 10^{-3}.

millimicron. Nanometre.

milling. Pulverisation.

millipedes. *See* Myriopoda.

mimicry. *See* Batesian mimicry, Müllerian mimicry.

Minamata. A bay and town in Japan which in 1959 gave its name to a disease of the central nervous system caused by mercury poisoning and affecting the people consuming the fish and shellfish caught in Minamata Bay. The pollution was traced to the discharge of effluents from the Chisso factory, manufacturing acetaldehyde and vinyl chloride. The effluent accumulated as dimethyl mercury in the sediment in the Bay, a toxic organic mercurial compound, ingested by marine organisms. Between 1953 and 1960, 43 persons died and many more were incapacitated by 'Minamata disease'. The damage was two-fold, direct poisoning of the nervous system and similar teratogenic effects, the toxin being able to cross the placental barrier following ingestion by the mother. The tragedy was particularly severe because fishing was an important industry and the fish and shellfish was a staple protein food of the community.

mineral. A natural, solid, homogenous, inorganic substance with a chemical composition which is either fixed or else falls within a defined range.

minimal area. (of a species). The smallest area that is certain to include a particular species.

minimum factor. The factor that limits the distribution of forest by the intensity of its occurrence. The limit may be imposed by a high intensity or a low one.

minnow reach. *See* river zones.

Miocene. Refers to a sub-division of the Cenozoic Era. The Miocene usually ranks as an Epoch, and follows the Oligocene. The Miocene is thought to have lasted from 26 to 7 million years before the present. Miocene also refers to rocks deposited during this time: these are called the Miocene Series.

miogeosyncline. A geosyncline with a relatively thin sequence and no volcanic rocks lying close to a craton.

mispickel. An old name for the mineral arsenopyrite.

Mississippian. A system used in North American stratigraphy equivalent to the Lower Carboniferous. *See* geological time.

mist. Liquid droplets smaller than 10 microns in diameter generated by condensation. (Meteor.) Suspension of water droplets in the atmosphere that reduces visibility to between 1 and 2 kilometres. *See* fog, smog.

mistral. A dry, cold, northerly wind that blows down the Rhône valley and into the Gulf of Lyons. Similar winds may occur wherever the movement of a cold air mass is obstructed by mountains. They blow through gaps in the mountains, sometimes for days on end, and may be associated with very deep air currents. If strong such winds can be accompanied by falling pressure. The cold northerly wind that blows over Northern Italy in winter is called the maestrale, and the similar cool westerly wind of Corsica and Sardinia is called the maestro. Other names include gregale of Euroclydon (in Greece); tramontana; etesian winds; pampero (in Argentina); and levanter (which blows through the Strait of Gibraltar). *See* southerly buster, föhn winds, bora.

mites. (Acarina) *See* Arachnida.

mitochondria (chondriosomes). Membrane-bounded particles present in the cytoplasm of all eucaryotic cells. Mitochondria contain enzyme systems responsible for providing energy in the form of ATP.

mitosis. (karyokinesis). The normal process by which a cell nucleus divides into two daughter nuclei, each having an identical complement of chromosomes. The following stages are recognised. (1) The prophase, during which the chromosomes (each consisting of two chromatids) condense and become visible; (2) The metaphase, in which the nuclear membrane usually disappears, and a gelatinous elipsoid structure (the spindle), consisting of a number of protoplasmic threads, is formed. The chromosomes become attached to spindle threads while lying in the equatorial plane. (3) The anaphase, during which the chromatids separate and move toward opposite poles of the spindle. (4) The telephase in which the two sets of chromosomes form new nuclei, each surrounded by a membrane. *See* meiosis.

mixed economy. An economic system in which some components operate according to free market forces (*see laissez-faire*) though today these are usually modified, while other components are subject to central control as in a planned economy.

mixed forest. *See* forest.

mixing length theory. Theory used to calculate the behaviour of molecules in a fluid involved in a turbulent exchange. The mixing length is the average distance perpendicular to the main flow covered by the mixing particle, a fluid in turbulent exchange being displaced in a direction perpendicular to its direction of flow. *See* mean free path.

mixotroph. An organism that is both heterotrophic and autotrophic (e.g. insectivorous plant).

mm. Million. Used in connection with brl or cf, in figures for hydrocarbon reserves to signify million barrels or million cubic feet.

Mn. manganese.

Mo. Molybdenum.

mobile belt. An elongated region of the Earth's crust undergoing uplift and subsidence, earthquakes and volcanic activity, and probably folding and faulting as well, producing a mountain chain.

mock sun (sun dog). A bright spot seen in the sky at an angle of 22° to one or both sides of the sun, and sometimes faintly discoloured. They are caused by the refraction of light by ice crystals in the lower atmosphere.

mode. (Stat.) The value of the variate possessed by the greatest number of individuals.

model. A simplified description of a system, used as an aid to understanding the system. Models may be verbal or mathematical. Mathematical models are constructed from numerical values given to the components of the system and the relationships among components. If the model is sufficiently accurate to be approximately true, it can be used to predict the effect of changes within the system. Models may be constructed as initially-complex models, by starting with a replica and discarding non-essential details, or as initially-simple models, which begin with the most essential elements and add others as they are needed. Apart from mathematical models, models may be physical. *See* analogue, analogue computer, digital computer.

modified mercalli scale. Scale for measuring the intensity of earthquake shock, going from instrumental (1) to catastrophic (12). *See* Isoseismic line.

modularism. The technique of making whole structures more permanent by making their component sub-structures less permanent. Thus, a modular building consists of a permanent framework into which interchangeable modules can be fitted to be removed and replaced later. The pen fitted with disposable ink cartridges is based on this principle: the framework has a long life, the sub-structure a short life.

Moho. Mohorovičic discontinuity.

Mohole. A project, now abandoned, whose aim was to drill through the Earth's crust to the Moho and beyond. *See* JOIDES.

Mohorovičić discontinuity. A seismic discontinuity within the Earth below which shock waves from the focus of an earthquake travel with increased velocities. This was first proposed in the early 20th century by Yugoslav seismologist A. Mohorovičić and is now accepted as a major structural divide in the Earth, the part above the discontinuity being the 'crust' and the part below, down to the core of the Earth, being the 'mantle'.

Mohs' hardness scale. System in which minerals are classified in order of increasing hardness, so that the hardness of any particular substance may be expressed by a number between 1 and 10. The hardest substance is diamond (10).

mol. mole.

molasse. An association of predominantly non-marine clastic sediments, including breccia conglomerate, arkose, sandstone, and shale, formed during the Miocene as a result of the rapid erosion of the Alps. The term molasse, which is contrasted with flysch, has been used for many similar deposits elsewhere.

mole (mol). The SI unit of amount of substance, being that amount which contains as many elementary units as there are atoms in 0.012 kg of carbon-12. The elementary units must be specified. 1 mole of a compound has a mass equal to its molecular weight in grams.

molecular diffusion. The spontaneous inter-mixing of different substances by molecular movement, giving uniform concentrations. *See also* diffusion, eddy diffusion.

mole drain. *See* soil drainage.

mollic. Applied to dark, organic-rich surface soil layers, high in calcium and magnesium and with a strong structure so that soil remains 'soft' when dry. *See* soil classification.

mollisols. *See* soil classification.

mollusca. A large phylum of soft-bodied, unsegmented, mainly aquatic animals, generally with calcareous shells. Locomotion is usually by means of a large, muscular 'foot'. *See* Amphineura, Cephalopoda, Gastropoda, Lamellibranchiata, Scaphopoda.

molluscicide. Chemical used to kill molluscs, e.g. metaldehyde, and copper sulphate, used to control the water snail *Limnaea trunculata*, the vector of the liver fluke parasite in sheep.

molten carbonate. US process for removing sulphur dioxide from flue gases using a scrubber to absorb the sulphur dioxide, which is converted to hydrogen sulphide from which sulphur is extracted, and an electrostatic precipitator to remove fly ash.

molybdenite. The mineral molybdenum sulphide MoS_2, from which most of the world's molybdenum is produced as a by-product of mining porphyry copper deposits and also from the geologically analogous porphyry molybdenum deposits. The most important use of molybdenum is in alloys.

molybdenum (Mo). Element. A hard, white metal, which occurs as molybdenite and is used in special steels and alloys. It is an essential micronutrient. AW 95.94; At.No. 42; SG 10.2; mp 2620°C.

moment of momentum. *See* angular momentum.

momentum. *See* acceleration.

monazite. A phosphate mineral of the 'rare earth' metals: cerium (Ce), lanthanum (La) and thorium (Th), (Ce, La, Th)Po_4. Monazite is a rare accessory mineral in granites and pegmatites, but is concentrated in some placer deposits (particularly black sands), from which it is obtained as a by-product of mining for titanium minerals and zircon. The rare earth metals are used in alloys, and as catalysts, and thorium is used in nuclear reactors. Monazites in pegmatites have been used extensively in determining the radiometric ages of the rocks.

monera. Procaryotic organisms, comprising the bacteria and blue-green Algae.

monitoring programme. A programme designed to measure, quantitatively or qualitatively, the level of a substance over a period of time. Experiments and demonstrations are monitored in this way. Such programmes are often used to provide information on the distribution through space or time of pollutants so that effective measures can be developed to limit their harmful effects, should the results of the programme indicate that remedial measures are necessary.

monocarpic. Applied to plants which flower only once during their lives.

monocaryon. Monokaryon.

monoclimax. The theory that within a regional climatic type a single climax formation will develop. *See* polyclimax.

monocline. A localised zone of steeply dipping beds in a region of horizontal to low dip.

monocotyledon. *See* cotyledon.

Monocotyledoneae. *See* hormone weed killers. One of the two classes of flowering plants (Angiospermae). The embryo has only one cotyledon. The leaves are usually parallel veined. The vascular bundles in the stem are scattered and do not contain cambium (*see* meristem). Flower parts are usually in threes or multiples of three. Most monocotyledons are small plants (e.g. grasses, lilies, orchids) but a few (e.g. palm trees) are large. The class includes very important food-producing plants (*see* Graminae, Cocos, Musa).

monocytes. *See* blood corpuscles.

monoculture. The cultivation of a single crop year after year, to the exclusion of all others.

monoecious. Applied to organisms bearing both male and female reproductive organs in the same individual. In flowering plants the term is restricted to individuals with separate male and female flowers on the same plant. *See* dioecious.

monogamy. (Zool.) Permanent pair bonding; among species that form permanent pair bonds both parents are generally involved in the rearing of young.

monogenetic. *See* provenance.

monohybrid inheritance. The inheritance of one pair of alleles. The monohybrid ratio indicates the proportions of different types of individuals in the F_2 generation produced by crossing individuals differing in respect of a single pair of alleles. This ratio is 3:1 if one character is dominant.

monokaryon (monocaryon). A fungus hypha or mycelium made up of cells each containing one haploid nucleus. All the nuclei are identical. *See* dikaryon.

monomer. Chemical compound consisting of single molecules. *Cf.* polymer.

monomolecular layer. Layer having the thickness of a single molecule, such as oils on water, which may prevent the natural transfer of substances from the liquid to the air.

monophagous. Applied to an animal which feeds on only one kind of food, for instance the Black Hairstreak Butterfly caterpillar (*Strymonidia pruni*) which feeds only on Blackthorn (*Prunus spinosa*).

monophyletic. Applied to a natural, closely related group of species or other taxa (*see* taxis) all having a common origin. *See* polyphyletic.

monophyodont. Having only one set of teeth during a lifetime. *See* diphyodont, polyphyodont.

monosaccharide. *See* carbohydrate.

monosome. An unpaired chromosome. A single X chromosome is normally present in the heterogametic sex of some organisms. Monosomes occur abnormally in many aneuploids.

monotocous (uniparous). Producing young singly. *See* polytocous, ditokous.

Monotremata (Prototheria, Ornithodelphia). A sub-class of mammals which lay eggs and have many primitive, reptilian features. They are represented today only by the Australian Duck-billed Platypus (*Ornithorhynchus*) and the Spiny Anteaters *Tachyglossus* (Echidna) of Australia and *Zaglossus*, of New Guinea.

monovular twins. Monozygotic twins.

monoxenic (ametoecious) parasite. *See* parasitism.

monozygotic twins (monovular twins, uniovular twins, identical twins). Twins derived from a single fertilized egg. They are genetically identical, so always of the same sex. *See* dizygotic twins, polyembryony.

monsoon forest. *See* forest.

Monte Carlo method. (Stat.) This method proceeds by the construction of an artificial stochastic model of the process upon which sampling experiments are formed.

montmorillonite. A clay mineral with a layer lattice structure, formula $Al_4(Si_4O_{10})_2(OH)_4.xH_2O$. Characteristically it has a variable water content and the clay will swell as water is taken in between the layers. Fuller's Earth consists mainly of montmorillonite, but the purest form of montmorillonite is found in bentonite, formed by the weathering of volcanic ash.

moor (moorland). An open area of acid peat, usually high-lying and occupied by heathers (*Erica, Calluna*), sedges and certain grasses, (e.g. Purple Moor Grass, *Molinia*).

mor (mar, raw humus). An acidic, crumbly humus layer, which is clearly marked off from the mineral soil beneath. It is poor in animal life, especially earthworms. *See* mull.

Moraceae. A family of tropical and sub-tropical dicotyledonous plants, mainly trees and shrubs with latex. The economically important genera include mulberry (*Morus*), cultivated for its fruit and for leaves used as food for silkworms; *Castilloa*, a source of rubber; *Brosium*, which yields bread-nut and a milky drink from the latex; *Artocarpus*, the source of breadfruit; *Broussonetia*, the inner bark of which is used as paper and cloth; and *Ficus*.

Moraine. (*a*) Ground moraine. A relatively flat-topped deposit of boulder clay which is exposed after the retreat of a glacier or ice sheet (*b*) Terminal (end) moraine. A ridge-like accumulation of glacial drift (usually boulder clay) formed during a lull in a glacial retreat.

morbidity rate. The incidence of sickness in a population. *See* mortality rate.

morphactins. A group of synthetic plant growth-regulating substances which cause effects such as dwarfing, bushiness and inhibition of germination.

mophallaxis. (*a*) The gradual development into a particular form. (*b*) The remodelling of tissue during the regeneration of missing parts in an animal.

morphology. The study of the form, structure and origin of organisms or of the Earth's physical features. *See* geomorphology.

mortality rate. The incidence of death in a population. *See* morbidity rate.

mortar. Construction material used to hold together bricks or stones. Mortars can be made from a mixture of sand and cement, or pure hydrated (slaked) lime with or without the addition of volcanic ash (pozzolana), coal or wood ash, crushed bricks or any fired clay in powder form. Masonry cement is a mixture of Portland cement, an air-entraining plasticizer and an inert filler.

morula. An early animal embryo, before the blastula stage. It consists of a solid ball of cells.

mosaic. (*a*) (Genet.) Synonym for chimaera. *See* gynandromorphism. (*b*) (Bot.) A patchiness in the green colour of plants, caused by virus diseases. (*c*) (Zool.) A mosaic egg is one in which various cytoplasmic areas give rise to definite structures in the embryo, so that if an area of the egg is damaged, the corresponding structure will not develop correctly. (*d*) (Ecol.) A vegetation pattern in which two or more community types are interspersed (e.g. grassland with patches of scrub).

moss. *See* Musci, Bryophyta, gemmation, bog.

moss animals. *See* Polyzoa

moss layer. *See* field layer.

mother-of-pearl clouds. Nacreous clouds.

motor pattern. 'What an animal does', i.e. that aspect of its behaviour that can be observed easily, involving the movement of muscles (which includes making sounds). Motor patterns provide the information used in the compilation of ethograms.

mould. (*a*) Any fungus which produces a superficial growth of mycelium, e.g. *Penicillium notatum*, the source of antibiotic penicillin. (*b*) Humus-rich soil (e.g. leaf-mould). *See* mildew.

moult. (*a*) *See* ecdysis (*b*) The periodic shedding of hair (e.g. the spring moult in mammals when the thick winter coat is lost) or feathers. Most birds moult in the autumn. Ducks also moult in the summer and become flightless for a period, owing to the loss of their large wing feathers. The drab moult plumage of the drake is called eclipse plumage.

Mount Agung. (Gunung) Volcano in Indonesia which erupted violently in March 1963, emitting large amounts of particulate matter and gases, whose behaviour and effects were monitored. Large quantities of sulphur dioxide, or possibly sulphates reached a height of about 18 km, spread over the world within about 6 months, lasted for several years and produced twilight displays of colours. The temperature of the lower equatorial stratosphere increased by about 6° or 7°C shortly after the eruption, probably because of the presence of particles, which may have increased the normal stratospheric particle content about tenfold. It is assumed that the Agung eruption was typical of tropical volcanic eruptions, such as those of Mount Tambora and Krakatau. *See* natural pollutant, sulphate, sulphur dioxide.

326

mountain waves (lee waves). Waves, usually a few miles in length, set in motion when stable air (*see* static stability) moves over a mountain ridge. The formation of such waves is more likely if the air is more stable at lower than higher levels. The vertical velocity of the air within the waves may be 300 m per minute or more, making the downdraughts a hazard to aircraft. Clouds formed in the wave crest indicate the vertical extent of the waves, which may be several times the height of the mountains. With waves of large amplitude, an eddy with reverse motion may form, called rotor-flow.

Mount Tambora. Volcano in Indonesia which, in 1815, produced what is believed to have been the most violent eruption in historical times, reducing the height of the mountain from 4000 to 2800 m, and emitting large quantities of gases and particles into the atmosphere. *See* natural pollutant, sulphate, sulphur dioxide, Mount Agung.

mp. Melting point.

mucoid feeding. Feeding by means of mucus which is extruded from the mouth, then ingested together with small particles that have stuck to it. A method employed by some molluscs.

mudstone. A lithified argillaceous sedimentary rock which is not fissile (in contrast to shale).

mulch. A material (e.g. straw, paper or plastic) used to protect the surface of the soil from erosion. Where the mulch is composed of biodegradable material it becomes incorporated into the topsoil to become humus.

mull. (mild humus) A loose, crumbly humus layer, mingling with the mineral soil beneath, and produced where decay is rapid. Its acidity is low, and it contains a rich fauna, including abundant earthworms. *See* mor.

Müllerian mimicry. The resemblance of one species of animal equipped with warning coloration and a defence (e.g. distastefulness) against predators to another similarly protected species. Both gain from the resemblance, as predators learn to avoid both species after attacking one. *See* Batesian mimicry.

multiple factors. (polygenes) A group of genes which combine to control the development of a single character (e.g. stature in Man). A wide range of variation is produced by numerous combinations of these genes.

multivoltine. Applied to organisms which produce several generations in a year.

municipal waste. Substances discarded by private households, offices, shops, etc. as unusable. Generally collected by a local authority for disposal by dumping,

327

sanitary landfill, composting or pyrolysis. It is used as a fuel for power production in large incinerators. Typically composed of paper, organic matter, plastics, metals and non-metallic minerals (ash). Historically, the content of plastics has increased rapidly and ash content has decreased.

Musa. Huge, tropical monocotyledonous herbs, with large oval leaves. The bananas (e.g. *Musa sapientum*) and the plantain (*M. paradisiaca*) are widely cultivated, producing a very high yield of food per acre. Manila hemp is obtained from the leaf stalks of *M. textilis*.

Musci (mosses). A class of Bryophyta. Mosses have prostrate or erect stems containing conducting tissue. Their leaves are usually only one cell thick except in the midrib region. Mosses occur in moist and dry terrestrial habitats and in water. *Sphagnum* is an important constituent of bog. *See* gemmation.

muscle. Highly contractile tissue in an animal. (*a*) Striped (striated, voluntary, skeletal) muscle consists of elongated, multinucleate fibres, whose cytoplasm is made up of longitudinal fibrils, with alternating bands of different composition. Striped muscle contracts rapidly, and is mainly concerned with moving skeletal parts in vertebrates and arthropods. *See* actomyosin. (*b*) Smooth (plain, involuntary) muscle consists of individual spindle-shaped cells. It performs slow, often sustained contractions, and is found in the walls of organs such as the gut and blood vessels. Similar muscle is found in many invertebrates (e.g. molluscs, worms).

mushroom rocks. Sandblasted rocks sculpted into mushroom shapes by the wind carrying sand at a height of under two metres. *See* corrasion.

mussels. *See* Lamellibranchiata.

mustard gas. Dichlorodiethyldisulphide, one of the first chemical warfare agents used in the First World War.

mutagen. An agent (e.g. X-rays, gamma rays, mustard gas, TCDD) which induces mutation.

mutation. A sudden change in the chromosomes of a cell. Most mutations are changes in the DNA of individual genes. Others are alterations in the structure of number of chromosomes (*see* polyploid, aneuploid). Mutations may occur during gamete formation, when they produce an inherited change, or in body cells (somatic mutation). Most mutations are harmful. Evolution occurs through natural selection of mutations. *See* mutagen.

mutualism. Once a synonym for symbiosis, but used variously as meaning (*a*) a type of symbiosis in which neither partner is essential for the well-being or life of the other; (*b*) both symbiosis and commensalism; (*c*) any association between organisms of different species, including parasitism.

m.y. Millions of years.

mycelium. The vegetative part of a fungus, consisting of a mass of minute filaments (hyphae).

mycetocytes. Specialized cells in an insect's body which contain symbiotic micro-organisms. Sometimes the cells are localized in special organs (mycetomes). *See* symbiosis.

mycetome. *See* mycetocytes.

mycetophagous. Applied to an animal which feeds on fungi.

Mycetozoa (Myxomycetes, Protomyxa). *See* Myxomycophyta.

Mycophyta (fungi). A large group of organisms (*see* Protista) including toadstools, mildews, yeasts, etc. They are either unicellular or composed of masses of fine filaments (hyphae), and reproduce by means of spores. None of them contains chlorophyll, so all are heterotrophic. Many (e.g. mildews, smuts, rusts) cause plant diseases, a few cause animal diseases (e.g. ringworm). Some fungi cause decay of food and timber, but many are beneficial because of the large part they play in the decomposition of organic matter in the soil. Some antibiotics are produced by culturing fungi (e.g. penicillin from *Penicillium*). A few fungi (e.g. yeasts) are used in the preparation of food and in brewing. *See* Ascomycetes, Basidiomycetes, Phycomycetes, *Fungi imperfecti*, agarics, mildew, yeasts.

mycoplasmas (pleuropneumonia-like organisms, PPLO). Procaryotic organisms without cell walls, usually considered to be very small bacteria. Some are saprophytic, others cause diseases in plants and animals.

mycorrhiza. The symbiotic association of the root of a higher plant with a fungus. In an ectotrophic mycorrhiza (e.g. heath, pine trees) the fungal mycelium covers the outside of the roots and in an endotrophic one (e.g. orchids) the fungus grows inside the cells of the root cortex.

mycotrophic. Applied to plants which possess a mycorrhiza.

myiasis. Infestation by parasitic larvae of Diptera (e.g. warble-fly).

mylonite. A very compact welded rock flour produced at the base of thrusts.

myoglobin. A protein which is found in muscle, and whose function is the temporary storage of oxygen. A molecule of haemoglobin consists of four myoglobin-like molecules.

myosin. *See* actomyosin.

Myriopoda. Centipedes and millipedes, terrestrial Arthropoda which have many similar, leg-bearing segments. Centipedes (Chilopoda) are carnivorous, with the first pair of appendages modified as poison claws and the body flattened dorso-ventrally. Millipedes (Diplopoda) are mainly vegetarian, with cylindrical bodies and two pairs of legs to each apparent segment.

myrmecochore. A plant whose reproductive structures are dispersed by ants (e.g. violets, whose seeds bear oil bodies (elaiosomes) attractive to ants).

myrmecophily. (*a*) Living with ants. Some animals (e.g. the caterpillers of the Large Blue Butterfly, *Maculinea arion*) live inside ants' nests. Some plants are inhabited by ants. See Myrmecophyte. (*b*) Pollination by ants.

myrmecophyte. A plant which, by means of special structures, shelters ants and may provide food for them (e.g. *Acacia sphaerocephala* has thorns which are inhabited by ants, which feed from nectaries and food-bodies on the leaves.

Mysticeti. *See* Balaenoidea.

Myxobacteria. A small group of bacteria, many of which produce 'fruiting bodies', formed by the gliding together of rod-shaped vegetative cells.

myxomatosis. A virus disease affecting rabbits and occasionally hares. It is endemic in South America, where it is transmitted by mosquitoes. Myxomatosis was deliberately introduced into Australia in 1950. The disease somehow reached Britain in 1953, and by 1955 had virtually wiped out the rabbit population in many areas. This led to startling changes in the vegetation (*See* downland vegetation). The rabbit populations in both Australia and Britain have gradually recovered since the advent of myxomatosis, and many more rabbits now survive an outbreak. In Britain the vector of the disease is the rabbit flea.

Myxomycetes. Mycetozoa. *See* Myxomycophyta.

Myxomycophyta (Slime fungi, Slime moulds). A group of organisms including the Mycetozoa (Myxomycetes, Protomyxa), intermediate between fungi and protozoa, commonly occurring as saprophytes on decaying vegetable matter. The vegetative stage usually consists of a multi-nucleate, amoeboid mass of protoplasm (a plasmodium), which can ingest food. Reproduction is by means of spores produced in plant-like organs (sporangia). A common species is the bright yellow Flowers of Tan (*Fuligo septica*).

Myxophyceae. *See* Cyanophyta.

N

N. (*a*) Nitrogen (*b*) Newton.

n. Nano-.

Na. Sodium.

nacreous (mother-of-pearl) clouds. Thin clouds which occur at heights of 20 to 30 km above the surface and are seen most often in Scandinavia in winter when the sun is several degrees below the horizon. A temperature of about -80°C is believed to be required for their formation. Their lifetimes are not known, because conditions for good visibility do not last for long. The low stratospheric temperature associated with them, and their seasonal and geographical variations, supports the hypothesis that they are produced when local saturation occurs. The required temperature may be reached partly as a result of mountain waves. It has been suggested that the injection of water vapour (e.g. from aircraft exhausts) in the stratosphere might modify high-level cloud formation by permitting nacreous clouds to form over a wider area by raising the frost point temperature.

Namurian. The third oldest stage of the Carboniferous System in Europe.

nanism. Stunting of plant growth occurring, for instance, in trees at the limit of their range on mountains.

nannoplankton. The smallest of the phytoplankton.

nano- (n). Prefix used in conjunction with SI units to denote the unit x 10^{-9}.

nanometre (nm). One-thousandth of a micron, or 10 ångströms.

nappe. (*a*) A faulted, overturned fold. (*b*) A large body of rock that has moved forward at least a mile by recumbent folding and/or thrusting.

NASA. National Aeronautics and Space Administration.

nastic movement. A plant response which is independent of the direction of the stimulus (e.g. drooping of the leaves of *Mimosa pudica* when touched (seismonasty); daily opening and closing of flowers; 'sleep movements' of leaves). Nastic response to illumination is photonasty, and that to temperature is thermonasty. Response to alternation of day and night, related to changes in both illumination and temperature, is nyctinasty.

natality rate. Birth rate.

National Aeronautics and Space Administration (NASA). US Government agency responsible for space exploration and for the launching and management of satellites (e.g. Earth Resources Technology Satellite).

National Environmental Policy Act (NEPA). *See* Council on Environmental Quality.

national nature reserve. *See* nature reserve.

national park. In British planning, an extensive tract of country (designated under Acts of Parliament as a national park by the Countryside Commission) which, by reason of its natural beauty and the opportunities it affords for open air recreation, shall be preserved and enhanced for the purpose of promoting its enjoyment by the public.

native element. An element occurring in the uncombined state as a mineral. The twenty native elements can be divided into those in which the native element is the major, or the only, ore mineral: carbon (graphite and diamond), sulphur, gold, and the platinoid metals (platinum, palladium, rhodium, iridium, ruthenium, and osmium), and the other eleven elements whose major source is from compounds: iron, nickel, copper, arsenic, selenium, silver, antimony, tellurium, mercury, lead and bismuth.

natric. A clayey layer of soil enriched with sodium.

natural frequency. The frequency at which a system oscillates freely after excitation.

natural gas. A mixture of gases, both hydrocarbons and non-hydrocarbons, found in nature, and often associated with deposits of petroleum. The principal component of natural gas is usually methane (CH_4) and the term is often restricted to such a gas, although natural gases composed predominantly of carbon dioxide and hydrogen sulphide are known. Under subsurface reservoir conditions, some natural gases are liquids. Natural gas is either transported by pipeline or, refrigerated, in a liquefied state (liquefied natural gas, LNG) in LNG carriers.

natural increase. The rate of population growth calculated by subtracting the number of deaths from the number of births (or vice versa if the population is decreasing).

naturalists' trusts (county trusts for nature conservation). In Britain, bodies of naturalists, organised roughly on a county basis, which seek to conserve natural habitats by acquiring and managing natural reserves, and offering advice and education regarding the preservation of wildlife in Britain. The first to be formed was the Norfolk Naturalists' Trust, in 1926. The English and Welsh Trusts are sponsored by the Society for the Promotion of Nature Reserves, and the Scottish Wildlife Trust serves the whole of Scotland.

natural pollutant. A substance of natural origin that may be regarded as an environmental pollutant when present in excess (e.g. volcanic dust, sea salt particles, ozone formed photochemically or by lightning, products of forest fires). *See* sulphur dioxide, sulphate, Mount Agung.

natural resource ecosystem. A natural ecosystem in which one element is of use to Man.

natural selection. *See* Darwin.

Nature Conservancy Council (NCC). The official body, established in Britain by Act of Parliament in 1973, to be responsible for the conservation of flora, fauna, geological and physiographical features throughout Great Britain. It establishes and maintains National Nature Reserves (*see* nature reserves), gives advice and education on nature conservation, and carries out research into related topics. It is financed by the Department of the Environment. The NCC replaced the Nature Conservancy, formed in 1949.

nature-nurture. The controversy over the categorisation of behaviour into that which is innate (instinctive) and that which is learned. *See* eugenics, instinct, heritability, Lamarck.

nature reserve. In British planning, an area of land and/or water managed primarily to safeguard the fauna, flora and physical features which it contains. Reserves which are declared and managed by the Nature Conservancy Council (on behalf of the Department of the Environment) are called national nature reserves. Those designated by local planning authorities (in consultation with the Nature Conservancy Council) are called local nature reserves. County Naturalists' Trusts and other voluntary bodies are often associated with the management of local nature reserves and, in addition, have established a large number of reserves of their own.

nauplius. A free-swimming larva, typical of many Crustacea, with an egg-shaped, unsegmented body, three pairs of appendages, and a single eye.

nautical mile. Defined originally as the average length of one minute of latitude (which varies slightly with longtitude). Now defined as 6080 feet. The international nautical mile is 1852 metres (6076.12 feet).

Neanderthal man. *See* Homo.

nearctic. *See* zoographical regions.

NEF. Noise exposure forecast.

negative price. An economic concept which imputes a price to, for example, industrial effluents.

neighbourhood noise. Term used by the British Noise Advisory Council to denote noise from any source that may cause disturbance and annoyance to the general public in their homes or going about their business. The term does not include industrial noise as experienced by workers.

nekton. The strongly-swimming animals of the pelagic zone of a lake or sea (e.g. fish).

Nemanthelminthes. A phylum used by some taxonomists which comprises the Nematoda and Nematomorpha.

nematicide. Chemical used to control nematode worms (e.g. certain dinitro, organochlorine and organophosphorus compounds used to kill eelworms infesting potatoes, sugar beet, etc.).

nematoblast (cnidoblast, thread cell). A type of cell, found in large numbers in most Coelenterata, which holds a bladder (nematocyst) containing a hollow, coiled thread which is shot out when the cell is stimulated, and becomes attached to the animal's prey. Some threads contain poison and can penetrate, so exerting a paralysing effect.

Nematocera. A group of Diptera with long, many-jointed antennae, including crane-flies and Culicidae.

Nematoda (roundworms). A very large group (usually considered a phylum) of unsegmented worms, without a coelom, usually pointed at both ends. Some are free-living, in the soil, in water, and even in liquids such as vinegar. Many are parasites (e.g. potato root and sugar beet eelworms, pinworms of vertebrates). *See* Filaria, Hookworms.

Nematomorpha (Gordiacea, Horsehair worms, Threadworms). A group of extremely long (up to about 90 cm), slender, freshwater and marine worms, resembling Nematoda. The larvae parasitise insects or Crustacea. The Nematomorpha are regarded by some taxonomists as a separate phylum. *See* Aschelminthes, Nemathelminthes.

Nemertea (nemertini). Proboscis or ribbon worms. A small phylum of flattened, unsegmented, mainly marine animals, with a large, eversible proboscis, used for catching prey. Nemertea resemble Platyhelminthes in their general body organisation.

Nemertini. Nemertea.

Nene (Hawaiian Goose). *Branta sandivicensis. See* Anseriformes.

Neocomian. A subdivision (usually ranking as a stage) of the Cretaceous System, and comprising the Berriasian or Ryazanian, and the Valanginian, Hauterivian and Barremian.

Neo-Darwinism. The current theory of evolution, which combines Darwin's theory of evolution by natural selection with modern concepts of genetics.

neogea. *See* zoographical regions.

Neogene. Refers to the later of the two periods of the Cenozoic Era, and includes the Miocene and Pliocene Epochs, and also refers to the rocks formed during this time. In some usages, Neogene also includes the Pleistocene and Holocene (or Recent). The terms Palaeogene and Neogene are restricted mainly to European usage.

Neognathae. A large group including all present-day birds except for the Palaeognathae.

Neolithic (Late Stone Age). Applied to early agricultural societies which obtain some of their food by the cultivation of domesticated crop plants and/or the husbanding of domesticated livestock. Neolithic peoples used polished stone tools, although many present day peoples whose agriculture is at a neolithic stage of development use metals, generally imported in ready-made form. *Cf.* Palaeolithic, Mesolithic, agricultural revolution.

neoplasm. A tumour, often malignant.

neossoptiles. Nestling feathers. *See* teleoptiles.

neotony. The temporary or permanent persistence of pre-adult stages or structures in animals. In extreme examples paedogenesis occurs, as in an axolotl, a salamander which breeds in the larval form. Neotony has probably played a part in the evolution of Man, as many adult human features (e.g. reduction of hair, large head) resemble those of foetal apes.

neotropical region (neogea). *See* zoographical regions.

neotype. A specimen chosen to replace a type specimen when all the original material is missing. *Cf.* paratype, lectotype.

NEPA. National Environmental Policy Act. *See* Council on Environmental Quality.

nephridium. An osmo-regulatory and excretory organ typical of many animals (e.g. Platyhelminthes, Annelida, Cephalochordata) consisting of a tubule open to the exterior at one end. The other end is often closed, ending in hollow flame cells (solenocytes) containing cilia.

neritic. Applied to the parts of the sea lying over the continental shelf, usually less than 200 metres deep. *See* oceanic, littoral zone, sublittoral zone.

nerve cell. Neuron.

nerve net. A network of neurons making up a simple nervous system in Coelenterata. In animals with central nervous systems nerve nets may exist in some parts of the body.

ness. Promontory or headland.

net primary aerial production. *See* production.

net primary production. *See* production.

net production rate. *See* production.

net reproduction rate. The average number of female babies that will be born to a newly-born female during her lifetime, if current rates of birth and death continue. If the net reproduction rate is larger than unity, the population will grow, or decline if it is less than unity, although the age structure of the population may postpone the increase or decrease.

neuron (neurone, nerve cell). The fundamental unit of a nervous system. Each neuron consists of a nucleated nerve cell body from which project fine processes (axon, dendrites) along which nerve impulses pass. The places where the processes make contact with those of other neurons are called synapses. *See* CNS, nerve net, ganglion, autonomic nervous system, acetylcholine.

Neuroptera. An order of predatory insects (Endopterygota) whose members possess two similar pairs of wings covered in a delicate network of veins. They include lacewings, alder-flies, snake-flies and ant-lions. Some of the larvae are aquatic, others (e.g. lacewing) eat large numbers of aphids and other pests.

neurula. A vertebrate embryo at the stage in which the brain and spinal cord are beginning to form.

neustron. Organisms associated with the surface film of water. Infraneustronic animals (e.g. mosquito larvae) suspend themselves from the underside of the film; supraneustronic species (e.g. pond-skaters) live on the upper side of the film.

névé (firn). Snow that has survived a summer melting and become granular. An intermediate stage in the conversion of snow to glacer ice.

new red sandstone. The red beds of Permo-Triassic age in Europe. The Old Red Sandstone is of Devonian age. The vast majority of red beds in Devonshire are Permo-Triassic.

Newton (N). The derived SI unit of force, being the force required to give a mass of one kilogram an acceleration of one metre per second per second. Named after Sir Isaac Newton (1642–1727). The unit of pressure is the newton per square metre (Nm^2), sometimes called the pascal. *See* bar, atmosphere.

NGO. Non-governmental organisation.

niacin. Nicotinic acid.

niche. (*a*) The specific part of a habitat occupied by an organism. (*b*) The role an organism plays in the ecosystem. In different communities, very different organisms may play comparable roles (e.g. on coral islands the 'earthworm niche' is filled by land crabs).

niche diversification. Alpha diversity.

nickel carbonyl. An intermediary compound ($Ni(Co)_4$) in the Mond process for extracting nickel from the impure metal by direct action of carbon monoxide (CO) on the finely divided metal. Nickel carbonyl is formed, a volatile substance, b.p. 43°C, and the gas decomposes at 200°C into pure nickel and carbon monoxide (which can then be used again). Nickel carbonyl was the first metal carbonyl to be discovered (Ludwig Mond, 1890). It is very toxic, explosive, a respiratory irritant, and possibly a carcinogen.

nick point. A sudden steepening of gradient in a river bed, usually over resistant rock. It is produced by an intersection of new and old graded profiles of a river when a previous base level has been altered.

nicotine. An alkaloid extracted from tobacco leaves and used as an insecticide to control aphids and other insects on horticultural and fruit crops. It acts as a contact poison and a fumigant, and although poisonous to mammals, it loses its toxicity rapidly after application.

nicotinic acid (niacin, p-p factor). An important vitamin of the B group, found in malt extract, yeast and liver, and synthesised by many micro-organisms. It plays a vital role in respiration. Deficiency in Man causes pellagra.

nidation. Implantation.

nidicolous. Applied to birds (e.g. blackbird) which are helpless and often naked when hatched, and are fed for some time in the nest. *Cf.* nidifugous.

nidifugous. Applied to birds (e.g. Mallard) which are well developed when hatched, with a downy covering and a store of yolk remaining inside the body. They are able to leave the nest at once. *Cf.* nidiculous.

nife. Acronym for nickel and iron, the presumed major components of the core of the Earth.

Nightjars. Caprimulgidae.

night soil. Human excrement, which used often to be collected by night before sewerage systems provided for continuous removal.

nitrate. A salt or ester of nitric acid (HNO_3).

nitrate bacteria. *See* nitrification

nitrification. The conversion by aerobic soil bacteria (nitrifying bacteria) of organic nitrogen compounds into nitrates, which can be absorbed by green plants. Dead organic matter is broken down into substances such as ammonia which combines with calcium carbonate. Ammonium carbonate is oxidized to nitrite by nitrite bacteria (*Nitrosomonas*), then nitrites are converted to nitrates by nitrate bacteria (*Nitrobacter*). *See* nitrogen cycle.

nitrilotriacetic acid (NTA). Chemical compound sometimes proposed as an alternative to phosphates as a builder in detergents. Its adoption has been delayed because it may afford no great environmental advantage.

nitrite bacteria. *See* nitrification.

nitrobacter. *See* nitrification.

nitrogen (N). Element. An odourless, invisible, chemically inactive gas that forms about 80° of the atmosphere. It is an essential macronutrient for plants and so is used in fertilisers. The main natural source is Chile saltpetre, but industrially, atmospheric nitrogen is fixed (e.g. by the Haber process) to form ammonia (NH_3) and ammonium (NH_4) compounds. AW 14.0067; At. No. 7. *See* nitrogen cycle, nitrogen fixation.

nitrogen cycle. The circulation of nitrogen atoms through ecosystems, brought about mainly by living organisms. Inorganic nitrogen compounds (mainly nitrates) are synthesised into organic compounds by autotrophic plants. These die or are eaten by animals and the organic nitrogen compounds in their bodies and excretory products eventually return to the soil or water. Nitrogen fixation augments the supply of organic nitrogen. Nitrifying bacteria convert organic nitrogen into nitrates (*see* nitrification) which are again available to green plants. Denitrifying bacteria convert some of the nitrates to atmospheric nitrogen. *See* denitrification, root nodules.

nitrogen fixation. The conversion of atmospheric nitrogen into organic nitrogen compounds. This enriches the soil and is carried out by certain bacteria and blue-green algae. Some nitrogen-fixing bacteria (e.g. *Azotobacter* and *Clostridium pasteurianum*) are free living in the soil. Others live symbiotically in the root nodules of Leguminosae. *See* nitrogen cycle.

nitrogen oxides. Oxides formed during combustion, by the oxidation of atmospheric nitrogen at high temperatures. Oxides include nitrous oxide, nitric oxide, nitrogen dioxide, nitrogen pentoxide, and nitric acid.

nitrosamines. Yellow oils which form naturally in the environment by the reaction of aqueous nitrous acid (HNO_2) with secondary amines. They can be formed in the stomach from nitrites (derived from food) and secondary amines. Following experiments with dimethylnitrosamine, other nitrosamine and nitro-sämide compounds have been found to be active carcinogens producing tumours in more sites than most known carcinogens.

Nitrosomonas. *See* nitrification.

nm. Nanometre.

NNI. Noise and number index.

No$_x$. Nitrogen oxides.

node. *See* antinode.

nodum. A general term for a unit of vegetation, whatever its rank (e.g. association, society, consociation).

noise. Random sound or electromagnetic radiation, not wanted by the observer.

noise and number index (NNI). An index for the measurement of disturbance by aircraft noise, developed in Britain and used widely since. The index is based on a social survey carried out in the vicinity of London's Heathrow Airport.

noise criteria. Sets of curves that relate sound levels in octave bands to speech interference and acceptability for particular applications, often a working environment.

noise exposure forecast (NEF). A technique for predicting the subjective effect of aircraft noise on the average person, expressed in NEF units, and sometimes drawn on maps by lines connecting points of equal NEF value. The estimate is based on values assigned to a range of factors that are considered (e.g. frequency of aircraft movements and their distribution by day and night, aircraft weights and flight profiles, etc.).

noise level. Sound level.

noise rating curves (numbers). Sets of curves that relate levels of sound in octave bands to acceptability for particular applications. When an octave-band analysis is plotted on a graph of noise rating curves, the number of the curve which is reached by the level in one or more bands is the Noise Rating Number (NRN) of the noise. The same result can be reached mathematically. dBA units (*see* decibel) are used more frequently than NRN.

noise reduction coefficient (NRC). The average of the absorption coefficients of a surface or material at 250 Hz, 500 Hz, 1 KHz and 2KHz.

noise zoning. The classification of areas according to use, taking particular account of noise levels, so that land use planning can contain noisy activities within areas where they will cause minimum disturbance to residents.

nominum conservandum. An organism's scientific name which is retained by agreement, although it may be at variance with the rules of nomenclature.

non-abstractive use. *See* abstractive use.

non-governmental organisation (NGO). Voluntary body, usually with an international membership, which is accorded official status by the UN, enabling it to attend certain meetings in an observer or consultant capacity, and to provide information and views to UN committees.

non-identical twins. Dizygotic twins.

non-parametric. (Stat.) Those statistical tests which are independent of a specific form of distribution.

non-renewable resource. A natural resource that in terms of human time scales is contained within the earth in fixed quantity and therefore can only be used once in the foreseeable future. This includes the fossil fuels and it is extended to include mineral resources and sometimes ground water, although water and many minerals are renewed eventually. The distinction is drawn according to time scales. If a resource will only be renewed over a geological time span, it is considered non-renewable.

normal distribution. Gaussian distribution.

normal fault. A fault in which the fault-plane dips to the downthrown side.

Northern Lights. Aurora borealis.

North West European Megalopolis (the Golden Triangle). An area of North West Europe extending from the Ruhr in the Federal Republic of Germany, west into France as far as Paris, north to include most of the Netherlands, Luxembourg and Belgium, and reaching an apex in the Lancashire conurbation (Manchester–Liverpool) in Britain. Within this roughly triangular area, population densities and industrial activity are very much greater than they are outside the area. There is a tendency for the Megalopolis (so termed by the EEC) to exert a strong economic attraction, drawing more industry and population into itself, so causing developmental problems outside the area. Most of the EEC's most severe pollution problems occur within the region, which includes or is traversed by the Danube, Rhine, Elbe, Meuse, Moselle, Seine, Weser, Thames, Severn and other major rivers.

notochord. A longtitudinal, elastic supporting rod, consisting of vacuolated cells surrounded by a sheath, which lies beneath the central nervous system, and is present at some stage of development in all chordates. A notochord is present in adult Acania and Agnatha, but in most vertebrates it is more or less obliterated by the development of the vertebral column.

notogea. *See* zoographical regions.

nouméite. Granierite.

novel protein foods. High protein foods that are not derived from livestock products, or that are animal in origin but converted into food for humans or livestock by novel processes. For human consumption, most novel proteins are derived from soya, other leguminous crops, or cereals, but for livestock feeds the wastes from food processing, from abbatoirs, from livestock excrement, or from hydrocarbons used as a substrate for the cultivation of unicellular organisms are also immediately or potentially available. Proteins produced from micro-organisms (e.g. yeasts) are called single cell proteins (scp). Novel proteins may be used directly as a powder, as 'meat extenders' in traditional livestock products (e.g. sausages, stews, meat pies, etc.) or may be spun to produce long fibres, or extruded through a perforated surface, then textured, coloured and flavoured to produce analogues of meat (e.g. textured vegetable protein, TVP).

noy. A unit of noisiness related to the perceived noise level in perceived noise decibels (PNdB). *See* decibel.

NPK. Nitrogen, Phosphorus, and Potassium fertiliser. *See* macronutrients.

NRC. Noise reduction coefficient.

NRN. Noise rating numbers.

NTA. Nitrilotriacetic acid.

nuclear fission. *See* fission.

nuclear fusion. A nuclear reaction in which light atomic nuclei are fused to form heavier nuclei with the release of energy, as radiation, including heat. Fusion reactions occur in the explosion of hydrogen bombs, and are believed to supply the energy of stars. Nuclear fusion is the principle underlying the thermonuclear (fusion) reactor. *See* fusion reactor.

nuclear reactor. A device for generating heat by the controlled fission of atoms of uranium 235 or by the fusion of light atoms (*see* fusion reactor). In a fission reactor, a moderator (often water, heavy water or carbon as graphite) is used to slow down escaping neutrons emitted by the unstable U-235 atomic nucleus, without absorbing them, so reducing them to speeds at which they are more likely to bombard other uranium atoms, causing further fissions and releasing more neutrons. Control rods (of cadmium or boron) absorb neutrons and so slow the reaction and are raised or lowered into or out of the reactor pile to regulate the energy output. The energy is released as heat and other radiation. The heat is carried away by a coolant that passes through the reactor core. Individual reactor systems differ in the coolants, moderators and types of fuel they use. A fission reactor will continue to function until its fuel is exhausted, though the life of the fuel is extended by the production, in all reactors, of some plutonium, which is also consumed. A breeder reactor uses no moderator. Its core is surrounded by a blanket of uranium-238 (the more common natural isotope of uranium, and in its natural form of no value as a nuclear fuel) which is bombarded by fast neutrons and so converted into plutonium, which can be used as a fuel in conventional fission reactors. Breeder reactors are often called 'fast' because they utilise fast neutrons. *See* liquid-metal fast breeder reactor, critical mass.

nucleic acid. *See* DNA, RNA.

nucleolus. A compact structure inside a cell nucleus, concerned with the synthesis of ribosomal RNA.

nucleo-protein. A compound made up of a protein and nucleic acid. *See* DNA, RNA.

nucleotide. *See* DNA, RNA.

nucleus. Of cells. A membrane-bounded body containing the chromosomes and one or more nucleoli present in eucaryotic cells.

nuclide. The nucleus of an isotope.

nuée ardente. An incandescent cloud of gas and lava violently erupted from some volcanoes, which produce acidic magma. Nuées ardentes flow downhill at great speeds, with the material buoyed up by expanding heated volcanic gases and trapped air, and do great damage (e.g. the destruction of St. Pierre in 1902).

nullisomic. Applied to cells showing a kind of aneuploidy in which both members of a homologous pair of chromosomes are absent from the diploid set.

numerical abundance. The number of individual plants of a species which are present in an area.

numerical forecast. A weather forecast made using a computer to calculate the expected events using fundamental laws of physics and mechanics as the basis of the method.

nut. A dry, indehiscent, single-seeded fruit, usually with a woody wall (e.g. acorn, hazel nut). *See* Achene.

nutation (circumnutation). A slow, spiral growth movement in a plant, caused by the continual alteration in the position of the most actively growing part of the organ.

nutrient budget. A budget drawn up in terms of nutrients for a living system.

nutrient stripping. The removal of nutrients from a substance. This may be the removal of nutrients from sewage to reduce rates of eutrophication in receiving waters, or it may be the incidental removal of nutrients from food during processing.

nyctinasty. *See* nastic movement.

nymph. A young stage of a exopterygote insect (e.g. Mayfly, dragonfly, earwig). This resembles the adult in many ways, but its wings are absent or incompletely developed, and it is sexually immature.

O

O. Oxygen.

OA. Oxygen absorbed.

oak. *Quercus.*

oasis. *See* arid zone.

oat. *Avena sativa. See* Graminae.

oblanceolate. (Bot.) Inverted lanceolate.

obligate parasite. *See* parasitism.

obliterative countershading. Obliterative shading.

obliterative shading (obliterative countershading). A very common type of animal coloration in which the dorsal surface is darkest and there is a gradual lightening towards the ventral side. This tends to obliterate the solid appearance of the animal, affording very effective camouflage.

obsidian. A glassy acidic igneous rock with a conchoidal fracture and with a composition similar to rhyolite.

occluded front. When a cold front overtakes a warm front, the boundary at the surface divides the cool air ahead of the warm air from the cold air behind it. The warm air is raised off the ground and is occluded, and the front is called an occlusion.

oceanic. Applied to parts of the oceans deeper than 200 m, i.e. seaward from the continental shelf. *Cf.* neritic.

oceanic rise. *See* ridge.

oceanic trench. A narrow, elongated depression at the ocean-continental margin characterised by large negative gravity anomalies (*see* isostasy) and seismic activity. According to modern theories of plate tectonics this marks the site of a destructive plate margin, where an oceanic plate plunges down into the mantle beneath an adjacent plate.

ocellus. A simple light receptor, composed of a small number of sensory cells and a single lens, found in many invertebrates. *See* eye.

ochre. Natural red and yellow pigments composed of limonite.

octane number. A parameter defining the quality of petrol. High octane fuel has better anti-knock properties than low octane fuel. The octane number can be increased by further refining or by the addition of tetraethyl lead.

octave. The interval between two sounds one of which has a frequency twice that of the other.

Octopoda. Cephalopod molluscs with eight tentacles and no shell (e.g. octopus, argonaut). *See* Cephalopoda, Decapoda.

octroi. Tax or duty raised on goods entering a town or area. Used in Europe in former times, and still used in India.

Odonata (dragonflies). An order of large, often highly coloured insects (Exopterygota) whose members have two similar pairs of membranous wings. The adults are strong fliers, preying on insects which they seize while on the wing. The nymphs are aquatic and predatory, with the lower lip (labium) modified as an extensible, prehensile 'mask'.

Odontoceti (Delphinoidea). Toothed whales, including sperm whale, killer whale, narwhal, porpoises and dolphins. Their prey is much large than that of the Balaenoidea.

OECD. Organisation for Economic Cooperation and Development.

oestrogens (estrogens). Natural or synthetic hormones which stimulate the growth and activity of the reproductive system and the development of secondary sexual characters in female vertebrates. Oestrogens are secreted by the ovary, the placenta and, in small amounts, by the adrenal gland and the testis.

oestrous cycle (estrous cycle). A reproductive cycle occurring during the breeding season in many female mammals, in the absence of pregnancy. One phase of the cycle is oestrus ('heat'), during which ovulation occurs and there is an urge to mate. Oestrogen and progesterone levels vary according to the phase of the cycle, which is controlled by hormones secreted by the pituitary gland. The menstrual cycle of some primates (Man, apes and old-world monkeys) is a modified oestrous cycle in which there is no well marked oestrous phase, and destruction of the uterine lining produces bleeding.

offensive industry. Industry that is noxious or offensive by virtue of the processes it employs, the materials it uses, its products, or the wastes it discharges.

ohm. (Ω) The derived SI unit of resistance, being the resistance between two points of a conductor when a constant difference of potential of one volt applied between the points produces in the conductor a current of one ampere. Named after Georg Ohm (1787–1854).

oil, crude. Crude oil.

oilfield. An area underlain by one or more oil (or gas) pools which are adjacent or overlapping in plan view, and which are related to a single geological feature which is either structural or stratigraphic. The size of an oilfield refers to the volume of recoverable oil (the 'reserves') which is given either in barrels, or tonnes (depending on the gravity of the oil, 1 tonne = 6.5 to 7.5 barrels).

oil film. *See* monomolecular layer.

oil-pool. *See* pool.

oil-shale. A black, brown or green shale more or less impregnated with kerogen, and from which gaseous and liquid petroleum can be obtained by destructive distillation to 300°C to 400°C. With increasing kerogen content, oil-shales grade imperceptibly through boghead coals into cannel coals.

oil slick. Oil, discharged naturally or by accident or design, floating on the surface of water as a discrete mass carried by wind, currents and tides.

okta. *See* cloud amount.

Old Man's Beard. *Clematis vitalba.*

old red sandstone. ORS.

Olea Europea. Olive. *See* Oleaceae.

Oleaceae. A family of dicotyledonous shrubs and trees, some of which are of economic importance. They include ash (*Fraxinus*), the olive (*Olea europea*) which yields oil from its fruit (a drupe) and lilac (*Syringia vulgaris*).

Oligocene. Refers to a subdivision of the Cenozoic Era. The Oligocene usually ranks as an epoch, and follows the Eocene. The Oligocene is thought to have lasted from about 38 to 26 million years before present. Oligocene also refers to rocks deposited during this time: these are called the Oligocene Series.

Oligochaeta. An order of terrestrial and freshwater Annelida, whose members (e.g. earthworm) have few bristles (chaetae), no muscular, lateral projections (parapodia) along the body, and no specialised head appendages. They are hermaphrodite, and the eggs develop inside a cocoon. Earthworms are very important in the maintenance of soil fertility, because their tunnels aerate and drain the soil, and some species continually bring up to the surface fresh soil containing humus. There may be up to three million earthworms per acre. In western Europe they bring up about 6 kg of soil per square metre per year to the surface.

oligophagous. Applied to an animal which feeds on only a few kinds of food.

oligosaprobic. Applied to a body of water in which organic matter is decomposing very slowly, and in which the oxygen content is high. *Cf.* mesosaprobic, polysaprobic, catarobic, saprobic classification.

oligotropic. Applied to freshwater bodies which are poor in plant nutrients and are therefore unproductive. Their waters are clear, and do not become depleted of oxygen. A deep lake may be classed as oligotrophic despite being productive in its surface layers, if its hypolimnion is large enough never to become depleted of oxygen. *Cf.* eutrophic, mesotrophic, dystrophic.

Olive (*Olea europea*). *See* Oleaceae.

olivine. A group of rock-forming silicate minerals ranging in composition from Mg_2SiO_4 to Fe_2SiO_4. Olivines occur mostly in basic and ultrabasic rocks, and are easily altered by weathering or hydrothermal processes to serpentine minerals.

ombrogenous (ombrotrophic, ombrophilous). Living in wet climates.

ombripholous. Ombrogenous.

ombrotrophic. Ombrogenous.

Onager (*Equus hemippus onager*). A wild ass of south west Asia hunted for meat in Palaeolithic times.

onchocerciasis. A disease that is prevalent in Equatorial Africa and the American tropics, which causes impaired vision leading eventually to blindness if not treated, and fibrous tumours, or nodules, most commonly on the head and shoulders. The disease is caused by the parasitic roundworm (filariae) *Onchocerca volvulus*, transmitted to Man by several species of Black Fly.

onchosphere (oncosphere). An early stage in the life history of a tapeworm, consisting of a spherical, chitinous shell containing a six-hooked (hexacanth) embryo. It develops from the egg, and precedes the cysticercus stage.

oncosphere. Onchosphere.

one-crop economy. The economy of a nation, or region, that is largely dependent on a single activity, so making it highly vulnerable to changes affecting the production or marketing of a limited range of commodities. Thus the Arab OPEC countries are dependent on oil revenues for most of their foreign exchange, Iceland is heavily dependent on fishing, and in Central America, a number of countries (esp. Honduras, Costa Rica and Panama, the so-called 'banana republics') derive about one-third of their exports earnings from the production and export of bananas.

onion weathering. Sphaeroidal weathering.

ommatidium. *See* eye.

omnivore. An animal which eats both plant and animal food (e.g. badger, Man).

ontogenesis. Ontogeny.

ontogeny (ontogenesis). The sequence of development during the whole life history of an organism.

Onychophora (walking worms). A small group of terrestrial, worm-like, many-legged animals, usually regarded as Arthropoda, but with a thin, unjointed cuticle and other Annelid-like characters (*see* Annelida). A single living genus, *Peripatus*, lives under stones and dead bark in warm climates.

onyx. A form of chalcedony.

oocyte. A cell which gives rise by meiosis to an animal ovum. During this process a primary oocyte usually gives rise to one large secondary oocyte and a cell with very little cytoplasm (polar body). The second meiotic division produces the ovum and another polar body.

oogamy. The production of two kinds of gamete. The female gamete (egg) is large and non-motile, the male one small and motile.

oogenesis (ovogenesis). The cell divisions and changes resulting in the formation of an ovum. *See* oocyte.

oogonium. (*a*) (Bot.) The female sex organ of some fungi and algae, inside which large, non-motile female gametes (eggs, oospheres) are formed. (*b*) (Zool.) A cell in an ovary which gives rise to oocytes.

oolite. A sedimentary rock composed predominantly of concentrically layered sphaeroidal grains (or ooliths). Ooliths forming at the present day are calcareous, and are found in tropical regions where currents roll around sand-size debris in evaporating shallow water. Some ancient ooliths are dolomitic, or siliceous, or composed of various ores of iron.

oosphere. *See* oogonium.

oospore. A resting spore with a thick wall, formed from an oosphere (*see* oogonium) after fertilisation.

open burning. The outdoor burning of wastes (e.g. lumber, sawdust, scrapped cars, textiles, etc.) and the burning of open waste dumps. It produces atmospheric pollution inevitably. *Cf.* incineration.

open-cast mining. The working of coal seams that lie close to the surface by stripping off the overburden rather than by tunnelling. *Cf.* strip mining.

open cut mining. Strip mining.

open community. A community which is readily colonised by other organisms because some niches remain unoccupied.

open country. In British planning, an area consisting wholly or predominantly of mountain, moor, down, cliff, foreshore, woodlands, rivers or canals. (As defined in the Countryside Act 1968, S. 16).

open hearth furnace. A reverberatory furnace containing a cup-shaped hearth, for melting and refining pig iron, iron ore, and scrap for steel production. A large amount of dust from ore and other materials and splashings from slag are carried away in the noxious waste gas stream.

operon. A group of genes which control the synthesis of one enzyme system.

Ophidia (Serpentes). *See* Squamata.

Ophiuroidea (Brittle Stars).· A class of Echinodermata whose members have flattened, star-shaped bodies with arms sharply marked off from the disc. They move mainly by muscular movement of the arms, and eat small Crustacea, molluscs and detritus. *See* Asteroidea.

Opiliones. Phalangida.

Opisthobranchia. *See* Gastropoda.

opportunity cost. The real cost of satisfying a want, expressed in terms of the cost of sacrificing an alternative activity. If capital (or labour) earns a particular sum in some activity other than that in which it is engaged, then that is the opportunity cost of the present activity, which may or may not be a net cost when present actual income is set against it.

optical activity. The power of a substance to rotate the plane of polarised light transmitted through it. All living systems show this property, as do all asymmetric molecules.

optimum population. The number of individuals that can be accommodated within an area to the maximum advantage of each of them. The concept may be interpreted ecologically or economically.

Opuntia. A genus of fleshy-stemmed Cactaceae, usually with small, fleshy leaves which drop off early. *Opuntia vulgaris* (Prickly Pear) has become a troublesome weed since its introduction into Australia. Some species are used for hedging, others as food for cochineal insects. The fruits of some are edible.

OR. Orientation response.

oranogenesis. Oranogeny.

oranogeny (oranogenesis). The formation of organs during the development of an individual organism.

Orchidaceae. A very large, cosmopolitan family of monocotyledonous plants. All of them are perennial herbs, with tuberous roots, vertical stocks or rhizomes. Epiphytic species, often with aerial roots, are an important feature of tropical vegetation. A few species are saprophytic (e.g. Bird's Nest Orchid, *Neottia nidus-avis*). Mycorrhizae occur almost invariably. Pollination is usually by insects, to which pollen masses (pollinia) become attached. The pods of the genus *Vanilla* are used for flavouring. Many orchids are cultivated for their flowers.

order. *See* classification.

Ordovician. Refers to the second oldest period of the Palaeozoic Era, usually taken as beginning some time between 500 and 530 million years before present. Ordovician also refers to the rocks formed during the Ordovician Period: these are called the Ordovician System, which is divided in the UK into five series (the Arenigian, followed by the Llanvirnian, Llandeilian, Caradocian, and Ashgillian). The Ordovician probably lasted 65 to 70 million years. Geologists in most other countries place the Tremadocian in the Ordovician, but in the UK the Tremadocian is placed in the Cambrian. The Ordovician is zoned using graptolites.

Ord River Irrigation Scheme. A scheme to construct a dam in the Ord River, Western Australia, to divert the river for irrigation use. The scheme aroused opposition but was eventually implemented and began to supply irrigation water in 1972.

ore. A mineral or mineral aggregate, more or less mixed with gangue, which can be worked and treated at a profit.

ore-body. A mass of ore which is capable of being worked. Metalliferous ore bodies can be classified into (*a*) igneous segregations formed by magmatic differentiation; (*b*) intrusive bodies (e.g. pegmatite, diamond pipes, and hydrothermal veins); (*c*) metasomatic bodies; (*d*) sedimentary ore-bodies (e.g. banded iron-stones, laterites, and placers.

ore-mineral. A mineral from which a useful metal may be extracted.

organelles. Formed structures within cells, based on a limited number of structural forms, each of which consists of a number of large molecules arranged in a pattern. *See* cell.

organic farming (biological husbandry). Farming without the use of artificial fertiliser or pesticides, according to principles laid down by Sir Albert Howard, Lady Eve Balfour and others, and as interpreted in Britain by the Soil Association.

Organisation for Economic Cooperation and Development (OECD). Intergovernmental institution concerned with a wide range of issues that have economic or developmental implications. The members, all developed countries, are Australia, Austria, Belgium, Canada, Denmark, Finland, France, W. Germany, Greece, Iceland, Ireland, Italy, Japan, Luxembourg, Netherlands, New Zealand, Norway, Portugal, Spain, Sweden, Switzerland, Turkey, UK, USA, and Yugoslavia which enjoys a 'special status' entitling it to participate in many OECD activities.

organochlorine. An organic compound containing chlorine, sometimes called a chlorinated hydrocarbon. Many have biocidal properties and were used as the active ingredients for pesticides with a high persistence, due to their chemical stability and low solubility in water. Pesticide applications include DDT, aldrin and lindane. These have proved very effective in the control of insect pests, esp. insect vectors of disease in tropical areas, but some insects have become immune to them. They persist in sub-lethal doses in animals, where they accumulate in fatty tissue. Low concentrations of residues may be increased as they pass up the food chain and toxic levels may thus occur in the bodies of predators, causing death or sterility (e.g. in some birds). Because of these side effects the use of many of these insecticides has been limited in the UK and in other countries. Other uses include PCBs. *See* endosulfan (thiodan), chlordane, endrin, heptachlor.

organomercury fungicides. Poisonous compounds used as sprays or seed treatments to control fungus diseases in crops.

organophosphorus. Group of non-persistent insecticide compounds of varying vertebrate toxicity, which appear to inhibit the action of cholinisterase, an enzyme which cancels chemical messages within the nervous system. Because they break down rapidly, danger to wildlife is of short duration compared with that caused by organochlorine insecticides. *See* parathion, malathion, disulfoton, fenitrothion, demiton-S-methyl.

organotin fungicides. *See* fentin.

oriental region. *See* zoographical regions.

orientation response (OR). The physiological response to sudden changes in the immediate environment (e.g. an animal may prick up its ears in response to a new sound, dilate its pupils in response to a new sight, etc.). The response represents the defensive mechanisms of an animal reacting to novelties that might threaten it.

original (prototype). In modelling (*see* model), the original is that which the model seeks to describe or simulate.

ornithischia. *See* dinosaurs.

ornithodelphia. Monotremata.

ornithophily. Pollination by birds (e.g. humming birds).

orogenesis. *See* diastrophism.

orogenic belt. An elongated region of intensely deformed rocks formed during an orogeny. Young orogenic belts are characterised by fold mountains, but, with age, erosion strips off the higher levels to expose metamorphic rocks (*see* metamorphism) and batholiths.

orogeny. A period of mountain building, involving the formation of folds and faults, metamorphism and igneous intrusion. Most orogenies affected geosynclines and lasted tens of millions of years, with several maxima. The major orogenies affecting Europe were the Caledonian, Hercynian and Alpine.

orographic lifting. *See* convection.

orpiment. *See* arsenic.

ORS (old red sandstone). The continental facies of the Devonian System.

ortet. An organism which gives rise to a clone. *See* ramet.

orthogenesis (orthoselection, directive evolution). An evolutionary trend caused by the inherent tendency of a group to develop in a particular way.

orthograde. Applied to animals (e.g. higher apes and Man) which walk erect, on two legs.

Orthoptera (Saltatoria). Grasshoppers, crickets and locusts. An order of terrestrial insects (Exopterygota) with biting mouthparts, narrow, thickened forewings covering membranous hind wings, and hind legs usually enlarged for jumping. Most are poor fliers and some are wingless. *See* Acrididae.

orthoquartzite (quartzarenite, silicaceous sandstone, quartzose sandstone). A sandstone composed of detrital quartz grains cemented by silica such that the silica content (i.e. grains plus cement) is at least 90% (95% in some classifications). Synonyms include, loosely, quartzite, but this term also includes metaquartzite.

orthoselection. Orthogenesis.

Orycteropodidae. Tubulidentata.

Oslo Convention. *See* Intergovernmental Maritime Consultative Organisation.

osmo-regulation. Control of the water content of an organism. *See* homoiosmotic, poikilosmotic.

osmosis. The passage of water through a semi-permeable membrane (i.e. one which is permeable to water but not to solute) from a solution of low concentration (the hypotonic solution) into one of a higher concentration (the hypertonic solution). This tends to continue until the solutions are of equal concentration (isotonic). Because cell membranes are more or less semi-permeable, osmosis plays a part in the movement of water in living organisms. *See* osmotic pressure.

osmotic pressure. The pressure that develops when a pure solvent is separated from a solution by a membrane which allows only solvent molecules to pass through it.

ossicle, auditory. *See* ear.

Osteichthyes. Bony fishes, comprising the Actinopterygii and Choanchthyes. *Cf.* chondrichthyes.

osteolepid. Having an outer skin of armoured bone-like scales. The osteolepid Crossopterygii, from which the Amphibia are descended, lived in freshwater swamps and seas during the Devonian and Carboniferous. They had an armour of cosmoid scales and functional lungs, probably assisted by gills, and paired pectoral and pelvic fins with bony elements similar to elements in primitive amphibian limbs.

Ostracoda. Small Crustacea with a bivalve carapace enclosing the body, and reduced trunk limbs. They live in fresh and salt water, and may be pelagic or bottom-dwellers. *Cypris* is common in ponds.

otocyst. Statocyst.

otolith. *See* ear, statocyst.

outbreeding. Exogamy.

outcrop. (*a*) The exposure of a rock body or a rock surface, such as a fault or joint (*b*) The total area, esp. as shown on a geological map, in which a rock body or rock surface lies at the Earth's surface, though possibly concealed by a thin superficial deposit such as soil. The verb is to outcrop, or to crop out.

outfall sewer. A pipe or conduit which transports sewage, raw or treated, to a final point of discharge.

outwash deposits. A stratified deposit of material eroded by glaciers and reworked by melt-water. Outwash deposits are also known as fluvioglacial deposits, and are one component of glacial drift.

outwash plain. A body of outwash deposits which form a broad plain beyond the

354

edge of their associated ground moraine.

ovarian follicle. *See* follicle.

oven. *See* kiln.

overburden. The material lying above mineral deposits that must be removed in order to work the deposits.

overstorey (overwood). The upper layer in woodland, usually trees forming a canopy.

overtone. *See* Harmonic.

overturn. The circulation of water and nutrients in a lake when the thermal strata become mixed by the wind during the spring and autumn. *See* Thermal stratification.

overturned fold. A fold in which one limb has been tilted beyond the vertical, so that both limbs dip in the same direction. Such a fold will have an inclined axial plane.

overwood. Overstorey.

oviparous. Applied to animals (e.g. frogs) that lay eggs in which the embryos are at the most only slightly developed. *Cf.* viviparous, ovoviparous.

ovogensis. Oogenesis.

ovotestis. An organ present in some hermaphrodite animals (e.g. snails) which produces both eggs and sperms.

ovoviparous. Applied to animals (e.g. vipers) which retain their eggs within the maternal body until embryonic development is complete. During all or most of its development, the embryo is separated from the mother's body by the egg membranes *Cf.* viviparous, oviparous.

ovulation. The release of a ripe egg from an ovary.

ovule. The structure which develops into a seed after fertilisation. *See* embryo sac.

ovum. *See* gamete.

Owls. Strigiformes.

ox-bow lake. A crescent-shaped lake left in the site of a former river meander.

Oxfordian. A stage of the Jurassic System.

oxic. Applied to a soil layer from which much of the silica that was combined with iron and alumina has been leached (*see* Leaching), leaving sesquioxides or iron and alumina and 1:1 lattice silicate clays. *See* soil classification, soil horizon.

oxidase. An enzyme (a dehydrogenase) which catalyses oxidative reactions in which hydrogen is removed from a substrate and combines with free oxygen.

oxidation. The combination of oxygen with a substance, or the removal of hydrogen from it or, more generally, any reaction in which an atom loses electrons (e.g. the change of a ferrous ion Fe^{++} to a ferric ion Fe^{+++}). The removal of oxygen from a substance, or the addition of hydrogen to it, or, more generally, any reaction in which an atom gains electrons, is called reduction.

oxidation pond. A shallow lagoon or basin in which waste water is purified by sedimentation, aerobic and anaerobic treatment.

oxidative phosphorylation. The combination of ADP with phosphate to form ATP. Respiration provides the energy needed for this reaction.

oxisols. *See* soil classification.

oxygen (O). Element. An odourless, invisible gas that forms approximately 20% of the atmosphere and is essential to most forms of life. It is the most abundant of all elements in the Earth's crust, occurring in rocks, water and air. Industrially it is used for welding and metal-cutting by mixing with other gases (e.g. acetylene, hydrogen) to produce high-temperature flames. AW 15.9994; At. No. 8.

oxygen absorbed (OA). The amount of oxygen, in parts per million of water, absorbed by a sample of water from acidic permanganate at 27°C in four hours.

oxygen sag curve. A line drawn on a graph to trace the dissolved oxygen content of water over a period of time, to measure the capacity of the water to purify itself following the discharge into it of an effluent.

oxytocin. A pituitary hormone which stimulates uterine contractions and ejection of milk in mammals.

oysters. *See* Lamellibranchiata.

ozone. O_3. A form of oxygen in which the molecule consists of three atoms, rather than two. It is very reactive chemically, and an irritant to eyes and respiratory tissues. It occurs naturally as about 0.01 parts per million of air, 0.1 ppm being considered toxic. It is formed by the recombination of oxygen following its ionisation by electrical discharges or certain forms of radiation, esp. ultraviolet light.

ozone layer. A layer of the atmosphere, about 20 to 50 km above the surface, which contains ozone produced by ultraviolet radiation.

ozonometer. Instrument for measuring the amount of ozone present in the air.

ozonosphere. Ozone layer.

P

P. Phosphorus.

p. Pico-.

P$_1$(parental generation). The parents of the F$_1$ generation in a genetical experiment. For the allelomorphs being observed, all the males of the P$_1$ generation are identical to each other, and so are all the females.

package treatment plant. A transportable unit for sewage treatment capable of achieving a desired quality of effluent. It is sometimes used as a temporary measure prior to the installation of permanent treatment facilities.

paedogamy. Autogamy.

paedogenesis. Reproduction in animals which have become sexually mature while still otherwise in a young stage. For instance, insufficient thyroid secretion leads to reproduction in the axolotl, which is the tadpole stage of certain salamanders (e.g. *Ambylstoma*). *See* neotony.

paedomorphosis. The evolutionary process by which youthful characters are introduced into the line of adults. *See* neotony, paedogenesis.

PAH. Polycyclic aromatic hydrocarbons.

pahoehoe. Hawaiian term describing a basaltic lava with a smooth, billowy or ropy surface. *Cf.* Aa.

palade granules. Ribosomes.

palaearctic. *See* zoographical regions.

palaeo-. Prefix meaning ancient.

Palaeocene. The lowest division of the Cenozoic Era. The Palaeocene usually ranks as an epoch, lasting from about 65 to 54 million years before present. Palaeocene also refers to rocks deposited during this time: these are called the Palaeocene Series. Some authors do not recognise the Palaeocene, and consider the Eocene to extend all the way to the Cretaceous.

palaeoecology. The study of past environments, their flora and fauna, and the interrelationships among them.

Palaeogene. Refers to the earlier of the two periods of the Cenozoic Era and includes the Palaeocene, Eocene and Oligocene Epochs. Palaeogene also refers to rocks formed during this time. The terms Palaeogene and Neogene are restricted mainly to European usage, and some authors restrict the Palaeogene to the Eocene and Oligocene only.

palaeogenesis. The persistence of normally embryonic characteristics into a later stage. *See* neotony.

Palaeognathae. A primitive group, comprising the flightless birds (ratites) and the tinamous (Tinamiformes) from Central America and South America which can fly well but resemble the ratites in skull structure. *See* Neognathae.

Palaeolithic (Early Stone Age). Applied to human societies that are pre-agricultural, obtaining food by hunting, fishing and gathering of wild plants. The domestication of crop plants and animals probably began about 11000 years BP, and so this date marks the beginning of the transition from Palaeolithic to Neolithic culture in the regions affected, although other peoples continued to live by hunting and gathering for much longer and some survive to the present day. Historically, palaeolithic peoples used unpolished stone tools. *Cf.* Mesolithic.

palaeomagnetism. The study of geomagnetic induction in the geological past of the Earth, by determining the magnitude and direction of magnetism in iron compounds and rocks at the time of their formation. Although rocks are comparatively weak magnets, they will preserve their magnetism for hundreds of millions of years and information can be deduced from them regarding variations in the position of the ancient geomagnetic pole. Recent work with contemporaneous rocks dating back 4.5 million years has shown that the Earth's geomagnetic field has undergone periodical polarity reversals, i.e. an effective switch between North and South Poles.

palaeontology. The study of life forms of the past.

Palaeozoic. One of the eras of geological time, occurring between the Pre-cambrian and the Mesozoic, and comprising the Cambrian, Ordovician, Silurian, Devonian, Carboniferous (replaced by the Mississippian and Penn-sylvanian in North America) and Permian Periods. Palaeozoic also refers to the rocks formed during this time. The Lower Palaeozoic comprises the Cambrian, Ordovician and Silurian, and the Upper Palaeozoic comprises the Devonian, Carboniferous and Permian.

paleo-. Palaeo- (North American spelling)

Paleosere. The eosere of the Palaeozoic Period.

palingenesis. The recapitulation of adult stages of ancestors during the early development of the descendants (e.g. the formation of gill pouches in the embryos of mammals). The recapitulation theory, or biogenetic law ('ontogeny repeats phylogeny') originally put forward by Haeckel is now largely discredited, although the embryonic stages of related organisms do resemble each other.

Palmae (Arecaceae). The palms, a family of tropical and subtropical monocoty-ledonous plants, mostly with very large, fan-shaped or feathery leaves. Some are rhizomatous, a few are climbers with thin, reed-like stems, and others have tall stems (reaching 60 m in some species) capped by a crown of leaves. The fruit is a berry (e.g. a date) or drupe (e.g. coconut). The palms are of great economic importance, yielding food, fibres, oil, timber, etc. *Phoenix dactylifera* is the date palm. Species of *Metroxylon, Caryota* and *Arenga* yield sago from their pith and palm sugar from their sap. *Areca* is the source of the betel-nut. The fruit of *Elaeis* yields palm oil, used for lubrication, and wax is obtained from the leaves of *Copernicia. See* Cocos.

palmate. (Bot.) Applied to leaves with three or more lobes or leaflets that radiate from the centre, like the fingers of a hand.

Paludicola. (*a*) Marsh-dwelling animals. (*b*) A group of free-living freshwater flatworms (Turbellaria).

palynology (pollen analysis). The study of pollen and spores. Pollen grains have walls of a highly indestructible organic material, and fossil pollen is used in stratigraphic correlation (particularly of coal and oil-bearing strata) and as environmental indicators, to determine the composition of past plant communities.

pampas. *See* grassland.

pampero. *See* mistral.

359

PAN. Peroxyacetyl nitrates.

panclimax (panformation). Two or more related climaxes or formations of similar climate, life forms and genera or dominants. Panclimaxes may arise from common origins in an eoclimax.

Pandanaceae. A family of tropical monocotyledonous plants, mainly confined to coasts and marshes. They have long, narrow leaves and tall, usually twisted stems, supported by aerial roots. Some screwpines (*Pandanus*) yield edible fruits (e.g. Nicobar breadfruit) and the leaves are used for weaving and thatching.

panemone. A vertical axis windmill consisting of two or more concave or flat surfaces designed to rotate around the axis. The spinning signs used for advertising purposes at many petrol stations are panemones, as are anemometers and Savonius rotors.

panformation. Panclimax.

Panhonlib group. Terms applied to ships registered in, and flying the flags of, Panama, Honduras and Liberia. These are substantial fleets, but many of the shipowners are nationals of other countries. The *Torrey Canyon*, which was wrecked off the Scillies UK, in 1967, spilling 117000 tons of oil, was registered in Liberia, flew the Liberian flag, was owned and chartered by US nationals, and was manned by an Italian captain and crew.

panicle. (Bot.) A branched inflorescence.

Pantopoda (Pycnogonida, Sea Spiders). A small class of marine Arachnida with much elongated legs, and an extra (fifth) pair of legs in front, used in males for carrying the eggs. Many species are ectoparasites (*see* parasitism) or sea anemones.

pantothenic acid. A vitamin of the B group, which is essential for the formation of a co-enzyme in many organisms, including vertebrates and some bacteria.

Pantotheria. A group of Jurassic mammals which probably gave rise to both marsupials and placentals.

Papaw. *Carica papaya.*

Papilionaceous. Applied to flowers that are characteristic of the pea family. *See* Leguminosae.

Papilonoidea (Rhopalocera). Butterflies. *See* Lepidoptera.

Papveraceae. A family of mainly herbaceous, dicotyledonous plants, mostly with conspicuous, nectarless flowers which are visited by insects for their pollen. Opium is obtained from the poppy *Papaver somniferum* by cutting notches in the half-ripened capsules, and collecting the hardened exudation of latex.

paraclimax. Subclimax.

paraffin oil. Kerosene.

paralic. Refers to sediments formed on the landward side of a coast in an area subject to marine incursion. Paralic is also used for the environment of such an area, and is contrasted with limnic.

parameter. (Statistical). A characteristic figure for a total population (e.g. mean height of all English women).

paramorph. A variant within a species.

paraquat. Organic contact herbicide used to control many broad-leaved weeds and grasses. It is poisonous and accidental ingestion of large amounts, usually by children, has caused a number of deaths, but used correctly it is rapidly adsorbed on soil particles and so inactivated.

parasematic coloration. Coloration (e.g. eye, patterns on the wings of some butterflies) which protect an animal by directing the attack of a predator towards a non-vital part of the body.

parasite. *See* parasitism.

parasitism. A close association between living organisms of two different species, during which one partner, the parasite, obtains food at the expense of the other, the host. Many animals and a few flowering plants (e.g. Dodder) are parasites. Pathogenic bacteria and fungi are parasitic on plants and animals and viruses parasitize plant, animals and bacteria. The term is sometimes used in a wider sense, to include commensalism, mutualism and symbiosis. Ecoparasites are each adapted to a specific host or closely allied group of hosts. Ectoparasites live for long or short periods on the outside of their hosts. Endoparasites live inside their hosts. Hemiparasites (e.g. Yellow Rattle) obtain only part of their food supply from their hosts. Clepto-parasites (Klepto-parasites) (e.g. skuas) are animals which habitually steal food caught by animals of another species. Brood parasites (e.g. cuckoos, 'homeless' bees) are animals in which the rearing of the young is entrusted to animals of a different species. Hyperparasites (Superparasites) are parasites living on or in other parasites. Obligate parasites can live only parasitically. Facultative parasites can live as saprophytes but under certain circumstances live parastically. Xenoparasites are either ones infecting hosts not normal to them, or are only capable of invading an injured organism. Autoecious parasites spend their whole life cycle on an individual host. Heteroecious (Heteroxenous) parasites need more than one host to complete their life cycle, or are not host-specific. Metoecious (Metooxenous) parasites are not host-specific. Ametoecious (Monoxenic) parasites are host-specific. Tropoparasites are obligate parasites during some part of their life cycle, but live non-parasitically for the rest of the time (e.g. the Swan Mussel, *Airodonta cygnea*, is parasitic as a larva in freshwater fishes).

parasitoid. An animal with a mode of life intermediate between parasitism and predation (*see* predator) because the parasitoid ultimately kills its host (e.g. Ichneumons (Hymenoptera) which lay eggs inside the larvae or eggs of other insects).

parasympathetic nervous system. *See* autonomic nervous system.

parathion (thiophos). The first organophosphorus insecticide to be used widely in British agriculture, being introduced in 1944. Used to control aphids and other insects, mites, eel-worms and woodlice. It is very poisonous to mammals and birds, but breaks down rapidly after application.

parathyroid glands. Organs lying adjacent to or inside the thyroid gland in vertebrates other than fish. They secrete a hormone which controls the levels of calcium and phosphorus in the blood.

paratonic (aitogenic). Applied to plant movements (e.g. tropisms, nastic movements) which occur in response to external stimuli. *See* Autonomic.

paratype. (*a*) Any specimen cited along with the type specimen(s) in the original description of a species. (*b*) All the external factors which affect the manifestation of a genetic character. (*c*) An abnormal type within a species used, for instance, about bacterial colonies.

paraxonia. Artiodactyla.

parenchyma. (*a*) Loose, soft packing tissue, consisting of living cells. It occurs extensively in plants (e.g. in pith, cortex) and in animals which lack a coelom (e.g. flatworms). (*b*) The cells specific to an animal organ, as distinct from its other components such as blood vessels and connective tissue.

Parazoa. Multicellular animals whose bodies are at a very low level of organisation, and which are usually excluded from the Metazoa. Sponges (Porifera) are the only examples.

parental generation. P_1.

Parker Morris. A set of standards to which all public and much private house building in Britain is required to conform. The standards were first proposed in *Homes for Today and Tomorrow* (the *Parker Morris Report*) commissioned by the then Ministry of Housing and Local Government and published in 1961. The standards govern such matters as floor space (including storage space), heating, basic fittings, outside appearance and gardens.

Parrots. Psittaciformes.

parthenocarpy. The development of seedless fruit without fertilisation (e.g. in banana).

parthenogenesis. The production of offspring by the development of unfertilised ova. Parthenogenesis occurs naturally in some plants (e.g. dandelion) and animals (e.g. aphids), the egg nucleus usually being diploid. Parthenogenesis can sometimes be induced artificially by cooling and other techniques applied to the haploid ova of other animals. *See* apomixis, merogony.

parthenogenetic merogony. *See* merogony.

partial population curve. The population density of a given developmental stage plotted against time.

particle. A small, discrete mass of solid or liquid matter (e.g. aerosol, dust, fume, mist, smoke, and spray, each of which has different properties).

Pascal. *See* Newton.

Passeriformes (Passerines). Perching birds and song birds. A very large order of birds containing about half the known species and occupying many land habitats. They have three toes in front and one long one behind. Often the display behaviour is complicated, the nests elaborate, and the males have a well-developed song. They include the larks, swallows, wagtails, pipits, shrikes, orioles, starlings, waxwings, crows, wrens, accentors, warblers, flycatchers, thrushes, tits, treecreepers, finches and buntings.

Passerines. Passeriformes.

pasteurisation. Partial sterilisation (named after its originator, Louis Pasteur (1822–95), widely used to destroy many pathogenic bacteria in food without drastically altering its flavour. Milk is pasteurized by being heated for half an hour at 62°C, after which it is cooled rapidly.

pasture. *See* grassland, Graminae.

patabiont. An animal which spends its life in forest floor litter.

patacole. An animal which lives transiently in forest floor litter.

paternoster lakes. Strings of lakes, which are dammed by moraines or rock-bars, in glacially-eroded valleys.

pathogen (pathogenic organism). An organism responsible for the transmission of communicable diseases (e.g. cholera, amoebic dysentry, bacillary dysentry, malaria, yellow fever, typhoid fever, etc.)

pathological waste. Waste that could contain pathogens and whose disposal might therefore threaten public health. It includes hospital wastes, some laboratory wastes, and any other infected materials.

patina. Desert varnish.

patoxene. An animal which occurs by accident in forest floor litter.

pattern intensity. The extent to which density of a pattern varies from place to place. If intensity is high, differences are pronounced and densely populated zones alternate with zones that are sparsely populated.

pavement. All the layers which together comprise a road. Generally there are four. (*a*) The sub-base, of unbound aggregate for drainage. (*b*) The road base, which bears most of the weight, made from aggregates that may be bonded (bound together, usually with bitumen) or unbonded. (*c*) The basecourse, generally made from asphalt, aggregate and bitumen, to provide a shaped surface. (*d*) the wearing course, made from asphalt, aggregate and bitumen to take the wear from vehicles and to protect the lower layers. Together, the basecourse and wearing course comprise the road surface. The road surface may be rigid (rigid pavement) or have a slight elasticity under traffic movement (elastic pavement).

P/B. A ratio of primary production to total biomass.

Pb. Lead.

PCB. Polychlorinated biphenyl.

peak sound pressure level. The value, in decibels, of the maximum sound pressure, as opposed to the root-mean-square or effective sound pressure.

pearly. *See* lustre.

pea soup fog. A particularly dark and dirty greenish fog characteristic of London in the decades before 1940, caused largely by domestic smoke, and sometimes having a relatively clear layer close to the ground giving the impression of nocturnal darkness in daytime.

peat. Organic soil, often many feet deep, composed of partly decomposed plant material. It forms under anaerobic conditions in waterlogged areas such as fens and bogs.

peck-order. The social hierarchy in a group of gregarious animals of one species.

pectin. *See* carbohydrate.

ped. An individual, natural aggregate of soil particles.

pedalfer. A leached soil formed where moisture and drainage is sufficient to remove soluble constituents. *See* pedocal.

pediment. An eroded wash-slope at the foot of a mountain in an arid or semi-arid region.

pedocal. A non-leached soil where moisture and drainage have been insufficient to leach away soluble constituents. An alkaline soil, containing a horizon of accumulated carbonates. *See* pedalfer.

pedogenesis. The origin and development of soils.

pedogenic. Applied to effects which are caused by edaphic (soil) factors.

pedology. The study of the morphology and distribution of soils.

pedon. The smallest vertical column of soil containing all the soil horizons present at that point. *See* polypedon.

pegmatite. Igneous bodies of coarse-grain usually found as dykes associated with a larger body of finer-grained plutonic rock. The grain sizes are relative: crystals ten metres long are found in some pegmatites. Pegmatites crystallised at a late stage from the magma, when it was enriched in volatiles, and some are sources of minerals of rare elements such as tantalum, lithium, niobium, and the rare earths.

Pekilo process. A process for treating starch and sulphite wastes from the food industry to produce protein by continuous fungal cultivation under aseptic conditions in stirred fermentation vessels. *See* novel protein foods.

Peking Man. *See* Homo.

pelagic. Applied to organisms of the plankton and nekton which inhabit the open water of a sea or lake. *Cf.* benthos, littoral, demersal.

Pelecaniformes. An order of large, aquatic, fish-eating birds, with all four toes webbed. They include the gannets, which feed by swimming under water or diving from great heights, the pelicans, which scoop up fish in their large beaks while swimming, and the cormorants, which dive from the surface. Most species nest in colonies.

Pelecypoda. Lamellibranchiata.

pelitic. Argillaceous; clayey. The term is used chiefly to refer to metamorphic (*see* metamorphism) rocks formed from argillaceous rocks.

pellagra. Deficiency disease caused by lack of nicotinic acid.

peltier effect. Electrical effect associated with the temperature difference between functions of dissimilar metals.

pelton wheel. *See* impulse turbine.

pene-. Prefix meaning almost. Hence penecontemporaneous, peneplain, etc.

peneplain. The logical end product of river erosion and landscape evolution. A low-lying surface, approaching base-level, literally 'almost a plain'.

penetrative convection. *See* convection.

Penguins. Impennae.

penicillin. *See* antibiotic.

Penicillium notatum. The mould from which penicillin is derived. *See* antibiotic.

Pennsylvanian. A system used in North American stratigraphy equivalent to the Upper Carboniferous. *See* geological time.

pentadactyl limb. The type of limb evolved as an adaptation to terrestrial life in vertebrates. The basic skeletal pattern, often much modified by loss or fusion of parts, consists of a large upper bone (humerus or femur) articulating with two parallel bones (radius and ulna or tibia and fibula). The distal end of the limb consists of several small 'wrist' or 'ankle' bones (carpals or tarsals) articulating with five 'palm' or 'sole' bones (metacarpals or metatarsals), which in turn articulate with five sets of 'finger' or 'toe' bones (phalanges).

Pentastomida. Elongated, worm-like parasites found in the nasal passages of predatory mammals. The larvae, which live mainly in herbivorous mammals, resemble parasitic mites, so the group is usually included in the Arachnida.

pentlandite. A sulphide mineral composed of iron and nickel combined in varying proportions (Ni, Fe)$_9$S$_8$, and formed by the separation of an immiscible sulphide liquid from a basic magma. Pentlandite is one of the two major ores of nickel, the other being garnierite.

pentose. *See* carbohydrate.

peptidase. An enzyme which breaks down peptides and often proteins (e.g. pepsin, secreted by the stomach lining in vertebrates).

peptides. Compounds composed of two or more amino acids formed by the condensation of the -NH$_2$ group of one of the acids and the carboxyl group of another, resulting in the -NH-CO-peptide linkage. *See* polypeptides.

peptones. Breakdown products of proteins, consisting of polypeptides. *See* peptides.

perceived noise level (PNdB). The sound pressure level between one-third of an octave and one octave of random noise at 1000 Hz which 'normal' people consider equally noisy to a sound of interest. $PNdB = 40 + 10 \log_2 Noy$.

percentage area method. Method for determining the distribution and density of a species. A square grid divided by wires into 10 cm or 5 cm squares is placed on the ground and the percentage of the area covered by each species and the percentage of bare ground are estimated.

percentage cover. The mean number of first contacts in each hundred quadrats.

percentage frequency. The mean number of contacts in each hundred quadrats.

percentage of vegetation. The importance of a species, expressed as its percentage of all individuals of all species recorded.

percolating filter (trickling filter). A bed of inert material through which water is trickled in order to purify it. The bed offers a very large surface area that becomes covered with aerobic micro-organisms fed by nutrients carried in the water.

perdominant. A dominant species that is present in all, or nearly all, the associations of a formation.

perennial. An individual plant which continues to live from year to year. In herbaceous perennials the aerial parts die down in the autumn, leaving an underground structure (e.g. bulb, corm, rhizome, tuber) to overwinter. In woody perennials (e.g. trees) there are permanent, woody, aerial stems from which the plant makes new growth each year, so many of these plants attain great size. *Cf.* annual, biennial, ephemeral.

perfect gas. An ideal gas. A theoretical concept of a gas that would obey the gas laws exactly, and would consist of perfectly elastic molecules, the volume occupied by the molecules and the forces of attraction between them, being zero or negligible.

perianth. *See* flower.

pericarp. The part of a fruit formed from the ovary wall (e.g. the flesh of a berry, the shell of a hazel nut).

pericline. An anticline which plunges in both directions.

peridotite. A group of coarse-grained, ultrabasic igneous rocks consisting essentially of olivine, with or without other ferromagnesian minerals, and with no feldspar.

periglacial. The area surrounding the limit of glaciation and subject to intense frost action.

period. A major, worldwide unit of geological time corresponding to a stratigraphic system. The geological periods since the Precambrian are usually grouped into the Palaeozoic, Mesozoic and Cenozoic Eras or, collectively, the Phanerozoic Eon. Periods are subdivided into epochs, which are subdivided into ages.

periodic. Applied to phenomena which repeat in identical form after regular intervals of time.

periodic table of elements. (*See* table).

peripheral nervous system. The part of the nervous system other than the central nervous system. It consists mainly of nerves, each of which is made up largely of a bundle of nerve fibres (long, fine processes of neurons).

periphyton. The organisms attached to underwater rooted plants. *Cf.* benthos, plankton.

Perissodactyla. The odd-toed ungulates (hoofed mammals). These have the

Table of elements in alphabetical order

Element	Symbol	Atomic number	Relative atomic mass	Element	Symbol	Atomic number	Relative atomic mass
Actinium	Ac	89	(227)	Mercury	Hg	80	200.59
Aluminium	Al	13	26.9815	Molybdenum	Mo	42	95.94
Americium	Am	95	(243)	Neodymium	Nd	60	144.24
Antimony	Sb	51	121.75	Neon	Ne	10	20.179
Argon	Ar	18	39.948	Neptunium	Np	93	237.0482
Arsenic	As	33	74.9216	Nickel	Ni	28	58.71
Astatine	At	85	(210)	Niobium	Nb	41	92.906
Barium	Ba	56	137.34	Nitrogen	N	7	14.0067
Berkelium	Bk	97	(247)	Nobelium	No	102	(254)
Beryllium	Be	4	9.01218	Osmium	Os	76	190.2
Bismuth	Bi	83	208.9806	Oxygen	O	8	15.994
Boron	B	5	10.81	Palladium	Pd	46	106.4
Bromine	Br	35	79.904	Phosphorus	P	15	30.9738
Cadmium	Cd	48	112.40	Platinum	Pt	78	195.09
Caesium	Cs	55	132.9055	Plutonium	Pu	94	(242)
Calcium	Ca	20	40.08	Polonium	Po	84	(210)
Californium	Cf	98	(251)	Potassium	K	19	39.102
Carbon	C	6	12.011	Praseodymium	Pr	59	140.0977
Cerium	Ce	58	140.12	Promethium	Pm	61	(147)
Chlorine	Cl	17	35.453	Protactinium	Pa	91	231.0359
Chromium	Cr	24	51.996	Radium	Ra	88	226.0254
Cobalt	Co	27	58.9332	Radon	Rn	86	(222)
Copper	Cu	29	63.546	Rhenium	Re	75	186.2
Curium	Cm	96	(247)	Rhodium	Rh	45	102.905
Dysprosium	Dy	66	162.50	Rubidium	Rb	37	85.47
Einsteinium	Es	99	(254)	Ruthenium	Ru	44	101.07
Erbium	Er	68	167.26	Samarium	Sm	62	150.4
Europium	Eu	63	151.96	Scandium	Sc	21	44.9559
Fermium	Fm	100	(253)	Selenium	Se	34	78.96
Fluorine	F	9	18.9984	Silicon	Si	14	28.086
Francium	Fr	87	(223)	Silver	Ag	47	107.868
Gadolinium	Gd	64	157.25	Sodium	Na	11	22.9898
Gallium	Ga	31	69.72	Strontium	Sr	38	87.62
Germanium	Ge	32	72.59	Sulphur	S	16	32.06
Gold	Au	79	196.9665	Tantalum	Ta	73	180.9479
Hafnium	Hf	72	178.49	Technetium	Tc	43	98.9062
Helium	He	2	4.0026	Tellurium	Te	52	127.60
Holmium	Ho	67	164.9303	Terbium	Tb	65	158.9254
Hydrogen	H	1	1.008	Thallium	Tl	81	204.37
Indium	In	49	114.82	Thorium	Th	90	232.0381
Iodine	I	53	126.9045	Thulium	Tm	69	168.9342
Iridium	Ir	77	192.2	Tin	Sn	50	118.69
Iron	Fe	26	55.847	Titanium	Ti	22	47.90
Krypton	Kr	36	83.80	Tungsten	W	74	183.85
Lanthanum	La	57	138.9055	Uranium	U	92	238.029
Lawrencium	Lw	103	(257)	Vanadium	V	23	50.9414
Lead	Pb	82	207.2	Xenon	Xe	54	131.30
Lithium	Li	3	6.941	Ytterbium	Yb	70	173.04
Lutetium	Lu	71	174.97	Yttrium	Y	39	88.9059
Magnesium	Mg	12	24.305	Zinc	Zn	30	65.37
Manganese	Mn	25	54.9380	Zirconium	Zr	40	91.22
Mendelevium	Md	101	(256)				

weight-bearing axis of the foot passing through the third toe. In horses the first and fifth digits are absent, and the second and fourth are incomplete (splint bones). In the rhinoceros all the feet have three digits, and in the tapir the front feet have four and the hind feet three digits. *See* Artiodactyla.

perknite. *See* ultrabasic.

permafrost. Ground that is permanently frozen. The water in the near-surface layers may melt in summer, with resulting solifluction on slopes. Associated with permafrost are such features as polygons, stone stripes and pingos.

permeability. The capacity of a rock for transmitting fluid. The degree of permeability depends on the size and shape of the pores in a rock as well as the extent, size, and shape of the connections between the pores. Permeability is measured in darcys or millidarcys. Reservoir rocks usually have permeabilities ranging from a few millidarcys to several darcys. The opposite of permeable is impermeable.

Permian. Refers to the youngest period of the Palaeozoic Era, usually taken as beginning some time between 275 and 290 million years BP and lasting about 55 million years. Permian also refers to the rocks formed during the Permian Period: these are called the Permian System, which in Eastern Europe is divided into two series, the Lower and Upper Permian. The Lower Permian is divided into four stages (Asselian, followed by the Sakmarian, Artinskian and Kungurian) and the Upper Permian is divided into two stages (the Kazanian followed by the Tartarian). The evaporite facies of the Permian in north-west Europe is historically divided into the Rotliegendes and the Zechstein.

Permo-Trias. The Permian and Triassic Systems considered together.

peroxisomes. Membrane-bounded particles found in the cytoplasm of plant cells. Peroxisomes contain oxidases and catalase and are probably the site of photorespiration.

peroxyacetyl nitrates (PAN). A component of photochemical smog and the primary cause of smog injuries to plants, many of which are sensitive to concentrations of 0.05 parts per million or less.

persistence. The capacity of a substance to remain chemically stable. This is an important factor in estimating the environmental effects of substances discharged into the environment, certain highly toxic substances (e.g. cyanides) having a low persistence, while other less toxic substances (e.g. many organochlorine insecticides) may cause more serious effects because of their high persistence.

pervious. (*a*) Permeable by virtue of mechanical discontinuities, such as joints, fissures, and bedding-planes in the rock. The opposite of pervious is impervious. (*b*) Permeable (US usage). *See* permeability.

pesticide. Chemical agent, most often artificial, that kills insects and other animal 'pests'. Sometimes used as a general term embracing insecticides, herbicides, fungicides, nematicides, etc. Some pesticides, e.g. DDT have had widespread disruptive effects among non-target species.

petiole. Leaf stalk.

petroleum. A naturally occurring material composed predominantly of hydrocarbons, in a solid, liquid or gaseous state (or a combination of these states) depending on the nature of the compounds and reservoir conditions. Petroleum can be subdivided into natural gas, condensate, crude oil and bitumen.

petrology. The study by all available means of the origins, alteration, present conditions and decay of rocks and minerals.

PFR. Prototype fast reactor.

pH. The measure of the acidity or alkalinity of a substance, based on the number of hydrogen ions in a litre of the solution, and expressed in terms of pH $= \log_{10}\left(\dfrac{1}{H^+}\right)$ where H^+ is the hydrogen ion concentration. Pure water dissociates slightly into hydrogen and hydroxyl ions $(H^+ + OH^-)$ the concentration of each type of ion being 10^{-7} mole per litre, so the pH of water is $\log_{10}\dfrac{1}{10^{-7}} = 7$, and this is taken to represent neutrality on the pH scale. Smaller values are acidic, larger values are alkaline.

phacolith. A lens-shaped igneous intrusion in the crest of or trough of a fold.

Phaeophyceae. Phaeophyta.

Phaeophyta (Phaeophyceae, Brown Algae). A division of algae, mostly marine and benthic, common in the intertidal zone (*see* shore zonation). Phaeophyta contain a brown pigment (fucoxanthin) which masks the chlorophyll and other pigments present, producing a brown to olive-green colour. Some species are filamentous and minute, others (e.g. bladder-wracks and kelps) are very large, and differentiated into a basal holdfast and a stem-like portion bearing ribbon-like or leaf-like fronds. Brown seaweeds are used as manure and fodder in some coastal districts, and a few species are edible.

phages. Bacteriophages.

371

phagocyte (phagocytic cell). Macrophages and granular leucocytes (*see* blood corpuscles) which ingest bacteria, tissue cells (usually dead), protozoa and small particles of matter. In simpler forms of life this is a method of feeding; in higher forms it counters infections from invading organisms.

phagocytic cell. Phagocyte.

phagocytise. To consume a particle by engulfing it.

phagocytosis. The process, involving flowing movements of the cytoplasm, by which a cell ingests particles from its surroundings. *See* blood corpuscles, macrophages.

Phalangida (Opiliones, Harvestmen). Long-legged, predaceous Arachnida, distinguished from spiders by the absence of silk glands and by their undivided bodies.

Phanerogamia. Spermatophyta.

phanerophyte. *See* Raunkiaer's Life Forms.

Phanerozoic. The period of time from the beginning of the Cambrian Period to the present day. *See* geological time.

phase. A measure of whether a periodic phenomenon is in time with another, usually measured in radians ($360° = 2$ radians). If one sine wave is at its maximum when another is at its minimum, the two are said to be $180°$ out of phase. *See* angular frequency.

Phasmida. Stick and leaf insects. An order of mainly tropical insects (Exopterygota), many of them wingless, closely resembling twigs or leaves. Some have no known male sex and reproduce by parthenogenesis. Phasmida were formerly included in the Orthoptera.

phellem. *See* cork.

phellogen. *See* cork, meristem.

phenocopy. A non-inherited variation, caused by environmental influence, which produces an effect like that of a known gene mutation.

phenocryst. A relatively large crystal set in a finer groundmass in an igneous rock, forming a porphyritic texture.

phenology. The study of the timing of recurring natural phenomena (e.g. the

372

flowering of particular plants, the arrival or departure of migrating birds, etc.) with particular reference to climatological observations.

phenotype. The outward expression of the genes. The characteristics displayed by an organism as determined by the interaction of its genetic constitution and its environment. *Cf.* genotype.

phenylalanine. Amino acid with the formula $C_6H_5CH_2$ CH. (NH_2). COOH. Molecular weight 165.2.

pheromone. A substance released by an animal into its surroundings which influences the behaviour or development of others of the same species (e.g. sexual attractants released by female moths; queen-bee substance, which prevents the development of other queens).

philoprogenetive. Prolific. Producing many offspring.

phloem. Vascular tissue through which synthesised food products are transported in Pteridophyta and seed plants. The characteristic element of phloem is the sieve-tube, a longtitudinal row of elongated, living cells, which communicate by means of perforations in their connecting walls.

Pholidota (Manidae). Pangolins or scaly ant-eaters of Asia and Africa, formerly included in the Edentata, but now classified as a separate order. Some are burrowing, others are able to climb trees. Like other ant-eaters, they have no teeth, an elongated snout, a long tongue and large claws, but their skin is covered with horny scales, interspersed with hairs.

phon. A unit of loudness. Phon $= \log_2$ Sone.

phoresy. The transport of one animal by another of a different species (e.g. certain nest-dwelling pseudoscorpions (arachnids) temporarily attach themselves, by means of their claws, to birds and insects, and are transported to new feeding grounds).

Phoronidea. A small phylum of marine, worm-like animals, which superficially resemble Polyzoa. Unlike these, Phoronidea possess a blood system and haemoglobin.

phosgene. A very toxic gas ($COCl_2$), used as a weapon in the 1914–18 war. It interferes with the nervous system and causes inflammation of the lungs. Given off by many degreasing agents (e.g. trichloroethylene) when they are heated, also produced by metal plating processes and by the combustion of some organic chlorine compounds (e.g. polyvinylchloride) if the temperature of combustion is high and carbon monoxide is present.

phosphagens. Organic phosphates (e.g creatine phosphate in vertebrates) which provide a store of phosphate for building up ATP during muscular contraction.

phosphate. Essential nutrient (*see* macronutrients) for plants, supplied naturally from the weathering of rocks and artificially from the mining of rock phosphate, an insoluble evaporite. Rock phosphate is processed by treatment with sulphuric acid to produce 'superphosphate', containing some soluble phosphorus pentoxide (P_2O_5) fertiliser, and superphosphate can be treated further, with phosphoric acid, to increase the P_2O_5 content, as 'triple superphosphate'. Gypsum is a by-product. Phosphates are also used industrially. They can cause eutrophication when discharged, mainly in sewage. Phosphate is fairly immobile in soils and contamination of water from phosphate fertiliser is not severe. *See* phosphorus cycle.

phosphate rock. A sedimentary rock rich in the mineral apatite, $Ca_5(PO_4)_3OH$, which is the only important source of phosphorus. Phosphate rock, which is relatively insoluble, is treated with dilute acid (*see* phosphate) to produce a more soluble material containing a higher percentage of such water-soluble compounds as $Ca(H_2PO_4)_2$.

phospholipid. *See* lecithin.

phosphorite. Phosphate rock. The term is variously used to refer to all phosphorus-rich deposits, or only to lithified rock phosphate where the distinction is made with the usually unlithified pebble phosphates.

phosphorus (P). Element. Occurs in several allotropic forms (*see* allotropy), the most common being white phosphorus, a waxy white, very poisonous, very inflammable solid (mp 44°C) or as red phosphorus, a dark red, non-poisonous, not very inflammable powder. It is found in nature only in the combined state, mainly as phosphate rock. It is used in fertilisers, detergents, and matches, and is essential to life (*see* macronutrients). AW 30.9738; At. No. 15.

phosphorus cycle. The circulation of atoms of phosphorus, brought about mainly by living organisms. Phosphorus is an essential constituent of DNA and RNA and is involved in the ADP-ATP (see ATP) process whereby energy is made available to cells. The death and decomposition of living tissues returns phosphorus to the soil in a form available to plants, so continuing the cycle.

photic zone. The surface waters of a sea or lake, penetrated by sunlight. This zone includes the euphotic and dysphotic zones.

photochemical. Applied to chemical reactions brought about by energy supplied to substances by light.

photochemical smog. *See* smog.

photogeology. The geological interpretation of aerial photographs. The photographs used are most commonly overlapping vertical views taken along a straight line of flight; these are then viewed stereoscopically. Satellite imagery, particularly from the ERTS and LANDSAT programmes is being used increasingly.

photon. A quantum of electromagnetic radiation with zero rest mass and energy equal to the product of the frequency of the radiation and Planck's constant. Photons are generated when a particle possessing an electrical charge changes its momentum, when nuclei or electrons collide, and when certain nuclei and particles decay. In some contexts photons are regarded as elementary particles.

photonasty. *See* nastic movement.

photoperiodicity. Photoperiodism.

photoperiodism (photoperiodicity). The response of plants or animals to the relative duration of day and night. The timing of the breeding season in many vertebrates and the flowering period of many plants is regulated by day length. For instance, the chrysanthemum is a 'short-day' plant, flowering when the nights are long, whereas lettuce is a 'long day' plant. *See* phytochrome.

photophosphorylation. The formation of ATP from ADP and phosphate, utilising light as the source of energy. *See* photosynthesis.

photorespiration. Respiration activated by light, characteristic of certain plants (e.g. wheat, sugar beet), but not of others (e.g. maize, sugar cane). The two types of respiration follow different biochemical pathways. *See* peroxisomes.

photosynthesis. The synthesis by living cells of organic compounds from simple inorganic compounds, using light energy. In green plants the absorption of light by chlorophyll initiates photochemical reactions ('light reactions') in which oxygen is released from water and light energy is converted into chemical energy by the formation of ATP. This is photophosphorylation. Subsequent 'dark reactions' result in the reduction of carbon dioxide (using hydrogen which originated in the water) to carbohydrates, utilising the energy stored in ATP. This may be summarised as:

$$CO_2 + H_2O \xrightarrow[\text{by chlorophyll}]{\text{light energy absorbed}} > \text{Carbohydrate} + O_2$$

Amino acids are synthesised by the combination of intermediate products, with elements such as nitrogen, derived from mineral salts. Most of the oxygen in the atmosphere is derived from photosynthesis. Photosynthetic bacteria, however, do not produce oxygen, because they take the hydrogen necessary for the reduction of carbon dioxide from organic compounds or inorganic sources other than water. Heterotrophic organisms depend on the supply of organic material produced by chlorophyll-containing plants during photosynthesis, and so are dependent on sunlight as their ultimate source of energy.

phototaxis (heliotaxis). *See* taxis.

phototrophic. Applied to organisms which obtain energy from sunlight. *See* photosynthesis, autotrophic, chemotrophic.

phototropism (heliotropism). (*a*) (Zool.) Once a synonym for phototaxis (*see* taxis). (*b*) A growth response of plants in which the stimulus is light. Most stems are positively phototropic, and will curve towards a light source; some roots (e.g. the aerial roots of Ivy) are negatively phototropic. Growth curvatures are under the control of auxins.

photovoltaic effect. The production of an electromotive force by radiant energy, commonly light, falling on the junction of two dissimilar materials, such as a metal and a semi-conductor.

phreaticolous. Applied to organisms which live in subterranean fresh water.

phreatic water. Water below the water table in the zone of saturation, i.e. groundwater.

phreatophyte. A plant with roots long enough to reach the water table.

Phthiraptera. A term sometimes used to cover the biting lice (Mallophaga) and sucking lice (Anoplura).

phycobilins (biliproteins). Blue and red pigments present in blue-green and red algae. Like the chlorophyll also present in these plants, phycobilins absorb light energy.

Phycomycetes. A class of fungi whose members usually show a well-marked sexual phase, with the production of thick-walled resting spores. The hyphae are mostly without cross walls. The group includes parasites which cause white rust of Cruciferae (*Cystopus*), potato blight (*Phytophthora*) and fungus diseases of fish (e.g. *Saprolegnia*). Saprophytes include *Mucor*, the common black pin mould which grows on food such as stale bread.

phyllite. An argillaceous metamorphic (*see* metamorphism) rock with shiny cleavage planes due to the presence of flaky minerals. The individual flaky minerals are too small to be seen with the naked eye: with increasing grain size phyllite grades into mica-schist. Phyllites are formed under low temperature regional metamorphism.

phylloclade. Cladode.

phyllode. A flattened, expanded petiole (leaf stalk), taking the place of a reduced leaf blade (e.g. in many species of *Acacia*).

phylogeny. The evolutionary history of a group of organisms, as distinct from the development of an individual organism.

phylum. *See* classification.

physiocracy. The rule of nature, as this was understood by the Physiocratic School which flourished in France in the 18th Century and which laid the foundations for modern economics, exerting considerable influence on Adam Smith. The Physiocrats believed that the only true source of wealth is the land, since it is only natural biological processes that produce goods from nothing: manufacturing industry merely alters the form of pre-existing materials and so adds nothing. Thus the products of agriculture should be prized highly and those of manufacture considered as of less worth. All living organisms, including Man, are governed by natural laws, which have universal applicability and with which man-made laws should conform, so that the purpose of a ruler is only to educate his people regarding natural laws, to foster their observance, and to remove hindrances to their observance. It was the Physiocrats who coined the phrase '*laissez faire, laissez passer*'. *See laissez-faire.*

physiographic succession. That component in a succession in which the composition of communities is related to topographical features.

physiological specialisation (biological specialisation). The occurrence of a number of genetically distinguishable forms within a species, which differ from one another in their biochemical characters but not in their structure. In disease-producing organisms (e.g. black stem rust of wheat) this is of great importance, because differences in physiology cause differences in pathogenicity towards different varieties of the host species. This complicates the breeding of resistant varieties of crops.

Physopoda. Thysanoptera.

phytoalexins. Substances poisonous to fungi and produced by plants when they are infected by fungus diseases.

phytobenthos. *See* benthos.

phytochrome. A plant pigment which becomes activated by a certain wavelength

377

of light, and in its activated form initiates growth, germination, flowering, etc. In the absence of the red light, the phytochrome becomes de-activated or lost. Phytochrome regulates photoperiodism.

phytocoenosis. The assemblage of plants inhabiting a particular area.

phytogeocoenosis. A plant community and its physical environment.

phytogeography. The study of the distribution of plant species in relation to climate, geography and history.

phytokinins. Cytokinins.

phytophagous. Applied to animals that feed on plants. Herbivores.

phytoplankton. *See* plankton.

phytoplankton bloom. Algal bloom.

phytosociology. The study of plants in the widest sense and including the study of all phenomena that affect their lives as social units. It is thus the study of plant communities and so covers much of the field of ecology, including the composition of communities, their morphology and structure, their development and change, the relationships among species and between species and their environment, and the classification of communities.

phytotoxic. Poisonous to green plants.

phytotron. A building used for growing plants under controlled environmental conditions.

pica. The habit in children of eating non-food materials.

Picidae (woodpeckers). A family of birds (order Piciformes) specialised for climbing and wood boring. The bill is strong and sharply pointed for digging out insects and making nest holes in trees, and the long, sticky, protrusible tongue is used for extracting insects from holes. The tail is used to support the bird as it climbs.

pico- (p). Prefix used in conjunction with SI units to denote the unit $\times 10^{-12}$.

picrite. *See* ultrabasic.

piedmont glacier. *See* glacier.

piezoelectric effect. The interaction between electrical and mechanical stress-strain variables in a medium, converting energy from one form to another.

piezometer. *See* piezometric level.

piezometric level. (*a*) The level to which water from a confined aquifer will rise under its own pressure in a borehole. (*b*) The level to which water will rise in a piezometer (a tube inserted into the ground so that the lower end, with a permeable tip, is at a position from which pre-pressure measurements are required).

pigeon's milk. Cropmilk.

piliferous. (Bot.) Having hair.

pillow lava. Lava having the shape of pillows piled on top of each other. The lava can be of any composition from acid to basic, but all recent pillow lavas appear to have been formed in water.

Pinaceae. A family of coniferous trees and shrubs which includes firs (*Abies*), pines (*Pinus*), cedars (*Cedrus*) and the deciduous larches (*Larix*). Pinaceae yield resins, turpentine, and bark for tanning, but are most important because of their easily-worked timber, for which the pines in particular are cultivated on a vast scale.

pingo. A more or less conical hill of unconsolidated material found in a region of permafrost.

pinnate. Applied to plants bearing compound leaves consisting of more than three leaflets arranged in two rows on a single stalk. *Cf.* bipinnate.

pinnatifid. Applied to leaves that are lobed pinnately (see Pinnate), but divided completely into leaflets.

Pinnipedia. Seals, walruses, sea-lions, etc. An order of carnivorous aquatic mammals with fin-like limbs, formerly included in the Carnivora.

pinocytosis. The ingestion of drops of liquid by cells.

pinworms. *See* Nematoda.

Pisces (fish). *See* Acanthodii, Chondrichthyes, Osteichthyes, Placodermi, Cypselurus.

pisciculture. Aquaculture.

pisé de terre. Rammed earth.

pistil. *See* flower.

pistillate. Applied to flowers which have carpels but not stamens, and so are female. *Cf.* staminate.

pitch. (*a*) (Acoustic) An aural assessment of sounds so that they can be ordered in a scale from low to high. Pitch is determined mainly by frequency but also by sound pressure and waveform. (*b*) The angle with which a propeller meets the air or water. Most propeller-driven aircraft, and many advanced aerogenerators, have variable pitch propellers. (*c*) A black, carbonaceous, non-crystalline material which is solid at normal temperatures, but flows on slight warming. Occurs in nature, rarely; more commonly it is obtained as a residue from oil refining.

pitchblende. A naturally occurring mixture of uranium oxides, including uraninite, found in massive form.

pitchstone. A predominantly glassy acidic igneous rock. Pitchstone tends to have a duller lustre and a flatter fracture than obsidian.

Pithecanthropus erectus. *See Homo.*

Pithecoidea. Anthropoidea.

Pitot tube. Instrument for measuring the velocity or pressure of a gas or liquid. It is a tube with two openings, one into the moving fluid and one away from it. The difference in pressure between ends of the tube is related to the velocity of the fluid. In an aircraft, the airspeed indicator, pressure altimeter and vertical speed indicator are linked to a pitot tube carried outside the aircraft.

pituitary gland (pituitary body, hypophysis cerebri). A gland situated under the floor of the brain in vertebrates, which secretes a number of hormones, some of which control the activity of other endocrine organs. The activity of the pituitary gland is influenced by the brain. *See* anti-diuretic hormone, oxytocin, follicle-stimulating hormone, luteinising hormone, corticotropin, lactogenic hormone, thyrotrophic hormone, intermedin, growth hormone.

placenta. (*a*) (Bot.) The part of the ovary wall which bears the ovules. (*b*) (Zool.) An organ by which the embryo of mammals is attached to the uterus, and through which the exchange of oxygen, food, antibodies and waste products takes place, maternal and foetal blood vessels being in close contact. In placental mammals the embryonic tissues involved are the choron and usually the allantois. In most marsupials there is no placenta, but in a few the yolk sac or allantois develops a weak connection with the uterus. The placenta secretes progesterone and small amounts of oestrogen.

Placentalia (placental mammals, Eutheria). Mammals with a highly developed placenta. A sub-class containing most of the present-day mammals (*See* Monotremata, Marsupalia). *See* the orders Artiodactyla, Carnivora, Cetacea, Chiroptera, Edentata, Insectivora, Lagomorpha, Perissodactyla, Pholidota, Pinnipedia, Primates, Proboscidea, Rodentia, Sirenia, Tubulidentata, Hyracoidea, Dermoptera.

placental mammals. Placentalia.

placer. A sediment in which transportation (by wind, water or ice) has caused concentrations of heavy minerals (e.g. gold, cassiterite, rutile. etc), weathered out from the rocks or veins.

Placodermi. An extinct group of mainly Devonian fishes whose bodies were heavily armoured with bone.

plaggen. Soils with deep cultivation horizons. *See* soil horizons.

plagioclase. *See* feldspar.

plagioclimax. A stable plant community, in equilibrium under existing conditions, but which has not reached the natural climax, or has regressed from it, because of the operation of biotic factors such as human intervention (e.g. constantly grazed or mowed grassland; woodland managed as coppice). *See* preclimax, postclimax, subclimax, serclimax, proclimax.

plagiogeotropism. *See* geotropism.

plagiosere. A plant succession deflected from its normal course by biotic factors. This results in the formation of a plagioclimax.

plain muscle. *See* muscle.

Planck's constant (h). The universal constant relating the frequency of radiation (v) with its quantum of energy (E) ($E = hv$). The value of Planck's constant is 6.62559×10^{-34} joule seconds, 6.62559×10^{-27} erg seconds. Named after Max Planck (1858–1947).

plane wave. Wave in which the wave fronts are parallel with one another and at right angles to the direction of propagation.

plankton. Animals and plants, many of them microscopic, which float or swim very feebly in fresh or salt water bodies. They are moved passively by winds, waves or currents. The animals (zooplankton) are chiefly Protozoa, small Crustacea, and larval stages of molluscs and other invertebrates. The plants (phytoplankton) are almost all algae (e.g. diatoms, Pyrrophyta, desmids, etc.). The smallest organisms (e.g. diatoms) are called nannoplankton. All animal life in the open sea depends ultimately on the phytoplankton, which is the basis of food chains leading to fish, whales, sea birds, etc.

planned economy. An economic system controlled centrally by an authority responsible for all planning connected with production and distribution of goods. *Cf.* mixed economy, *laissez-faire*.

planning permission. The official permission required to carry out a particular development. Under British planning laws, administered by planning authorities ranging from the lowest tier of local government to the Secretary of State for the Environment, permission must be obtained for any development that would alter the use of a building or area of land, or that would have a major effect on the landscape or environment.

planosol. Soil with a compact clayey depositional layer developing into hardpan. *See* soil horizon.

plant. An organism characterised chiefly by a holophytic mode of nutrition dependent upon the possession of chlorophyll. Some plants, however, are parasitic (*see* parasitism) or saprophytic. Most have cellulose cell walls, are sedentary, and have branching bodies. The distinction between plants and animals becomes blurred in some groups (e.g. slime moulds, unicellular algae, Protozoa), which are now often regarded as part of a third kingdom, the Protista.

plantain. *See* Musa.

plantigrade. Applied to animals (e.g. Man) which walk with all of the lower surface of the foot (metacarpals or metatarsals and digits) on the ground. *Cf.* digitigrade, unguligrade.

plant louse. *See* Aphididae.

plasmagenes. Cytoplasmic particles or substances which can reproduce and pass on inherited qualities to daughter cells. Unlike genes, they are not inherited through the chromosomes of the gametes in a Mendelian way. Plasmagenes may be present in bodies such as plastids (plastogenes) or mitochondria.

plasmalemma. The cell membrane. In plant cells the term is applied only to the external membrane (ectoplast) and not to the one surrounding the vacuole. *See* tonoplast.

plasma-membrane (cell membrane). A very fine membrane, composed mainly of protein and fat, bounding the cytoplasm of cells. It isolates the cell, allowing its internal composition to differ from that of its surroundings, because it is selectively permeable, allowing water and fat molecules to pass readily across it, but preventing the passage of many substances (e.g. proteins and many ions). In active transport, substances are passed across cell membranes against concentration gradients. *See* osmosis.

plasmodemata. *See* cell wall.

Plasmodium. (*a*) A genus of protozoan parasites (Sporozoa) found in the blood of birds and mammals, causing malaria. The intermediate hosts are mosquitoes. (*b*) The vegetative stage of slime fungi (Myxo-mycophyta) consisting of an amoeboid, multinucleate mass of protoplasm. *See* coenocyte, synctium.

plasmogamy. Plastogamy.

plasmon. A cell's complement of plasmagenes.

plastics. Organic polymers of wide variety of types and uses (e.g. PVC, polyurethane, polythene). Commonly plastics fall into two groups: thermoplastic and thermosetting, depending upon their behaviour on heating.

plastids. Membrane-bounded bodies found in the cytoplasm of most plant cells (not in fungi, blue-green algae or bacteria). Leucoplasts are colourless, and store food materials. Chromoplasts (chromatophores) contain pigments (e.g. carotenoids and chlorophyll).

plastogamy (plasmogamy). The fusion of cytoplasm but not nuclei when cells come together to form a plasmodium.

plastogene. A plastid which functions as a plasmagene.

plateau basalt. An extensive flow, or series of flows, of basaltic lava which, because of differential erosion, forms a plateau. Most plateau basalts are also flood basalts. The terms are often used synonymously.

plate tectonics. The theory that the lithosphere is made up of several moving major rigid plates, plus some minor ones, whose edges are defined by belts of earthquakes, volcanoes, Benioff zones, oceanic trenches, island arcs, midoceanic ridges, and transform faults. Boundaries between plates can be of three major types:(*a*) constructive, where plates are moving apart with the production of new material, by sea-floor spreading along ridges; (*b*) destructive, where one plate descends beneath another along a subduction zone, most of the sediment being scraped off to form fold mountains, until eventually the ocean may close and a

383

continent collision result, (*c*) conservative, where plates slide past each other along transform faults. The mechanism causing the movement of plates is unknown, but is thought to be related to the Low Velocity Zone. The movements of ancient plates far back into the Precambrian have been postulated, with variable amounts of agreement. Although there are still unexplained and apparently contradictory features of the Earth on this model, the theory has revolutionised Earth science, and provided new ideas in the search for resources.

platform. A stable, flat area, e.g. a continental shelf, on which a thin sequence of sediments may accumulate.

platinoids. The platinum group of metals (platinum, palladium, rhodium, iridium, ruthenium and osmium) which occur native with nickel and copper sulphide as a result of differentiation in basic magma, and in placer deposits. The metals find uses as catalysts, in alloys and electroplating, and in applications such as electrical contacts, where corrosion or oxidation would be detrimental.

Platyhelminthes (flatworms). A phylum of flattened invertebrate animals, comprising the flukes (Trematoda), tapeworms (Cestoda) and free-living flatworms (Turbellaria). Platyhelminthes are triploblastic and have no coelom or blood system. The alimentary canal, when present, has a single opening, the mouth. The reproductive system is complicated and usually hermaphrodite.

playa. A lake which exists temporarily after rainfall, or its dried-up lake-bed. Playas are common features of some deserts and may contain evaporite deposits.

Plecoptera (stoneflies). A small, primitive order of insects (Exopterygota) whose members have two pairs of membranous wings. The larvae are aquatic and predatory. The adults are short-lived and do not feed.

pleciotropic. Applied to a gene which affects more than one characteristic in the phenotype.

Pleistocene. Refers to the older subdivision of the Quaternary. The Pleistocene, usually ranking as an epoch, was the time of the most recent glaciation, and is thought to have lasted from about 2 million years ago to about 10000 years ago. Some climatologists consider we are still living in an interglacial period of the Pleistocene. Pleistocene also refers to the rocks deposited during this time: these are called the Pleistocene Series.

pleomorphism. Polymorphism.

Plesiosauria (Sauropterygia). An order of large Mesozoic reptiles with long necks and two pairs of five-toed, paddle-like limbs. They were marine, mostly entirely aquatic, but some could probably also move about on land.

pleuriilignosa. Rain forest and rain bush.

pleurodont. *See* thecodont.

pleuropneumonia-like organisms. Mycoplasmas.

pleustron. A single-layered plant community floating on or in a water body (e.g. duckweeds on the surface of a pond).

Pliensbachian. A stage of the Jurassic System.

plinthite. *See* hardpan.

Pliocene. Refers to a subdivision of the Cenozoic Era. The Pliocene, usually ranking as an epoch, follows the Miocene, and is thought to have lasted from 7 to 2 million years before present. Pliocene also refers to rocks deposited during this time: these are called the Pliocene Series.

pliomorphism. Polymorphism.

plug. A comparatively small, steep-sided igneous body which is roughly circular in plan. Some plugs are in-filled and solidified feeder channels for volcanoes and are composed of lava or pyroclastic rock.

plume. Chimney effluent composed of gases alone or gases and particulates. The form of the plume depends on turbulence in the atmosphere. Descriptions of plumes include looping, coning, fanning, fumigating and lofting.

plumule. (*a*) (Bot.) The terminal bud in the embryo of a seed plant. (*b*) (Zool.) A down feather.

plunge. The plunge of a fold is the angle between the axis of the fold and the strike of the axial plane.

pluteus. A stage in the development of Echinoidea and some Ophiuroidea, formed from the coeloblastula (*see* coelom, blastula) with the development of skeletal rods and limb-like projections and the disappearance of cilia. Plutei metamorphose into adults without attaching themselves to a substrate.

pluton. A large igneous intrusion which solidified at depth. The term is more rarely used to refer to any igneous intrusion of unknown configuration or to an intrusion which does not fit any other category (such as batholith, stock, boss, laccolith, sill, dyke, etc.).

plutonic. (*a*) Refers to a rock, or to a large igneous intrusion which crystallized at depth. Plutonic rocks are contrasted with hypabyssal and with volcanic rocks. (*b*) Loosely, coarse-grained. (*c*) Refers to any process involving heat deep within the Earth's crust (e.g. migmatisation).

plutonium. Pu. A transuranic element, different isotopes of which are produced by nuclear reactions. $^{349}_{94}$ Pu is produced in nuclear reactors and has a half-life of 24400 years. It is also used in nuclear weapons, one kilogram having an energy equivalent of about 10^{14} joules. At. No. 94.

PNdB. Perceived noise level in decibels.

pneumatolysis. The alteration of rock, both the igneous body and country rock, by volatiles produced during the consolidation of magma. Kaolinisation, tourmalinisation, greisening, and serpentinisation are examples of pneumatolysis, which is itself one type of metasomatism.

pneumatophores. Erect, aerial root branches produced by some plants (e.g. mangroves such as *Avicennia* and *Sonneratia*) which grow in water or tidal swamps. Air spaces inside the roots communicate with the atmosphere by means of pores.

pneumoconiosis. Incapacitating disease caused by retention of inhaled dusts in the lung, reducing lung function. Occurs among workers in the mining industries, esp. in coal mining.

Podicipedidae (Grebes). A family of diving birds (order Podicipediformes or Colymbiformes) with lobed toes. Grebes nest on floating vegetation. They are clumsy on land and have weak and hurried flight. Courtship displays are often elaborate.

podzol (podsol). An acid soil where aluminium and iron oxides and hydroxides have been leached from the upper layers (A horizon) and redeposited in the lower layers (B horizon), leaving a surface of raw humus over an upper pale horizon and a denser, darker horizon beneath. *See* leaching, soil horizons.

podzolisation (podsolisation). The process by which iron and aluminium oxides and hydroxides are leached from upper layers of soils and precipitated in the zone of deposition beneath. *See* leaching, soil horizons.

poikilosmotic. Applied to animals (e.g. many marine invertebrates) which have body fluids with an osmotic pressure which varies according to that of the surrounding water. *Cf.* homoiosmotic.

poikilothermy. 'Cold bloodedness'. The possession of a body temperature which varies, approximating to that of the surroundings. All animals except birds and mammals are poikilotherms. Their rate of metabolism slows down in cold conditions, so many of them have to hibernate. *Cf.* homoiothermy.

point bar. Sand or gravel deposit built up by rivers on the inner curve of a meander.

Polar desert soil. *See* tundra soil.

polje. A feature of karst landscape consisting of a closed basin with an irregular outline, and usually, steep walls. Surface water disappears down a swallow hole after traversing the flattish floor of the polje.

poll. (*a*) To remove the branches from the main trunk of a living (unfelled) tree. (*b*) To remove the horns from cattle.

pollard. (*a*) Tree from which the branches have been removed, leaving a main trunk from which new growth develops at the top. (*b*) Animal whose horns have been removed.

pollen. Microspores (*see* heterospory) of seed-producing plants. Each pollen grain contains a much reduced male gametophyte (see Alternation of generations). Pollen grains are transferred by wind, insects, water or birds to the ovules (in Gymnospermae) or to the stigmas (in flowering plants), where pollen tubes containing male nuclei grow out and penetrate the embryo sac.

pollen analysis. Palynology.

polluter pays principle. *See* externalities.

pollution. The direct or indirect alteration of the physical, thermal, biological or radioactive properties of any part of the environment in such a way as to create a hazard or potential hazard to the health, safety or welfare of any living species.

pollution accretion. *See* accretion.

polyandry. A breeding system in which one female mates with several males. It is uncommon, but occurs occasionally in certain primate societies (e.g. Howler monkey).

Polychaeta (bristle worms). An order of mainly marine annelid (*see* Annelida) worms, whose members have numerous bristles (chaetae), borne on flat outgrowths (parapodia) of the body segments. Their heads bear sense organs and tentacles. Males and females are separate, and fertilization is external. Some (e.g. ragworms) are free-swimming, others (e.g. tubeworms, lugworms) live in tubes or burrows.

polychlorinated biphenyl. A group of closely related chlorinated hydrocarbons (organochlorines) whose principal use has been as liquid insulators in high-voltage transformers. Their use is being reduced owing to evidence of their persistence and toxicity in the environment.

polyclimax. A number of climaxes occurring within a climatic region, probably owing to the subordination of climatic factors by edaphic ones. *Cf.* monoclimax.

polycyclic aromatic hydrocarbons (PAH). A group of chemical compounds, including benzopyrene, dibenzopyrene, dibenzoacridine, which are carcinogenic to man.

polyembryony. The development of more than one embryo from a single fertilised egg. In some insects (e.g. parasitic hymenoptera), many hundreds of embryos may develop from one egg. *See* monozygotic twins.

polyenergid. Applied to a cell nucleus containing several sets of chromosomes which have been produced as a result of repeated division within the intact nuclear membrane (endomitosis). *See* polyploid.

polygamy. (*a*) (Bot.) The production of male, female and hermaphrodite flowers on the same or different plants. (*b*) (Zool.) The possession of more than one mate.

polygenes. Multiple factors.

polygenetic. *See* provenance.

polygon. In topography, a unit of polygonally-patterned ground surface found in regions of permafrost, and also in areas of alternating flooding and dessication, such as playas. Polygons vary from a few millimetres to several tens of metres in diameter, and have several modes of origin. Polygons in regions of permafrost grade into stone stripes on hill tops.

polygoneutic. Applied to an animal which produces several broods in one season.

polygordius worms. *Archiannelida. See* Annelida.

polygyny. An animal breeding system involving one male and several females. There are two main forms, lek and harem formation, in which a male defends a group of females with which it has exclusive breeding rights.

polymer. Substance formed by the joining together of simple basic chemical units in a regular pattern. Commonly based on the carbon system, or on silicon and oxygen atoms.

polymorphism (pleiomorphism, pleomorphism). (*a*) The occurrence, usually in the same habitat and within an interbreeding population, of several distinct forms making up one species. The different forms are present in fairly constant proportions. Examples are some colour varieties in plants; feeding, stinging and reproducing polyps in the Portugese-Man-of-War (*Physalia*); colour varieties in certain butterflies (e.g. *Heliconius* spp. of the tropical Americas); human blood groups; queen, drone and worker castes in the honey bee. (*b*) The occurrence of different forms in the same individual at different stages in the life history (e.g. medusa and polyp in some coelenterates). (*c*) (Geol.) The occurrence of separate minerals having the same chemical composition but different physical structures and properties.

polymorphonuclear leucocytes. *See* blood corpuscles.

polymorphs. *See* blood corpuscles.

polyp. A sedentary form in coelenterates, living singly (e.g. sea anemone) or in colonies (e.g. corals). The typical type of polyp (hydranth) has a tubular body, attached at one end, and with a mouth surrounded by tentacles at the other. In some colonies there is more than one type of polyp (e.g. some corals have different forms to catch and to digest the prey). Some polyps produce medusae.

polypedon. A collection of pedons.

polypeptides. A chain of three or more amino acids joined together by the peptide linkage. Polypeptide chains may consist of hundreds of amino acid units. Proteins consist of polypeptide chains cross-linked together in a variety of ways.

polyphagous. Applied to an animal which feeds on many kinds of food.

polyphyletic. Applied to a group of organisms which do not all share a common origin. The group must therefore be an artificial one, the similarities of its members resulting from convergent evolution (*see* convergence). The phylum Polyzoa is thought to be polyphyletic. *Cf.* monophyletic.

polyphyodont. Applied to an animal (e.g. a shark) which has a continuous succession of teeth. *Cf.* diphyodont, monophyodont.

polyploid. Applied to organisms or cells which have three or more times the haploid number of chromosomes. Polyploid individuals are sterile when crossed with normal (diploid) ones. Polyploidy is rare in animals, but fairly common in flowering plants. *Cf.* allopolyploid, autopolyploid, allotetraploid, polyenergid, tetraploid, triploid.

polyribosome. *See* ribosomes.

polysaccharide. *See* carbohydrate.

polysaprobic. Applied to a body of water in which organic matter is decomposing fast, and in which the free oxygen is either exhausted or is present in very low concentrations. *Cf.* mesosaprobic, oligosaprobic, catarobic, saprobic classification.

polysome. *See* ribosomes.

polytocous. (*a*) Producing many young at a time. *Cf.* monotocous, ditokous. (*b*) fruiting repeatedly, or caulocarpous.

polytopic. Applied to an organism or group of organisms which occurs in more than one area. *See* discontinuous distribution.

polytypic. Applied to a species which has a variety of forms living in different parts of its range. *See* cline.

polyvinylchloride (PVC). One of the most common plastics, used in clothes, furniture, records and containers. Produced from the monomer vinyl chloride. *See* Minamata.

Polyzoa (Bryozoa, Moss animals). *See* mats, corallines. Small, colonial animals, all aquatic but mostly marine, superficially resembling coelenterates, but far more complex, feeding by means of ciliated tentacles. Each individual has a horny, gelatinous or calcareous case, and the colony may be encrusting, bushy or very like a seaweed in form. The Polyzoa are often divided into two phyla, Endoprocta and Ectoprocta.

pome. A 'false fruit', such as an apple or pear, in which the fleshy part develops not from the ovary but from the receptacle (axis) of the flower.

pool. A body of oil or gas, or both, occurring in a separate reservoir and under a single pressure system. The oil or gas occurs with water within the pores or fissures in the reservoir rock.

population. A group of individuals considered without regard to inter-relationships among them.

population dynamics. The study of changes in population densities, i.e. changes in the number of individuals in a specified area.

population ecology. The study of the factors that effect the number of individuals of a particular population present in a specified area over a period of time.

Porifera (sponges). A phylum usually regarded as comprising a subkingdom, the Parazoa, separate from all other multicellular animals. Sponges are aquatic (freshwater as well as marine) and sessile. Their bodies have a single, usually much-branched body cavity, communicating with the surrounding water by numerous pores. The cavity is lined by collared cells bearing flagella, by means of which currents of water containing food particles are passed through the body. Most sponges have horny, calcareous or silicious skeletons.

porosity. The percentage of pore space in a rock. Porosity in sedimentary rocks ranges from less than 1 % to over 50 %, and depends on the sorting, angularity, and packing of the grains, as well as the degree of cementation of the rock. The porosity of an original sediment may be increased by leaching during diagenesis, or it may be decreased by compaction and cementation. A porous rock need not necessarily be permeable: both chalk and clay have very high porosities (above 30 %), but neither is permeable. The opposite of porous is non-porous. *See* permeability.

porphyritic. Refers to a texture of igneous rocks in which large crystals (phenocrysts) are set in a finer ground mass.

porphyry copper. Large, low-grade, disseminated hydrothermal deposits of copper ore (chiefly chalcopyrite) in finely broken up intrusions of porphyritic, silica-rich igneous rock. Porphyry copper deposits appear to have been formed during the last 170 million years in areas of widespread volcanism, and are being worked opencast at grades of 0.5 % copper, providing about half of the world's annual copper requirement.

Portland cement. A mixture of ground limestone and clay, invented in 1756 by John Smeaton and patented in 1834 by John Aspdin, who considered it as strong as the Jurassic Portland stone of Dorset, used in many buildings at the time. Portland cement is the cement used most commonly today. Its essential ingredients are calcium carbonate ($CaCo_3$), silica (SiO_2) and alumina (Al_2O_3).

Portlandian. A stage in the Jurassic System.

positive feedback. *See* feedback.

postclimax. According to F. E. Clements, a stable plant community whose composition reflects more favourable (e.g. cooler or moister) climatic conditions than the average for a region. *Cf.* preclimax, subclimax.

postclisere. The series of formations that arises when the climate becomes wetter. *See* clisere, postclimax.

potamobenthos. *See* benthos.

potassium (K). Element. A silvery-white, soft, very reactive metal, widely distributed in the form of salts. It is an essential nutrient (*See* macronutrients) and is used in fertilisers. AW 39.102; At.No. 19; SG 0.86; mp 62.3°C.

potassium sulphite (or **sodium sulphite**) **process.** A Japanese process for removing sulphur dioxide from flue gases, using a wet scrubber to produce recoverable sulphur dioxide.

Potato. *See* Solanaceae.

potential climax. The climax that will replace the existing climax should the climate change. *Cf.* postclimax, preclimax.

potential instability. *See* conditional instability.

potential temperature gradient. The difference between the adiabatic lapse rate and the actual lapse rate in the lower atmosphere. If the gradient is zero (i.e. the two are the same) a mass of rising gas will have a constant buoyancy. If the gradient is negative the buoyancy of the gas will increase with height, if the gradient is positive buoyancy will decrease until a point where it reaches zero.

Poza Rica incident. An air pollution incident which occurred in 1950 at Poza Rica, Mexico, in which some 22 people died and 320 were made ill following the discharge into the atmosphere of large quantities of hydrogen sulphide. The discharge resulted from failure of equipment at an oil refinery sulphur recovery unit, and the gas was trapped beneath an inversion.

pozzolana. Volcanic ash found near Pozzuoli, Italy, and used since Roman times as an ingredient in some cements and mortars.

P-P factor. Nicotinic acid.

PPLO (pleuropneumonia-like organisms). Mycoplasmas.

prairie. *See* grassland.

prairie soil (brunizem). Grassland soil, which is dark brown, mildly acidic at the surface, over a leached layer. There is a brownish subsoil grading down to the parent material, with little or no calcium carbonate. It is found in generally cool to warm temperate climates. *See* soil horizon, soil classification.

pre-adaptation. The possession of characteristics which provide an advantage for an organism when it is exposed to new conditions (e.g. the lobed fins of Devonian Choanchthyes,) from which the legs of Amphibia were evolved).

Precambrian. Refers to geological time before the beginning of the Cambrian Period, or else to rocks formed during that time. Precambrian rocks are grouped into many local stratigraphic subdivisions. Precambrian time, also called the Cryptozoic, is sometimes divided into the Archaean Era followed by the Proterozoic Era.

precipitation. (*a*) (Chem.) Formation of solid particles in a solution. Generally the settling out of small particles. *See* electrostatic precipitation. (*b*) (Met.) The settling out of water from cloud, in the form of dew, rain, hail or snow, etc.

Precipitous Bluff. A geological feature in south-west Tasmania, threatened by plans to exploit the limestone deposits it contains. The top 300 metres of the Bluff are composed of dolerite overlying sedimentary rocks including beds of limestone.

preclimax. According to F. E. Clements, a stable plant community whose composition reflects less favourable (warmer or drier) climatic conditions than the average for a region. *Cf.* subclimax, postclimax.

preclisere. The series of formations that arises when the climate becomes drier. *Cf.* postclisere, clisere, postclimax, preclimax.

predator. An animal which kills other animals for food. A secondary consumer in a food chain. A predator preys externally on other animals, usually destroying more than one individual, whereas a parasite (*see* parasitism) often lives in or on a single host without killing it.

preferential species. Species that are present in several communities but that are predominant or more vigorous in one particular community.

presbycousis (presbyacusis). Hearing loss due to advanced age, usually at high frequencies.

presence indicator. Species whose presence is taken to indicate the existence of a particular environmental factor.

pressurised-water reactor (PWR). A light-water nuclear reactor which uses enriched uranium as a fuel, light water as a moderator, and as a coolant light water held in the liquid state under a pressure of 150 atmospheres, contained inside a vessel of welded steel.

prevailing wind. The direction from which the wind in a particular area blows more frequently than any other.

Prickly Pear. *Opuntia vulgaris. See* Opuntia.

Pridolian. The youngest series of the Silurian system in Europe. In the UK, this series is frequently called the Downtonian, and is placed at the bottom of the Devonian System.

primary air. Air admitted to a furnance or incinerator during the first part of the firing cycle, i.e. with the fuel.

primary minerals. Minerals formed directly from the cooling magma and persisting as the original mineral even in sedimentary rocks (e.g. quartz). *Cf.* secondary minerals.

primary production. *See* production.

primary rock. Rock produced directly from the magma. *Cf.* secondary rock.

primary succession. *See* succession.

primary treatment. A process to remove most floating and settleable solids in waste water and so reduce the concentration of suspended solids.

primates. A primarily arboreal mammalian order, comprising the Anthropoidea (monkeys, apes and Man) and the more primitive Prosimii (lemurs, lorises and tarsiers). Primates have large cerebral hemispheres and forwardly-directed eyes, and their thumbs (and sometimes also big toes) are apposable.

prisere. A natural plant succession beginning on a bare area and culminating in a climax.

probe. Tube inserted for sampling or for measuring pressures at a distance from the actual collection or measuring apparatus. Probes are commonly used for chimney or duct sampling. *Cf.* Pitot tube.

Proboscidea. An order of herbivorous placental mammals including the two present-day species of elephant and fossil forms (e.g. mammoths and mastodons). Modern elephants are characterised by their enormous size, stout legs, long tusks and trunks, and three sets of large grinding molars, developed in a series, only two pairs being in use at any one time. The earliest known proboscidean (*Moeritherium*) was only 60 cm high, and had more complete dentition, small tusks and probably a short snout.

Proboscis Worms. Nemertea.

procaryotic (prokaryotic). Applied to organisms or cells whose genetic material (filaments of DNA) is not enclosed by a nuclear membrane, and which do not possess mitochondria or plastids. Bacteria and blue-green algae are the only

394

procaryotic organisms. *Cf.* eucaryotic.

Procellariiformes (Tubenoses). An order of oceanic pelagic birds including petrels, albatrosses, fulmars and shearwaters. They have external, tubular nostrils, hooked beaks and long, narrow wings. They nest in colonies on remote shores.

Prochordata. Protochordata. *See* Chordata.

proclimax. According to terminology introduced by F. E. Clements, any plant community resembling the climax in stability of permanence, but lacking the proper sanction of the existing climate. The term includes Subclimax, Serclimax, Preclimax, Postclimax, Disclimax.

producer. *See* food chain.

production. (*a*) Gross production rate. The rate of assimilation shown by organisms of a given trophic level. (*b*) Gross primary production. The assimilation of organic matter or biocontent by a grassland community during a specified period. (*c*) Net primary production. The biomass or biocontent incorporated into a plant community during a specified period of time. (*d*) Net aerial production. The biomass or biocontent incorporated into the aerial parts (leaf, stem, seed and associated organs) of a plant community. (*e*) Net production rate. The assimilation rate (gross production rate) minus losses of matter by predation, respiration and decomposition. (*f*) Primary production. The total quantity of organic matter newly formed by photosynthesis.

production ecology. The study of biomes in terms of the production and distribution of food, and hence the flow of energy, within them.

production line. A manufacturing process that breaks fabrication into simple, discrete elements, each performed by one person or group of persons, in order to increase individual productivity.

production rate. The number of organisms formed within an area during a given period of time.

production residues. Wastes resulting from production and distribution. *Cf.* consumption wastes.

productivity. (*a*) Primary productivity is the amount of organic matter made in a given time by the autotrophic organisms in an ecosystem. Net productivity is the amount of organic matter produced in excess of that used up by these organisms during respiration, and represents potential food for the consumers of

395

the ecosystem. *See* food chain, production. (*b*) (forestry). The total timber yield per annum. The Productivity Rating Index is the expected yield per unit area divided by the standard yield for the same area, and expressed as a percentage.

productivity rating index. *See* productivity.

product standard. A standard calculated for products, esp. such consumable products as foods and detergents, that aims to ensure a particular quality of effluent discharged from their production or use.

profundal zone. The zone of a lake (deep water and bottom) lying below the compensation depth (*see* compensation point).

progesterone. A hormone secreted by the corpus luteum of the ovary and by the placenta in mammals. It is responsible for preparing the uterus for the implantation of the embryo(s), and inhibits ovulation during pregnancy.

Project Independence. A proposal by the US Government in January, 1974, whose aim was to develop America's own energy resources so that by 1980 the country would be self-sufficient in energy and so invulnerable to pressures from overseas suppliers of oil. The Project required the relaxation of many restrictions imposed in earlier years to improve the quality of the environment. Recently, it appears to have had little importance in US energy policy considerations.

prokaryotic. Procaryotic.

prolactin. Lactogenic hormone.

proline. Amino acid with the formula $NH.(CH_2) CH.COOH$. Molecular weight 115.1.

pro-natalist. *See* anti-natalist.

propagule (propagulum). Any part of a plant (e.g. seed, cutting) capable of forming a new individual when separated from the original plant.

propagulum. Propagule.

prophage. *See* lysogenic bacterium.

propolis. Plant resin used by honey bees for sealing crevices in the hive.

proprioceptor. A sense organ (e.g. balancing organ of the inner ear; receptors in joints, muscles or blood vessel walls) by which an animal receives information regarding its position or the movements and other changes going on inside its body. Using this information, the animal automatically coordinates its movements. The term does not cover sense organs which detect substances introduced into the alimentary canal or respiratory system. *Cf.* interoceptor.

Prosobranchia (Streptoneura). *See* Gastropoda.

protandrous. Applied to flowers (e.g. Rose-bay Willow-herb, *Chamaenerion angustifolium*) whose anthers mature before their carpels, and to hermaphrodite animals which produce first sperm and then eggs (e.g. some nematode worms). *Cf.* protogynous, dichogamy.

proteins. Very large and complex nitrogenous organic compounds, made up of many amino acid molecules linked to form one or more chains. Proteins are the basic constituents of living organisms, and form an essential part of the food of heterotrophs. The possible number of combinations of amino acids is immense, so very large numbers of proteins exist, and each species has proteins peculiar to itself. *See* enzyme, globulins, actomysin, haemoglobin, cytochrome, antibody, antigen.

proteoclastic enzymes. Proteolytic enzymes.

proteolytic enzymes (proteoclastic enzymes). Enzymes that have the power to decompose or hydrolyse proteins.

Proterozoic era. *See* Precambrian.

prothallus. The gametophyte (*see* alternation of generations) in vascular plants, which is much reduced, compared with that in non-vascular plants. In the lower plants (Thallophyta) the gametophyte is dominant, often forming by far the largest part of the whole plant. In the vascular plants the sporophyte has roots, stems, leaves and predominates in the life history of the plant, the gametophyte, or prothallus being reduced to a small but independent generation bearing the sex organs.

prothrombin. *See* clotting of blood.

Protista. A term now used by some authorities to denote a kingdom comprising all the simple organisms, i.e. bacteria, fungi, slime moulds, Algae and Protozoa. The term was originally used only for unicellular organisms. *Cf.* animal, plant.

proto-. Prefix denoting first, or original (from Greek *protos*, meaning 'first').

Protochordata (Prochordata). *See* Chordata.

protogynous. Applied to flowers (e.g. figwort, *Scrophularia*) whose carpets mature before their pollen is shed. *See* protandrous, dichogamy.

protomyxa. Mycetozoa. *See* Myxomycophyta.

Protoneolithic. Mesolithic.

protoplasm. The living matter of a cell, comprising both the nucleus and the cytoplasm.

protoplast. The protoplasm, as distinct from the non-living cell wall of a plant cell.

protore. A primary mineral deposit from which an economic ore may be formed by enrichment (e.g. leaching) or secondary enrichment. The term is also used to denote a deposit which may become economically workable with technological change or price increases.

protosoil. Early soil.

Prototheria. Monotremata.

prototroph. A strain of micro-organisms which has no nutritional requirements other than those necessary to the organism in its natural state. *Cf.* auxotroph.

prototype. (*a*) The first full-scale working model of a new technological device, used for testing under operating conditions. (*b*) Original.

prototype fast reactor. The British 250 megawatt fast breeder reactor built at Dounreay, Caithness, Scotland. *See* nuclear reactor.

Protozoa. A sub-kingdom (and phylum) comprising the animals whose bodies are not divided into cells (i.e. they are unicellular, or, according to some authorities, non-cellular). *See* Ciliophora, Flagellata, Rhizopoda, Sporozoa, Metazoa, Parazoa.

provenance. The terrane or parent rock from which the clasts (see Clastic) of a sediment are derived. Sediments derived from one source are termed monogenetic, whereas those from several sources are polygenetic.

psammitic. Arenaceous. The term is chiefly used to refer to metamorphic rocks formed from arenaceous rocks.

psammon. The organisms inhabiting the water lying between grains of sand on the shore of a lake.

psammophyte. A plant which grows in sandy soil.

psammosere. The stages in a plant succession beginning in sandy soil.

pseudaposematic coloration. Batesian mimicry.

pseudepisematic characters. Lures (e.g. the filamentous dorsal fin forming the fishing line of the angler fish *Lophius*) which enable certain animals to catch their prey.

pseudogamy. The development of an ovum when it is stimulated by the entry of a male gamete whose nucleus does not fuse with the egg nucleus. This occurs in some seed plants and some nematode worms.

pseudokarst. Karst-like terrain, not on limestone, and whose rough topography is due to causes other than solution.

pseudopodium. A temporary protrusion of cytoplasm from the surface of a cell (e.g. *Amoeba*, some white blood cells), serving for locomotion or ingestion of particles.

psilomelane. A mixture of manganese oxides, psilomelane is an important source of manganese, and is formed at or near the surface as a secondary mineral.

Psilophytales. An order of extinct Pteridophyta common in the Devonian Period. They were small green plants with horizontal rhizomes and erect, forked branches, which in some species bore numerous small leaves. They reproduced by spores formed in terminal sporangia.

psilopsida. A modern term covering Psilophytales and Psilotales. *See* Tracheophyta.

Psilotales. An order of Pteridophyta, probably related to the fossil order Psilophytales, now represented by a few small plants (*Psilotum* and *Tmesipteris*), mostly epiphytes found in tropical and subtropical areas.

Psittaciformes (parrots). An order of birds found mainly in warm climates and living amongst trees. They are mainly vegetarian, and some use the powerful beak to break open hard nut shells. Most are hole-nesting.

Psocids. Psocoptera.

Psocoptera (Psocids). Barklice, booklice, dustlice. An order of small, soft-bodied insects (Exopterygota) with long antennae and biting mouthparts, some of which are wingless. They live among vegetation, paper or dried materials, feeding on fungi and other organic matter. Some species harbour sheep tapeworms.

psychosocial stressors. Stimuli suspected of causing diseases which originate in social relationships, affecting the organism through the higher nervous processes.

psychosomatic. Term applied to physical effects caused or influenced by the mind, including many diseases.

psychosphere. Human thought and culture as an environmental phenomenon. Some writers have suggested that the psychosphere be used to complement concepts of lithosphere, biosphere, hydrosphere and atmosphere. *See* technosphere.

psychrometer. An instrument for measuring dry-bulb and wet-bulb temperature.

psychrometry. (*a*) The measurement of the humidity of the atmosphere. (*b*) The thermodynamics of air and water vapour mixtures, esp. as applied to air conditioning.

psychrophilic micro-organism. A micro-organism whose optimum temperature for growth lies below 20°C. *Cf.* mesophilic, thermophilic.

Pteridium aquilinum. Bracken.

Pteridophyta. A division of plants, mainly terrestrial, including the present-day ferns (Filicales), clubmosses (Lycopodiales), horsetails (Equisetales) and Psilotales. Pteridophyta have proper roots, stems and leaves, and a well-developed vascular system. Asexual spores are produced on sporophylls which often resemble foliage leaves, or are grouped to form cones. The gametophyte (*see* Alternation of generations) is a small, green thallus (prothallus). Fossil orders include the Psilophytales and Sphenophyllales.

Pteropsida. A modern term covering ferns and seed plants. *See* Tracheophyta.

Pterosauria (Pterodactyla). Flying reptiles of the Mesozoic Period, whose membranous wings were supported mainly by the elongated fourth 'fingers'. In 1971 a complete specimen of a pterosaur was discovered in Upper Jurassic rocks in Kazakhstan. This species, *Sordes pilosus*, had a thick, furry coat, suggesting that it was warm-blooded, and that the Pterosauria were not true reptiles.

pteroylglutamic acid. Folic acid.

Pterygota (Metabola). All insects except those comprising the Apterygota. Some Pterygota are wingless (e.g. lice, fleas), but these are thought to have evolved from winged forms. *See* Endopterygota, Exopterygota.

Pu. Plutonium.

pubescent (Bot.) Covered with soft, short hairs.

Puccinia graminis. *See Berberis vulgaris.*

puddingstone. A popular term for conglomerate, now usually restricted to particular strata such as the Eocene 'Hertfordshire Puddingstone'.

puffins. *See* Alcidae.

Pulmonata. *See* Gastropoda.

pulses. Food crops of the family Leguminosae, including the pea (*Pisum sativum*), lentil (*Lens culinaris*), lablab (*Dolichos lablab*), butter bean (*Phaseolus lunatus*), chick pea (*Cicer arietinum*), black gram (*Phaseolus mungo*), green gram (*Phaseolus aureus*), etc., which form important protein foods in many countries.

pulverization (milling, shredding). An intermediate step in refuse disposal in which refuse is reduced to small particles, so reducing the volume.

pumice. A highly vasicular acid pyroclastic rock.

pumped storage. A system for generating peak load electricity by turbines turned by a fall of water from a reservoir filled by pumps lifting water from a lower level. During base-load periods, when demand for electrical power is low, water is lifted from the lower to the higher level, and at times of peak demand the water is released. In certain situations wind power can be used for the pumping operation.

pumping station. An installation for raising sewage to a higher elevation, comprising a wet-well, a pump, and a pressure pipe. The term is also used to describe similar installations for pumping mains water supplies.

pupa (chrysalis). The stage between larva and adult in an endopterygote insect. The pupa appears from the outside to be quiescent, but inside the remodelling of the insect's body is going on. The term 'chrysalis' is often used only for the pupae of butterflies. *See* metamorphosis.

Purbeckian. The uppermost stage of the European Jurassic System.

401

pure line. A succession of generations homozygous for all genes. Pure lines are achieved by intensive inbreeding.

pure tone. A sound whose waveform is sinusoidal.

putrefaction. The anaerobic decomposition or organic matter with incomplete oxidation and the production of noxious gases.

putrescible wastes. Wastes of animal or vegetable origin which are degraded bacteriologically.

puy. A volcanic plug, esp. those in the Auvergne.

PVC. Polyvinylchloride.

P-wave. The primary wave reaching a seismograph from an earthquake. P-waves are compressional and travel through the crust, mantle and core. *Cf.* S-wave.

PWR. Pressurized-water reactor.

pycnium. Spermagonium.

Pycnogonida. Pantopoda.

pyramid of numbers. Biotic pyramid.

pyrethrum. Insecticide prepared from the flowers of *Chrysanthemum cinerariae-folium*. It breaks down rapidly, does not harm plants, is not very toxic to vertebrates, and so has a minimal effect on wildlife.

pyridoxine (vitamin B6). A vitamin of the B group which is essential for the formation of a co-enzyme in many organisms, including birds, mammals and some bacteria.

pyrite. An iron sulphide mineral, FeS_2. Pyrite, commonly known as Fool's Gold, is a widespread mineral. It is used in the manufacture of sulphuric acid, and because cobalt substitutes for iron in the crystal-lattice, pyrite is also an important source of cobalt, which is used in the iron industry.

pyroclastic. Refers to material (tephra) blown out by an explosive volcanic eruption and subsequently deposited. Pyroclastic deposits include tuffs, ignimbrites and some agglomerates.

pyrolusite. One of the major ore minerals of manganese, pyrolusite, MnO_2, occurs in sedimentary rocks or as a residual deposit concentrated by leaching.

Pyrolusite is also one of the manganese minerals in deep sea manganese nodules, also occurring commonly as dendritic forms on joint surfaces. Manganese is an almost ubiquitous alloying metal.

pyrolysis. Destructive distillation.

pyrometry. The art of measuring high temperatures.

pyroxene. A group of rock-forming silicate minerals with a wide range of compositions. The general formula approximates to $A \, B \, Si_2O_6$, where A is usually magnesium, iron, calcium or sodium, and B is usually magnesium, iron or aluminium, and the silicon can also be partly replaced by aluminium. Pyroxenes occur in igneous rocks being characteristic of basic and ultrabasic varieties, and in some metamorphic (*see* metamorphism) rocks. Augite, basically a calcium magnesium aluminosilicate, is the best known pyroxene.

Pyrrophyta (Fire algae). A group of mainly unicellular algae, most of which have flagella. They constitute a large part of the plankton in seas and inland waters. The most important members of the Pyrrophyta, the Dinophyceae (Dinoflagellata), are often included in the Protozoa (Flagellata), some being without chloroplasts. When abundant, Dinophyceae cause luminescence and 'red water' in the sea, and 'mussel poisoning' in Man and some sea birds.

Q

Q_{10} **(temperature coefficient).** The increase in the rate of a process brought about by raising the temperature 10°C. In many living systems, within certain limits, the rate is doubled or more than doubled for each 10°C increase.

QFE. *See* altimeter.

QO_2. The oxygen uptake of an organism, expressed in microlitres per milligram (dry weight) per hour.

quadrat. A sampling area (often one metre square) used in studying the composition of an area of vegetation. *See* quadrat method.

quadrat, major. A quadrat that includes all the more important species as well as a number of the less important ones.

quadrat method. An intensive study of the environment within a circumscribed area in order to gain a comprehensive knowledge of the wider area. *See* vegetation study.

quantum. Any observable quantity is quantised when its magnitude, in some or all of its range, is restricted to a discrete set of values. If the magnitude of the quantity is always a multiple of a definite unit, then that unit is called the quantum. The concept can be applied in many fields.

quartz. A crystalline mineral with the composition SiO_2, quartz is characteristic of acid igneous rocks, and because of its resistance to chemical weathering and its lack of cleavage is very abundant in many sedimentary and metamorphic rocks. Pure quartz (rock-crystal) is water clear, but slight impurities cause the well-known varieties which include amethyst (purple), rose quartz (pink), citrine (yellow) and cairngorm (brown).

quartzarenite. Orthoquartzite.

quartzite. A rock composed almost entirely of silica which is either a metamorphic (*see* metamorphism) metaquartzite or a sedimentary orthoquartzite. Quartzites are extensively quarried for aggregate.

quartzose sandstone. Orthoquartzite.

quartz-porphyry. Porphyritic microgranite.

quassia. A 'natural' insecticide extracted from the roots and timber of the tropical genus *Simaroubaceae*, used since 1825 as a fly-killer and later for aphid control. It is still used to control plum sawfly and as a bird and mammal repellant.

Quaternary. Refers to geological time since the end of the Pliocene, i.e. to Pleistocene and Holocene time. Quaternary, which also refers to rocks formed during this time, is ranked as either an era, following the Cenozoic Era, or a Period of the Cenozoic Era, following the Tertiary Period.

Queensland arrowroot. Achira.

Quercus. (Fagaceae). The oaks, evergreen or deciduous trees or shrubs which produce nuts (acorns) born in 'cups'. Many species yield bark for tanning and valuable timber. The bark of *Quercus suber* is cork. Oaks are the dominant trees in natural woodland throughout much of Britain. The two native species are *Q. robur*, the widespread Common Oak, with long-stalked acorns, and *Q. petraea*, the Durmast Oak, characteristic of northern and western areas, with short-stalked acorns.

quicklime. *See* calcium oxide.

quicksilver. *See* mercury.

R

Ra. radium.

rabbits. *See* Lagomorpha.

race. (*a*) A loose term used to refer to microspecies, permanent varieties or particular breeds. (*b*) A rhizome.

raceme (Bot.) An unbranched inflorescence whose individual flowers are stalked.

rachion. The line on a lake shore where wave action causes the most disturbance.

rachis (Bot.) The main axis of an inflorescence.

rad. (*a*) A unit used in measuring the amount of ionizing radiation absorbed by living tissues. One rad equals 100 ergs of energy per gram of tissue. *See* Curie, Rem. (*b*) Symbol for radian.

radar. *See* echo.

radial drainage. River systems forming a radial pattern. Typical of high mountain areas or systems on volcanic cones.

radially symmetrical. Actinomorphic.

radian (rad). The supplementary SI unit of plane angle, being the angle subtended at the centre of a circle by an arc equal in length to the radius of the circle. 2π radians $= 360°$. 1 rad $= 57.296°$

radiation. *See* heat transfer, adaptive radiation, alpha-, beta-, gamma-, X-rays, nuclear reactor, fission, nuclear fusion.

radiation fog. *See* fog.

radiation window. A band in the radiation spectrum extending from 8.5 to 11.0 microns in wavelength in which little absorption by water vapour occurs. In clear skies radiation from the ground in this band can escape to space but the remainder is absorbed.

radical (radicle) (Chem.) A group of atoms present in a series of compounds which maintains its identity through chemical changes that affect the rest of the molecule, but which generally is incapable of independent existence.

radicle. (*a*) The part of an embryo in a seed which develops into a root during germination. (*b*) *See* radical.

radioactive waste. Wastes produced by nuclear power generation that are radioactive and hence hazardous. Radioactive wastes fall into two categories: low-level wastes, whose half-life is short and which can be discharged safely after storage for a period long enough to permit them to decay; and high-level wastes, with a long half-life, for which permanent safe disposal methods and sites must be found. The main long-lived fission products require isolation for about 1000 years, but some (e.g. zirconium-93, iodine-129, caesium-135, americium-241 and technetium-99) remain biologically dangerous for 1 million to 100 million years.

radioactivity. Property exhibited by unstable isotopes of elements which decay, emitting radiation, principally alpha, beta and gamma particles, which are biologically harmful.

radiolaria. *See* Rhizopoda.

radiolarian ooze. A deep-sea deposit composed chiefly of the siliceous tests of radiolaria.

radiometric age. The time, measured in years before the present, that it has taken for a particular ratio of 'daughter' to 'parent' atoms to be formed by the radioactive decay of the parent atom. (*See* radioactivity). This assumes a closed system, the absence of native daughter atoms in the original material, and depends on the unique half-life of particular radioactive nuclides, (i.e. the time it takes for half the radioactive atoms present to decay). Of the many radioactive nuclides in nature four have been found most suitable for determining the radiometric age of rocks. Uranium-238, (half-life 4510 m.y.) decaying to Lead-206; Uranium-235, (half-life 713 m.y.) decaying to Lead-207; Potassium-40 (half-life 47000 m.y.) decaying to Strontium-87. Carbon-14 is a rare isotope occurring in living plants and animals and continually renewed in the atmosphere, with a low half-life of 5730 ± 40 years and has been used for dating organic material younger than 45000 years ± 5000 years by determining the ratio of carbon-14 to other carbon in the sample. When a living organism dies it ceases to assimilate carbon-14 and that present decays relatively rapidly. Carbon-14 dating has proved a very useful tool for archaeologists and students of recent Earth history, but has proved unreliable for dating material younger than 1000 B.C. *See* dendrochronology.

radiosonde. *See* balloon.

radium (Ra). Element. A very rare metal, a naturally occurring radioactive element. Its most stable isotope, $^{226}_{88}$ Ra, has a half-life of 1620 years. At. No. 88; SG 5: mp 700°C.

radius of curvature. *See* curvature.

radon (Rn). Element. Formerly known as niton. A naturally occurring radioactive gas, the immediate breakdown product of radium and chemically inert. It has a half-life of 3.825 days. At. No. 86.

rainbow. A circular arc of coloured light which displays the colours of the spectrum (violet, indigo, blue, green, yellow, orange, red) with red on the outside. The centre of the arc is opposite to the sun, the rainbow never describes more than a half circle and the higher the sun the smaller the arc. Rainbows are caused by the reflection and refraction of light in water droplets in such a way that the light emerges split into its spectrum colours at an angle of 42° to the direction of the sun's rays, so producing an arc with a radius of 42°. The formation of a rainbow requires water droplets to be falling and sun to be shining simultaneously, so that they occur in showery weather.

rain forest. *See* forest.

rainmaking. *See* artificial rain.

rainout. The removal of particulate matter from the atmosphere by formation of water droplets on the particles which act as condensation nuclei, followed by rain. This is generally a more effective removal mechanism than washout.

raise (rise). In mining, a small tunnel excavated upwards from a drive or level.

raised beach. Beach material left above the present high water mark when sea level was higher relative to the land than at present. *See* raised beach platform.

raised beach platform. Beach platform above the present high water mark, cut by waves at a time when sea level was higher relative to the land than at present. This may be due both to eustatic changes in sea level and raising of the land due to isostatic readjustment. *See* isostasy.

raised bog. *See* bog.

ramet. An individual member of a clone. *See* ortet.

rammed earth (*pisé de terre*). Building technique using subsoil, preferably rather sandy and sometimes with the addition of a stabiliser of cement or other substance, rammed into wooden shutters or made into blocks in a press. *See* adobe.

random noise. A fluctuating quantity (of sound, electromagnetic radiation, etc.) whose amplitude distribution is Gaussian.

range management. The planning and management of the use of grazing land (range) in order to sustain maximum livestock production consistent with the conservation of the range resource.

rank. The stage reached by coal in the course of its carbonification. The chief ranks of coal, in order of increasing carbon content, are lignite, sub-bituminous coal, bituminous coal, and anthracite.

Ranunculaceae. A family of dicotyledonous plants, mostly herbaceous perennials of northern temperate and arctic regions. Their floral structure shows primitive features (e.g. in the spiral arrangement of many free stamens and carpels in the buttercup), but many species (e.g. Water Crowfoot, *Ranunculus aquatilis*) are highly adapted to specialised habitats. Many (e.g. *Clematis, Delphinium, Anemone*) have showy flowers, and are cultivated in gardens. Most Ranunculaceae are acrid, and they are often very poisonous because of the presence of alkaloids. Some have been used in medicine (e.g. *Aconitum* for analgesic). Buttercups (*Ranunculus repens, R. acris*) are poisonous weeds, but are normally avoided by grazing animals.

rape. *See* Brassica.

rapid hardening cement. *See* cement.

Raptores. Falconiformes.

rarity. A species is rare if it occurs only once in 20 samples taken within a 64 sq m area, or if it is known to be present in a community but fails to occur in samples taken at random.

Ratitae (Ratite birds). Flightless birds with reduced wings and sternum, long legs and curly feathers. Many are large. They include ostriches (Africa and south west Asia), Rhea (South America), Emu and Cassawary (Australasia), moas (extinct birds of New Zealand) and kiwis (New Zealand). *See* Palaeognathae.

Rattite birds. *See* Ratitae.

rattomorphia. The tendency to extrapolate into human situations information derived from observations of animal behaviour.

Raunkiaer's Life Forms. A classification of plants based on the type of organs (often buds) which they possess to survive unfavourable periods, and the position of these organs in relation to soil level. (1) Phanerophytes are woody plants (e.g.

trees, shrubs) in which the perennating parts are more than 25 cm above ground level. (2) Chaemaeophytes are woody plants in which the perennating parts are above the ground but below the 25 cm level. (3) Hemicryptophytes are herbaceous plants in which the perennating parts are at soil level, often protected by dead portions of the plant. (4) Geophytes are herbaceous plants in which the perennating parts are below ground level. (5) Hydrophytes are herbaceous plants in which the perennating parts lie in water. (6) Helophytes are herbaceous plants of marshes, in which the perennating parts lie in mud. (7) Therophytes are herbaceous plants which survive unfavourable periods as seeds. Cryptophytes comprise the classes (geophytes, helophytes, hydrophytes) in which the perennating parts are covered by soil or water.

raw humus. Mor.

rayl. *See* characteristic impedence.

Rayleigh number. The non-dimensional number representing the interplay of forces tending to produce and subdue buoyant convection. Thus: Ra $= \dfrac{g\beta h^4}{kV}$ where g is gravity, $\beta = \dfrac{1}{\rho}\dfrac{\delta\rho}{\delta z}$ where ρ is density and z is height, h is the depth of the layer of fluid, k is the thermal conductivity and V is the kinematic viscosity. ρ may be replaced by $\dfrac{\alpha}{T}\dfrac{\delta T}{\delta z}$ where T is the absolute temperature and α the coefficient of thermal expansion.

Rayleigh scattering. The scattering of visible light by air molecules in a predictable and isotropic way. Particles of more than about 0.1 micron radius scatter light in a more complicated way, much of it in a forward direction (Mie scattering) relatively independently of the wavelength. They also absorb sunlight to some extent.

razorbills. *See* Alcidae.

realgar. *See* arsenic.

recent (Geol.) Holocene. *See* geological time.

receptacle. *See* flower.

receptor. A sense organ; *See* exteroceptor, interoceptor, proprioceptor, chemo-receptor, ear, eye.

recessive. A recessive character is one of a pair of contrasted characters which is not developed in a heterozygous individual (i.e. in the presence of the dominant gene).

recharge. Process of renewing underground water by infiltration during wet seasons. *See* artificial recharge.

recharge well. *See* artificial recharge.

recumbent fold. A fold with a horizontal, or nearly horizontal, axial plane.

recycling. The recovery and re-use of materials from wastes.

Red algae. Rhodophyta.

red beds. An assemblage of sedimentary rocks formed in a highly oxidising environment so that the iron present is in the ferric state. Red beds are probably indicative of an arid continental environment, and the term is often used not only to describe the red marls, shales, and sandstones of the New Red Sandstone, but associated breccias and evaporites as well.

red book. A collection of all the available date on species threatened with extinction, maintained by the International Union for Conservation of Nature and Natural Resources.

red clay. A soft deep-sea deposit rich in iron oxides and found at depths of over 200 fathoms (about 4 km). It is made up of wind-blown dust (including volcanic particles), manganese nodules, dust from meteors and meteorites, and insoluble organic remains such as sharks' teeth, whales' earbones etc.

Red Pepper. *See* Solanaceae.

reducer (decomposer). Organisms (e.g. bacteria and fungi) which break down dead organic matter into simpler compounds.

reduction. *See* oxidation.

reduction division. Meiosis.

reductionism. The belief that a system may be understood by a detailed examination of its components, based on dissection and analysis. *See* holism, vitalism.

Redwood. *See* Taxodiaceae.

refection (autocoprophagy). An animal's habit of eating its own faeces, practised by some herbivorous animals (e.g. rabbit).

reflex. A simple form of animal behaviour in which a stimulus evokes a specific automatic and often unconscious response (e.g. touching a hot object causes immediate withdrawal of the affected part of the body). The response depends on the existence of an inborn nervous pathway, the reflex arc, in which nerve impulses pass from a sense organ along nerve fibres to the central nervous system, then along other nerve fibres to an effector organ (e.g. muscle, gland) which brings about the response. *See* conditioned reflex.

refrigerant. Substance which is suitable as the working medium of a cycle of operations producing refrigeration, e.g. liquid ammonia, Freon.

refugium. An area which has escaped great changes undergone by the region as a whole, and so often provides conditions in which relic colonies can survive (e.g. a driftless area (nunatak) which escaped the effects of glaciation because it projected above the ice).

reg. A desert region with a surface of pebbles. Reg is chiefly used in Algeria, whilst serir is the term used in Libya and Egypt. Reg is contrasted with erg and hamada.

regional park. Area intended primarily for recreation use and offering a wide range of recreation facilities, generally larger and more varied than a country park and intended to serve a wider catchment area, but smaller than a national park and with less emphasis on the landscape value.

regional sea. UN term used to designate seas that are land-locked, or whose waters mix only slowly with those of the oceans so that they are especially susceptible to pollution from coastal states, e.g. the Mediterranean, Baltic, Caribbean, Persian Gulf, Malacca Straits, etc.

regolith. Loose, unconsolidated and broken rock material covering bedrock.

regosol. Weakly developed soil. *See* soil classification.

regression. (*a*) (Statis.) The dependence of one variable upon another independent variable. Regression analysis seeks to find both the degree and mathematical form of the dependence which can be expressed in a regression equation. (*b*) (Ecol.) The destruction of vegetation (e.g. by fire, grazing, etc.) and subsequent colonisation at a lower level (e.g. the replacement of forest by grasses following the destruction of the trees). (*c*) (Geol.) Marine Regression: lowering of sea level relative to the land, or retreat of the sea by build-out of terrigenous sediments. *See* transgression.

regressive. Refers to a body of water or sediments associated with a lowering of sea level.

Reinluft-process. German process for removing sulphur dioxide from flue gases by adsorption with peat coke and lignite charcoal in a moving bed reactor at 140°C, with up to 90% efficiency. The reagent and sulphuric acid are recovered at 400°C.

rejuvenation. An interruption in the evolution of a landscape by tectonic movements, altering its form and presenting a new landscape for the agents of wind and water to sculpt. *See* tectonic.

relative abundance. The measure of the abundance of a species indicates the degree to which it is able to maintain itself under prevailing conditions of environment and competition.

relative frequency. The mean number of contacts per 100 quadrats.

relative transpiration. The rate of transpiration per unit area from the plant surface, divided by the rate of evaporation from an equivalent area of open water surface under similar climatic conditions.

releaser stimulus. A stimulus which initiates an instinctive behaviour pattern in an animal. If the stimulus comes from a member of the same species it is called a social releaser.

relic (relict). Applied to an organism or group of organisms representing the surviving remnants of a population which was formerly more widespread or characteristic of the area. *See* Refugium.

relict. Relic.

rem. The unit dose of ionizing radiation which gives the same biological effect as that due to one roentgen of X-rays. The name means: 'Roentgen Equivalent Man'. *See* curie, rad, roentgen.

rendzina soils. Dark grey or black organic surface layers developed over soft, light calcareous material derived from chalk, limestone or marl. *See* soil classification, soil horizons.

rentalism. That economic system, or part of an economic system, based on the short-term renting of goods rather than on private ownership. Rentalism tends to be characteristic of highly industrialised, affluent, mobile societies.

replica. In modelling (*see* model) a replica is a complete reconstruction of the original in all its structural and functional details, and so cannot be called a model. Since it possesses all the features of the original, not just the essential ones, it offers no advantage over the original as a subject for study.

412

reproduction curve. The relationship between the numbers of a given stage in generation $(n+1)$ plotted against the numbers of that stage in generation (n).

reptiles. Reptilia.

Reptilia (reptiles). A class of essentially terrestrial vertebrates, whose modern representatives are poikilothermic and scaly-skinned, with embryos protected by an amnion and an allantois. Most lay large, shelled eggs. Reptiles were the dominant land vertebrates in the Mesozoic Era. *See* Chelonia, Squamata, Crocodilia, Rhynchocephalia, Therapsida, Dinosaurs (Ornithischia, Saurischia), Pterodactyla, Ichthyosauria, Plesiosauria.

reserves. *See* resource.

reservoir. (*a*) Natural or artificial lake for the storage of water for industrial and domestic purposes and for regulation of inland water-way levels. Service reservoirs store water for domestic supply purposes under cover and regulate diurnal fluctuations in demand. Impounding reservoirs provide storage to cover seasonal or year to year variations in inflow. Such reservoirs may supply water for domestic or industrial use or for regulating water levels in rivers and canals (feeder reservoir). Some impounding reservoirs storing water for domestic use contain relatively pure water which needs little further treatment before being piped direct to a supply network. (*b*) (Geol.) A natural, underground container for fluids, e.g. water, crude oil, or natural gas.

reservoir rocks. Rocks forming an underground reservoir for fluids such as water, natural gas, and crude oil. Reservoir rocks are characterised by high porosity and permeability, and are often sandstones, limestones or dolomites.

residual oil. Residue from the refining of crude oil to produce petroleum gas, petrol, diesel oil, lubricants, paraffin and asphalt, etc. Generally sold for burning.

resinous. *See* lustre.

resonance. When a system is vibrating as a result of forced excitation at a certain frequency, a diminution of the amplitude of vibration caused by altering the frequency of the exciting force indicates that the system is in resonance.

resonant. Capable of being excited into resonance.

resonant frequency. Frequency at which resonance occurs.

resource. A means that is available for supplying an economic want (e.g. land, labour, etc.) Minerals and fossil fuels are described as stock, resource, or reserve. The stock of a substance is the total amount of that substance contained in the environment, much of which will be inaccessible or unprocessable by present-day technology. That part of the stock which could be used under specified social, economic and technological conditions is called the resource, and estimates of resources change with economic (e.g. alterations in prices), social and technological changes (e.g. a new technology might increase the quantity of a resource). The reserves are that part of the resource that can be exploited with current technology under current economic and social conditions. Reserves are further sub-divided into proven, probable and possible recoverable reserves, all of which have been identified, but some of which will be paramarginal or submarginal, and undiscovered reserves which are either hypothetical resources or specified resources, their location being unknown or known respectively.

respiration. (*a*) External respiration. Process whereby oxygen is taken into an organism and carbon dioxide is given out. This often involves breathing, or respiratory movements, e.g. pumping air into and out of lungs or passing water across gills. (*b*) International respiration (tissue or cell respiration). The chemical reactions from which an organism derives energy, e.g. in living cells glucose is oxidised to carbon dioxide and water, liberating 3.74 Calories/gramme. This oxidation is carried out through a complicated series of oxidation-reduction reactions (e.g. Kreb's cycle, citric acid cycle) under the control of enzyme systems (e.g. oxidases, dehydrogenases).

respiratory enzymes. *See* dehydrogenase, oxidase.

respiratory pigment. A substance contained in the blood and capable of combining reversibly with oxygen, so carrying it from the respiratory organs to the tissues. *See* haemoglobin, haemocyanin.

respiratory quotient (R.Q.). The ratio of the volume of carbon dioxide given off to the volume of oxygen used up by an organism during respiration. R.Q. varies according to the kind(s) of food being oxidised, and indicates whether respiration is aerobic or anaerobic.

rest mass. The mass of a body when it is at rest relative to the observer. Mass varies with velocity, which becomes important as speeds (e.g. of atomic particles) approach the speed of light.

retarded timing (TC). Technique for improving fuel combustion in internal engines, so reducing the formation and emission of pollutants.

retting. Process in the extraction of certain vegetable fibres (e.g. flax, jute). The plant stem is immersed in water (in a pond, or on grass to be covered with dew 'dew retting') and the outer sheath is partially decomposed by bacteria. When washed and dried, the outer parts become brittle. Scutching breaks them into short pieces without damaging the fibre, and they are hackled by being drawn through a series of metal combs of increasing fineness until all the straw is removed and only the fibres remain. In flax preparation, the short fibres are called 'tow' and the longer fibres 'line'.

reverberation. Sound at a point, which is increased by multiple reflections from surrounding surfaces and persists after the source ceases to emit sound.

reverberation time. The time required for reverberant sound of a given frequency to decay by 60 dB after the source ceases to emit sound waves. Commonly calculated by measuring the first 30 dB decay and then extrapolating.

reverberatory furnace. Furnace where the material heated is not mixed with the fuel. Its roof is heated by flames and the heat radiated down to the material.

reverse fault. A fault in which the fault-plane dips to the upthrown side.

reverse osmosis. Industrial process for removing salts and other substances from water by forcing water through a semi-permeable membrane under a pressure that exceeds the osmotic pressure so that the flow is in the reverse direction to normal osmotic flow. Reverse osmosis is used to desalinate brackish water and has been tested experimentally for the purification of water polluted by sewage effluent.

Rhaetian. A Stage of the Triassic System.

rheophyte. A plant which grows in running water.

rheotaxis. *See* taxis.

rhizobium. *See* root nodules.

rhizoid. A hair-like organ of attachment found in liverworts, mosses, fern prothalli, some algae and some fungi. Rhizoids may be formed from one or several cells.

rhizome (rootstock). An underground stem, bearing buds and scale leaves, which last for more than one season and usually serve for both vegetative propagation and perennation. Many rhizomes (e.g. in iris) are stout, containing large amounts of stored food. The term 'rhizome' is often reserved for horizontal underground stems (e.g. in mint), and 'rootstock' is often confined to more or less erect structures (e.g. in primroses).

rhizomorph. A root-like strand of hyphae occurring in certain fungi (e.g. Honey or Bootlace Fungus, *Armillaria mellia*), serving to spread the fungus and transport food materials.

415

Rhizophoraceae. A family of tropical trees, many of which (e.g. *Rhizophora Bruguiera*) are mangroves. *Rhizophora* spp. develop many roots from the stems and branches to support the plants in the unstable mud. The seeds germinate while on the trees, and thus avoid being buried during their early development.

rhizoplane. The micro-environment of a root surface. *See* rhizosphere.

Rhizopoda. A class of mainly free-living, freshwater and marine Protozoa, whose members usually feed and move by means of pseudopodia. Rhizopoda may be entirely soft (e.g. *Amoeba*) or have shells or skeletons. The calcareous shells of marine, planktonic groups of Foraminifera (e.g. *Globigerina*) on falling to the bottom form an important constituent of deep-sea oozes, and have contributed towards the formation of chalk. The siliceous skeletons of Radiolaria similarly contribute to the formation of marine oozes and chert.

rhizosphere. The part of the soil immediately surrounding roots. Roots alter the nutrient status of the soil close to them by absorbing minerals and releasing other substances. This leads to an increase in the numbers of micro-organisms, and often alters the relative porportions of the different kinds of micro-organisms present.

Rhodesia Man. *See Homo.*

Rhodophyceae. Rhodophyta.

Rhodophyta (Rhodophyceae, Red Algae). A large, diverse, mostly marine group of algae. Most are red, owing their colour to the presence of large amounts of phycoerythrin, which masks the chlorophyll and other pigments which they contain. Some species are microscopic, others membranous or filamentous, and often much branched. Red seaweeds are relatively small compared with the brown ones (Phaeophyta) and grow at greater depths or in rock pools too shady for the brown to thrive. Calcareous species are common, especially in tropical seas, where they often play a part in the formation of coral reefs. Some yield agar, others (e.g. laverbread, *Porphyra*; carragheen or Irish moss, *Chondrus crispus* and *Gigartina stellata*) are edible.

Rhopalocera (Papilonoidea). Butterflies. *See* Lepidoptera.

Rhus. Genus of trees and small shrubs of the Anarcardiaceae family, whose popular name is sumac. *R. toxicodendron* is N. American poison ivy; *R. vernicefera* yields lacquer. Other species also grown for ornament.

Rhynchocephalia. A small order of reptiles with many primitive features. The only living representative is the burrowing, lizard-like *Sphenodon* (the Tuatara of New Zealand), which has a pineal eye in the roof of the skull.

Rhynchota. *See* Hemiptera.

rhyolite. A fine-grained to glassy acidic igneous rock with a similar mineralogical and chemical composition to granite. Many rhyolites are porphyritic and some show flow-structure. Glassy rhyolites are termed obsidian or pitchstone, and many have spherulites (spherical masses of radiating crystals) indicating partial devitrification. Rhyolites form near-surface intrusions and lava-flows, although rhyolitic flows are not extensive at the present time: many ancient rhyolites have proved to be ignimbrites.

ria. A drowned river valley caused by changes in sea level relative to the land after the river has worn its channel down to base level. *See* eustatic.

ribbon worms. Nemertea.

riboflavin (lactoflavin). A vitamin of the B group needed by animals (including Man) and some bacteria for the formation of certain respiratory co-enzymes. Important sources for Man are liver, yeast, eggs and green vegetables. Some bacteria living in the human intestine are able to synthesise riboflavin.

ribonucleic acid. RNA.

ribose. *See* carbohydrate.

ribosomes (palade granules). Minute bodies, too small to be visible with a light microscope, present in the cytoplasm of all organisms, and often attached to the endoplasmic reticulum. Ribosomes consist of RNA (which is synthesised in the nucleoli of eucaryotic cells) and protein. They are the site of protein synthesis (see RNA), groups of ribosomes (polyribosomes, polysomes, ergosomes) probably being associated with single molecules of messenger RNA.

rice. *Oryza sativa. See* Graminae, Wild rice.

richardson number. The ratio of the stabilising force due to stratification to the destabilising effect of shearing motion, equal to $g\beta/(\delta u/\delta z)^2$ where $\beta = \dfrac{1\,\delta\rho}{\rho\,\delta z}$, ρ being the density, z the height, and u the horizontal fluid velocity. It must be less than $\frac{1}{4}$ if a stream is to be dynamically stable. Named after Lewis Fry Richardson (1881–1953), the first person to attempt to forecast weather mathematically.

Richter scale. *See* earthquake.

rickets. *See vitamin D.*

rickettsias. A group of very small bacteria which cause severe diseases such as human typhus fever. Rickettsias are found inside the cells of arthropods such as ticks, lice and fleas, and are transmitted to mammals by the bites of these parasites. The vector of typhus is the human louse (*Pediculus humanus*).

ridge. Oceanic ridge. A submarine, broadly bilaterally symmetrical ridge with sloping sides sometimes but not necessarily in the middle of an ocean. Ridges are the site of earthquake and volcanic activity and of higher heat flow than the crust. According to the theory of Plate Tectonics they form the boundary between lithospheric plates where material is added to the plate edge by the process of sea floor spreading. Some ridges are known as rises. There are also currently inactive ridges.

rift valley. A linear valley bounded by normal faults. The longest rift valley is developed along the axes of oceanic ridges. The East African Rift Valley is postulated as an incipient spreading plate margin. *See* Plate tectonics.

Right Whale. *See Balaenoidea.*

rigid pavement. *See* pavement.

rime. Ice deposited on solid surfaces caused by supercooled droplets (as in freezing fog) that freeze immediately on contact, so building up layers of ice sometimes to considerable thickness.

riparian. Land bordering sea shore, lake or river.

rip current. A localised, strong outgoing surface or near-surface current returning water seaward against the incoming surf.

rise. Raise.

river capture. This occurs when a river system, in eroding the slopes providing its own drainage, pushes the watershed between its own drainage basin and an adjacent one further back until a gap is breached and the river with the lower channel captures the headwaters of the other river, thus enlarging its own drainage basin. The river thus captured is said to be beheaded and the point of diversion is known as the elbow of capture.

river terrace. The remains of alluvial flood plains of a river when the river channel ran at a higher level than at present. Raised sea levels of former times led to a higher base level down to which a river would cut its channel, but as sea levels descended relative to the land the river would cut a deeper channel, making a new, lower flood plain and leaving behind stepped alluvial terraces at the valley sides.

river zones. For biological purposes, rivers are divided into four zones: (*a*) Headstream or Highland Brook which is often small, may be torrential (i.e. with water flowing at 90 cm per second or more), temperature conditions may vary widely and fish may be absent. (*b*) The Troutbeck is larger and more constant, though it may still be torrential. Trout may be the only permanent fish. (*c*) The Minnow Reach is still fairly swift, but patches of silt and mud collect in sheltered spots. Higher plants can gain a foothold and the fish population is more varied. In Europe the Minnow Reach is sometimes known as the Grayling Zone. (*d*) Lowland Reach is slow and meandering and coarse fish commonly appear. In Europe this is known as the Bream Zone. Some workers divide the Lowland Reach into two.

RMS value. Root-mean-square value.

Rn. Radon.

RNA (ribonucleic acid, ribodenucleic acid). An RNA molecule is a single, long, unbranched chain of nucleotides. Each nucleotide is a combination of phosphoric acid, the monosaccharide ribose, and one of the four nitrogenous bases uracil, cytosine, adenine or guanine. The bulk of the cell's RNA occurs in the ribosomes. RNA is responsible for translating the structure of DNA molecules of the chromosomes into the structure of proteins, as follows: Messenger RNA is built up on a length of DNA (a cistron) by specific base-pairing (adenine will link only with thymine or uracil, and guanine only with cytosine). The messenger RNA molecule then moves from the chromosome to the ribosomes. Here a specific polypeptide molecule is built up. Its amino acids are put in place by smaller molecules of RNA, transfer RNA, of which there is a different kind for each amino acid. Each transfer RNA molecule becomes temporarily attached at one end to an amino acid molecule, and at the other, by a distinctive set of three nucleotides, to a site on a messenger RNA molecule determined by specific base pairing. In this way the polypeptide chain is assembled, its sequence of amino acids being determined by the sequence of nucleotides in the messenger RNA and ultimately in the DNA of the chromosomes. RNA forms the genetic material of some viruses.

road base. *See* pavement.

roche moutonnée. A hillock or jointed rock with one end smoothed by abrasion by ice-borne debris, and the other end with a stepped surface produced by the plucking out of blocks by the glacier. The smoothed (upstream) surface commonly shows glacial striations.

rock. Any naturally formed aggregate or mass of mineral matter.

rock crystal. Pure quartz.

rock-flour. Finely-comminuted rock debris, especially that produced by the grinding effect of glaciers. Rock-flour is a major constituent of boulder clay; the fine fraction of boulder clay is not rich in clay minerals when fresh, except when the rock-flour was derived from argillaceous rocks.

rock phosphate. Phosphate rocks.

rock-salt. Halite.

rod. (*a*) Specialised cell of the eye consisting of an inner and an outer segment. The outer segment abuts on the eye wall and consists of a stack of thin discs to which molecules of visual pigments are attached. It is probable that the first events in vision occur at these discs, from which effects are transmitted to the inner segment, connected to the outer by a cilium and with mitochondria lying close to the cilium. The inner segment narrows behind the region containing the mitochondria to become a fibre with a swelling containing the nucleus, with a synapse linking it to the bipolar cell and thence to the optic nerve. (*b*) Straight stems of willow (osier) used in basket making. (*c*) An obsolete measure of length (also called pole and perch) equal to 5.5 yards (5 m).

Rodentia (rodents). An order of placental mammals which includes rats, mice, voles, squirrels, hamsters, porcupines, beavers, coypu etc. Rodents have teeth adapted for gnawing, with one pair of large, chisel shaped persistently-growing incisors in each jaw, and no canines. The smaller species have a high reproductive potential, and often become pests. They constitute a staple food for many carniverous mammals, birds and reptiles. *See* Lagomorpha.

rodents. Rodentia.

Roentgen. The amount of X or gamma radiation which will produce ions carrying one electrostatic unit of electricity of either sign in 1 cc of dry air. Named after Wilhelm Konrad Roentgen (1845–1923). *See* curie, rad, rem.

root. The part of a vascular plant which is usually underground, serving for anchorage and absorption of water and minerals. A few plants (e.g. certain tropical epiphytic orchids) have green aerial roots, resembling stems or even leaves, and having a photosynthetic function. Roots differ from stems principally in the absence of leaves and buds, and in the arrangement of the vascular tissue in a central core. The tip of an underground root is covered by a protective layer of cells (the root cap). A short distance behind this is a zone of root hairs, tubular outgrowths of the epidermal cells which provide an extensive absorbing surface.

root-mean-square (RMS) value. The effective value of a fluctuating quantity. Calculated by squaring the values, averaging them, then extracting the square root.

root nodules. Gall-like swellings containing nitrogen-fixing bacteria (Rhizo-bium) on the roots of Leguminosae and a few other plants (e.g. Alder, *Alnus*; Bog Myrtle, *Myrica*). The bacteria enter the roots from the soil (different strains infecting different species of plant) and cause the formation of the nodules. Here they multiply rapidly, using carbohydrates from the host plant, but supplying it with organic nitrogen compounds formed during nitrogen fixation. Ultimately the nodule disintegrates, so bacteria are returned to the soil. Cultivation of leguminous crops such as clover, increases the nitrogen content of the soil, especially if the crop is ploughed into the ground. *See* nitrogen cycle, nitrogen fixation.

rootstock. Rhizome.

rorqual. *See* Balaenoidea.

Rosaceae. A cosmopolitan family of dicotyledonous herbs, shrubs and trees, which includes many species with edible fruits. The fruits of *Prunus* spp. (plum, peach, cherry, almond, etc.) are drupes, and that of *Rubus fruticosus*, the bramble, is an aggregation of small drupes. The fruits of *Fragaria*, the strawberry, *Malus* the apple, and *Pyrus*, the pear, are 'false fruits', with the fleshy parts formed from the receptacle (axis) of the flower and not from the ovary. *See* Crataegus.

rose quartz. Quartz made pink by slight impurities.

rotenone. *See* derris.

Rotifera (Wheel animalcules), A phylum of minute animals, the smallest of the Metazoa. Rotifers move and feed by means of a crown of cilia, and they have no true coelem. Most live in fresh water, a few in the sea, and they are very resistant to dessication.

Rotliegendes. The lower division of the Permian System in north-west Europe.

rotor-flow. *See* mountain waves.

rottenstone. A much weathered but still coherent rock resulting from the leaching out of one or more components. Most rottenstones are calcareous: fossiliferous sandstones often weather to rottenstones.

roughness, aerodynamic. *See* aerodynamic roughness.

roughness height. *See* aerodynamic roughness.

roundworms. Nematoda.

Royal Society for the Protection of Birds (RSPB). Society which has worked since 1889 to protect birds in Britain and has prevented the loss of various breeding species (e.g. osprey). It maintains a number of bird reserves.

r. q. Respiratory quotient.

rubber. *See* Euphorbiaceae, Moraceae, Ficus, Hevea, Manihot.

Rubiaceae. A very large, mainly tropical family of dicotyledonous plants, including trees, shrubs and herbs. Most have conspicuous insect pollinated flowers. The family includes the bedstraws (*Galium*) and Madder (*Rubia*). Genera of economic importance are *Coffea*, whose seeds yield coffee, and *Cinchona*, trees native to the Andes, from the bark of which alkaloid drugs such as quinine are extracted.

ruby. Corundum.

rudaceous. *See* rudite.

ruderal. Applied to plants which inhabit old fields, waysides or waste land.

rudite. A sedimentary rock with grains larger than 2 mm in diameter. Such sedimentary rocks are termed Rudaceous. *See also* arenite, argillite.

Ruminantia (ruminants). Even toed ungulates (Artiodactyla) which chew the cud (ruminate). The stomach is typically four-chambered, and there are no upper incisor teeth. Cattle, sheep, goats, deer, antelope and giraffes are members of this group.

ruminants. Ruminantia.

runner. *See* stolon.

run-off. Water from rain or snow that runs off the surface of the land and through streams and rivers.

Rust fungi (Uredinales). An order of parasitic Basidiomycetes with complicated life histories, many of which are important plant pathogens. *Puccinia graminis* causes black stem rust in cereals, and uses Barberry (*Berberis vulgaris*) as an intermediate host. *Uromyces fabae* causes broad bean rust. *See* physiologic specialisation.

rutile. A major ore mineral of titanium, with the formula TiO_2. Rutile is widespread in small amounts in many rocks, and is produced, often with ilmenite, another titanium mineral, from placer deposits.

Ryazanian. *See* Neocomian.

rye. *Secale cereale.* An important cereal crop that was probably first domesticated in areas away from the main wheat-growing centres where it thrived as a weed among wheat crops, so leading early farmers to develop it as an alternative to wheat. *See* Graminae.

S

S. Sulphur.

s. Second.

saccharine disease. Concept proposed by T. L. Cleave and others of a single degenerative disease caused by an environmental factor: the over-consumption of refined carbohydrate foods. The disease has many manifestations including appendicitis, peptic ulcer, diabetes mellitus, obesity, diverticulitis, coronary thrombosis, femoral thrombosis, gall stones, varicose veins, haemorrhoids, peridontal disease, dental caries, etc.

saccharoidal. (**sucrosic** in North American usage). Having the appearance of sugar. A term used to describe an equigranular texture of white, or nearly white, rocks such as some marbles and some dolomite-rocks.

saccharomyces. See yeasts.

saddle-reef. An ore-body occupying the space between relatively competent beds in the hinge-zone of a fold. Some gold-quartz veins form saddle-reefs.

Sahel (Sahelian region). Area of semi-arid lands bordering the southern Sahara and covering all or part of: Mauretania, Mali, Upper Volta, Niger, Chad, Senegal, Ghana, Cameroon, Nigeria, and the Central African Republic. The name is derived from the Arabic word for 'shore'.

St. David's series. The second oldest series of the Cambrian System.

Salicaceae. A family of dicotyledonous trees and shrubs, mainly of northern temperate regions, including *Salix* (willows) and *Populus* (poplars). The flowers are grouped in male and female catkins, borne on different plants. Poplars are rapidly-growing trees cultivated for timber which is used in matches, packing cases and paper pulp. White Willow yields wood for cricket bats, and osiers are used in basket making.

423

Salientia. Anura.

salinity. The degree of concentration of salt solutions, determined by measuring the density of the solution using a salinometer, a type of hydrometer designed for the purpose.

salinometer. *See* salinity.

salivary gland chromosome. megachromosome.

salpingectomy (tubal ligation). The operation for the sterilisation of a female mammal in which the abdomen is opened and the fallopian tubes are removed, so preventing the ova from reaching the uterus.

salt. (*a*) A chemical compound formed when the hydrogen ion of an acid is replaced by a metal, or, together with water, when an acid reacts with a base. Salts are named according to the metal and the acid from which they are derived. (*b*) Common salt (sodium chloride, NaCl). *See* halite.

saltation. (*a*) A leaping movement of sedimentary grains that are too heavy to be carried entirely by wind or water, but are bounced and rolled along the ground or stream bed by turbulence, thereby inducing the movement of other grains with which they impact. (*b*) (Ecol.) A sudden change.

Saltatoria. Orthoptera.

salt-dome. A structure resulting from the upward movement of evaporites. Oil and gas fields are frequently associated with salt-domes, which are usually roughly circular to lozenge-shaped in plan and often several thousand feet in depth.

salt-field. An area underlain by workable deposits of salt.

samara. A winged achene, e.g. a sycamore fruit or an ash 'key'.

sand. *See* arenite.

sand dollar. Echinoidea.

sand-dune. Mounds or ridges of sand blown by prevailing winds and exhibiting a long windward slope and a steeper lee slope. Unless stabilised (e.g. by vegetation) dunes will migrate in the direction of the prevailing wind as sand is blown up the windward slope and down the slip-face of the leeward slope. In coastal regions, as a dune migrates landwards it may be followed by a second dune as new material is blown in from the beach.

sandstone. A sedimentary rock consisting of cemented clastic grains predominantly between 2 mm and 0.06 mm in diameter.

Sanitary landfill. Dump of domestic refuse, compacted on site and covered regularly by a layer of earth. Micro-organisms decompose the organic part of the refuse. This is engineered burial of refuse, but in many places the term is synonymous with a rubbish dump or waste tip. *See* land-fill.

Santonian. *See* Senonian.

sapphire. Corundum.

saprobe (saprobiont). An organism which feeds on dead or decaying organic matter.

saprobic classification (saprobien classification). A classification of river organisms according to their tolerance of organic pollution. a) The polysaprobic group (including sewage fungus, 'bloodworms', and the Rat-tailed Maggot (*Eristalsis tenax*) can live in grossly polluted water in which decomposition is primarily anaerobic. b) The alpha mesosaprobic group (including the water louse *Asellus*) can tolerate polluted water where decomposition is partly aerobic and partly anaerobic. c) The beta mesosaprobic group (including Canadian Pond Weed, *Elodea canadensis*, some caddis-fly larvae, *Trichoptera,* the eel, *Anguilla anguilla*, and the Three-pined Stickleback, *Gasterosteus aculeatus*) can tolerate mildly polluted water. d) The oligosaprobic group (including stone-fly nymphs, *Plecoptera,* and the river trout, *Salmo fario*), are restricted to non-polluted water, which may contain the mineralized products of self-purification from organic pollution. *See* biotic index.

saprobien classification. Saprobic classification.

saprobiont. Saprobe.

saprolite. An ancient soil, with rock deposits showing a gradual transition above hard bedrock, through deeply weathered and leached zones which show the same structure as the bedrock, to a massive weathered zone and clay, beneath the modern soil.

saprobel. An amorphous material formed by the slow, anaerobic decomposition of planktonic remains, which may subsequently be converted to petroleum compounds after compression under accumulated sediment by processes that are not well understood.

saprophagous (saprozoic). Applied to an animal (e.g. House Fly larva) which feeds on dead or decaying plant or animal material. *See* saprophyte.

425

saprophyte

saprophyte. Organism (plant or Protista) which obtains food in solution from the dead or decaying bodies of other organisms. Many fungi and bacteria are saprophytes, so are a few Protozoa (e.g. *Euglenia, see* Euglenophyta) and flowering plants (e.g. Bird's Nest Orchid, *Neottia*). Saprophytes carry out the essential process of breaking down organic matter into simple substances such as carbon dioxide and nitrates, which are then available for synthesis by autotrophic organisms into new organic matter. *See* carbon cycle, nitrogen cycle.

saprozoic. Saprophagous.

sapwood. (*a*) The outer layers of wood, which contain living cells. As well as providing mechanical support, sapwood conducts water and stores food. It is generally lighter in colour and less resistant to decay than heartwood. (*b*) A term applied to trees with no clear distinction between sapwood and heartwood.

Sargasso Sea. A roughly elliptical, fairly still area in the central North Atlantic (between about 20° and 35°N and 30° and 70°W) that lies inside a clockwise current of the Gulf Stream on the west side and other currents on the east. Floating matter, esp. from the south-west, converges on the area, which contains large quantities of Gulfweed (*Sargassum*) and supports many animals typical of the littoral zone, some of which are found nowhere else. The sea is particularly susceptible to pollution, esp. from oil and floating debris. It was reported in 1969 by workers from the Woods Hole Oceanographic Institution, that oil globules were more plentiful in the Sea than the Gulfweed.

sarsen. A boulder of hard sandstone occurring on the surface on the chalk downs of southern England. Sarsens are probably remnants of a widespread Eocene deposit. Some have been erected to form stone circles, such as Stonehenge, although the inner circles at Stonehenge itself are formed from dolerite imported from the Prescelly Mountains in Wales.

satellite. Object that orbits around a larger one. Artificial satellites orbiting the Earth are used for communications, the gathering of military intelligence, the monitoring of weather and other environmental phenomena, etc. *See* Earth Resources Technology Satellite.

satin spar. Fibrous calcite or fibrous gypsum.

saturated adiabatic lapse rate. Wet adiabatic lapse rate.

saturation. A relative humidity of 100 %, measured by comparing the difference in readings between a dry bulb and wet bulb thermometer. In unsaturated air, water evaporating from the wet bulb will lower its reading. In saturated air there will be no evaporation from the wet bulb, the water on the bulb will be at the same temperature as the surrounding air, and the wet and dry bulb readings will be

426

identical. The amount of water that air can contain as vapour varies with temperature (e.g. from 2 g/cu m at $-10°C$ to 51 g/cu m at $40°C$).

Sauria. Lacertilia. *See* Squamata.

Saurischia. *See* Dinosaurs.

Sauropsida. A term originally used to cover all living and extinct reptiles and birds. It is now applied to birds and living reptiles, together with fossil reptiles which have a similar skull structure (dinosaurs, pterodactyls), but it excludes the extinct mammal-like reptiles. *See* Theropsida.

Sauropterygia. Plesiosauria.

savannah. *See* grassland.

Savonius rotor. A vertical axis windmill with high inertia and low efficiency, but producing a more continuous output than conventional aerogenerators. *See* darrieus generator, panemone.

saw-flies. *See* Hymenoptera.

saxatile. Saxicolous.

saxicoline. Saxicolous.

saxicolous (saxatile, saxicoline). Applied to organisms which live among rocks (e.g. saxifrages).

Sb. Antimony.

scabrous. Rough to the touch.

scale insects. Plant bugs (Hemiptera), members of the family Coccidae. Some are harmful parasites of trees and shrubs (e.g. *Mytilapsis pomorum*, the Apple Scale), others yield cochineal and shellac. Female scale insects are wingless, and remain fixed to their food plant, covered with a scale formed from cast-off exoskeletons.

Scaphopoda (tooth shells, tusk shells). A small class of marine, burrowing molluscs, with curved, tapering tubular shells, open at both ends, and prehensile tentacles around the mouth, which are used to obtain food.

scattering. *See* backscatter, Mie scattering.

scheelite. A mineral exhibiting fluorescence which is a tungstate of calcium,

$CaWO_4$. Scheelite is a major source of tungsten, and commonly occurs with wolfram and cassiterite in contact metamorphic (*see* metamorphism) or hydrothermal deposits adjacent to acid igneous bodies, and in placer deposits. Tungsten is mainly used in the manufacture of tungsten carbide (WC) and for alloy steels.

schist. A medium to coarse-grained metamorphic (*see* metamorphism) rock with subparallel arrangement of flaky or acicular minerals. Many schists are micaceous and have an undulose cleavage. Schists are named according to their most prominent minerals (e.g. hornblende-schist, garnet-mica-schist). Greenschists are chlorite-schists and blueschists have the blue amphibole glaucophane.

Schistoma (Bilharzia). Genus of blood flukes (*Trematoda*) which infest mammals, including Man, causing much disease in Africa, Asia and South America. The larvae develop in various snails and enter the human body with drinking water or by burrowing through the skin.

schistosity. A texture of fairly coarsely crystalline metamorphic (*see* metamorphism) rocks caused by the subparallel arrangement of flaky or platy minerals. Schistosity is shown by schists and some gneisses.

schizocarpic. Applied to a dry fruit formed from two or more united carpels, which splits when mature into parts (mericarps), each of which represents the product of a single carpel (e.g. Hollyhock fruit).

schizogenesis. Reproduction by cell division. *See* schizogony.

schizogony. Asexual reproduction by multiple fission, during which a single cell (schizont) produces many smaller cells (schizozoites). This occurs in many Sporozoa.

Schizomycophyta. *See* bacteria.

schizont. *See* schizogony.

schizozoites. *See* schizogony.

sciophyll. Skiophyll.

sciophyte. Heliophobe.

sclereids (stone cells). *See* sclerenchyma.

sclerenchyma. A type of supporting tissue found in plants. It consists of cells whose walls are much thickened with cellulose or lignin. These cells occur singly or in groups, and when mature usually contain no living protoplasm. They are of two kinds, fibres, which are much elongated, and sclereids (stone cells), which are not. Sclereids occur often in seed coats and fruits (e.g. forming gritty particles in pear flesh). *See* collenchyma.

sclerophyll. Plants with tough or leathery evergreen leaves (e.g. holly, pine).

scleroproteins. Stable, fibrous proteins (e.g. collagen, keratin) present in skeletal, protective and connective tissues of animals.

sclerotium. A hard, compact mass of fungal hyphae, capable of remaining dormant for long periods when conditions are unfavourable for growth. *See* ergot.

Scolytus scolytus. Elm bark beetle. *See* Ulmus.

scoria. Pyroclastic material, ejected by a volcano, which is dark in colour, basic in chemical composition, and full of vesicles. Scoria is at least partly glassy. Scoria between 4 mm and 32 mm in diameter are called cinders. Scoria is contrasted with pumice.

Scorpion. Scorpionidea.

scorpion flies. Mecoptera.

Scorpionidea (scorpions). *Viviparous Arachnida* with the posterior segments forming a flexible tail bearing a terminal sting, which is used to paralyse their larger prey (mainly insects, spiders and other scorpions). Scorpions date from the Silurian, and the earliest were marine.

scp. Single cell protein.

scramble competition. Applied to a situation in which a resource is shared equally among competitors.

scree. Slopes of frost-shattered rock material.

Scrophulariaceae. A cosmopolitan family of dicotyledonous, mainly herbaceous plants, including *Veronica, Verbascum* and *Antirrhinum*. Some members of the family (e.g. Eyebright, *Euphrasia* and Yellow Rattle, *Rhinanthus*) are hemiparasites which absorb food from the roots of grasses. Most Scrophulariaceae are poisonous. The drug digitalis, obtained from the foxglove (*Digitalis*) is used to treat heart complaints.

scrub. (*a*) A type of vegetation dominated by shrubs and containing no tall trees. Scrub may be the natural climax, but in Britain it is often a stage in the succession, giving way to woodland. (*b*) (Forestry). A scrub forest is an inferior growth of shrubs and small or stunted trees.

scrubber. Apparatus used in sampling and in gas cleaning, in which gas is passed through a space containing wetted packing or spray. The liquid is used to remove solid or liquid particles or some gases from the carrier gas stream.

scurvy. Deficiency disease caused by lack of ascorbic acid. *See* vitamin.

scutching. *See* retting.

Scyphomedusae. Scyphozoa.

Scyphozoa (Scyphomedusae). A class of coelenterates in which the polyp stage is absent or inconspicuous, and the medusa is large and complex. The group consists mainly of the free-swimming jellyfishes, saucer- or bell-shaped animals with many tentacles around the margin, and a mouth at the end of a 'stalk' on the underside. *See* Actinozoa, Hydrozoa.

Se. Selenium.

Sea Beet. *See* Beta vulgaris.

sea breeze. *See* breeze.

sea cows. Sirenia.

sea cucumbers. Holothuroidea.

sea floor spreading. The concept that the new material from the Earth's mantle is being added to the Earth's crust at the site of oceanic ridges and thus 'spreading' some ocean floors. According to the modern theory of plate tectonics these are sites of constructive plate margins. Supporting evidence for sea floor spreading comes from palaeomagnetism (see also Magnetic anomaly), from seismic evidence, heat flow studies along oceanic ridges and radiometric dating. (Oceanic rocks and sediments have been shown to be geologically relatively young and progressively younger towards oceanic ridges). It is estimated, for example, that the Atlantic is growing at the rate of about 4 cm per year, 2 cm per year being added to each ocean ridge flank.

sea fog. Literally, fog at sea. It is formed typically in moist tropical air in warm sectors of cyclones in early summer (June–July) over colder areas of sea. In many coastal areas it is formed over land, drifts out to sea, where it remains by day, often crossing the coast in a sea breeze in the afternoon.

sea lilies. Crinoidea.

sea-mount. An isolated, usually conical, submarine mountain rising at least 1000m above the sea-floor. Sea-mounts occur on the continental rises, abyssal plains and in the trenches. Most rise abruptly from the bottom and are associated with large magnetic anomalies, and submarine eruptions and are thought to be of volcanic origin.

seasons. Divisions of the year according to regular changes in the character of the weather and, in the European languages, named by association with the annual life cycle of plants, winter being the period of dormancy, spring of sowing, summer of ripening, and autumn of harvesting. In the higher latitudes, day-length changes, related to progress toward a single maximum and a single minimum average temperature, regulate plant growth. Conventionally, in the northern hemisphere, winter is taken to cover the months December to February, spring March to May, summer June to August, and autumn September to November, the seasons being reversed in the southern hemisphere. Actual weather experienced is often unseasonal, however, so that the divisions are purely arbitrary.

sea spiders. Pantopoda.

sea urchins. Echinoidea.

second. (*a*) One-sixtieth of one minute, which is one-sixtieth of one degree of arc. (*b*) The SI unit of time, equal to the duration of 9192631770 periods of the radiation corresponding to the transition between two hyperfine levels of the ground state of the caesium-133 atom. One-sixtieth of a minute.

secondary enrichment. The natural enrichment of an ore by material of later origin deposited from descending aqueous solutions. Commonly, the added material has been derived from an overlying oxidised deposit. The enrichment may involve, (*a*) the infilling of voids by further deposition of a pre-existing mineral or by deposition of a new mineral, or (*b*) the replacement of an ore mineral by a mineral richer in the valuable constituent.

secondary minerals. Those minerals formed after consolidation of cooling magma by reactions within still cooling rock or with circulating groundwater; as a result of metamorphic activity; minerals formed by weathering of igneous rock; during diagenesis of sedimentary rock (*see* authigenic). *Cf.* primary minerals.

secondary pollutant. A pollutant formed in the environment by the combination or reaction of other (primary) pollutants. *See* synergism.

secondary rock. Rock formed by alteration or pre-existing rock. Sedimentary and metamorphic rock fall into this category. *Cf.* primary rock.

secondary sexual characters. Characteristics, other than those of the actual reproductive organs, which differ according to the sex of the animal. They are produced as a result of hormones secreted by endocrine organs, including the testis or ovary. Examples are the antlers of a stag, the facial hair of a male human, and differences in colour and size in the two sexes in birds.

secondary succession. *See* succession.

secondary thickening. The formation of secondary vascular tissue (xylem, phloem) as a result of the activity of cambium in the stems and roots of woody perennials. Most of the woody part of a tree is the product of secondary thickening. *See* annual rings.

secondary treatment. A process to remove the amount of dissolved organic matter in waste water, and to reduce further the suspended solids.

secretion. (*a*) The passage of a substance elaborated by a cell, particularly that of a gland, to the outside of the cell. Usually the substance is passed through the plasma-membrane, often against the concentration gradient, but sometimes the cell breaks down to liberate its contents. (*b*) A substance made and passed out by a cell (e.g. nectar in flowers, hydrochloric acid in the stomach of vertebrates).

sedentary. Untransported. As applied to soils, those weathered from underlying rock or formed by the accumulation of organic matter *in situ*.

sedimentary rock. Rock formed by the accumulation of material in water, or from the air or from glaciers. Sedimentary rocks can be formed from fragments of pre-existing rocks (i.e. clastic rocks such as sandstone, shale, arkose, and bioclastic limestone); from organic remains (e.g. peat, coal and some limestones); from chemical precipitates (i.e. the evaporites such as gypsum and halite); or from fragments blown out of volcanoes (i.e. the pyroclastic rocks such as tuffs.) Characteristically, sedimentary rocks have bedding planes, and some contain fossils.

sedimentary structures. Various structural features within beds of sedimentary rock and on bedding planes, which are used (*a*) to work out the environment of deposition, (*b*) the upper side of the bed as originally deposited, (*c*) to determine palaeocurrent directions. Sedimentary structures within beds include cross-bedding, graded bedding and imbricated pebble beds; and on bedding planes: rain prints, ripple marks, dessication cracks and various structures associated with turbidites such as flute casts eroded by turbulent flow, and tool marks made by moving objects.

sedimentation. The separation of an insoluble solid from a liquid in which it is suspended by settling under the influence of gravity or centrifugation. The settling of particles by gravity.

seed. A structure derived from a fertilized ovule, and consisting of an embryo plant and its food store surrounded by a protective coat (testa). The food is either stored in the seed leaves (cotyledons) or outside the embryo in the endosperm.

seed plants. Spermatophyta.

seepage tank. *See* septic tank.

segmentation. (*a*) Cleavage. Repeated cell divisions following fertilization and resulting in the zygote of a plant or animal becoming an early embryo. (*b*) Metameric segmentation (Metamerism). The repetition of a series of fundamentally similar segments (metameres) along the length of many animals. This is particularly obvious in the Annelida and Arthropoda. In the earthworm, the segments are visible externally, and each possesses a similar pattern of blood vessels, nerves, muscles, excretory organs, etc. Metameric segmentation is usually much modified, for instance by the development of the head. Vertebrate embryos clearly show metameric segmentation in muscular, skeletal and nervous systems.

Segregation, Law of (Mendel's first law). The law of segregation states that only one pair of contrasted characteristics (allelomorphs) may be represented in a single gamete. Segregation is a result of the behaviour of chromosomes during meiosis.

seiche. An oscillation in the level of the water of a lake or inland sea. A strong wind can pile up the warm surface layer of water (epilimnion), which slides back over the lower layer (hypolimnion) when the wind drops, then continues to oscillate for some time. Seiches can also occur as a result of temporary local fluctuations in the water table.

seif. A longtitudinal sand-dune parallel to the wind direction.

seismograph. An instrument for recording shock-waves generated by any transient disturbance of the Earth, whether natural (e.g. an earthquake) or man-made (e.g. an explosion). *See* isoseismal line.

seismonasty. *See* nastic movement.

selection pressure. A measure of the effects of natural selection on the genetic composition of a population.

selective species. A species found most usually in a particular community, but also, rarely, in other communities.

selenite. A colourless, transparent variety of gypsum.

selenium (Se). Element. A non-metal that exists in several allotropic (*see* Allotropy) forms. The silvery-grey crystalline solid ('metallic') form varies in electrical resistance on exposure to light and is used in photo-electric cells and solar cells as well as in photographic light meters (*see* photovoltaic effect). It is also used in the manufacture of rubber and some glasses. It occurs in nature as metallic selenides together with the sulphides of the metals. AW 78.96; At. No. 34; SG 4.81; mp 217°C.

selenodont teeth. Cheek teeth with crescent-shaped ridges (cusps) on the crown. This type of tooth is characterisitc of ruminants (e.g. cow). *Cf.* bunodont, lophodont.

self-fertilisation. The fusion of gametes from the same individual or from ramets of a clone. *Cf.* cross fertilization.

self pollination. The transference of pollen from a stamen to the stigma of the same flower, or of another flower in the same plant. *Cf.* cross pollination.

sematic coloration. *See* aposematic coloration.

semi-permeable membrane. A membrane through which a solvent, but not certain dissolved or colloidal substances may pass. *See* colloid, osmosis.

Senonian. A subdivision (usually ranking as a stage) of the Cretaceous System, and comprising the Coniacian, Santonian and Campanian.

sepal. *See* flower.

septic tank. A watertight sedimentation tank for sewage in which solids settle and are decomposed anaerobically. The liquid effluent may be passed from this tank into the ground, or into a seepage tank in which it is filtered through sand or gravel before release. A well designed septic tank rarely requires emptying. *Cf.* cesspool.

septum. (Bot.) A partition.

Sequoia. *See* Taxodiaceae.

seration. A series of communities within a formation or ecotone (e.g. the series of communities that may stretch across a valley from one hill top to the next).

serclimax. According to F. E. Clements, a stable plant community which persists at a stage before the subclimax in the seral sequence.

sere. (*a*) A series of plant communities resulting from the process of succession. *See* hydrosere, lithosere, xerosere, plagiosere, prisere, psammosere, clisere, subsere. (*b*) Any stage in a plant succession.

series. (Geol.) In stratigraphy, the major subdivision of a system. *Cf.* stage.

serine. Amino acid with the formula $CH_2OH.CH.(NH_2).COOH$. Molecular weight 105.1.

serir. *See* reg.

Serpentes. Ophidia. *See* Squamata.

serpentine. A group of silicate minerals with the approximate formula $Mg_3(Si_2O_5)(OH)_4$. Serpentine is the main alteration product of pyroxene and olivine. Chrysotile, a fibrous serpentine, is one of the minerals worked for the manufacture of asbestos. Nickeliferous serpentine, an important mineral, is called garnierite.

serpentinite. A rock consisting of serpentine minerals formed by metasomatism of various ultrabasic rocks. Serpentinites are cut, turned and polished as ornamental objects.

serule. A miniature succession composed of minute or microscopic organisms which usually is engaged in the breaking down of organic matter to its simpler constituents.

serrulate. Having a serrated edge. Applied to leaves whose edges have many small teeth pointing forward (i.e. toward the tip). *Cf.* crenate.

service reservoir. *See* reservoir.

sesquioxide. An oxide containing three oxygen and two metallic elements. In soil these are chiefly iron and alumina, Fe_2O_3; Al_2O_3.

sessile. (Bot.) Not stalked.

seston. The organisms (bioseston) and non-living matter (abioseston) swimming or floating in a water body.

seventh approximation. United States Department of Agriculture Soil Survey's seventh draft, advancing towards an ideal soil classification, and issued in 1960.

435

Seveso. Village near Milan, Italy, which suffered as a result of the accidental discharge of TCDD following an explosion in July, 1976 at the Icmesa factory which was manufacturing the herbicide 2,4,5-T. The population of Seveso was evacuated and vegetation had to be destroyed. A similar, but much smaller incident, involving TCDD occurred several years earlier in Derbyshire, UK, and as a result of the Seveso incident the Derbyshire factory was closed temporarily.

sevin. Carbaryl.

sewage fungus. A pale, slimy, bacterial deposit which often occurs in water subject to organic pollution.

sex chromosomes. Chromosomes which are responsible for determining the sex of an animal. There is a homologous pair (two X chromosomes) in the nuclei of the homogametic sex, and a dissimilar pair (X and Y) or an unpaired X chromosome in the nuclei of the heterogametic sex. The X chromosome is usually larger than the Y chromosome and, unlike it, contains numerous genes which show sex-linkage. All of the gametes produced by the homogametic sex, but only half of those produced by the heterogametic sex, contain an X chromosome.

sex determination. *See* sex chromosomes.

sex-linkage. The occurrence of a character more frequently in one sex than in the other because the gene controlling that character is carried only on the X chromosome (*see* Sex chromosomes). For instance, in Man, the male has a single X chromosome, and the effects of a recessive gene carried on this chromosome cannot be masked by a dominant allelomorph. This results in many more men that women manifesting sex-linked recessive characters such as haemophilia and red-green colour blindness.

sexual reproduction. Reproduction which involves the fusion of two haploid nuclei to form a zygote. The nuclei are often those of male and female gametes.

sexual selection. A type of selection which Darwin suggested might bring about the evolution of secondary sexual characters such as brilliant plumage and courtship behaviour in birds. Selective mating, caused because a female chooses only males with certain characteristics, would reslult in these characteristics being transmitted to the offspring, to the exclusion of features possessed by rejected males.

SG. Specific gravity.

SGHWR. Steam-generating heavy-water reactor.

shade plant. Heliophobe.

shale. A sedimentary rock, made up largely of clay, which splits along bedding planes.

shear stress. In fluid flow, shear stress is the force per unit area tending to produce shear (e.g. wind shear in the atmosphere). This stress is generated at surfaces as a result of viscosity and turbulence.

sheep. *See* Bovidae.

sheep month. The amount of feed or grazing needed to maintain a mature sheep or a ewe and its suckling lamb in good condition for 30 days. This is usually considered to be equivalent to one-fifth of a cow month. *See* animal unit.

shelterbelt. Windbreaks, usually stands of trees, planted so as to provide nearly continuous protection against the wind.

shield. (Geol.) A major continental block, usually of Precambrian igneous and metamorphic (*see* Metamorphism) rocks, which has been stable over a relatively long period of Earth history, and which has undergone only faulting or gentle warping.

shield volcano. A central vent volcano with slopes of very low angle built up of flows of very fluid basaltic lava. Also called a Hawaiian type volcano.

shipworms. Teredo.

SHM. Simple harmonic motion. *See* sine wave.

shock wave. Int the atmosphere, a travelling wave in which thermodynamic properties of the air change suddenly to a different value.

shore zonation. The division of the sea shore into zones, each of which supports a characteristic fauna and flora. (*a*) The splash zone lies above the extreme high water level of spring tides, but may sometimes be drenched with spray. Small periwinkles and channel wrack (*Pelvetia canaliculata*) extend up into this zone. (*b*) The upper shore lies between the average high tide level and the extreme high water level of spring tides, so is only covered occasionally, and then for short periods. The number of species inhabiting this zone is relatively small. (*c*) The middle shore lies between the average low tide level and the average high tide level. It is an extensive zone, mostly covered by the sea twice a day. A large number of speices inhabits this zone. (*d*) The lower shore extends from the extreme low water level of spring tides up to the average low tide level. It is only uncovered occasionally, and for short periods. Some species typical of the area below tide levels extend into this zone. (*e*) The sublittoral fringe lies below the extreme low water level of spring tides, and is never uncovered. The shallow water here has a greater range of temperature than the open sea. The upper part of the large oar weeds (*Laminaria*) typical of this zone on rocky shores may project above the surface at low water.

short-day plants. Plants that develop normally only when the photo-period is less than a critical maximum.

short horned flies. Brachycera.

shredding. Pulverisation.

shrub. A low-growing, woody perennial, which, unlike a small tree, branches from the base. The term is often restricted to plants less than about 6 m in height. *Cf.* coppice.

shuttle box. A device used in behavioural experiments with animals, consisting of a box in which the animal 'shuttles' from one end to the other in response to a stimulus.

Si. Silicon.

sial. The discontinuous upper part of the crust of the Earth. 'Sial' is derived from *si*licon and *al*uminium. Tending to be replaced by the less committal term 'Upper Crust'.

siblings (sibs). The offspring of the same parents.

sibling species (species pair). Species which are very similar or even indistinguishable in appearance, but are distinct genetically and do not interbreed (e.g. Chiff-chaff and Willow Warbler, which are chiefly distinguishable by their song). Sibling species are probably examples of recent speciation.

sibs. Siblings.

siccicolous (xerophilous). Drought resistant. *See* Xerophyte.

siderite. (*a*) A meteorite composed entirely of metal. (*b*) A carbonate mineral with the formula $FeCO_3$, often found as a gangue mineral, in sedimentary iron ores, as nodules in clay, and as a cement.

Siegenian. The second oldest stage of the Devonian System in Europe.

Sierra Club. Society devoted to conservation. Founded in 1892, the Club works in the US and in other countries to restore the quality of the natural environment and to maintain the integrity of ecosystems.

sieve tube. *See* phloem.

silcrete. *See* hardpan.

silencer. A sound-absorbing duct for exhaust systems, as on a car.

Silesian. Upper Carboniferous. The Silesian usually ranks as a series, and comprises the Westphalian and Stephanian Stages.

silica. SiO_2. Silica refers either to a mineral of that composition (e.g. quartz, chalcedony etc.), or to the silicate mineral content of a rock, which is usually expressed chemically as the weight percentage of silica.

silicaceous sandstone. Orthoquartzite.

silicate mineral. Silicates are the major group of compounds in the crust. Silicates are based on SiO_4 tetrahedra joined by the oxygen (which also bonds to the metal cations), and are classified on the arrangement of the tetrahedral units. The silica tetrahdra can be independent (e.g. olivine), or in rings (e.g. beryl, mica, and talc), or in chains (e.g. pyroxene) or bands (e.g. amphiboles), or in a three-dimensional framework (as in quartz and feldspar), or two SiO_4 can share one oxygen. Aluminium can replace the silicon in some of the tetrahdra. The properties of the mineral are closely related to the atomic structure.

silicon (Si). Element. A non-metal that occurs as a brown powder and as dark grey crystals. In nature it occurs in various silicate minerals and as silica. It is used in glass manufacture and in alloys and in silicone compounds, used in lacquers, resins, lubricants and water-repellant finishes. AW 28.086; At. No. 14; SG 2.42; mp 1420°C.

silicosis. Type of pneumoconiosis caused by the inhalation of silica dust.

silicula. *See* Siliqua.

siliqua. A type of dry fruit found in the Cruciferae (e.g. *Cheiranthus*, wallflower), consisting of an elongated capsule, formed from two united carpels, which opens from below, exposing the seeds attached to a central framework. A silicula (e.g. the fruit of *Capsella bursapastoris*, Shepherd's Purse) is similar, except that it is short and broad.

silk. Very fine, strong protein threads secreted by various insects (e.g. Lepidoptera, Embioptera, Coleoptera, Trichoptera) and spiders. The silk glands of the silkworm (*Bombyx*) are modified salivary glands. Embioptera have silk glands in the front legs, and those of spiders are situated in the abdomen. Silk is produced as a fluid, which is extruded through tubular spinnerets, and hardens on contact with the air. It is used in the construction of egg and pupal cocoons, webs, nets, 'balloons', etc.

silky. *See* lustre.

sill. A tabular igneous intrusion which is concordant (i.e. it follows a bedding-plane in sedimentary rocks or the foliation in metamorphic (*see* metamorphism) rocks). Sills which cut across from one horizon to another are described as transgressive.

silt. Sediment made up predominantly of grains between 0.06 mm and 0.004 mm in diameter.

Silurian. Refers to the third oldest period of the Palaeozoic Era, usually taken as beginning some time between 435 and 460 million years before present. Silurian also refers to the rocks formed during the Silurian Period: these are called the Silurian System, which is normally divided into four series (Llandoverian, Wenlockian, Ludlovian and Pridolian or Downtonian). The Silurian probably lasted 30 million years. The Downtonian is placed by some geologists in the succeeding Devonian System. The Silurian is zoned using graptolites.

silva. The aggregate of forest trees of an area.

silver (Ag). Element. A white, rather soft metal, extremely malleable and ductile and an excellent conductor of electricity. It occurs as the metal and as argentite, horn silver and other compounds. It is used in jewellery and coinage, etc. and its compounds are used in photography. AW 107.87; At. No. 47; SG 10.5; mp 960.5°C.

silver fish (Thysanura). *See* Apterygota.

sima. The lower part of the crust of the Earth. 'Sima' is derived from *si*licon snd *ma*gnesium. Tending to be replaced by the less committal term 'Lower Crust'.

simazine. Organic soil-acting herbicide used to control germinating weeds in many crops and for total weed control on waste land, paths, etc. It is a long-lasting chemical which may prevent regrowth for as long as a year.

similarity. In fluid motion, similarity is a property of certain flows in which the flow pattern remains constant when the linear scale is changed. Examples are the momentum jet, and the buoyant thermal, in which the development is conical and all of the flow properties are proportional to the distance from the origin.

simoom. sirocco.

simple interest. Linear growth. *See* exponential growth.

Sinanthropus pekinensis. *See* Homo.

Sinemurian. A stage of the Jurassic System.

sine wave. Simple harmonic motion. The path traced by a point which oscillates along a line about a central point, so that its acceleration towards the centre is always proportional to its distance from the centre. On a graph, tracing position against time, a sine wave may be described by the formula:

$$y = r \sin 2\pi \left(\frac{t}{T} - \frac{x}{\lambda} \right)$$

where y = the distance of the point from the centre, t = time, r = amplitude, T = the period of the wave, λ = wavelength, x = the distance the point has travelled from the centre in time t.

single cell protein (scp). Protein derived from unicellular organisms (e.g. bacteria, yeasts) grown on a hydrocarbon substrate (e.g. crude oil, methanol, cellulose, or wastes from food processing). Scp may be used directly as a 'meat extender', but in general it is used in livestock feedingstuffs as a protein additive. *See* novel protein foods.

sinistral fault. A wrench fault, or a fault with a considerable component of strike-slip motion, in which the distant block shows the relative displacement to the left when viewed across the fault-plane.

sink. (*a*) (Geol.) Swallow-hole (*b*) (Biol.) Receptacle, or receiving area, for materials translocated through a system (e.g. the oceans are the sink into which many water-borne pollutants are drained; the seed of a plant is the sink in which nutrients are stored). (*c*) In air pollution, a place or mechanism associated with the removal of air pollutants from the atmosphere. Examples are rainout of particles; soil as an absorber of carbon monoxide; oxidation of sulphur dioxide to form sulphur trioxide and thence sulphate particles, etc.

sink hole. (*a*) A depression in marshy, flat land where water collects. (*b*) Swallow hole.

sinter plant. An industrial plant in which metal particles are compressed into a coherent body under heat, but at temperatures below the melting point of the metal. Glass, ceramics and certain other non-metals may also be sintered. Sinter plants can be a source of particle emission.

sinusoidal wave. Sine wave.

Siphonaptera. Aphaniptera.

Siphonopoda. Cephalopoda.

Siphunculata. Anoplura.

Sipunculoidea. *See* Annelida.

Sirenia (sea cows). An order of large aquatic herbivorous placental mammals, with flipper-like front limbs, vestigal hind limbs, a horizontal tail fin and little hair. Modern forms are the Manatee of the Atlantic and the Dugong of the Pacific and Indian Oceans. Sea cows live along the coasts and in rivers in tropical and subtropical regions.

sirocco. A warm, dry wind that occurs in the Mediterranean area in spring, when it brings air from the central Sahara across sea that is much cooler than the desert. Siroccos are known locally as khamsin (in Egypt and Malta), leste (in Madeira and North Africa), simoom (in north-east Africa and Arabia), and leveche (in south-east Spain). Siroccos bring oppressive weather and often cause damage to vegetation.

site class. The productivity of a forest, measured in cubic metres of wood per hectare per year, calculated for a rotation of 100 years.

Site of Special Scientific Interest (SSSI). In British planning, an area of land which, not being land for the time being managed as a statutory nature reserve, is (in the opinion of the Nature Conservancy Council) of special interest by reason of its flora, fauna or geological or physiographical features. The local planning authority must notify the Conservancy of any proposed development for the area, so that its implications for the scientific interest may be assessed. As defined by the Countryside Commission.

site-type. A group of communities that corresponds to a particular set of site characteristics.

SI units (Système International d'Unités). An internationally agreed system of units. The seven basic units are the metre (m), kilogram (kg), second (s), ampere (A), kelvin (K), mole (mol) and candela (cd), with the radian (rad) and steradian (sr) added as supplementary units. Derived from these units are the hertz (Hz), newton (N), joule (J), watt (W), coulomb (C), volt (V), farad (F), ohm (Ω), weber (Wb), tesla (T), henry (H), lumen (lm) and lux (lx). *See* each unit under its own name.

skarn. An impure limestone or dolomite which has undergone thermal metamporphism and metasomatism, and usually contains calcium silicates and borosilicates. Many skarns are worked for sulphide minerals and manganese silicates.

skeletal muscle. *See* muscle.

skiophyll (sciophyll). A plant with dorsiventral leaves.

Skiophyte. Heliophobe.

slag. Non-metallic residue from the smelting of metallic ores, generally as a molten mass floating on the molten metal.

slaked lime. Calcium hydroxide.

slash-and-burn. Swidden farming.

slate. An argillaceous metamorphic (*see* metamorphism) rock with well developed cleavage but little recrystallisation. Slates are formed under low grade regional metamorphism. Slates subjected to thermal metamorphism develop new minerals in small patches, so producing spotted slates, or long rods of chiastolite (a variety of andalusite): with increased temperature the cleavage is destroyed and a hornfels results. Slates are used for roofing, walling and hedging, and as an inert filler.

sleeping sickness. Disease caused by *Trypanosoma gambiensis* and *T. rhodesiensis,* and transmitted by the tsetse fly (*Glossina*). *See* Cyclorrhapha, Trypanosomidae.

sleet. Precipitation of water and ice simultaneously.

slickenside. A polished and scratched surface of a fault plane.

slides. Rockslides occur when massive whole rocks slide downhill along sloping bedding planes or joints. They may be very rapid and catastrophic. *See* slump.

slime fungi. Myxomycophyta.

slime moulds. Myxomycophyta.

slope of front. A warm front has a surface (the boundary between the two air masses) with a relatively shallow slope, around 1:100, while cold fronts are steeper, at 1:50 over most of their length, steepening considerably near the surface.

sludge. (*a*) Thick mud, often greasy. (*b*) The suspended solid matter in industrial effluent or sewage, after partial drying.

slump. The movement downhill of a unit of weathered rock or regolith over a definite surface of failure, often lubricated by water. A slump may also be caused by water undercutting the base of a slope or by the faulty design of the cut of an embankment. If the surface of the failure is spoon-shaped, the slump is called a rotational slump.

smelting. The extraction of a metal from its ore by a process involving heat, generally reducing the oxide of the metal with carbon, in a furnace, or the roasting or calcination of sulphide ores. Smelting works emit grit and, where sulphide ores are processed, sulphur dioxide, as well as emissions from the combustion of fuels.

Smith, Adam (1723–1790). Scottish economist who advocated free trade and, in *The Wealth of Nations*, introduced the concept of 'the invisible hand'. Smith was much influenced by the school of Physiocrats (*see* Physiocracy), but divorced their main concept from its exclusive concern with agriculture and applied it to manufacturing industry.

smog. Originally, a contraction of 'smoke' and 'fog', which characterised air pollution episodes in London, Glasgow, Manchester and many other cities. The word was coined by H. A. Des Voeux in 1905. The Great London Smog of 1952 casused 4000 excess deaths and led to the enactment of the Clean Air Acts of 1956 and 1968, which provided for the banning of smoky fuels within specified areas. The word smog has since been applied to other air pollution effects not necessarily connected with smoke, such as 'Los Angeles Smog' , which arises from nitrogen oxides and hydrocarbons emitted by motor vehicles, and the photochemical action of sunlight. *See* London Smog Incidents.

smoke. An aerosol of minute solid or liquid particles (most less than 1 micron in diameter) formed by the incomplete combustion of a fuel. In air pollution it is mainly associated with the burning of coal.

smoke stack. Chimney.

smooth muscle. *See* muscle.

smudge pot. A smoke generator used to create an artificial fog with the object of preventing a ground frost or frost in a shallow layer of air which might damage fruit, etc., by deepening the layer of air cooled by radiation at night.

smut fungi (Ustiginales). An order of parasitic Basidiomycetes with black spores, the cause of many plant diseases. *Tilletia caries* infects wheat, causing stinking smut disease, or bunt.

Sn. Tin.

snow line. The altitude above which snow lies throughout the year. It varies from place to place depending on summer temperature, wind, snow amount, steepness of slopes, etc.

snow storm. Intense precipitation of ice crystals. In synoptic meteorology, a snow storm is classed as 'heavy' if it exceeds 4 cm/h.

soaring. The art of gaining height to remain airborne in motorless or non-flapping flight by the use of updraughts and upslope flow, practised by birds of prey and glider pilots.

social releaser. *See* releaser stimulus.

sociation. A plant community of homogenous composition so that each part is characteristic of the whole.

socies. A group of subdominant plants in a stage of succession which precedes the climax.

society. A group of plants forming a minor climax community within a consociation, and in which the dominant species is different from that of the consociation.

Society for the Promotion of Nature Reserves (SPNR). Society founded in 1912, which owns and manages nature reserves in Britain, and sponsors the Naturalists' Trusts movement.

sodium (Na). Element. A soft, silvery-white metal. It is very reactive, tarnishes rapidly in air, and reacts violently within water to produce sodium hydroxide and hydrogen gas. Compounds are abundant and distributed widely, the most common being common salt (sodium chloride). It is an essential micronutrient and is used as a coolant in breeder reactors (see Liquid metal fast breeder reactor). AW 22.9898; At. No. 11; SG 0.971; mp 97.5°C.

sodium bicarbonate. *See* bile.

sodium chlorate. Translocated and soil-acting herbicide used for total weed control on land not intended for cropping. It can persist in the soil for months after application, and makes plant residues inflammable when dry. It is not highly poisonous to mammals.

sodium sulphite process. *See* potassium sulphite process.

softwoods. Trees belonging to the Gymnospermae or their timber. Almost all commercial softwoods are Conifers. These do not possess vessels in their wood.

445

Cf. hardwoods.

soil. (*a*) (Biol.) Weathered, unconsolidated surface material in which plants anchor their roots and from which they derive nutrients and moisture. (*b*) (Eng.) Loose unconsolidated material that can be moved without blasting. (*c*) (Geol.) Any material weathered *in situ*, thus including older weathered deposits that might be parent material for the present soil. *See* surface deposits, soil classification, soil horizons.

soil-acting herbicides. Herbicides which are applied to the soil and are absorbed by weeds before the emergence of aerial parts from the soil (e.g. Diuron, Linuron, Simazine, Diallate, Triallate, sodium chlorate).

soil association. (*a*) (US usage) A group of related soil types. Soil series (UK usage) (*b*) British-based voluntary organisation with world-wide connections and affiliations, concerned with the promotion of organic farming.

soil classification. There are two main systems of classification. (*a*) Based on the concept of zonality: i.e. that commonly found soils can be associated with particular regions or zones of climate. There are three main orders, divided into sub-orders, world groups, series, and types. (1) Order I Zonal. Belts of soil that correspond roughly with climatic zones (e.g. Tundra soils, Podzols (Podsols), Grey brown podzols, Brown forest soils, Black earth soils (chernozems), Prairie soil, Chestnut and brown soils, Desert soils, Laterite soils, etc.) *See* each soil under its own name. (2) Order II Intrazonal. Soils where local conditions are more important than climate in development (e.g. Rendzinas, Gley soils, etc.) See each soil under its own name. (3) Order III Azonal or Skeletal. Young soils (e.g. Blown sand, River alluvium, material in screes and shingles, etc.) See each under its own name. (b) The US Seventh Approximation (1960) gives ten major soil orders based on the present state of development of the soils. The orders are divided into sub-orders, great groups, sub-groups, families, and soil series: (1) Entisols. Young soils, without horizon development, which occur in all climates (e.g. lithosols). (2) Vertisols. Clay-rich soils that swell and crack in seasonally wet and dry environments, thus mixing or inverting horizons (e.g. Grumosols). (3)' Inceptisols. Young soils with horizons weakly developed, occurring in variable climates, including Brown earths and Tundra Soils. (4) Aridisols (Aridosols). Soils of desert and arid regions with generally mineral profiles. The accumulation of salt, gypsum and carbonate is common. (5) Mollisols. Grassland soils characterised by a thick, organic-rich surface layer, covering a wide variety of profiles with strong structural development, including the brown, chestnut, chernozem, prairie (brunizem), red prairie, rendzina and brown forest soils of the earlier definitions. (6) Spodosols. Podzolised soils with a diagnostic iron oxide and/or organic-enriched B horizon underlying an ashy grey, leached layer in the A horizon, associated with a cool and cool-humid climate and a forest or heath vegetation cover. (7) Alfisols. Relatively young, acid soils characterised by a clay-enriched B horizon, commonly occurring beneath deciduous forests and associated with humid, sub-humid, temperate and subtropical climates. (8) Ultisols. Deeply weathered, red and yellow clay-enriched soils, associated with humid temperate to tropical climates. (9) Oxisols. Tropical and subtropical soils, intensely weathered and including most lateritic and bauxitic soils of earlier definitions. (10) Histosols. Organic soils developed largely by the accumulation of organic matter in a waterlogged site. There is no major climatic distinction, and the term includes bog soils, half bog and peat soils of earlier definitions. *See* soil horizons.

soil creep. The slow movement of soil under the force of gravity. Soil creep produces such phenomena as terminal curvature of planar structures in underlying rock, and deformed fence-lines and bulging walls.

soil drainage. The removal of surplus water from the soil. Naturally, water will drain to the ground water, but where natural drainage is insufficient for agricultural or other purposes, the rate of drainage can be increased. Ditches dug along the upper boundary of a sloping field will reduce the amount of water entering the field from land higher up the slope. Mole drains are narrow tunnels made in the subsoil by a cylindrical implement attached to a blade. When dragged through the soil, the slit made by the blade closes, but the tunnel remains open. Tile drains are made from unglazed clay as pipes, each about 25 cm long, laid in trenches dug to receive them on a bed of, and packed around with, small stones. Tile drains are laid end to end in straight lines, either running parallel to the slope to form a series of channels, or running parallel to the slope and with side branches laid in a 'herring bone' pattern. Tile drains are sometimes improvised using pieces of corrugated iron, gorse, or other woody material.

soil-flow. Solifluction.

soil horizons. Distinctive, successive layers of soil produced by internal re-distribution processes (*not* by sequential sedimentary deposition). Conventionally the layers have been divided into A, B and C horizons. The A horizon is the upper layer, containing humus and is leached and/or eluviated of many minerals. The B horizon forms a zone of deposition and is enriched with clay minerals and iron/aluminium oxides from the A layer. The C layer is the parent material for the present soil and may be partially weathered rock, transported glacial or alluvial material or an earlier soil. Sometimes a D layer is quoted, which is the massive rock underlying the soil layers, and an O layer to designate the fresh organic litter on the surface of the ground. These layers have been subdivided by the use of numbers to indicate minor differentiation within the horizons: A_0 fresh organic litter; A_1 organic-rich A layer; A_2 leached A layer; A_3 A layer grading into B; B_1 B layer grading into A; B_2 depositional layer; B_3 B horizon grading into C; C weathered parent material. Latterly, and coming into favour, is a more detailed nomenclature for the sub-layers, using lower case letters to follow the capital letter which designates the kind of horizon (e.g. a ploughed or cultivated surface horizon (e.g. A_p);$_b$ buried horizon (e.g. A_b);$_g$ waterlogged, gleyed layer (e.g. A_g);$_e$ leached, acid horizon (e.g. A_e);$_h$ accumulation of organic matter (e.g. B_h);$_t$ enriched with clay (e.g. B_t);$_{ca}$ accumulation of calcium carbonate (e.g. B_{ca}, C_{ca});$_s$ enriched with translocated sesquioxides of iron and aluminium (e.g. B_s);$_{ir}$ accumulation of iron (e.g. B_{ir});$_{sa}$ accumulation soluble salts (e.g. B_{sa}, C_{sa}).

soil map. A map that shows the distribution of soil types in relation to other features of the land surface.

soil particle size. (Scale in mm)

	Old international	American
Coarse sand	2.00 – 0.2	2.00 – 0.2
Fine sand	0.2 – 0.02	0.2 – 0.05
Silt	0.02 – 0.002	0.05 – 0.002
Clay	< 0.002	< 0.002

See soil texture.

soil phase. Local variation of soil type or soil series based on some unusual condition of soil, e.g. stoniness, slope, salinity, etc.

soil profile. A vertical cross-section of soil horizons, not including the D layer.

soil series. The basic soil mapping unit composed of soils similar in structure, colour and depth, developed on a uniform or similar soil parent material. The surface layer may show differences. *See* soil types, soil phase.

soil type. A subdivision of a soil series based on the texture of the surface horizon. *See* soil horizon.

Solanaceae. A family of dicotyledonous herbs, shrubs and small trees of tropical and temperate regions. Economically important genera are *Nicotiana* (tobacco), *Capsicum* (red pepper) and *Solanum*. *S. tuberosum*, native to South America, is cultivated for its stem tubers (potatoes) and *S. lycopersicum* is cultivated for its fruit (tomatoes). *Atropa bella-donna* (Deadly Nightshade, Dwale) is very poisonous, containing the alkaloids atropine and hyoscyamine.

solanin. A naturally-occurring poison found in some members of the Solanaceae. *See* alkaloids.

solanum. *See* Solanaceae.

solar cell. Device for converting solar energy into electrical power using a semi-conductor sensitive to the photovoltaic effect. Solar cells are used on space satellites to power electronic equipment and may come to be used to provide energy on the Earth.

solar collector. A dark-coloured (ideally matt black) surface used to absorb solar heat. The heat is then transferred, usually to water that flows beneath the surface. Solar collectors can be used to heat water or to provide, less usefully, space heating.

solar constant. The energy flux per unit area due to solar radiation, which would be measured outside the Earth's atmosphere at the mean distance of the Earth from the sun. Its value is about 139.6 mW/sq cm.

solar farm. A suggested power utility, based in a desert and covering a considerable area, where large amounts of electrical energy would be generated from solar energy.

solar furnace. Device for generating large amounts of energy from solar radiation, using mirrors to focus heat rays so that intense heat is collected in a small area.

solarimeter. An instrument for measuring total radiation per unit area received on the ground.

solarisation. The inhibiting effect of very high intensities of light upon photo-synthesis.

solar power. The extraction of useful energy from solar radiation. The most successful examples so far are solar cells used in satellites, and solar collectors heating water. In domestic applications, solar heating is at a disadvantage because the annual cycle of solar radiation intensity is out of phase with the cycle of heating demand. Since it is solar heat that provides the energy to power weather systems, wind power is sometimes classed (e.g. in USA, but not in UK) as solar energy, and since the movement of ocean currents and waves is also caused partly by the weather, wave power may also be counted as derived partially from solar energy.

solar radiation. Electromagnetic waves emitted by the sun. In space outside the Earth's atmosphere, this radiation covers a wide range of wavelengths, but absorption in the stratosphere restricts the spectrum received at the ground to certain limited bands which include the range of visible light wavelengths.

sol brun acide. Acid forest soil with a strongly leached upper brown layer, but without much accumulated iron, aluminium or clay compounds in the de-

solenocyte

positional zone. *See* soil horizons.

solenocyte. *See* nephridium.

solfatara. A volcanic vent emitting primarily sulphurous and aqueous vapours. Solfataras are a variety of fumarole.

solifluction (solifluxion, soil-flow). The gradual down-slope movement of particles at the Earth's surface. Some authors restrict solifluction to processes controlled by freezing and thawing, and the production of such deposits as head and combe-rock.

solifluxion. Solifluction.

soligenous. Applied to a wet, peaty area in which the water table is maintained by a flow of water through the substratum.

solum. The soil mantle, comprising the organic layer, the leached layer, and the layer of deposition, but not including the parent material (i.e. the A and B horizons). *See* soil horizon.

solution. (Geol.) Water entering the ground encounters organic matter containing carbon dioxide and acid ions which become dissolved in the water. The acids thus produced dissolve mineral matter as they percolate downwards. Most commonly, a weak solution of carbonic acid (H_2CO_3) is formed. This is a solvent of some rock, particularly limestone, chalk and the cementing matrix of some sandstones.

soma. All the cells of an organism other than the reproductive cells.

somatic. Pertaining to the soma or to the body wall of an animal.

somatotrophic hormone. Growth hormone.

sonar. A device for locating underwater objects using sound waves which are reflected from the objects, in a way similar to the use of microwaves in radar. *See* echo.

sonde. A device for obtaining direct measurements of the condition of the atmosphere at various altitudes. It comprises a lifting device such as a balloon or rocket, and a set of transducers with either a recorder or more frequently a radio transmitter.

sone. Unit of loudness, designed to give a scale of numbers roughly proportional to the loudness. *Cf.* phon.

sonic boom. The transient noise heard by a stationary observer as an object (esp. an aircraft) passes nearby travelling above the speed of sound. The cause of the sound is the passage of a shock wave with its associated pressure pulses. For supersonic aircraft, the pressure pulse is of the order of 1.0 lbs/sq in (0.0705 kg/sq cm), but may fluctuate above and below this level. Strong sonic booms can damage buildings and break windows, while their sudden onset can be disturbing to people even at quite modest pressure levels. *See* SST, Mach number.

Sorghum. *Sorghum vulgare See* Graminae.

sorting. According to nature of transportation, wind or water will fractionate a homogenous rock debris into grains of similar sizes. The sorting of a sediment refers to the range of grain sizes within a particular sample. The term 'well sorted' has a different meaning in geology and engineering. To a geologist a well sorted sediment has a preponderance of just one grain size. To an engineer, 'well sorted grains' means an even mix of range of sizes and this is referred to as a graded aggregate.

sound level. The value of sound, in decibels. *See* loudness.

sound power level. The total energy per second emitted by a sound source, expressed in decibels.

sound pressure level. The effective (i.e. root-mean-square) value of pressure fluctuations above and below atmospheric pressure caused by the passage of a sound wave, expressed in decibels.

sound propagation. Sound waves propagate in gases in the form of small variations in pressure and displacement in the direction of propagation. A sound source is a generator of pressure fluctuations, which are communicated from each small element of the gas to the next by the acceleration of gas due to transient pressure gradients leading to compression of the adjacent gas and consequently higher pressures there. *See* speed of sound.

sound shadow. The acoustical equivalent of a light shadow.

sound, speed of. Speed of sound.

sound waves. Pressure and displacement waves in material media. From point sources in uniform media the waves are spherical and propagate radially. *See* sound propagation, speed of sound.

source rock. The geological formation in which petroleum or minerals originated.

sour gas. Natural gas containing much hydrogen sulphide (H_2S). The hydrogen sulphide is separated, and used as a source of sulphur. The opposite of sour gas is sweet gas.

southerly buster. Buster, southerly.

Soya (soybean). *Glycine max.* An annual herb belonging to the family Leguminosae, growing as an erect, bushy plant to a height of between 45 cm and 2 m, bearing rough, brownish hairs on its leaves and pods. The pods, up to 5 to 7 cm long, are constricted between the two to four seeds that each contains. The seeds (beans) may be green, brown, yellowish or black, and they are grown for their high protein content as food for livestock and humans. The plant also yields an oil used for cooking, as a salad oil, in the manufacture of margarine, and in the manufacture of soaps, plastics, paints and other products. Its young shoots are important in Chinese cuisine (bean sprouts), its seeds can be dried and ground to make a flour used to increase the protein content of other flours, it is used in ice cream manufacture, and a 'milk' extracted from the seeds is used in Chinese and Japanese cuisine and is recommended for invalids. Soy sauce is used widely as a condiment.

spadix. (Bot.) A spike bearing flowers, sometimes sunken, and enclosed in a spathe.

spathe. (Bot.) A large, sheathing bract.

speciation. The evolutionary process by which a new species is formed.

specient. An individual member of a species.

species. *See* classification.

species pair. Sibling species.

specific gravity. (SG). Relative density, being the ratio of the density of a substance at a particular temperature to the density of water at the temperature of its maximum density (4°C). Numerically, the SG is equal to the density in grams per cubic centimetre, but it is stated as a pure number.

spectrum of turbulence. The variation of turbulence intensity with frequency. Atmospheric turbulence may be thought of as being composed of oscillations in local velocity with a wide range of time and length scales (and therefore of frequency). The spectrum of these oscillations is the r.m.s. velocity at a given point and in a given direction per unit bandwidth centred on the various frequencies contributing to the motion.

specularite. Specular iron ore. Haematite in the form of metallic black crystals.

speed of sound. In air, the speed of propagation of sound waves is about 332 m/sec at 0°C. Generally, in a perfect gas, the speed of sound is given by: $a = \sqrt{(\gamma R T)}$, where γ is the ratio of specific heats C_p/C_v, R is the gas constant, and T is the absolute temperature.

spell of weather. A convenient concept in the study of persistence of weather types. It may be defined for the purpose of statistical studies as a period of consistent type of n days, n being 5 or 10 or a number sufficiently large to make the spell a notable event. Thus a spell of dry weather would be longer than a spell of fog.

spermagone. Spermogonium.

spermagonium. Spermogonium.

Spermaphyta. Spermatophyta.

spermatia. *See* spermogonium.

spermatocyte. (*a*) (Bot.) A cell which gives rise to a spermatozoid without cell division. (*b*) (Zool.) A primary spermatocyte gives rise by meiosis first to two secondary spermatocytes, then to four spermatids. These become spermatozoa, usually developing flagella.

spermatogenesis. The cell divisions and changes resulting in the formation of spermatozoids or spermatozoa.

spermatogonium. A cell in a testis which gives rise to spermatocytes.

spermatophores. Packets of spermatozoa produced by certain animals (e.g. newts, some Crustacea and Mollusca) with internal fertilisation.

Spermatophyta (Spermaphyta, phanerogamia, seed plants). Plants which produce seeds. The group comprises the Gymnospermae and the Angiospermae.

spermatozoid (antherozoid). A male gamete produced by lower plants (Algae, Bryophyta, Pteridophyta) and able to move by means of a flagellum.

spermatozoon (Pl. Spermatozoa). A male gamete produced by an animal. Most spermatozoa can move by means of a flagellum.

spermogonium (spermagonium, spermagone, pycnium). A hollow organ inside which non-motile male gametes (spermatia) are formed in some fungi (e.g. rusts).

Spermatia are transferred by the wind or insects to other hyphae.

sphaeroidal weathering (onion weathering). The weathering of a rock such that concentric layers separate off a less weathered core. Commonly seen in well-jointed rocks like basalts and dolerites in which water penetrates along the joints attacking each block from all sides.

sphagnicolous. Inhabiting bog moss (*Sphagnum*).

sphagniherbosa. Plant communities growing on peat and containing large amounts of bog moss (*Sphagnum*).

Sphagnum moss. *See* bog.

sphalerite. The mineral zinc sulphide, ZnS, also known as zinc blende. Cadmium commonly substitutes for zinc in sphalerite, which is the most important ore of both metals. Cadmium is mostly used in the electroplating industry and in pigments, whereas zinc is used principally in alloys for die-cast products and in anti-corrosion treatments for iron and steel. Sphalerite commonly occurs, with galena, in hydrothermal deposits, in sedimentary stratiform deposits, and in metasomatised (*See* metasomatism) limestones.

Spheniscidae. Penguins. *See* Impennae.

Sphenisciformes. An order that includes the family Spheniscidae to which all the penguins (*See* Impennae) belong. The penguins have no close living relative, but may be allied distantly to the petrels (Procellariformes).

sphenodon. *See* Rhynchocephalia.

Sphenophyllales. An order of fossil Pteridophyta which flourished during the Carboniferous and Permian Periods. They were shrubby or herbaceous plants with grooved stems and whorls of leaves. Spores were formed in terminal cones.

Sphenopsida. A group of vascular plants (Tracheophyta) comprising the horsetails (Equisetales) and Sphenophyllales.

spherulite. Spherical mass of radiating crystals. *See* rhyolite.

spiders. Araneida.

spike. (Bot.) An unbranched, elongated flower head, bearing flowers that have no stalks.

spikelet. A unit of the flower-head of a grass, usually bearing two glumes and one or more florets.

spilite. A sodium-rich basalt.

spindle. *See* mitosis, meiosis.

spindle attachment (centromere, kinetochore, kinomere). The part of a chromosome which becomes attached to the spindle during nuclear division.

spinel. A group of minerals with the general formula $A\,B_2\,O_4$, where A is magnesium, iron, zinc, manganese or nickel, and B is aluminium, iron or chromium. Spinels are formed in igneous and metamorphic (see Metamorphism) rocks and include chromite $FeCr_2O_4$, the main ore mineral of chromium, and magnetite Fe_3O_4. Most of the gem spinels are spinel *sensuo stricto*, $MgAl_2O_4$. Spinel can become concentrated in placer deposits.

spiracles. *See* stigma.

spirochaete (spirochete). A spirally twisted, unicellular bacterium which moves by undulating. Some species are free-living, others are parasitic, causing such diseases as syphilis.

spirochete. Spirochaete.

splash zone. *See* shore zonation.

splenic fever. Anthrax.

splice. To join together two pieces of rope or cable by interweaving their constituents strands to form a whole. Also splicing of wood to repair a fracture in furniture.

spodic. Applied to a soil layer enriched in sesquioxides and/or organic matter, with or without clay. *See* soil classification.

spodosols. *See* soil classification.

sponges. Porifera.

sporangium. A plant organ inside which asexual spores are formed. *See* Myxomycophyta, sporophyll.

spore. A reproductive body, consisting of one or several cells, produced by plants, bacteria and Protozoa. Spores are usually microscopic, and are often produced in vast numbers. They are widely dispersed and can effect a rapid sporogonium

increase in the population of a species. Some are thick-walled resting spores which can survive unfavourable conditions. In flowering plants, the spores are the pollen grain (microspore) and embryo sac (megaspore). *See* heterospory, homospory.

sporogonium. The spore-producing structure (sporophyte generation) of Bryophyta.

sporophyll. A leaf, or structure derived from a leaf, on which sporangia are borne. In some plants (e.g. Bracken) sporophylls are like ordinary leaves, but in other plants they are much modified. The stamens and carpels of flowering plants are sporophylls.

sporophyte. A generation in the life cycle of certain multicellular plants in which spores are produced that develop without fertilization into gametophytes, the alternate generation. *See* alternation of generations.

Sporozoa. A class of parasitic Protozoa which form large numbers of spores. Most Sporozoa have complicated life cycles involving alternation of sexual and asexual reproduction. Sporozoa cause coccidiosis and malaria (*See* plasmodium), red water fever in cattle (transmitted by a tick), and diseases in the silkworm and honey bee.

spotted slate. Slate with darker inclusions of new minerals, found in the outer part of a metamorphic aureole. Inclusions are usually small, spherical or ovoidal in shape, but may be rectangular and they are due to the recrystallisation of minerals by thermal metamorphism.

spray. Liquid droplets greater than 10 microns in size, created by mechanical disintegration processes. Spray is a source of salt particles in the atmosphere. The smallest drops are formed when tiny air bubbles break the surface. These remain airborne long enough for evaporation of their water to occur before they fall back to the sea. Drops in spray from wave crests are too large to make a significant contribution.

spring. *See* seasons.

Spring-tail (Collembola). *See* Apterygota.

sputnik. Soviet earth-orbiting satellite, launched in 1957, the first satellite whose success stimulated the US space research programme.

squall. A strong wind that begins suddenly, lasts for several minutes, then dies away rather more slowly. The term is used also to describe a kind of storm characterised by a series of squalls, i.e. a succession of wind gusts each lasting from a few seconds to a minute and with speeds that may be half as great again as the average wind speed. Windspeeds of 50 to 100 km/h are common in squalls and gusts of upto 160 km/h have been known. *See* Line squall.

squall line. Line squall.

Squamata (Lepidosauria). An order of reptiles comprising lizards (Lacertilia, Sauria), snakes (Ophidia, Serpentes) and some extinct groups. Snakes are limbless, elongated, and have a jaw arrangement enabling the mouth to be opened extremely widely. Locomotion is by lateral undulations of the body, helped by ventral scales attached to long, movable ribs. Lizards have a normal sized jaw gape, and only a few species (e.g. the Slow-worm, *Anguis fragilis*) lack limbs.

Sr. Strontium.

sr. Steradian.

SSSI. Site of Special Scientific Interest.

SST (supersonic transport aircraft). Civil aircraft designed to fly at speed in excess of the speed of sound over part of their route. Development of SSTs has proceded in the USA, the USSR and in the combined Anglo-French Concorde. SST development has caused great controversy, particularly with regard to the cost of the development programme and the likelihood of the aircraft being able to compete commercially with much cheaper, slower aircraft, although experience with Concorde operations has shown that a demand exists. Environmental objections arise from noise at take-off, sonic boom and the possible chemical reactions between its exhaust gases and atmospheric gases at its high operating altitude, although these fears have largely receded. *See* landing and take-off cycle, ozone layer.

subduction zones. According to the theory of plate tectonics, those belts of the Earth's crust and mantle in which lithospheric plates are descending into the mantle.

stabilisation. (Biol.) The increase of dominance that ends in a stable climax, produced by the invasion of species leading to the establishment of a population most completely fitted for prevailing conditions. This degree of adaptation to conditions ensures that the climax is permanent and can be changed only by a change in the conditions to which the population is fitted.

stabiliser. Trace chemical added to plastics to reduce degradation by oxidation and the action of ultraviolet radiation.

457

stable. *See* dynamic stability, static stability.

stable population. A population in which births and deaths are in balance, so that, discounting migrations, the size of the population remains constant.

stack. Chimney.

stage. (Geol.) In stratigraphy, a stage is a subdivision of a series, which itself is a subdivision of a system.

stagnation point. When a fluid flow divides to pass by an object, there exists on the upwind side a point (or in two dimensional flow a line) on the surface of the object at which the velocity of flow is zero. This is the stagnation point.

stalk. *See* flower.

stamen. *See* flower.

staminate. Applied to flowers which have stamens but not carpels, and so are male. *Cf.* pistillate.

stand. (*a*) (Ecol.) An aggregation of plants which is more or less uniform in species composition, age and condition, and is distinguishable from adjacent vegetation. (*b*) (Forestry) The amount (usually expressed as volume) of standing timber per unit area.

standing crop. The amount of living matter (usually expressed as biomass) present in a population of one or more species within a given area.

standing wave. In the atmosphere, a wave or distortion in the airflow whose shape is stationary relative to the surface (e.g. the mountain wave).

starch. *See* carbohydrate.

star fishes. Asteroidea. *See* Crown of Thorns Starfish.

static electricity. Electricity that is at rest, rather than flowing. If the electrons within a substance are at rest, there is no electrostatic force present in the substance to move them, i.e. there is no electrical potential between any two points within the substance. However, this substance may enter the electrostatic field of another substance, so creating a potential between the two substances and a consequent flow of current.

static reserve index. The length of time for which the known reserves of a resource will last if the rate at which they are used remains constant, calculated by dividing the amount of the reserve by the amount used during each time period. If the rate of use is increasing exponentially (*see* exponential growth), then the Exponential Reserve Index applies, calculated by comparing the reserves with a rate of usage that increases exponentially.

static stability. In the atmosphere, the condition in which small vertical displacements of a parcel of air cause gravitational restoring forces in the absence of horizontal wind. This will be the case if the lapse rate is greater than the adiabatic lapse rate. *Cf.* dynamic stability.

station. The geographical location in which an organism or community occurs, making no reference to environmental factors.

statocyst (otocyst, lithocyst). An organ of balance, present in many invertebrates. It consists of a fluid-filled sac containing one or more granules of lime or a similar substance (statoliths, otoliths, lithites) which shift about as the animal moves and stimulate the sensory cells lining the sac.

statolith. *See* statocyst.

steam fog. A fog which forms when cold air overlies sufficiently warm water. The air just above the surface is saturated and at the water temperature, so that on mixing with air at higher levels condensation may occur. The density profile is unstable so that free convection occurs and the fog appears as vertical streaks.

steam-generating heavy water reactor (SGHWR). A nuclear reactor that uses enriched uranium as a fuel, boiling light water as a coolant and heavy water as a moderator.

steel. Alloy of iron with small amounts of carbon. *See* Bessemer process.

steering of storms. The movement of storms that must be understood if their tracks are to be predicted. The most useful rule is that storms travel in the direction of the thermal wind, or with the wind at about the mid height of their circulation.

Stefan's law. States that the total radiation emitted by a black body is proportional to the fourth power of its absolute temperature. Named after the Austrian physicist Josef Stefan (1835–1893).

stele (vascular cylinder). The core or (in some ferns) network of vascular tissue in a root or stem. It is made up of xylem and phloem with accompanying ground tissue, and is surrounded by the epidermis, when this is present.

stem. That part of a vascular plant which bears leaves, buds, and often reproductive structures (e.g. flowers). The vascular bundles in a stem are scattered or arranged in a ring. Most stems grow above the ground, but rhizomes are subterranean, and their leaves are reduced to scales.

stenohaline. Able to tolerate only a narrow range of salinity (i.e. a small variation of osmotic pressure) in the environment. *Cf.* euryhaline.

stenothermous. Able to tolerate only a small variation of temperature in the environment. *Cf.* eurythermous.

stenotopic. Applied to organisms with a restricted distribution. *Cf.* eurytopic.

Stephanian. The youngest stage of the Carboniferous System in Europe.

steppe. An extensive, treeless plain. *See* grassland.

steradian (sr). The supplementary SI unit of solid angle, being that angle which encloses a surface on a sphere equal to the square of the radius of the sphere.

Sterculiaceae. A family of dicotyledonous, chiefly tropical herbs, shrubs and trees. The seeds of the tropical American genus *Theobroma* yield cocoa and cocoa butter. The tree *Cola* is the source of kola nuts, which contain caffein and are an important article of trade in West Africa.

stereo pictures. Two photographs of the same object taken from different positions (e.g. by an overflying aircraft which takes two pictures a few seconds apart). They are used to obtain a three-dimensional appreciation of a scene (e.g. to measure the height of buildings and hills). For visual use the baseline should be between one-fifth and one-twentieth of the distance of the object viewed.

stereotaxis. *See* taxis.

stereotropism. Haptotropism.

Stevenson screen. A standardised wooden container in which surface weather measurements are made. It is of louvred construction so that the air mass may pass freely through it, while thermometers are shielded from direct sunlight. It is placed so that the thermometers are 1.25m above ground, and is painted white all over. Designed by the engineer Thomas Stevenson.

STH. Growth hormone.

stick insects. *See* Phasmida.

stigma. (*a*) The terminal part of a carpel, which receives the pollen. (*b*) Eye

460

spot. A light-sensitive, pigmented organelle, found in many flagellate Protozoa and microscopic, motile green algae. (*c*) Spiracle. The external opening of a trachea (breathing tube) in an insect. (*d*) A small coloured area on the clear wings of some insects (e.g. dragonflies).

still. Apparatus for distilling, used to prepare alcoholic drinks, distilled water and other purified liquids.

still-process. German process for removing sulphur dioxide from flue gases with 70 to 80 % efficiency by absorbing the sulphur dioxide with brown-coal filter ash, hydrated lime or limestone, at 100 to 1300°C, with a surplus 10 % of surface water. The development of this process was interrupted and it has not been applied.

stimulus. A change in the internal or external environment of an organism which evokes a response in the organism, but does not provide energy for this response.

stipule. (Bot.) An outgrowth, which may be leaf-like or scaly, at the base of a petiole.

stochastic. (Stat.) Adjective implying a random element (e.g. a stochastic process is one in which some element of chance occurs).

stock. (*a*) A discordant igneous intrusion with an area of outcrop of a few square kilometres to tens of square kilometres. Some stocks are probably cupolas of batholiths. *See* boss. (*b*) *See* resource.

Stockholm conference. The UN Conference on the Human Environment.

stockwork. An ore deposit consisting of a large-scale mass of closely-spaced, narrow mineralised veins. Stockworks are commonly worked opencast.

stoichiometric. Pure. A chemical compound is stoichiometric when its component elements are present in the precise proportions represented by its chemical formula. A stoichiometric mixture is one that will yield a stoichiometric compound on reaction (i.e. it consists of elements in exactly the proportion required to yield a stoichiometric compound utilising all the components of the mixture).

stoker. Originally a man, now more commonly a machine for feeding coal or other combustible material into a furnace and supporting it there during combustion.

stolon. (*a*) (Bot.) A short-lived, horizontal stem, e.g. a strawberry runner, which produces roots and shoots, so vegetatively propagating the plant. The term is usually restricted to stems which creep above the ground. (*b*) (Zool.) 1) A root-like part of a colony of animals (e.g. some Hydrozoa) which serves for anchorage. (2) An outgrowth of the body of some sea-squirts, which produces new individuals by budding.

stoma (stomate). A pore in the epidermis of a plant, through which exchange of gases (water vapour, oxygen, carbon dioxide) takes place. Stomata are particularly abundant in leaves. Each pore is surrounded by two crescent-shaped, chlorophyll-containing guard cells. Absorption and loss of water causes alterations in the shape of these cells, leading to variations in the size of the stomatal aperture.

stomate. Stoma.

stone. A meteorite composed mainly of silicate minerals.

Stone Age. That period in the development of human societies in which tools and implements were made from stone, bone or wood, and no metals were used. The date of the period varies very widely from place to place and the term is often used loosely to describe the cultures of present-day pre-agricultural peoples, although these may (and often do) use imported metal artifacts. *Cf.* Palaeolithic, Mesolithic, Neolithic.

stone cells. Sclereids. *See* Sclerenchyma.

stoneflies. Plecoptera.

Stone Webster tonics process. Process for removing fly ash and sulphur dioxide from flue gases by electrostatic precipitation and pre-scrubbing with water, followed by the removal of the sulphur dioxide by absorption in water, then contacting the absorbent with sodium bisulphate to produce sodium sulphate and sulphur dioxide gas. Caustic soda and sodium bisulphate are recovered. The process is used in the USA.

stoneworts. Charophyta.

stony iron. A meteorite rich in both nickel-iron alloys and silicate minerals.

stope. *See* stoping.

stoping. (*a*) A process postulated to occur during igneous intrusion, particularly of granites, whereby magma forces its way along fissures in the country rock, so forcing blocks to fall into the magma chamber and be assimilated. (*b*) In mining, the method of winning ore from a steeply-inclined lode by driving horizontal tunnels (called drives) along the lode, then extracting the ore above or below the drive (called overhand or underhand stoping, respectively), thus forming caverns called stopes.

storm surge. An unusual variation in the amplitude of the tide, caused by atmospheric factors (e.g. wind and pressure gradients).

stoss. The direction from which ice has come. The opposite of 'lee', the side away from the direction of ice flow. 'Stoss-and-lee' topography is a landform showing rocks with smoothly abraded slopes on one side and broken steep slopes on the other.

stoss-and-lee topography. *See* stoss.

strain. (*a*) A subspecific group in which the organisms are not sufficiently different genetically from the rest of the species to form a variety. (*b*) The distortion of material, usually producing stress, i.e. internal forces. In an elastic solid the stress is proportional to the amount of the strain, in a viscous fluid to the rate of strain. Plastic and viscoelastic materials have more complex relationships.

strangers. Species that occur with a markedly greater frequency in areas other than those being examined.

strange species. Species that are rare and accidental intruders from another community, or relicts of an earlier community.

strata. (Ecol.) layers.

stratification. In the atmosphere, the forming of stable, horizontal layers which do not intermingle, due to a lapse rate less than the adiabatic lapse rate.

stratocumulus. Cumulus clouds in a layer, or a layer, often complete, of clouds generated by convection which occurs when convection is confined by an inversion at the top.

stratopause. The boundary between the stratosphere and the mesosphere, which is at a height of about 50 km.

stratosphere. The region of the atmosphere above the troposphere in which temperature increases with height. There is therefore little mixing. The stratosphere contains relatively large amounts of ozone formed by absorption of ultraviolet radiation.

stratovolcano (composite volcano). A volcano which emits both tephra and lava, and which builds a steep-sided cone.

streak. The colour of a powdered mineral. Streak is normally determined by drawing the mineral across an unglazed porcelain tile. The colour of the streak is far less variable than the colour of the mineral in a coarser form.

streak clouds. Fibrous patches of cloud elongated in the direction of the wind shear.

streamline. A line drawn through a fluid in motion which is always parallel to the local direction of flow.

stream-tin. Cassiterite occurring as detrital grains in alluvial deposits.

Strepsiptera (stylopids). An order of minute insects (Endopterygota) whose larvae are mostly parasitic on other insects such as plant bugs and bees, often rendering the host sterile. The adult females are grub-like and usually remain inside their hosts, but the males are free-living insects with vestigial fore-wings.

Streptomyces griseus. The source of streptomycin. *See* antibiotic.

Streptomycin. An antibiotic produced by the bacterium *Streptomyces griseus*. It is used in the treatment of diseases such as tuberculosis, and as a fungicide for the control of downy mildew on hops. *See* Actinomycetes.

Streptoneura. *See* Gastropoda.

striae. Glacial striation.

striated muscle. *See* muscle.

Strigiformes (owls). Birds of prey specialised for hunting at night. They have large heads, short necks, hooked beaks and powerful claws. The eyes cannot be moved in their sockets and the whole head swivels to compensate. There are large external ear flaps. They fly silently, probably detecting prey (small mammals, birds and insects) mainly by sound. Mostly nest in cavities.

strike. The direction of the line of intersection of an inclined rock surface (e.g. bedding plane, vein, joint, cleavage-plane, fault, schistosity) with a horizontal plane. Strike is at right angles to the direction of dip. Strike is also used, less precisely, to indicate the general trend of strata.

strike-slip fault (wrench, tear fault). A fault in which the relative displacement is parallel to the strike of the fault.

striped muscle. *See* muscle.

stripes. Linear arrangement of stones down a slope, found in regions of present or former permafrost. Stripes grade into polygons with a decrease of slope.

strip mining. The working of coal or an ore by removing the overburden to expose the ore-body which is removed and may be partly processed on site. The technique is used to obtain some coal and many other minerals, esp. where the rock contains so little of the required substance that conventional techniques of tunnelling along veins cannot be used. The effect on the landscape is considerable, with some minerals the pollution may be high, and restoration of the land when work has finished may be difficult.

strobilation. The repeated formation of similar structures (e.g. proglottides of a tapeworm, larvae of some jellyfish) which for a time remain connected, but after a period of development drop off the chain one by one. *See* segmentation.

strobilus. Cone.

stroma. (*a*) Connective tissue binding together the components of an organ in an animal. (*b*) A mass of fungal hyphae from which, or inside which, fruiting bodies are formed. (*c*) The colourless part of a chloroplast.

Strong, Maurice. *See* UNEP.

strontium. (Sr.) Element. A reactive metal that resembles calcium and can replace it in biological processes. It occurs as celestite and strontianite. The radioisotope strontium-90 is present in the fall-out from nuclear explosions and is a health hazard because it can lodge in bone. Strontium-90 has a half-life of 28 years. AW 87.62; At.No. 38; SG 2.6; mp 757°C.

structural terrace. Terrace.

structure. The spatial and other arrangements of species within an ecosystem. The structure takes account of the composition of the biological community, including species, numbers, biomass, life cycle and spatial distribution, quantity and distribution of the non-living materials (nutrients, water, etc.); the range, or gradient, of conditions such as temperature, light, etc.

strychnine. *See* Strychnos.

Strychnos. A genus of tropical, dicotyledonous trees and climbing shrubs (family Strychnaceæ) whose seeds yield the poison strychnine. The bark of *Strychnos toxifera* yields curare, used on poison arrows by South American Indians. The seeds of *S. potatorum* are used to purify water by precipitation.

style. *See* flower.

stylopids. Strepsiptera.

sub-base. *See* pavement.

subclimax. According to F.E. Clements: (*a*) The stage preceding the climax in a complete sere. (*b*) A plant community which is stabilised at the subfinal stage in the succession by edaphic or other factors which arrest development into the climatic climax. *Cf.* preclimax, postclimax, serclimax.

subcloud layer. The layer of air below cloud base, used sometimes to mean the shallow stable layers of air immediately beneath convection clouds, at others to mean all the air between cloud base and the Earth's surface.

subdominant. Species that may appear more abundant than the true dominant in a climax at particular times of the year (e.g. trees and shrubs are more conspicuous in a savannah than are the grasses that are the true dominants).

subdominule. Subdominant in a micro-community.

suberin. *See* suberisation.

suberisation. The laying down of suberin, the waterproof substance in the walls of cork cells. *See* cork, bark.

subharmonic. A harmonic of a frequency that is an integral number of times lower than the fundamental frequency in a periodic wave.

sub-imago. *See* Ephemeroptera.

sublimation. The change in a substance between the solid and the gaseous state without passing through the liquid state.

sublittoral fringe. *See* shore zonation.

sublittoral (sublitoral) zone. (*a*) The part of a lake which is too deep for rooted plants to grow. *Cf.* littoral. (*b*) The zone of a sea lying below the intertidal zone and extending to the limit of the continental shelf. The part of the neritic lying beneath the littoral zone. *See* shore zonation.

submetallic. *See* lustre.

subsequent. Refers to rivers whose courses follow channels cut by themselves into easily erodable rock formations.*Cf.* consequent.

subsere. (*a*) The series of plant communities making up the stages in a secondary succession (*b*) Any one of these stages.

subsidence. (Meteor.) In the atmosphere, subsidence refers to the gradual descent of an air mass. Nearer the ground, the rate of descent is lower and the flow spreads horizontally. Adiabatic compression occurs and the descending air is warmed, producing a stable lapse rate. This is called the subsidence inversion. Subsidence is usually associated with anticyclonic conditions.

subsidence inversion. High level inversion. *See* subsidence.

subsonic. A flow is said to be subsonic if the relative speed between all parts of the flow and any solid boundary is less than the local speed of sound in the fluid.

sub-species. A group of organisms genetically distinct from the rest of their species, which has arisen through partial or recent reproductive isolation. Sub-species are often regarded as incipient species still capable of interbreeding within their species.

substituent. *See* substitution product.

substitution product. (*a*) (Chem.). A compound obtained by replacing an atom or group of atoms by another atom or group of atoms within a molecule, the new atom or group being known as the substituent. (*b*) (Econ.) A product which enters a market in place of another, but which satisfies the same consumer want.

substrate (substratum). (*a*) The surface to which an organism is attached, or upon which it moves. (*b*) The material on which a micro-organism grows (e.g. culture medium, host organism). (*c*) The particular substance or group of substances which an enzyme activates.

substratum. (*a*) The soil layer beneath the solum. The substratum either conforms (horizon C) or is unconforming (horizon D). *See* soil horizons. (*b*) Substrate.

subsun. A bright spot of light seen where the sun is reflected in the horizontal upper surfaces of crystals in ice clouds below the observer.

subtropical high. A feature of the general circulation, the subtropical high is a region of high pressure found in latitudes intermediate between the tropical and temperate zones. The Azores high is in this region.

subtropical jet. A jetstream or wind maximum at the tropopause blowing from the west at around 30° latitude above places where there is predominantly a descent of air and a dry climate.

succession. The progressive natural development of vegetation towards a climax, during which one community is gradually replaced by others. A primary succession starts on sites (e.g. sand dunes, lava flows) which have not previously borne vegetation. A secondary succession is one which follows the destruction of part or all of the original vegetation of an area. A natural succession has two components, the physiographic, in which living organisms respond to topographical features, and the biotic, in which organisms react with one another. *See* sere.

successive percentage mortality (apparent mortality). The mortality in each developmental stage expressed as a percentage of the number alive at the beginning of the stage.

succulent. The leaves of a plant, or a plant, which has enlarged tissues that hold water or sugar.

sucrose. *See* carbohydrate.

sucrosic. Saccharoidal.

suffrescent. *See* suffruticose.

suffrutescent. Suffruticose.

suffruticose (suffrutescent). Applied to perennial plants which are woody at the base but herbaceous above, and do not die down to ground level in the winter. A distinction may be drawn between 'suffrescent' and 'suffruticose', the former referring to less woody plants, the latter to more shrub-like (fruticose) ones.

Sugar beet. *Beta vulgaris.*

sulfacid process. German process for removing sulphur dioxide from flue gases by absorption with activated coal (*see* activated carbon) in a fixed bed reactor at temperatures between 60 and 80°C.

sullage water. *See* combined sewer.

sulphate. The ion SO_4^{2-} formed when sulphur trioxide reacts with water to form sulphuric acid (H_2SO_4). This ion is often found in the atmosphere in other chemical compounds (e.g. ammonium sulphate ($2(NH_4)SO_4$), a major constituent of Teesside mist, caused by the combination in the air of sulphur trioxide and ammonia and water. Sulphate ions are also released in volcanic eruptions, together with sulphur dioxide, especially the eruptions of Mount Tambora (1815), Krakatau (1883) and Mount Agung (1963).

sulphate-resistant cement. *See* cement.

sulphite. The ion $SO_3{}^{2-}$ formed notably in the reaction of sulphur dioxide with water to form sulphurous acid (H_2So_3).

sulphur (S). The native element sulphur, deposited from volcanic vents and fumaroles and also found in sedimentary rocks, particularly with gypsum and limestone, and associated with salt-domes. Native sulphur (or brimstone) is the main source of sulphur for the sulphuric acid industry, followed by sour gas (natural gas containing hydrogen sulphide (H_2S) and pyrite. Sulphur is an essential plant macronutrient. AW 32.064; AT.No. 16; SG 2.07; mp 112.8°C; bp 444.6°C.

sulphur dioxide (S0$_2$). A constituent of products of combustion of a wide range of fuels, but particularly heavy fuel oil and coal. It is widely used as an index of the level of air pollution, but epidemiological research has so far failed to prove that it has harmful effects on humans in even the highest concentrations found in the atmosphere. Its importance lies mainly in its ease of measurement (by acidity) and its dependence on fuel consumption and atmospheric dilution, on which the levels of many other pollutants also depend. It is also released, with sulphates, in volcanic eruptions, esp. Mount Tambora (1815), Krakatau (1883) and Mount Agung (1963). *See* sulphur trioxide.

sulphuric acid (H_2SO_4). In the atmosphere, sulphuric acid is formed by the combination of sulphur trioxide with water to form a relatively stable mist of acid droplets. Sulphur is an essential macronutrient for plants and so atmospheric sulphuric acid contributes nutrients to soils. It also tends to increase soil acidity and in soils that are naturally acid this may require the use of lime to raise the pH. Some Scandinavian soils have been damaged in this way as a result of industrial air pollution originating in the UK, Germany (East and West) and Poland. *See* transfrontier pollution.

sulphur trioxide. In the atmosphere, sulphur trioxide is formed by the oxidation of sulphur dioxide in the reaction $2SO_2 + O_2 \rightarrow 2SO_3$. It is believed that this reaction may occur more readily in the surface of water droplets, leading immediately to the formation of sulphuric acid.

summer. *See* seasons.

Sundew. *See* Droseraceae.

sun dog. Mock sun.

Sunn hemp. *Crotalaria juncea.* A tropical crop grown for its fibre. It is unrelated to true hemp, *Cannabis sativa.*

sun pillar. A vertical streak of light above the sun, usually seen at sunrise or sunset. It is caused by reflection from ice crystals.

sun plant. Heliophyte.

sunrise. Defined in meteorology as the moment when the upper edge of the sun appears to rise above the apparent horizon on a clear day. *Cf.* sunset.

sunset. In meteorological convention, the moment when the upper edge of the sun appears to fall below the apparent horizon on a clear day. Effects of refraction cause the apparent position to be about 34′ above the true position.

sunshine. Solar radiation.

sunspot. A local disturbance on the visible surface of the sun which causes an increase in solar radiation and consequently alters the Earth's magnetosphere and ionosphere. The intensity of sunspot activity varies in an 11 year cycle and it appears possible that this may cause a similar cycle in the global meteorology, but as yet this has not been proved.

supercooled fog. *See* frozen fog.

supercooling. A metastable state of a liquid in which it remains in the liquid state although at a lower temperature than the freezing point. It often occurs in cloud droplets in the atmosphere.

superfluent. An animal species of similar importance to that of a subdominant plant species.

supergene. Refers to mineral deposits or processes involving downward percolating water or aqueous solutions. Secondary enrichment is a special case of a supergene process. *Cf.* hypogene.

superior image. An optical reflection at an inversion giving the appearance of 'castles in the air'. They are most common in winter over land when the ground is snow-covered and the weather anticyclonic, or over closed seas (e.g. the Mediterranean) in summer.

supernumerary bow. The additional rainbows seen inside the primary (42°) rainbow when the range of raindrop sizes is small, the order of the colours being the same.

superorganism. Epiorganism.

superparasite. *See* parasitism.

superphosphate. *See* phosphate, phosphate rock.

superposition. The arithmetical combination of two or more waves at successive instants or points.

supersaturation. A metastable state of moist air in which there is more vapour than that required to saturate. This can occur in cooling air if there are too few condensation nuclei to act as initiating sites for condensation of droplets. *See* saturation.

supersonic transport aircraft. SST.

supraneustronic. *See* neustron.

supraorganism. Epiorganism.

surface-active agent (surfactant). Substance that causes a liquid to spread more readily on a solid surface, mainly by reducing surface tension. Many detergents are surfactants.

surface analysis. The analysis of a surface chart, i.e. the identification of pressure systems, air masses, and fronts from surface observations.

surface deposits. Unconsolidated material covering bedrock, either weathered from bedrock in situ (residual) or weathered from one area and transported to another by wind (aeolian, e.g. sand dunes and loess), by water (e.g. alluvium), by ice (e.g. till or glacial drift), or by gravity (colluvium).

surface pressure. Normally, the atmospheric pressure inside a Stevenson screen (i.e. effectively at ground level).

surface pressure chart. A chart of surface atmospheric pressure plotted as isobars over a geographical area. It may include other surface parameters.

surface topography. *See* topography.

surface-water. *See* ground water.

surface wind. Conventionally defined as the wind at a height of 10 m above a flat and smooth piece of ground unaffected by obstructions.

surfactant. Surface-active agent.

surficial deposits. Surface deposits.

suspended sediment load. Particles small enough to be carried in suspension by moving water.

suspended solids. (*a*) Fine dust particles distributed as an aerosol in the air. (*b*) Small particles of solids distributed through water.

swallow hole (sink hole). A funnel-shaped hole in limestone caused by solution of the rock as rainwater with dissolved carbon dioxide from the soil drains through fissures and enlarges them. By continued solution the holes may be connected with vast underground caverns.

swamp. An area which is saturated with water for much of the time, but in which the soil surface is not deeply submerged. Used in a resricted sense, the term implies an area characterised by woody vegetation, but it is often used more widely to include marsh, bog, etc.

swans. *See* Anatidae.

S-wave. The secondary wave reaching a seismograph from an earthquake. S-waves are shear waves and so do not travel through liquids. The absense of S-waves beyond 105° of the arc from the epicentre of an earthquake indicated the fluid nature of the Earth's core. *Cf.* P-wave.

sweet gas. Natural gas containing little hydrogen sulphide (H_2S), in contrast to sour gas.

swidden farming (slash-and-burn farming). A system of primitive agriculture in which areas of natural vegetation are cleared by felling trees, cutting back shrubs and then burning off the stumps and herbaceous plants. The area is sown to crops for several years in succession and when yields begin to fall due to the cxhaustion of soil nutrients, or invading wild plants depress crops, the farmers move to a new area where the process is repeated. Thus each community moves every few years to a new site, eventually returning to the point from which they began.

swifts. Apodidae.

syenite. A coarse-grained, alkali, intermediate igneous rock characterized by the presence of alkali feldspar with or without feldspathoids and such fer-romagnesian minerals as biotite, hornblende, augite and more alkaline types. Syenites are the plutonic equivalents of trachytes.

sylvite. An evaporite mineral, with the composition KC1, which is one of the major sources of potassium, an essential fertiliser. *See* carnallite, macronutrients.

symba process. A process for treating starch wastes from the food industry using

Endomycopsis fibuliger fungus to hydrolyse the starch, then *Torula* yeast to produce proteins. *See* novel protein foods.

symbiont. A symbiotic organism. *See* symbiosis.

symbiosis. A close and mutually beneficial association of organisms of different species. The occurrence of cellulose-digesting protozoans in the guts of wood-eating cockroaches and termites is a symbiotic relationship, as the insects cannot digest cellulose unaided, and the Protozoa cannot live independently. *See* mycorrhiza, parasitism, commensalism, root nodules.

symmetrical fold. A fold in which both limbs dip away from the axial plane.

sympathetic nervous system. *See* autonomic nervous system.

sympatric. Applied to different species or sub-species whose areas of distribution overlap or coincide. *Cf.* Allopatric.

Symphyta. *See* Hymenoptera.

syn-. Prefix meaning with, together, along with, at the same time. Hence synorogenic, syntectonic, synsedimentary, syngenetic, etc.

synapse. *See* central nervous system, neuron, acetylcholine.

synapsis. The pairing of chromosomes during meiosis.

syncarp. A multiple fruit, made up of many small fruits united together.

synchorology. The study of the distribution ranges of plant communities, phytosociological regions, vegetation complexes, geographical complexes, dissemination spectra, and contemporary plant migration patterns.

synchronous satellite. A satellite that occupies such an orbit that its position remains fixed relative to a particular point on the Earth's surface. Such satellites are often used for communications.

syncline. A fold that is convex downwards. *Cf.* anticline.

synclinorium. A compound syncline in which the limbs are folded. The term is used for large-scale structures.

syncytium. A mass of protoplasm containing many nuclei, and enclosed by a single plasma membrane. The term is applied only to animal tissues (e.g. muscle fibres). *See* plasmodium, coenocyte.

syndynamics. The study of successional changes (*see* succession) in plant communities, their causes and trends.

synecology. The study of the relationships between communities of species and their environment. *Cf.* autecology.

synergism. (*a*) The combined effects of agents such as hormones or drugs when they act in the same direction on living systems. Synergism may result in an effect which is greater than the sum of the effects of the agents when they act individually. (*b*) The coordinated action of pairs or sets of muscles which contract together to produce a particular movement. *See* antagonism.

syngameon. A group of species between which hybridisation occurs.

syngamy. Sexual reproduction in organisms (e.g. aphids) which, during part of their life cycle, reproduce by parthenogenesis.

syngeneic. *See* isogeneic.

synmorphology. The study of the floristic composition, structure and minimal area of a community, its floristic homogeneity, and its extent in time and space, which aims to provide a comprehensive inventory of communities.

synnecrosis. Mutual depression or death.

synphylogeny. The study of the historical and evolutionary trends and changes within plant communities.

synphysiology. The study of metabolic processes of plant communities or species that are in constant competition with one another, including considerations of their water requirements, transpiration, assimilation and production of organic matter, the physiological effects of light, temperature, root exudates and other ecological factors.

synpiontology. The study of ancient patterns of distribution and migration of plant communities.

syntaxonomy. The classification of plant communities in relation to the taxonomy (*see* classification) of the flora, fauna, and Man.

syntype (cotype). Any specimen of a series used to designate a species when neither holotype nor paratype have been selected.

synusia. A group of plants, all of the same general form, occurring together in the same habitat (e.g. an aggregation of floating herbs, the tree layer in a wood).

Syringia vulgaris. Lilac. *See* Oleaceae.

system. (*a*) (Geol.) A sequence of strata deposited during a geological period. The systems were originally defined by geologists working in different regions in the 18th and 19th centuries. Some systems were established on distinctive rock types and others on distinctive faunal content, with boundaries usually chosen at distinct breaks in the geological record. Nowadays, the systems and the subdivisions (series, stage and substage) within them are recognised on the basis of distinctive fossils. (*b*) An assemblage or combination of things or parts forming a complex or unitary interacting whole.

Systéme International d'Unités. An internationally agreed system of units. *See* SI units.

systemic insecticides. Insecticides which are taken in and then translocated throughout a plant, rendering the whole surface lethal to insect pests, even if only a small area has been treated. These insecticides can be less harmful to wildlife than contact poisons because insects not actually feeding on the treated plants may escape their effects. *See* translocation.

T

T. (*a*) Trillion (in the American sense of one million million, 10^{12}). Used in conjunction with cf. in figures for natural gas reserves to signify trillions of cubic feet. (*b*) Tera- (*c*) Tesla.

taconite. Unleached banded ironstone.

taiga. Forest (usually coniferous) adjacent to arctic tundra.

tailings (debris). Those portions of washed ore that are considered too poor to be treated further.

talus. The slope formed by fallen rock debris or slide-rock at the foot of a cliff.

Tambora, Mount. Mount Tambora.

tangent arc. A coloured arc commonly seen at the highest point of a halo around the sun, which has opposite curvature to the halo when the sun is low. Tangent arcs may be seen at the bottom of a halo. They occur when there is a predominance of crystals with their axes horizontal. The circumferential arc is sometimes thought to be the very rare upper tangent arc of a 46° halo because it is in about the same position.

tangential velocity (circumferential velocity). The component of velocity along the tangent in a curved flow.

tapeworms. Cestoda.

tapioca. *See* Manihot.

tap root system. A root system with a prominent, vertical main root (tap root) bearing numerous small lateral roots. Some tap roots (e.g. carrot) become swollen with stored food. *Cf.* fibrous root system.

Tardigrada (bear animalcules, water bears). Minute Arthropoda found in damp moss, ponds, etc., which resemble Arachnida because of their four pairs of stumpy legs.

tarn. A lake in a cirque.

tar-sand. A surface or near-surface sand or sandstone containing a high percentage of very viscous natural hydrocarbons. Some such deposits probably represent the remains of oilfields exposed by erosion and from which the lighter hydrocarbons have evaporated.

taurocholate. *See* bile.

Taxaceae. A small family of trees and shrubs (Gymnospermae) including the yew (*Taxus*). In the yew the seed is surrounded by a fleshy red cup (aril) attractive to birds, which disperse the seeds. The leaves are very poisonous, and the wood is valuable. Some yews in English churchyards have girths of 9 to 10 m, and are about 1000 years old.

Taxales. An order of Gymnospermae containing a single family, Taxaceae. They are much branched trees and shrubs with solitary ovules each surrounded when ripe by a fleshy integument, the aril. The Common Yew (*Taxus baccata*) has red arils, valuable wood and very poisonous leaves.

taxis. A locomotory response of a cell or organism in which the direction of movement is oriented with relation to the stimulus (*cf.* kinesis). In chemotaxis the stimulus is a chemical (e.g. a pheromone emitted by a female moth to attract a male; cane sugar exuding from the female sex organ (archegonium) of a moss plant to attract spermatozoids). In phototaxis (heliotaxis) the stimulus is light, in geotaxis it is gravity, in thermotaxis it is heat, and in rheotaxis it is flowing movements in the immediate enviornment (e.g. water currents). In thigmotaxis (stereotaxis) the stimulus is touch, which sometimes causes inhibition of movement leading to the close contact of an organism with a solid surface.

Taxodiaceae. A small family of coniferous trees, some of which are enormous and yield valuable timber. The Redwood (*Sequoia sempervirens*) grows up to 102 m high and 8.5 m thick. The Mammoth Tree (*Sequoiadendron giganteum*) of California reaches 10.5 m in thickness, the tallest is 96 m, and the age of the largest is about 1500 years. The oldest Redwood has been dated as 3212 years old. *See* Bristle-cone pine.

taxon. Any named taxonomic group, whatever its rank (e.g. phylum, genus, variety, etc.).

taxonomy. The science of classification, usually restricted to mean the classification of plants and animals. The word is derived from the Greek *taxis*, 'arrangement', and *nomos*, 'law'. Taxonomy arranges plants and animals into hierarchical groups.

TC. Retarded timing.

TCDD ('dioxin', 2, 3, 7, 8, tetrachlorodibenzo-p-dioxin). A compound which is extremely toxic to plants and animals, very persistent, and capable of causing chromosome malformation. Traces of TCDD have found their way into food chains in some parts of the world. It may be formed during some methods of manufacturing the widely-used herbicide 2, 4, 5-T. *See* Seveso.

TDI. Tolylene di-isocyanate.

tea. *See* Camellia.

teak. *See* Tectona.

tear fault. Strike-slip fault

technological fix. A solution to a problem based on technology, often used in a pejorative sense as an apparent, or simplistic technological solution to a complex human problem whose benefits may be only cosmetic. *See* Cornucopian premises.

technosphere. That part of the physical environment built or modified by Man. *Cf.* psychosphere.

Tectona. *Tectona grandis* (family Verbenaceae) is the teak, a deciduous tree cultivated in India, Java, etc. for its very hard, durable timber, extensively used in shipbuilding. In India the wood is dried out by removing a ring of bark and sapwood from the standing tree, which dies and is left standing for two years before felling.

tectonic. Refers to deformation of the Earth's crust through warping, folding and faulting.

tectonic creep. Slight, and apparently continuous, movement along a fault.

Teesside mist. *See* sulphate.

tektite. A piece of natural glass, probably of extra-terrestrial origin, usually weighing a few grammes, and resembling obsidian but with a distinctive shape, such as that of a tear-drop, dumb-bell, or button. Tektites are found in groups often far from a volcanic source, and unlike meteorites they are geologically young, with ages of up to 30 million years.

teleoptiles. The feathers of an adult bird, including down feathers, contour feathers, quills and hair-like filoplumes. *Cf.* neossoptiles.

Teleosti. A large group of bony fishes containing all the present-day Actinopterygii except for a few primitive species (e.g. the sturgeon and gar-pike). *See* Clupeidae, Cyprinidae, Gadidae.

telolecithal (megalecithal). Applied to eggs (e.g. those of birds) which contain much yolk. *Cf.* microlecithal.

temperature coefficient. Q_{10}.

tendon. Connective tissue, associated closely with muscles, that transmits the mechanical force of muscle contraction to the bones, which function as levers.

tendril. A modified stem, leaf or part of a leaf used by climbing plants for attachment. Some tendrils twine (e.g. the modified leaflets of the Sweet Pea, *Lathyrus odoratus*) others end in adhesive pads (e.g. the branch tendrils of Virginia Creeper, *Ampelopsis veitchii*).

tensiometer. Instrument to measure the amount of water in the plant root area of a soil.

Teosinte. *Zea mexicana*. Wild maize, which still occurs in Central America and is the ancestor of domesticated maize, with which it interbreeds locally at random producing crops that display heterosis. *See* Graminae.

tepal. *See* flower.

tephigram. A diagram that shows the temperature and entropy at different levels in the atmosphere.

tephra. A collective name for all pyroclastic material ejected from a volcanic vent into the air. Tephra includes volcanic dust, ash, cinders, lapilli, scoria, pumice, bombs and blocks.

tera- (T). Prefix used in conjunction with SI units to denote the unit $\times$ 10^{12}.

teratogen. Substance that produces deformation of the foetus in the womb.

Teredo (shipworms). A genus of lamellibranch molluscs with long, slender bodies and much reduced shells, by means of which they bore into wood. They cause great damage to ships, piers, wharves, etc.

terminal curvature. The bending-over of the upper end of rock layers caused by creep of the overlying soil.

terminal (end) moraine. *See* moraine.

termite. Isoptera.

termiticole (termitiphile, termitophil). An organism which lives with termites inside their nests (e.g. certain fungi and insects).

termitiphile. Termiticole.

termitophil. Termiticole.

ternate. Arranged in threes.

terns. Laridae.

terrace. (Geol.) (*a*) A local flattening in an otherwise uniformly dipping series of beds. Otherwise known as a structural terrace, and the opposite of monocline. (*b*) River terrace.

terrapin. Chelonia.

terra rossa. Red earths developed on limestone due to accumulations of insoluble iron oxides.

terrigenous. Refers to sediments derived from the land.

terriherbosa. Herbaceous vegetation growing on dry land.

territory. An area which an animal or group of animals defends, mainly against members of the same species. *See* bird song.

Tertiary. Refers to that part of geological time from the end of the Cretaceous to the beginning of the Pleistocene. The Tertiary comprises the Eocene, Miocene, Oligocene, and Pliocene (with the prior addition of Palaeocene in most American usages). Tertiary also refers to rocks formed during this time. In some usages Tertiary is synonymous with Cenozoic, while in others Tertiary and Quaternary together make up the Cenozoic.

tertiary treatment. Advanced waste treatment.

tesla (T). The derived SI unit of magnetic flux density, being the density of one weber of magnetic flux per square meter. Named after Nikola Tesla (1870-1943).

testa. *See* seed.

testosterone. *See* androgen.

Testudinae. Chelonia.

tetraethyl lead. Compound of lead added to petrol to increase its octane number.

tetrahedrite. A copper-containing antimony sulphide material, $Cu_{12}Sb_4S_{13}$. Tetrahedrite is an important ore of both copper and antimony, and is found in hydrothermal veins.

tetraploid. Applied to a polyploid organism or cell with four times the haploid number of chromosomes *Cf.* allotetraploid.

tetrapoda. The essentially terrestrial vertebrates (Amphibia, reptiles, mammals) which characteristically possess two pairs of limbs.

texture. Size, shape and arrangement of particles making up a surface deposit, rock or soil (in soil the term generally applies only to size). The degree of sorting and rounding of the component particles of sediments is referred to as their textural maturity.

textured vegetable protein (TVP). *See* novel protein foods.

Th. Thorium.

thallium (Tl). Element. A white, malleable metal used in alloys. Its compounds are all extremely poisonous and are used in pesticides. AW 204.37; At. No. 81; SG 11.85; mp 303.5°C.

thalloid. Applied to plants (e.g. to one of the two classes of true liverworts (Hepaticae) the best known of which is *Marchantia polymorpha*) which have no leaves, being flattened ribbon or rosette-like structures whose thalli are from about 2 to about 25 mm in diameter (in Hepaticae).

Thallophyta. A large group of plants, comprising the primitive forms (algae, fungi, bacteria, slime fungi, lichens) in which the plant body (thallus) is organised in a simple way, without differentiation into root, stem and leaf. Many thallophytes are unicellular, but in some (e.g. brown seaweeds) the thallus may be very large and show some tissue differentiation.

thallus. A simple plant body which does not show differentiation into root, stem and leaves. Some thalli are unicellular. Multi-cellular ones are filamentous or ribbon-like, branched or unbranched. *See* Thallophyta.

thanatocoenosis. An assemblage of fossils consisting of the remains of organisms brought together after death. Thanatocoenosis is contrasted with biocoenosis.

Thea. Camellia.

thecodont. Having teeth set in sockets in the jaw bones as, for instance, in mammals. This is in contrast to the acrodont condition (e.g. in most bony fishes) in which the teeth are fused to the bones, and to the pleurodont condition (common in lizards and snakes) in which the teeth are attached by one side to the inner surface of the jaw bones.

theodolite. Instrument used for measuring horizontal and vertical angles, particularly in surveying.

Therapsida. A group of extinct reptiles which originated in the Permian Period and were ancestral to mammals. Mammal-like tendencies which evolved in the Therapsida include the differentiation of teeth for various functions, the turning of the limbs under the body, and enlargement of the brain case.

thermal metamorphism. Contact metamorphism.

thermal pollution. The raising of the temperature of part of the environment by the discharge of substances whose temperature is higher than the ambient (e.g. thermal pollution of rivers from the discharge of cooling waters). In fresh water, the amount of free dissolved oxygen tends to decrease with increasing temperature, so affecting living organisms.

thermal soaring. The art of circling in thermals to gain height, used by some birds such as gulls over land, vultures and other similar birds of prey, and incidentally by swallows and swifts feeding on insects carried up in thermals.

thermal stratification. The existence within a water body a succession of well-differentiated layers, each with a different temperature, lying at various depths, the coldest at the bottom. Inverse stratification occurs beneath ice. *See* epilimnion, hypolimnion, thermocline.

thermal wind. The increase in geostrophic wind with height due to horizontal temperature gradients. Its magnitude is about 3.6 km per second (7.2 knots) per km per 1°C per 100 km.

thermocline. The layer of water in a lake which lies between the epilimnion and the hypolimnion. Within the thermocline the temperature decreases rapidly with increasing depth (usually by more than 1°C for each metre). *See* thermal stratification.

thermodynamics, Laws of. O (Zeroth Law): Two objects are in thermal equilibrium when no heat passes between them when they are placed in contact with one another. 1. (First Law): Energy can be neither created nor destroyed, but it can be changed from one form to another (the Conservation of Energy). 2. (Second Law): When two bodies at different temperatures are placed in contact, heat will flow from the warmer to the cooler, and there is no continuous, self-sustaining process by which heat can be transferred from a cooler to a warmer body. Thus energy tends to become distributed evenly throughout a closed system (i.e. the entropy of a closed system increases with time). 3. (Third Law): The absolute zero (0 kelvin) can never be attained.

thermograph. Instrument for recording, as a line on a rotating drum, changes in temperature.

thermonasty. *See* nastic movement.

thermophilic micro-organism. A micro-organism which grows well at temperatures over 45°C (e.g. bacteria in hot springs, manure heaps and fermenting hay ricks). *Cf.* mesophilic, psychrophilic.

thermosphere. That part of the upper atmosphere in which temperature increases with height.

therophyte. *See* Raunkiaer's Life Forms.

Theropsida. A wide term which covers all the mammals and mammal-like extinct reptiles (e.g. Therapsida). These groups all have a similar skull structure. *See* Sauropsida.

thiamin (aneurin, vitamin B1). A vitamin of the B group which is needed by many animals (e.g. vertebrates, insects) for the formation of a co-enzyme concerned in carbohydrate metabolism. Some bacteria, fungi and Protozoa are able to synthesize thiamin. Rich sources are germ of cereals, leguminous seeds and yeast. Deficiency in Man causes beri-beri, a disease affecting the nerves, which is prevalent among populations which subsist on polished rice.

thigmotaxis. *See* taxis.

thigmotropism. Haptotropism.

thin section. A fragment of rock or mineral ground to a thickness usually of 0.03 mm, and mounted in glass as a microscopical slide. Nearly all minerals, except for most oxides and sulphides, are transparent at this thickness. Thin sections are usually viewed through a petrographic (petrological) microscope which has two polarising filters, one on each side of the specimen. The optical properties, as viewed through the microscope, are diagnostic of the rock from which the slice was taken.

thiocarbamate. *See* carbamate.

thiodan. Endosulfan.

thiophos. Parathion.

Third World. A term introduced in the late 1940s to describe the developing countries (*see* economic development). The world was divided into three: the First World comprised the countries whose economies were governed to a large extent by the free market; the Second World of those (socialist) countries whose economies are planned centrally. Only the term Third World remained in use, describing those countries whose economies were at a less advanced stage than the others. More recently, a Fourth World has been added by dividing the old Third World into those countries (e.g. the OPEC countries) that have natural resources of high value (Third World) and those developing countries (Fourth World) that have not. As the new Third World develops an industrial economic base, presumably the countries which comprise it will enter either the old First or Second Worlds, and the original three worlds will re-establish themselves, although by then it is probable that the term 'Third World' will have fallen into disuse.

thixotropy. An infinitely reversible property of certain clay minerals (e.g. montmorillonite). When mixed with water the mineral forms a gel and on agitation becomes a highly viscous fluid. This property has applications in civil engineering.

thorium (Th). Element. Dark grey radioactive metal used in alloys, filaments and in nuclear reactors. It occurs in monazite. AW 232.038; At. No. 90; SG 11.2; mp 1845°C.

thread cell. Nematoblast.

threadworms. Nematomorpha.

threonine. Amino acid with the formula $CH_3CHOH.CH.(NH_2).COOH$. Molecular weight 119.1.

threshold limit value (TLV). The concentration of an airborne contaminant to which workers may be exposed legally day after day without any adverse effect. Several defined TLV levels have been found later to cause adverse health effects.

threshold of audibility. The minimum sound pressure level at which a person can hear a sound of a given frequency.

threshold of pain (feeling). The minimum sound pressure level at which a person begins to experience pain in the ear from a sound of a given frequency.

threshold shift. An alteration, temporary or permanent, in a person's threshold of audibility.

thrips. Thysanoptera.

thrombin. *See* clotting of blood.

thrombocytes. *See* blood platelets.

thrombokinase. *See* blood platelets, clotting of blood.

throw. The vertical displacement between the upthrown and downthrown sides of a fault.

thrust. A low angle reverse fault.

thunderflies. Thysanoptera.

thymonucleic acid. DNA.

thyrocalcitonin. calcitonin.

thyroid. *See* thyroxin.

thyrotropic hormone (thyrotrophic hormone, TSH). A pituitary hormone which stimulates the secretory activity of the thyroid gland. *See* thyroxin.

thyroxin (thyroxine). An iodine-containing amino acid hormone, produced

484

along with a similar hormone (tri-iodo-thyronine) by the thyroid gland, which is situated in the neck region of vertebrates. Thyroid secretion increases the rate of oxidative processes and stimulates growth in young animals. Deficiency (due sometimes to lack of iodine in the diet) results in physical and mental stunting (cretinism in Man). In Amphibia, thyroid secretion initiates metamorphosis in the tadpole. *See* paedogenesis, calcitonin.

Thysanoptera (Physopoda, thrips, thunderflies). An order of minute, slender insects (Exopterygota) commonly found in flowers. They usually have two pairs of narrow, fringed wings. Most feed on plant juices, and may be present in such large numbers that they become serious agricultural pests, damaging peas, wheat and other crops. Some transmit plant diseases.

Thysanura. *See* Apterygota.

ticks (Acarina). *See* Arachnida.

tidal wave. Tsunami.

tidal power. Mechanical power, which may be converted to electrical power, generated by the rise and fall of ocean tides. The possibilities of utilising tidal power have been studied for many generations, but the only feasible schemes devised so far are based on the use of one or more tidal basins, separated from the sea by dams or barrages, and of hydraulic turbines through which water passes on its way between the basins and the sea. The world's largest tidal power plant is located on the estuary of the River Rance, in Brittany, France, completed in the 1960s, which generates 544 kwh per year. The disadvantages of tidal power generation are the very high capital cost of the dams or barrages, possible ecological disturbance, but the major limitation imposed by the tidal cycle can be overcome by generating energy from both the filling and emptying of the basins to provide base-load power, and by using base-load power at times of low demand to pump water to higher reservoirs, from which it can be released to meet peak load demands.

tile drain. *See* soil drainage.

Tiliaceae. A family of dicotyledonous plants, mostly trees and shrubs, of tropical and temperate regions. *Tilia*, a genus including the European limes and the American bass-wood, is visited by bees for the nectar secreted by the sepals, and is a valuable source of honey as well as of timber. *Corchorus* is an annual about 3 m high, grown in India and elsewhere, from which jute fibre is obtained by retting the stems in water.

till. *See* boulder clay.

tillite. Lithified boulder clay.

tilth. (*a*) Cultivation. (*b*) The depth of soil affected by cultivation. (*c*) The physical condition of the soil with reference to its suitability for cultivation. Soil with a good tilth is friable and porous, with a stable, granular structure.

tiltmeter. An instrument used to measure changes in slope of the ground surface. Tiltmeters measure vertical displacement and are used to indicate impending earthquakes or volcanic activity.

tin (Sn). Element. A silvery-white metal which is soft, malleable and ductile. At ordinary temperatures it is unaffected by air or water. It exists in three allotropic forms (*see* allotropy) and occurs as cassiterite. It is used for tin-plating and in many alloys. AW 118.69; At. No. 50; SG 7.31; mp 231.85°C.

tinstone. Cassiterite.

tissue. A group of cells with similar structure and function (e.g. muscle tissue in animals, cork tissue in plants).

tissue culture (explantation). A method of keeping cells alive after their removal from the organism, using a suitable medium containing the correct balance of salts, pH, oxygen and food, kept at the right temperature.

titration. An operation that forms the basis of volumetric analysis, in which a measured amount of a solution of one reagent is added to a solution of another until the action between them is complete, the second regent having been entirely consumed.

Tl. Thallium.

TLV. Threshold limit value.

Toad. *See* Anura.

toadstool. *See* Agarics.

Toarcian. A stage of the Jurassic System.

tobacco. *See* Solanaceae.

tocopherol. Vitamin E. *See* vitamin.

Tolba, Mostafa K. *See* UNEP.

tolylene di-isocyanate (TDI). A chemical of the di-isocyanate group, which causes chest irritation, leading to sensitisation, after which the victim cannot tolerate even unmeasurably small atmospheric concentrations, and suffers symptoms akin to acute asthma. A high concentration, in Britain, is one that exceeds the Threshold Limit Value of 0.02 ppm, and the National Institute of Occupational Safety and Health have proposed a maximum exposure of 0.02 ppm for not more than 20 minutes, and an 8 hour TLV of 0.005 ppm. Di-isocyanates are produced by the chemical industry and are used in the rubber industry.

tomato. *See* Solanaceae.

tombolo. A spit or bar joining an island to the mainland or to another island.

tone. Sound of definite pitch.

tonoplast. The plasma membrane surrounding the vacuole in plant cells. *See* plasmalemma.

tooth shells. Scaphopoda.

topography. A detailed description of the natural and artificial surface features of an area.

topotype. A population in one geographical region with characteristics that differ from those of another region. A topotype may be extraclinal if it does not fall within a geographical gradient or intraclinal if it is characteristic of a particular geographical gradient.

topset. A flat-lying sedimentary layer overlying the foresets of a delta.

tornado. *See* bath plug vortex.

tortoise. *See* Chelonia.

torula. *Torula thermophila*, a yeast that is grown on sugars to produce novel protein foods.

total energy system. District heating.

total population curve. The total population density of individuals of all stages plotted against time.

Tournaisian. The oldest stage of the Carboniferous System in Europe.

tow. *See* retting.

toxicology

toxicology. The study of poisons.

toxic waste. *See* industrial waste.

trace elements. Elements which are necessary in extremely small amounts for the proper functioning of metabolism in plants and animals. Most are probably constituents of enzymes. Higher plants need traces of copper, zinc, boron, molybdenum and manganese. 'Heart rot' of sugar beet is produced by lack of boron. Deficiency of cobalt causes disease in cattle and sheep. *See* micronutrients.

tracer. A substance mixed with or attached to a given substance to enable the distribution or location of the latter to be determined subsequently. Tracers can be physical, chemical, radioactive and isotopic.

trachea. (*a*) (Bot.) Vessel. A non-living, tube-like element of xylem, derived from a row of cylindrical cells by the more or less complete breakdown of adjacent end walls. The longitudinal walls of the vessels are thickened with deposits of lignin in spiral, annular or reticulate patterns, and the vessels often communicate by means of minute pits in these thickened walls. Vessels conduct water and dissolved mineral salts, and provide mechanical support. They occur mainly in flowering plants. *See* tracheid. (*b*) (Zool.) 1. The 'wind pipe' of vertebrates. 2. A branching, cuticle-lined tube in an insect, which conducts air from an external opening (spiracle) directly to the tissues.

tracheid. A non-living element of xylem formed from a single, elongated cell, with tapering ends. The walls are thickened with ligin. Tracheids usually communicate by means of pits in the walls. They conduct water and dissolved mineral salts, and provide mechanical support. The conducting tissue in the wood of conifers is entirely made up of tracheids, which also occur along with vessels in some parts of flowering plants.

Tracheophyta. A modern term for the vascular plants (Pteridophyta and Spermatophyta of the older classifications). The group comprises the Psilopsida, Lycopsida, Sphenopsida and Pteropsida.

trachyte. A fine-grained, alkali intermediate rock. Trachytes characteristically show trachytic texture, the parallel alignment of lath-like feldspar crystals. Trachytes are the volcanic equivalent of syenites.

trade, commercial and industrial waste. *See* industrial waste.

trade cumulus. Cumulus clouds in the tradewinds which carry up moisture, converting the air from dry polar into moist tropical.

trade inversion. The inversion, or stably stratified layer, which lies at the top of the tradewinds, which blow from ENE, and above which the wind is usually from a westerly point. The height increases as the air moves over warmer water. It may be at around 1 km off NW Africa, rising to 3 or 4 km as the air approaches the West Indies, where the convection breaks through to make showers, thunderstorms, and seasonal hurricanes.

tradewinds. *See* angular momentum.

Tragedy of the Commons. The title of an essay by Garrett Hardin, published in *Science*, based on *Two Lectures on the Checks to Population* by William Foster Lloyd (1794-1852) first published in England by the Oxford University Press in 1833, which Hardin republished in part in his *Population, Evolution, and Birth Control* (W. H. Freeman, 1964). Lloyd sought to rebut the concept of the 'invisible hand' by showing that any resource that is the property of all (a commons) may be over-exploited, since each user seeks to increase his profit from it. The profit accrues to the individual, but the cost is shared among all. Thus, for all users, the benefit of increasing use exceeds the costs until the shared resource is reduced, possibly suddenly, to a level that prevents any further use by any of the commoners. Hardin's essay became a classic during the late 1960s and was reprinted in many anthologies.

tramontana. *See* mistral.

transad. Closely related organisms which have become separated by an environmental barrier (e.g. the caribou of North America and the reindeer of Europe).

transduction. The transfer by bacteriophages of genetic material from one bacterium to another.

transect. A cross-section of an area, used as a sampling line for recording the vegetation.

transfer RNA. *See* RNA.

transformation. (*a*) The acquisition by a bacterium of genetic material (DNA) from a related strain. Characteristics such as resistance to antibiotics may be modified by growing bacteria in the presence of killed cells or extracts of other bacteria. (*b*) An inherited change in cultured animal cells produced by viruses or other agents.

transform fault. A strike-slip fault offsetting a spreading ridge axis. In the offset zone the motion along the fault plane is in the opposite direction to the displacement. Transform faults probably originate as irregularities in the of the crust at the initial separation of two plates. *See* plate tectonics.

transfrontier pollution. Pollutants produced in one country which cross international frontiers in water or air, and whose effects can be mitigated only as a result of international agreement, since damage is not caused in the country of origin. Examples include the pollution of certain European rivers (e.g. Danube, Rhine), regional seas, and air pollutants carried from industrial areas in Europe into Scandinavia (*see* sulphuric acid).

transgression. (Geol.) Marine transgression is the raising of sea level in relation to the level of the land.

transgressive. (*a*) (Ecol.) A species that overlaps from one community to another. (*b*) (Geol.) Refers to a body of water or sediments associated with a spreading of water over the land.

transgressive sill. *See* sill.

translocated herbicides. Herbicides which spread throughout a plant after absorption through the roots or leaves (e.g. 2, 4, 5-T, 2 4,-D, Dichlorprop, Dalapon, MCPA, Mecoprop, Barban, Asulam, Linuron).

translocation. (*a*) (Genet.) The breaking away of any part of a chromosome and its attachment to another chromosome. *See* mutation. (*b*) (Bot.) The transport of materials (e.g. mineral salts, organic materials) from one part of a plant to another, mainly via the xylem and phloem in higher plants. *See* hormone weed killers, systemic insecticides, translocated herbicides.

transmission loss. A measure of the sound insulation of a wall, partition or panel, equal to the difference in sound level on either side of the wall, etc., or, if the intensity of the waves on either side are expressed as a ratio of incident: transmitted, the transmission loss is equal to ten times the logarithm of the ratio.

transpiration. The loss of water vapour from a plant, mainly through the stomates and to a small extent through the cuticle and lenticles. Transpiration results in a stream of water, carrying dissolved mineral salts, flowing upwards through the xylem.

transpiration coefficient. The ratio of the dry weight of material produced to the amount of water transpired by a plant during a given period of time.

transuranic elements. Elements beyond uranium in the periodic table, i.e. elements with atomic numbers greater than 92. They do not occur in nature, but are produced by nuclear reactions and all are radioactive

trash (US usage). Municipal waste.

Traveller's Joy. *Clematis vitalba.*

travelling wave. A motion in which a fixed pattern of displacements, or wave, travels longitudinally while any one particle involved in the motion does not travel in the mean, but executes an oscillatory motion. In the atmosphere, examples are gravity waves, sound waves and atmospheric tides.

travertine. A form of tufa deposited by hot springs. Travertine is used as an ornamental stone.

tree. A large, woody perennial plant with a single stem (bole). The term is often restricted to plants at least 6 m in height. *See* hardwoods, softwoods, broadleaved trees, conifers.

tree layer. *See* layers.

trellis drainage. A river system resembling a trellis pattern and characteristic of areas of folded sedimentary rocks where tributaries cut channels through less resistant beds.

Tremadoc series. In the UK, the youngest series of the Cambrian System, but in most countries, placed at the bottom of the Ordovician System.

Trematoda (flukes). A class of parasitic flatworms (Platyhelminthes) with leaf-like bodies, hooks and/or suckers and thick cuticles. Some are ectoparasites (*see* parasitism) of aquatic animals, living, for instance, on the gills of fish. Others are endoparasites of vertebrates (e.g. the blood fluke *Schistoma*; the liver fluke of sheep, *Fasciola hepatica*). The various larval stages (miracidium, redia, cercaria) of endoparasitic flukes require at least one intermediate host, including a mollusc (e.g. the water snail *Limnaea truncatula* for *Fasciola hepatica*).

trench. *See* oceanic trench.

triage. An approach to the medical treatment of casualities following a catastrophe (e.g. thermonuclear attack) that exceeds the capacity of the services available. Casualties are divided into three categories: (*a*) those who will die, even with treatment; (*b*) those who will die without treatment, but recover with treatment; (*c*) those who will recover without treatment. Priority is given to categories b, c and a in that order. Some writers have suggested adapting the approach to the allocation of aid (e.g. food aid) to developing countries.

triallate (tri-allate). Soil-acting herbicide of the thiocarbamate (*see* carbamate) group used to control wild oat and black-grass in cereal, pea, bean and carrot crops. It can be irritating to the skin and is harmful to fish.

Triassic. Refers to the oldest period of the Mesozoic Era, usually dated as beginning some time between 225 and 245 million years ago, and lasting about 30 million years. Triassic also refers to the rocks formed during the Triassic Period: these are called the Triassic System, which in Europe is divided into three series, the Lower, Middle and Upper, which correspond approximately to the original threefold division into Bunter, Muschelkalk and Keuper. The three series are divided into stages: the Lower has one stage, the Scythian, the Middle two, the Anisian and Ladinian, and the Upper three, Karnian, Norian and Rhaetian. In the UK the Triassic is divided into Bunter, Keuper and Rhaetic.

tribe. (Biol.) A group of closely related genera within a plant family. For instance, within the grass family (Graminae) the tribe Festuceae contains *Festuca, Dactylis, Lolium, Poa* and other genera.

2, 4, 5-T (2, 4, 5-trichlorophenoxyacetic acid). Translocated hormone weed killer used to control woody weeds, and of great value in preventing regrowth of scrub after clearance. *See* translocated herbicides.

Trichoptera (caddis flies). An order of insects (Endopterygota) almost all of which have aquatic larvae. The adults have two pairs of wings covered with minute hairs, long, slender antennae, and reduced mouthparts, so that they rarely feed. Many caddis larvae build tubular cases of sand, snail shells or plant material, others build silk nets in which food is trapped. Trichoptera form an important item of food for freshwater fish.

trickling filter. Percolating filter.

Trilobita. A class of marine, Palaeozoic arthropods, with flattened, oval bodies divided longitudinally into three lobes, a single pair of antennae, and numerous similar forked (biramous) appendages. *See* index fossil.

triple superphosphate. *See* phosphate, phosphate rock.

Triploblastica. Animals whose bodies are made up of three layers of cells, ectoderm, mesoderm and endoderm (*see* germ layers). All Metazoa except coelenterates are triploblastic. *Cf.* Diploblastica.

triploid. Applied to a polyploid cell or organism with three times the haploid number of chromosomes.

tripton. The non-living matter suspended in a body of water. The abioseston. *See* seston.

trisomic. Applied to cells or organisms showing a kind of aneuploidy in which one kind of chromosome is represented three times instead of twice, as is normal. Mongolism in Man is caused by the presence of an extra chromosome (number 21).

tritium (T). Element. A radioactive isotope of hydrogen, with mass number 3 and atomic mass 3.016. It occurs in natural hydrogen as one part in 10^{17}, but it can be made artificially in nuclear reactors. Its half-life is 12.5 years. The nucleus of a tritium atom is called a triton. *See* fusion reactor.

triton. *See* tritium.

trochophore (trochosphere). A type of plantonic larva characteristic of many aquatic invertebrates (e.g. polychaete worms, some molluscs and Polyzoa). The larva swims by means of cilia, a ring of which encircles the mouth.

trochosphere. Trochophore.

trophallaxis. The exchange of food or secretions such as saliva between animals, esp. social insects (e.g. in ants and wasps the worker taking food to the grub receives in return a drop of saliva).

trophic level. (*a*) *See* food chain, energy flow. (*b*) The nutrient status of a water body, esp. as regards the levels of nitrate and phosphate. *See* dystrophic, eutrophic, mesotrophic, oligotrophic.

trophobiont. *See* trophobiosis.

trophobiosis. A type of symbiosis in which two organisms (trophobionts) of different species feed one another (e.g. some species of ant rear aphids on plant roots inside their nests, and feed on the honeydew secreted by the aphids).

trophogenic region. The region of a water body where organic material is produced by photosynthesis. *See* photic zone, compensation point.

tropical cyclone. A violent storm with a very small area of very low pressure at the centre around which the isobars are almost circular, very close together, and the winds are extremely violent. They may be 500 or 600 km in diameter, with the wind, cloud and general weather pattern similar all around them. At the very centre, the 'eye of the storm', winds are light, skies are clear or almost clear, and there is no precipitation. The term 'cyclone' is used in the Indian Ocean and Bay of Bengal, 'hurricane' in the Caribbean, 'typhoon' in the China Sea, and 'willy nilly' in Australia. *See* bath plug vortex.

Tropina. Single cell protein used as a feed supplement for livestock and made in

France by British Petroleum Ltd. from yeasts (*Candida* spp.) grown on paraffins (gas-oil containing waxes). After extracting the yeast-protein, the wax-free paraffins are returned for refining.

tropism. (*a*) A growth response by plants or sedentary animals, which produces curvatures whose direction is determined by the location of the stimulus. *See* nastic movement, geotropism, hydrotropism, phototropism, haptotropism. (*b*) (Zool.) Once a synonym for taxis.

tropoparasite. *See* parasitism.

tropopause. The boundary between the troposphere and the stratosphere. The height of the tropopause varies from day to day, but its average height over the Equator is about 17 km, decreasing to about 6 km over the Poles.

tropopause break. A phenomenon which occurs when a cyclone induces the folding over of the tropopause, with the tropopause higher on the warm side of a front by perhaps a few thousand metres.

tropophyte. A plant which lives under moist conditions for part of the year and under dry conditions for the rest of the time (e.g. a deciduous tree, which sheds its leaves for the dry season or for the winter, when water may not be available because it is frozen).

troposphere. The lowest level of the atmosphere, in which temperature decreases with height, bounded by the land or sea surface below and by the tropopause above.

troutbeck. *See* river zones.

truewood. *See* heartwood.

truncated spur. The steep side walls of the U-shaped trough formed by a glacier as spurs and projections are abraded away by the advancing ice.

Trypanosomidae. A family of parasitic Protozoa (Flagellata), some of which cause serious diseases in Man and domestic animals. Oriental sore and kala-azar are caused by *Leishmania*, which attacks the cells lining the blood vessels, and is transferred by flies of the genus *Phlebotomus*. *Trypanosoma gambiensis* and *T. rhodesiensis* cause sleeping sickness, and are transmitted to Man by the bite of the tsetse fly (*Glossina*). These species of *Trypanosoma* and *T. Brucei*, which causes African cattle sickness, are non-pathogenic in antelopes.

tryptophan. Amino acid with the formula $C_6H_4.NHC_2H.CH_2CH.(NH_2).COOH$. Molecular weight 204.2.

494

tsetse fly (Glossina). *See* Cyclorrapha, Trypanosomidae.

TSH. Thyrotropic hormone.

tsunami. A sea wave produced by a submarine earthquake or volcanic explosion. Tsunamis travel across oceans at a velocity of several hundred kilometres per hour, but with a low wave-height, but on reaching a shelving shore the energy is concentrated into a hugely destructive series of waves several metres high. Tsunami are misleadingly called 'tidal waves'. 'Tsunami' in Japanese means 'harbour wave'.

tuatara. *Sphenodon. See* Rhynchocephalia.

tubal ligation. Salpingectomy.

tubenoses. Procellariiformes.

tuber. A swollen, food-storing portion of an underground stem (e.g. potato) or root (e.g. Dahlia). Tubers are organs of perennation and vegetative propagation which last only one year, and do not give rise to those of the succeeding year. *Cf.* corm.

Tubulidentata (Orycteropodidae). An order of placental mammals containing only *Orycteropus*, the aardvark (earth pig) or African ant-eater. *Oryctoperus* lives on termites, has a long tongue and an elongated snout like other ant-eaters (Edentata and Pholidota), but possesses unique, peg-like teeth with tubes in the dentine.

tufa. A sedimentary rock composed of calcium carbonate or silica and deposited from aqueous solution by evaporation of circulating ground water, or water from springs or in lakes.

tuff. Lithified pyroclastic rock with a preponderance of fragments less than 2 cm in diameter.

tundra. Treeless arctic and alpine regions. Tundra may be bare of vegetation, or may support mosses, lichens, herbaceous plants and dwarf shrubs. *See* tundra soil.

tundra soil. A shallow soil with permanently frozen subsoil impeding drainage. Relatively little moisture is available. Where plants (lichens, mosses, herbs and shrubs) can colonize, organic matter accumulates and a dark peaty layer overlies an anaerobic greyish layer. Profiles generally are poorly developed and the soils grade between Polar Desert Soil of arid northern regions to Arctic Brown Earths in better drained, more humid upland regions. *See* soil classification.

tungsten (W). Element. A grey, hard, ductile, malleable metal which resists corrosion. It is used in alloys, as tungsten carbide for hard tools, and in electric lamp filaments. AW 183.85; At. No. 74; SG 19.3; mp 3410°C. *See* wolfram, scheelite.

Tunicata. Urochordata.

Turbellaria. A class of mainly free-living flatworms (Platyhelminthes) which move by means of cilia which cover the body. Some (e.g. *Planaria*) live in freshwater, others are marine and a few are terrestrial (e.g. *Bipalium kewense*, a tropical form reaching 30 cms in length, sometimes found in greenhouses).

turbidite. Deposits from a moving slurry of sediment and water (called a turbidity current or density current). Experimental turbidity currents produce sedimentary structures within each bed such as graded bedding and on the base of each bed and underlying bed such as grooves, flute casts, load casts and tool marks. These features are shown by many natural deposits from unlithified graded deposits in the abyssal plains to Precambrian sandstones.

turbidity. The cloudiness in a liquid caused by the presence of finely divided, suspended material.

turbine. A machine that converts the energy in a stream of fluid into mechanical energy by passing the stream through a system of fixed and/or moving fan-like blades, causing them to rotate. Turbines have wide uses in large scale power generation (usually employing steam-driven turbines), small-scale power generation (e.g. Pelton wheel), jet aircraft propulsion (gas turbines deriving power from a stream of heated air), marine engines, etc.

turbulence. An irregular movement of a fluid, in which, in general, no two particles of the fluid follow the same path. Most natural movement of fluids is turbulent rather than laminar and in air, turbulence is a major cause of mixing.

turgor pressure. The hydrostatic pressure within a plant cell exerted against the cell wall as a result of the intake of water by osmosis. Turgor pressure plays a large part in the mechanical support of plant tissues. When water loss exceeds intake, turgor pressure falls and wilting follows. Some plant movements such as seismonasty (*See* nastic movement) and the opening and closing of stomates are caused by variations in turgor pressure. The turgor pressure of vacuolated cells in the notochord of chordates, acting against the notochord sheath, creates a stiff, flexible supporting rod.

turions. Detachable winter buds formed by many water plants, enabling them to survive the winter.

turnip. *See* Brassica.

turn-over. (*a*) Overturn (*b*) The continuous, balanced process of generation and loss of cells or molecules in living systems. The turn-over time is the time needed for the replacement by turn-over of the cells or molecules equivalent to the total biomass of a population, or the time taken for an individual organism to mature, die and undergo decomposition.

Turonian. A stage of the Cretaceous System.

turtle. *See* Chelonia.

tuscacora rice. Wild rice.

tusk shells. Scaphopoda.

TV dinner. Meal whose protein content is derived from textured vegetable protein and from the vegetables present in the meal in their traditional form. The meal contains no meat. *See* novel protein foods.

TVP. Textured vegetable protein. *See* novel protein foods.

twister. (US usage) A tornado or waterspout.

type specimen (holotype). The original specimen from which the description of a new species is made. *Cf.* lectotype, neotype, paratype.

typhoon. Tropical cyclone in the China Sea.

tyrosine. Amino acid with the formula $C_6H_4OH.CH_2CH.(NH_2).COOH$. Molecular weight 181.2.

U

U. Uranium.

UKAEA. United Kingdom Atomic Energy Authority.

Ulex. Gorse, furze, whin. *See* Leguminosae.

Ulmus. Elms. Tall, deciduous trees (family Ulmaceae) with winged fruit, which yield valuable timber. *Ulmus procera* is the English Elm, abundant in hedgerows and woods in the south of England. Dutch Elm Disease, which has drastically reduced the elm population in Britain, is caused by the fungus *Ceratotomella ulmi*, and is spread by the Elm Bark Beetle, *Scolytus scolytus*. The disease has been endemic in Britain at least since 1927.

ultisols. *See* soil classification.

ultrabasic. A term applied to igneous rocks almost entirely composed of ferromagnesian minerals to the virtual exclusion of quartz, feldspar, and feldspathoids. Ultrabasic rocks are grouped into (*a*) peridotites (e.g. dunite and kimberlite), consisting essentially of olivine with or without accessory ferromagnesian minerals, and without feldspar; (*b*) perknites consisting essentially of ferromagnesian minerals other than olivine; (*c*) picrites, consisting of over 90 % ferromagnesian minerals and up to 10 % feldspar. Picrite is thus gradational into basic rocks such as gabbro. Purely ultrabasic intrusions are rare and occur late in an orogeny. Ultrabasic rocks are normally found in association with basic rocks in layered igneous intrusions: such intrusions often contain economic deposits of chromite, ilmenite, magnetite, and the platinum group of metals (platinum, iridium, osmium, palladium, rhodium and ruthenium).

ultramafic. Ultrabasic.

ultra-violet radiation. Very short wave electromagnetic radiation in the wavelength range of about 4×10^{-7} to 5×10^{-9} metres, which places it between visible light waves and X-rays. Much of the ultra-violet radiation reaching the Earth from the sun is 'filtered' in the ozone layer where it loses energy by powering chemical reactions, UV radiation affects photographic films and plates and in the skin of some animals, including Man, it acts on ergosterol to produce vitamin D (*see* vitamins). It is powerfully ionising, and excessive UV radiation is lethal to plants and animals. UV radiation can be produced artificially by mercury vapour lamps.

Umbelliferae. A cosmopolitan family of dicotyledonous plants, mostly herbs with stout, hollow stems. The inflorescence is usually umbrella-shaped, with many small flowers massed together, making them attractive to the insects which pollinate them. Many umbellifers are cultivated as food crops (e.g. *Daucus*, carrot; *Pastinacea*, parsnip; *Apium*, celery). Many are poisonous (e.g. *Conium maculatum*, Hemlock).

unconformity. A lack of continuity of sedimentary deposition between two beds in contact, and implying a gap in the geological record.

UNCTAD. United Nations Conference on Trade and Development.

undergrowth. The shrubs, saplings and herbaceous plants in a forest.

understorey. The lower layer of trees in a two-layered woodland (e.g. coppice under standards), or the lower storey of a two-storied high forest.

UNESCO. United Nations Educational, Scientific and Cultural Organisation.

United Nations Educational, Scientific and Cultural Organisation. United Nations agency, founded in 1945, to support and complement the efforts of member states to promote education, scientific research and information, and the arts and to develop the cultural aspects of world relations. It holds conferences and seminars, issues publications, promotes research and the exchange of information and provides technical services. It is financed by its own budget and also draws on funds pledged by member states to the UN Development Programme. Its headquarters are in Paris.

UNEP (United Nations Environment Programme). UN agency charged with the coordination of inter-governmental measures for environmental monitoring and protection. Formed after the 1972 UN Human Environment Conference. UNEP's first Executive Director was Mr Maurice Strong, who had been Secretary General of the 1972 conference. He was succeeded, in 1976, by Mr Mostafa K. Tolba. UNEP operates the Earthwatch Programme and funds Earthscan. Its headquarters are in Nairobi, Kenya.

Ungulata (ungulates). Hoofed mammals, including the Artiodactyla and Perrisodactyla.

ungulates. Ungulata.

unguligrade. Applied to mammals which walk on the tips of the digits, which end in hooves. cf. Digitigrade, Plantigrade.

uniformitarianism. The principle that processes active today also acted in the past producing the same results. Uniformitarianism is often glibly encapsulated in the statement 'the present is the key to the past.'

uniformity. The general tendency of an association to have a uniform distribution of component species within it, producing a uniformity of the association as a whole, so providing one of the most distinctive characteristics of that association.

union. Groups of plants that are ecologically significant within the total flora, because they usually appear together whenever environmental conditions are suitable, their requirements being similar or complementary.

uniovular twins. Monozygotic twins.

uniparous. Monotocous.

unisexual. (*a*) (Bot.) Applied to monoecious or dioecious plants with stamens and carpels in separate flowers. (*b*) (Zool.) Applied to an animal which produces either male or female gametes. *Cf.* hermaphrodite.

United Kingdom Atomic Energy Authority (UKAEA). Official body in the UK responsible for all aspects of atomic energy, and commercial and scientific uses of radioisotopes. The UKAEA is also involved in more general research into energy use and resources.

United Nations Conference on the Human Environment. A conference held in Stockholm in the summer of 1972, attended by most UN member governments, except for the USSR and the East European states, but including the People's Republic of China. *See* Declaration on the Human Environment, UNEP, Earthwatch programme.

United Nations Conference on Trade and Development (UNCTAD). A principal UN forum in which developed and developing countries meet to discuss matters relating to trade and development.

United Nations Research Institute for Social Development (UNRISD). UN agency, based in Geneva, engaged in social research, especially into the effects of development policies in developing countries.

unit membranes. Term used to describe all osmotically active cell membranes because they have a universal molecular architecture, regulating the quantity of materials being exchanged across it by altering its thickness, rather than by making any fundamental change in its composition. *See* osmosis, plasma membrane.

Universal Oil Product (UOP). Process for removing sulphur dioxide from flue gases using a wet scrubber followed by extraction of sulphur and the regeneration of the reagent. The process is used in the USA.

univoltine. Applied to an organism which produces only a single generation in a year.

UNRISD. United Nations Research Institute for Social Development.

unsettled weather. Weather that exhibits considerable changes from sunshine to cloud and rain within the passage of hours. It is typical of the westerly type in temperate climates.

unsorted. *See* sorting.

unstable. *See* dynamic stability, static stability.

UOP. Universal Oil Product.

updraught. Air rising in a convection current.

upper air contours. Upper air maps are drawn showing the contours of constant height of an isobaric surface rather than isobars at constant height because of greater ease of construction and simpler mathematical use.

upper shore. *See* shore zonation.

uraninite. The mineral oxide, UO_2 (uranium oxide), found in granites and pegmatites, in hydrothermal veins, and also in some sedimentary rocks. Uraninite is the chief ore of uranium.

uranium (U). Element. Naturally occurring radioactive, hard white metal, the natural element consisting of 99.28 % $^{238}_{92}U$, which has a half-life of 4.5×10^9 years, and 0.71 % $^{235}_{92}U$, which has a half-life of 7.1×10^8 years. Uranium-235 is used as a fuel in nuclear reactors. AW 238.03; At.No. 92; SG 19.05; mp $1150°C$.

uranium enrichment. The improvement of the properties of natural uranium by removing the diluent U-238, which comprises 99.28 % of the natural element in order to increase the concentration of U-235, which is able to sustain a nuclear chain reaction.

urea ($CO (NH_2)_2$). A soluble breakdown product of proteins, excreted by many vertebrates. It is also produced by some plants. *See* ureotelic, deamination.

urea (group of herbicides). Group of herbicides (e.g. Diuron and Linuron) which control weed seedlings by inhibiting photosynthesis. They persist for long periods in the soil, but are not very toxic to mammals.

Uredinales. Rust fungi.

ureotelic. Applied to animals whose main nitrogenous excretory product is urea. Fish, Amphibia and mammals are ureotelic. During their embryonic development dissolved urea diffuses away into the surrounding water or is removed in the maternal blood stream. *Cf.* uricotelic.

uric acid. A complex nitrogenous compound, an almost insoluble breakdown product of proteins and nucleic acids, excreted in large amounts by uricotelic animals and in small amounts by some other animals.

uricotelic. Applied to animals whose main nitrogenous excretory product is uric

501

acid. Birds, insects, terrestrial snails and most reptiles are uricotelic. These animals develop inside shells (see Cleidoic egg) and must therefore store their nitrogenous excretory product in an insoluble form. cf. Ureotelic.

Urochorda. Urochordata.

Urochordata (Urochorda, Tunicata). The sea squirts (ascidians) and their allies, a group of marine Chordata whose larvae are tadpole-like, with a notochord in the tail. Most adult urochordates are sedentary, with no notochord, a reduced nervous system, and an enormous perforated pharynx through which water, bearing food particles is passed by ciliary action. The body is surrounded by a horny or gelatinous coat (the tunic or test). Many sea squirts are colonial.

Urodela (Caudata). An order of Amphibia which includes the newts and salamanders. Urodeles have a well-developed tail, and two pairs of limbs of more or less equal sizes. Gills may persist in the adult (e.g. in the axolotl).

US AID. United States Agency for International Development. The major US agency concerned with aid to developing countries.

Usonia. Idealised city planned by Frank Lloyd Wright.

Ustiginales. Smut fungi.

utopianism. A form of planning based on the construction of ideal futures, together with assessments of their feasibility and strategies by which they may be realised.

UV light. Ultra-violet radiation.

V

V. Volt.

vacuole. A fluid-filled space inside the cytoplasm, more common in plant than in animal cells. The vacuole in many plant cells occupies most of the cell and contains cell sap. Many Protozoa contain contractile vacuoles which collapse periodically, expelling excess water from the cell. Food vacuoles of Protozoa and other animals are spaces where ingested food particles are digested. *See* Turgor pressure.

vadose. Refers to the region between the ground surface and the water table.

Vadose is contrasted with phreatic.

vadose water (gravitational water). Water moving in aerated ground above the water table. Literally 'wandering water'.

Vaiont. Area in Italy that was the site of one of the most severe man-made disasters in modern times. On the night of 9 October 1963, a sudden and very large and rapid rockslide completely filled the Vaiont reservoir causing large waves. Some 3000 people died. The disaster was caused by two weeks of heavy rainfall which raised the water level in the cavernous surrounding rocks, which were inherently weak and whose limestone beds and clay layers offered little frictional resistance to sliding. The high water level in the reservoir saturated part of the side, increased buoyancy and decreased friction further. No witness of the incident survived.

Valanginian. *See* Neocomian.

valine. Amino acid with the formula $(CH_3)_2CH.CH.(NH_2).COOH$. Molecular weight 117.1.

valley fog. Fog formed in air cooled on hillsides by radiation at night and frequently causing the air containing the fog to stagnate in a valley by day because the fog reflects a significant fraction of sunshine and cools by radiation at infra-red wavelengths.

valley glacier. *See* glacier.

valley train. A body of outwash deposits partly filling a valley.

valve. (*a*) Device for controlling the passage of a fluid through a pipe or tube; (*b*) (Zool.) Membranous structure that permits a liquid (e.g. blood) to flow in one direction only; (*c*) The shell of a mollusc or each complete part of the shell of a bivalve mollusc; (*d*) A segment of a dehiscent fruit; (*e*) Thermionic valve or tube: a system of electrodes arranged in an evacuated glass or metal envelope.

vanilla. *See* Orchidaceae.

variable. (Stat.) A quantitative characteristic of an individual (e.g. height or weight in man). *See* attribute, variate. Any quantity which can have more than one value.

variate. (Stat.) A quantity which can have one of a range of specific values, each with a specific probability.

variation. Differences which exist between individuals of the same species at corresponding stages in the life cycle. Variation is a reflection of the genetic constitution of the individuals, and of the environmental influences brought to bear upon them. (*a*) Continuous variations: (i) Relatively small variations, not due to mutation; (ii) Variations which show gradation from one extreme to another, with a preponderence of intermediate types (e.g. stature in Man). This type of variation is caused by the combined action of a number of genes, each having a small effect (*see* polygene). (*b*) Discontinuous variations: (i) Mutations; (ii) Variations in which the expression of a particular character differs sharply amongst individuals, and there is no range of intermediate types (e.g. sex differences). This type of variation is caused by the action of a few (often two) genes which exert a large effect.

variety. (*a*) A group of plants which differ distinctly from others within the same sub-species. (*b*) A term used loosely for a variation of any kind within a plant or animal species.

Variscan (Geol.) Usually understood today to refer to processes and products of the mountain-building in Europe from the Carboniferous to Triassic Period, inclusive. Originally Variscan was restricted to the Hercynian of central Europe, but usually now Hercynian and Variscan are interchangeable terms.

varves. Rhythmically laminated sediments where, due to special conditions, distinctive layers have been deposited in the course of a single year. (e.g. the coarse and fine sediments deposited after the annual melting of an ice sheet, or different types of sediment deposited in a lake bottom during summer and winter). These laminae can be counted like the annual rings of a tree and used for estimating the time taken for the entire sediment to be deposited.

vascular. Applied to tissues which conduct liquids (e.g. water, blood), and to plants which possess vascular tissue. *See* vascular bundle, Tracheophyta.

vascular bundle. A strand of conducting tissue in a higher plant, consisting mainly of xylem and phloem.

vascular cylinder. Stele.

vascular plants. The Tracheophyta. The higher plants, characterized by the possession of true roots, stems and leaves, which probably evolved with the invasion of dry land by plants that formerly lived only in the sea, and through whose tissues liquids are conducted.

vasectomy. Male sterilisation, in which the *vas deferens* is cut and the two ends sealed, so preventing the release (but not the manufacture) of sperm.

Vavilov, Nikolai Ivanovich (1887–1943). Russian plant geneticist who made a comprehensive study of the origin of cultivated plants and proposed 'centres of diversity' and 'centres of origin.' He argued that if the rate of natural mutation among plant species is constant through time, the longer a species occupies a particular site the greater the genetic diversity that will exist among its individual members. Thus, the fewer the variations, the shorter the time the species must have been present. From this he located a series of areas in all parts of the world, where particular species of crop plants are found in the greatest diversity. From this he attempted to argue that these centres of diversity were also centres of origin from which knowledge of the cultivation of those particular species had spread by a process of cultural diffusion. This conclusion could not be supported on the evidence, but Vavilov made a major contribution to knowledge of plant genetics. He was removed from all his posts as Lysenko rose in favour in the USSR, and is believed to have died in prison in Siberia.

vector (*a*). Any physical quantity that cannot be described completely without reference to a direction (e.g. velocity) (*b*) Organism (e.g. insect) which conveys a parasite from one host to another.

veering wind. *See* backing wind.

vegetation. The plants of an area considered in general, or as communities, but not taxonomically. The total plant cover in a particular area, or on the earth as a whole. The vegetation is a significant source of moisture in the atmosphere. Consequently grass and trees remain cool compared with bare rock, sand or dry earth, because of the latent heat absorbed. At the same time the base of a convection cloud is lowered.

vegetation study. (*a*) Qualitative. Study of the species and characteristics of the plants that make up a community; (*b*) Quantitative. Study of the distribution of particular plants.

vegetative propogation (reproduction). (*a*) Asexual reproduction in plants or animals. (*b*) Asexual reproduction in plants by means of part of the plant other than a spore (e.g. bulb, corm, tuber, rhizome). *See* gemmation.

vegetative reproduction. Vegetative propagation.

veil of cloud. A layer of almost textureless cloud which spreads characteristically across the sky ahead of the warm front of an advancing cyclone. It is caused by the slow ascent of air over a wide area, and halos are often seen around the sun in it.

vein. A tabular or sheet-like body of a mineral or minerals found in another rock body. Most veins of ore-minerals, also called lodes, lie along joints or fault planes.

veliger. A free-swimming larval stage of many aquatic molluscs, which feeds by means of cilia and has the rudiments of adult organs, including the shell.

velocity. Speed and direction. e.g. the wind velocity is the wind speed and the direction from which the wind blows (the reciprocal of the direction of flow).

vent. The opening through which a volcano ejects material.

vent-agglomerate. *See* agglomerate.

ventifact. A pebble faceted by abrasion by wind-blown sand. A dreikanter is a special form.

ventral (*a*) (Zool.) The part of an animal or organ at, or nearest to, the side normally directed downwards. In bipedal mammals the ventral surface is directed forwards. *Cf.* dorsal. (*b*) (Bot.) Adaxial. The surface of the leaf facing towards the stem.

venturi effect. The acceleration of a fluid stream as it passes through a constriction in a channel, e.g. a narrowing of river banks, a narrow nozzle on a pipe, also an aerofoil section. The acceleration is associated with a reduction in the pressure in the fluid. In the flow of air over an aerofoil this produces lift.

verfluent. A minute, or microscopic animal member of a biome.

vermes. An outmoded term covering all worm-like animals.

vermiculite. Naturally occurring vermiculite is a magnesium iron hydro-aluminosilicate, with water molecules held between the silicate sheets. The flaky natural material is passed momentarily through a furnace at 1000°C which brings the particle temperature to about 230°C. The steam produced causes the mineral to exfoliate to a low density material which is used in thermal and acoustic insulation, packaging, horticulture, and as an inert carrier for animal feedstuffs.

vernalisation. The bringing forward of flowering by subjecting a plant to abnormal temperatures or to treatment with gibberellin. e.g. spring sown 'winter' varieties of cereals, which would not normally flower the same year, can be made to produce a crop in the summer of the same year by exposing the germinating seed to temperatures just above 0°C for a few weeks. When sown in the winter these plants are subjected naturally to similar temperatures.

Vertebrata (Craniata). Animals with a skull and (except perhaps in some extinct forms) a vertebral column made of bone or cartilage. The vertebral column is developed only imperfectly in lampreys. Vertebrates form a sub-phylum of the

Chordata and comprise Agnatha, true fishes (Pisces), Amphibia, reptiles, birds and mammals.

vertical. Perpendicular to the surface of static water. A perpendicular line does not point exactly to the geometrical centre of the Earth, because the Earth is oblate due to its rotation. Minor variations are due to the pressure of large mountains or significant variations in the density of the Earth's crust.

vertisols. *See* soil classification.

vesicle. (*a*) A small, rounded or pipe-like cavity in a fine-grained or glassy volcanic rock, formed by the trapping of a gas-bubble as the lava solidified. Where these cavities have been filled with secondary minerals they are known as amygdales. (*b*) (Medicine) Blister. (*c*) (Biol.) A small bladder or other hollow structure.

vesicular. Possessing internal hollow structures. (vesicles)

vessels (cell) vessel elements). One of the two principal types of tracheary elements in plants, the other being the tracheids. Whereas the tracheid is an imperforate cell with a continuous primary wall, a vessel element has perforations or openings in the end walls of the cell. Vessel elements are combined into long pipe-like units, the vessels, through which liquids pass, via the perforations, from element to element.

vestigial. Applied to a structure, function or behaviour pattern which has diminished during the course of evolution, leaving only a trace (e.g. the veriform appendix of Man; the much reduced 'wing' bones of the Kiwi).

vibration. Forced oscillation.

vicariad (vicarious species). A member of a group of closely related species which are ecologically equivalent and whose distribution is allopatric, for instance the European Reindeer and the Caribou of North America.

vicarious species. Vicariad.

virement. The diversion of funds to purposes for which they were not allocated, especially the spending of surplus moneys remaining after a funded project has been completed.

virga. Trails of ice crystals falling from a cloud.

virion. A mature virus.

viruses. Sub-microscopic agents infecting plants, animals and bacteria, and unable to reproduce outside the tissues of the host. A fully formed virus consists of nucleic acid (DNA or RNA) surrounded by a protein or protein and lipid coat. The nucleic acid of the virus interferes with the nucleic acid synthesising mechanism of the host cell, organising to produce more viral nucleic acid. Viruses cause many diseases (e.g. mosaic diseases of many cultivated plants, myxomatosis, foot-and-mouth disease, common cold, influenza, measles, poliomyelitis). Many plant viruses are transmitted to insects (e.g. aphids), some by eelworms. Animal viruses are spread by contact, droplet infection or by insect vectors (e.g. yellow fever virus is transmitted by the mosquito *Aedes aegypti*). *See* bacteriophage.

Visean. The second oldest Stage of the Carboniferous System in Europe.

vitalism. The belief that life originates in a vital principle that is distinct from chemical and physical processes and that is not susceptible to examination by analytical techniques. *See* reductionism, holism.

vitamin. An organic food substance needed in very small amounts by heterotrophic organisms. Vitamins play essential roles in metabolism (e.g. as constituents of enzyme systems) and their deficiency causes diseases. Vitamin A (axerophtel) is a fat-soluble substance required by vertebrates and present in liver, milk and animal fats. Carotene is converted by animals into Vitamin A. Deficiency causes impaired activity of epithelia (e.g. the lining of the respiratory passages and alimentary canal), drying of the cornea, and night blindness. The vitamin is a constituent of one of the pigments in the retina. Vitamin B is a complex of water-soluble substances. *See* thiamin, riboflavin, pyridoxine, nicotinic acid, folic acid, biotin, cobalamine, pantothenic acid. Vitamin C is ascorbic acid. Vitamin D (calciferol) is a group of fat-soluble substances (sterols) present in animal fats and required by vertebrates. In Man, vitamin D is synthesised by the action of ultra-violet light from a precursor present in the skin. The vitamin is essential for the metabolism of calcium and phosphorus, and deficiency in children causes rickets. Vitamin E (Tocopherol) is the anti-sterility vitamin required by vertebrates. It is fat-soluble and present in seeds and green leaves. Deficiency results in abortion in females and sterility in males. Vitamin F is a complex of essential fatty acids (including linoleic, linolenic and arachidonic acids) needed by various animals and possibly by Man. Deficiency in Man may lead to certain kinds of eczema. Vitamin K is required by mammals and birds for the clotting of blood. Some of Man's requirements may be provided by bacteria living in the alimentary canal. It is possible that a further substance, 'vitamin P' (for permeability) affects the permeability of capillary blood vessels, so assisting in the control of bleeding.

Vita-Soy. A high protein beverage marketed in Hong Kong. *See* novel protein foods.

vitelline membrane. *See* egg membranes.

Vitis. Vines. Dicotyledonous climbing plants, (family Vitidaceae) in which the tendrils coil round supports or force their way into crevices and cement themselves there. *Vitis vinifera* is the common cultivated species, yielding grapes for wine making and the production of dried fruits (currants, raisins, sultanas). *Vitis aestivalis* (Summer Grape) and *V. labrusca* (Fox Grape) have been introduced into Europe from North America, as they are more resistant to the plant louse *Phylloxera*, a troublesome pest in vineyards.

vitreous. Glassy; vitreous is used in descriptions of lustre.

viviparous. (*a*) (Bot.) Applied to seeds (e.g. mangrove), which germinate within the fruit, or to flowers which proliferate vegetatively and do not form seeds, as in bulbous rush (*Juncus bulbosus*). (*b*) (Zool.) Applied to animals in which the embryos develop within the maternal body, deriving nutriment from it (usually by means of a placenta and not being separated from it by egg membranes. *Cf.* oviviparous, oviparous.

volatile. Having a low boiling or subliming pressure, i.e. a high vapour pressure, e.g. ether, camphor, chloroform, benzene.

volatiles. Elements and compounds dissolved in magma which would, at atmospheric pressure, have been gases at the temperature of the magma.

volcanic dust. The finest pyroclastic ash ejected by an explosive eruption. Volcanic dust sometimes travels great distances in the upper atmosphere, causing spectacular sunsets. *See* Mount Agung.

volcanic rock. (*a*) Rock formed by volcanic action at the Earth's surface (i.e. includes lava, and, usually, pyroclastic rocks). With this meaning, volcanic is contrasted with hypabyssal and plutonic (*b*) Rock formed by volcanic action plus rock formed by the associated intrusive activity. With this meaning, volcanic rocks include some, if not all, rocks classified as hypabyssal by most authors. (*c*) Loosely, fine-grained and/or glassy. The first usage is the most frequent.

volcano. A vent or fissure through which magma, gases and solids are ejected from the Earth's crust. The shape of the mountain built up by central vent eruptions seems to depend on the viscosity, gas content and rate of extrusion of the magma. Volcanoes often erupt in different manners within a short space of time, but the types of eruptions have been named after volcanoes characteristically erupting in that particular way: (1) Hawaiian: production of very mobile lava with very little explosive activity, apart from lava fountains, produces very broad low-angle cones (called shield volcanoes). (2) Strombolian: frequent small eruptions inter-spersed with evolution of more acid magma produces a steeper-sided composite cone of interbedded lava and tephra. (3) Vulcanian: more infrequent eruptions than Strombolian, coupled with more viscous magma. (4) Vesuvian: very explosive eruptions occur whenever the plug of viscous lava in the vent is blown out (particularly violent eruptions are called Plinian, after Pliny who lost his life in one). (5) Peléan: extrusion of rhyolite spines and the liberation of nuées ardentes.

Volgian. A Stage of the Jurassic System.

volt (V). The derived SI unit of electric potential, being the difference of potential between two points on a wire conducting a constant current of one ampere when the power dissipated between the points is one watt. Named after Alessandro Volta (1745–1827).

voluntary muscle. *See* muscle.

Volz photometer. Instrument used in air pollution monitoring to make low precision measurements of the intensity of direct sunlight at wavelengths defined by narrow-band interference filters. Readings can be taken by comparatively unskilled observers, and the instrument does not require completely cloudless skies. The Volz photometer is thus suitable for the widespread collection of data. *See* LIDAR.

vortex. A flow of a fluid in which the streamlines form concentric circles. There are two kinds of vortex, forced and free, depending on whether torque is applied externally. e.g. a forced vortex can be made in a liquid contained in a vessel and stirred with a paddle; a free vortex occurs when the liquid is allowed to leave the cylinder through a small hole in the bottom, in which case the force is provided by gravity acting on the fluid. Vortices occur when a solid body moves through a fluid unless the body is designed (e.g. in an aerofoil) to avoid creating them.

vugh. A cavity in a rock, usually with a lining of well-formed crystals.

vulture. General name for about 20 species of birds of prey that are distributed widely in tropical and subtropical regions, especially the latter and that are divided into two distinct groups: Old and New World. The Old world species form a sub-family, Aegypiinae, of the hawk family, Accipitridae. The New World species constitute the family Cathartidae and include the largest living land bird capable of flight, the Andean condor (*Vultur gryphus*). Vultures depend on thermal soaring (mostly in anabatic winds) whereby they remain airborne in search of carrion. They have a wing form suited to low airspeed and narrow, circling flight, with low span: chord ratio and separated primaries at the wing tips.

W

W. (*a*) Watt (*b*) Tungsten.

wad. An amorphous mixture of several hydrous manganese oxide minerals commonly occurring in the oxidised zone of ore deposits, and also in bog and shallow marine deposits.

waders. Charadriidae.

wadi. Arroyo.

wake. A disturbance of the air behind a moving object, or a similar flow behind a fixed object in an airstream. Wakes in the atmosphere are usually characterised by strong turbulence, low mean velocity relative to the object, and may exhibit downwash.

walking worms. Onychophora.

Wallace, Alfred Russell (1823–1913). An English naturalist who, jointly with Darwin, published the first work advancing the theory of evolution by natural selection, and who developed the idea of dividing the world into zoogeographical regions. 'Wallace's Line' runs south east of the Philippine Islands through the Macassar Straits, and between Bali and Lombok. It divides the Australian Region, where the characteristic mammals are marsupials and monotremes from the Oriental Region, where all the mammals are placentals.

Wallace's Line. A line drawn by A. R. Wallace separating the distinct faunas of the Oriental and Australian zoogeographical regions. The line follows a deep-water channel running between Borneo and Celebes, and between Bali and Lombok.

Warfarin (200 coumarin). Poison used to kill rodents such as rats and mice. It acts by reducing the ability of the blood to clot, so that internal bleeding occurs. Animals which eat rodents killed by Warfarin are not harmed. Strains of rat resistant to Warfarin have appeared.

warm-bloodedness. Homoiothermy.

warm front. The boundary between two air masses of differing temperatures, moving so that the warmer air is advancing into, and rising over, the colder air. The passage of a warm front over a fixed point is preceded by falling pressure and steady rain, followed by a sudden increase in temperature, veering wind and clearing skies.

warm rain. Rain falling from water clouds as opposed to ice clouds.

warm ridge. A ridge of high pressure (e.g. ahead of the advancing front of a depression) in which clear skies give warm, sunny weather.

warm-season plant. A plant which grows mainly during the warmer part of the year, usually late spring and summer.

warm sector. The region between the cold and warm front associated with a depression.

washout. The removal of aerosols in a layer of the atmosphere by impact with raindrops falling from above.

waste. Any substance, solid, liquid or gaseous, for which no use can be found by the organism or system that produces it, and for which a method of disposal must be devised.

water balance. It is assumed that in the long term a water budget will balance, and the water balance lists the ingoing and outgoing water substance.

water bears. Tardigrada.

water budget. A budget of the incoming and outgoing water from a region, including rainfall, evaporation, runoff, seepage, and with perhaps special attention to ablation (evaporation) of snow, evapotranspiration from vegetation, dew, or other special aspects relevant for special interests (e.g. agriculture).

water cloud. A cloud in which the particles are water droplets in the liquid phase.

water fleas. *See* Branchiopoda.

waterfowl. Anatidae.

Water hyacinth (*Eichhornia crasspipes*). A freshwater plant of tropical regions that has become a noxious weed in many parts of the world. Highly prolific, and reproducing mainly vegetatively, it can double its numbers in 8 to 10 days in water at a temperature of 10°C or more, provided nutrients are present. It was introduced to North America in 1884 by visitors to the New Orleans Cotton Exposition, who brought specimens from Venezuela. NASA has experimented with the utilisation of water hyacinth to provide fertiliser and feedingstuffs for livestock and the hippopotamus and *Sirenia* (dugongs and manatees) have been suggested as animals that might be farmed in order to clear choked waterways while also providing food.

water injection (WI). Technique for reducing the formation, and so the emission, of pollutants from internal combustion engines. Research has shown that WI is the most efficient way of reducing pollutant emissions, but it requires vehicles to carry water tanks as large as fuel tanks, the water may freeze, and conventional anti-freeze compounds produce exhaust pollutants, and there may be long-term corrosion in the engine.

water meadow. *See* meadow.

watershed. A divide separating two catchment areas.

water spout. A cyclonic storm similar to a tornado (*See* bath plug vortex) which occurs over water and forms a dense funnel-shaped cloud by entraining water droplets from the surface.

water-table. *See* ground-water, meteoric water.

water vapour. Water vapour enters the atmosphere from evaporation from the sea and lakes, and from damp vegetation. It is responsible for most of the absorption of radiation that occurs in the atmosphere. It is removed by condensation (dew or cloud formation) leading often to precipitation, and is thus a vital element in the determination of weather quality. *See* hydrological cycle.

watt (W). The derived SI unit of power, equal to one joule per second, which is the energy expended per second by an unvarying electric current of one ampere flowing through a conductor the ends of which are maintained at a potential difference of one volt. Named after James Watt (1736–1819).

waveband. *See* band.

wave cloud. A cloud situated in the crest of a mountain or lee wave, and as a consequence almost stationary, with condensation of cloud at the upwind edge and evaporation in the descending air at the downwind edge. Wave clouds usually have a characteristically smooth outline (*see* whale-back cloud) often showing iridescence.

wave cyclone. A depression which forms at a wave or kink in a front. *See* frontal wave.

wave front. A theoretical surface composed of points at which the phase of a wave is the same.

wave hole. A hole in a layer of cloud caused by the descent of the air and evaporation of cloud in the trough of a mountain or lee wave.

wavelength. The distance between the crests of a sine wave or, more correctly, the perpendicular distance between two wave fronts in which the phases differ by one period. The wavelength of a sound wave is equal to the speed of sound divided by the frequency.

wave soaring. The technique used by glider pilots of soaring in the upslope side of a lee or mountain wave.

Wb. Weber.

wearing course. *See* pavement.

weathering. The mechanical and chemical breakdown of rocks by exposure to water, the atmosphere or organic matter, with no transport involved, except by gravity. *See* erosion.

weathering series. A sequence of common primary silicate minerals in order of their resistance to weather: olivine→ pyroxene→ hornblende→ biotite→ muscovite→ feldspars (calcium→ sodium→ potassium)→ quartz (most resistant).

weather map. A chart on which meteorological variables are plotted over an extensive geographical area for a particular time. The most common form is the surface pressure chart on which isobars are shown.

weather modification. The artificial stimulation of rain, prevention of hail and tornadoes, which has been attempted by seeking to interfere with the mechanisms that cause freezing and/or agglomeration of cloud particles into fallout. Success has been restricted to occasional minor modification of clouds and is unlikely on a significant or economic scale because of the efficacy of natural mechanisms. In view of the insuperable task of mounting a definitive statistical test of a weather modification technique, claims of success are likely to continue.

weather radar. A rader carried by ships or aircraft, designed to detect unfavourable weather on the route by means of reflection of radar waves from precipitation.

weather ship. A ship equipped with meteorological instruments providing routine observations at a fixed station at sea. It may also be used for research purposes.

Weber (Wb). The derived SI unit of magnetic flux, being the flux which, linking a

514

circuit of one turn, produces in it an electromotive force (EMF) of one volt as it reduces to zero at a uniform rate in one second. Named after Wilhelm Weber (1804–1891).

weight of the atmosphere. The cause of atmospheric pressure. It is about 10.3 tonnes per square metre, or 5.3×10^{15} tonnes for the whole atmosphere.

Weismannism. The theory advanced by the German biologist August Weismann (1834–1914) of the continuity of the germ plasm and the non-inheritance of acquired characters (*See* Lamarck). According to this theory the reproductive cells are not influenced by the body cells because they are set aside early in development. Weismann emphasised that the body (soma) is a product of the germ cells and not vice-versa, and that effective changes could only take place through the germ cells.

Wellman Lord process. US process for removing sulphur dioxide from flue gases, using electrostatic precipitators to remove fly ash and a wet scrubber. The process produces concentrated sulphur dioxide gas which is used to produce sulphuric acid or sulphur. The process is about 90% efficient.

Wenlockian. The second oldest series of the Silurian System in Europe.

westerlies. Belts of wind which occur in latitudes between 40° and 60° in which south-west winds predominate in the northern hemisphere and north-westerlies in the southern. *See* general circulation.

westerly type. A type of weather prevalent in temperate latitudes in which the winds at all latitudes in the troposphere are from a westerly point. It gives variable weather with passing cyclones and moving anticyclones.

Westphalian. The fourth oldest stage of the Carboniferous System in Europe.

wet adiabatic. A process (or on the tephigram a line representing a process) in which air saturated with water is expanded or compressed without external heat addition. In the atmosphere this is usually caused by changes in altitude.

wet adiabatic instability. A condition in the atmosphere in which vertical movements of moist air tend to increase. Such a situation will occur if the actual lapse rate of rising (or descending) air is greater than the adiabatic. The behaviour of moist air is more complex than that of dry air since the cooling of the air may cause the condensation and loss of water, with a consequent warming. Thus the critical lapse rate for descending moist air is the wet adiabatic lapse rate only if the air retains sufficient water to remain saturated.

wet adiabatic lapse rate (saturated adiabatic lapse rate). The drop in temperature

515

of a moving parcel of saturated air per unit increase in height in adiabatic ascent (i.e. there is no exchange of heat between the parcel and the surrounding air). The saturated adiabatic lapse rate is much lower than the dry adiabatic lapse rate, being about one-third of its value at 300K, about two-thirds at 273K, and about 95% at 240K.

wet bulb. A thermometer bulb maintained wet with distilled water, usually by means of a muslin wick. The temperature indicated by this bulb can be used together with the dry bulb to give a measure of humidity. *See* absolute humidity.

wet bulb temperature. The lowest temperature to which air can be cooled by evaporating water into it, the air supplying the heat for evaporation.

wetland. An area covered permanently, occasionally, or periodically by fresh or salt water up to a depth of 6m (e.g. flooded pasture land, marshland, inland lakes, rivers and their estuaries, intertidal mud flats).

wet scrubber. An absorption tower used to remove polluted gases from a waste gas stream by contact with a liquid (e.g. hydrogen chloride being absorbed in water).

wetting agent. *See* surface-acting agent.

whaleback cloud. A lenticular wave cloud so named by seamen in high latitudes. The cloud is situated in strong winds over islands and steep coasts and the name is derived from the smooth shape of the top.

whales. *See* Balaenoidea (Baleen whales), Cetacea.

wheat. *See* Graminae.

wheel animalcules. Rotifera.

whin (gorse, furze). Ulex. *See* Leguminosae.

whirling psychrometer. A psychrometer in which wet and dry bulb thermometers are mounted on a pivot about which they can be swung by hand to provide an airflow past the wet bulb so that the humidity of the air next to the bulb is always close to the ambient value.

whirlwind. A small, near-vertical vortex which forms in conditions of exceptionally strong convection (e.g. in deserts, where it may form a dust devil, or near large fires).

white ant. Isoptera.

white bicycles. Bicycles, painted white, that were used experimentally to provide free transport in several cities, including Amsterdam and Oxford, and that were provided officially by the UN at the 1972 Stockholm Conference. *See* Witkar.

white fish. Marine fish with white flesh, esp. flatfish and species belonging to the cod family (Gadidae.).

white horizontal arc. A rare arc seen in sunshine on clouds of ice crystals which have both horizontal and vertical faces which act as mirrors. The arc passes through the observer's shadow.

white matter. The part of the central nervous system of vertebrates which consists mainly of nerve fibres and generally lies outside the grey matter.

white noise. Random noise that has equal energy at every frequency in a particular band.

WI. Water injection.

Wildfowl Trust. Society founded in 1946 which promotes the study and conservation of wildfowl (ducks, geese and swans). It maintains several collections in England and breeds wildfowl from all over the world, including the Hawaiian Goose (Nene) which was saved from extinction and returned to the wild. The Wildfowl Trust has also established extensive wetland reserves in Britain. *See* duck decoy pond.

Wild Rice (Indian, Tuscacora rice). *Zizania aquatica.* An annual aquatic grass (*see* Graminae) native to eastern North America, which grows to a height of about 4m, with a panicle up to 60 cm long bearing pendulous male spikelets on its spreading lower branches and appressed female spikelets on the erect upper branches. The seed is used as food for waterfowl and for humans and is reputed to have a higher protein and vitamin content than true rice.

willy-nilly. Severe tropical cyclone occurring in Western Australia.

wind classification. A system designed to emphasise the forces mainly responsible for the characteristics of wind. Geostrophic winds have a balance between pressure gradient and the coriolis force and blow along the isobars. Gradient winds are a modification in which the curvature of the isobars and of the flow is important, as in a cyclone. Isallobaric winds are caused by rapidly changing pressure patterns. Katabatic winds (cold downslope) and anabatic (warm upslope) winds are shallow and local, as are antitriptic (friction dominated) winds which are exemplified by katabatic ravine winds and cold outflows from storms. Land and sea breezes are antitriptic but the coriolis force becomes important with time. The thermal wind is a geostrophic component.

517

The ageostrophic component is due to acceleration or friction. *Cf.* berg winds, mountain waves.

wind drag on the sea. The basic cause of most ocean currents, reacting with temperature gradients and the Earth's rotation. The drag of the sea also significantly reduces the wind speed, esp. in the tradewinds where air moving towards the Equator is accelerated towards the west. The influence of the sea is important because of its large area and the greater persistence of wind direction than over land.

wind erosion. The removal of material from the land or from buildings by the action of the wind. The mechanisms include straightforward picking up of dust by the airflow, and the dislodging or abrasion of surface material by the impact of particles already airborne.

wind measurement. The measurement of wind may be achieved by the direct tracking of balloons, optically or by radar. Doppler radar may be used to determine the velocity of rain or other airborne objects, their horizontal motion being attributed to wind. The displacement of clouds as seen by satellite photography may be used, provided clouds can be satisfactorily identified as moving with the wind.

windmill. Machine with a rotor that is moved slowly by the wind to produce mechanical power, used originally to mill grain and pump water. *See* aerogenerator, Savonius rotor, Darrieus generator.

wind power. Mechanical or electrical power generated by a windmill using the kinetic energy of the wind as the primary energy source. Windmills may take various forms including the traditional large horizontal axis mill with 2, 3 or 4 blades, and vertical axis types.

wind profile. The variation of wind characteristic (e.g. mean speed, direction, turbulence level) with altitude.

wind rose. A diagram summarising the frequencies of winds of different strengths and directions as measured at a specified point over an extended period of time. The most common form consists of a circle with radius proportional to the frequency of calms, from which a number of bar symbols radiate, one for each wide direction band, with lengths proportional to the frequency of occurrence. The bars are often subdivided to represent the contributions from different wind strength bands.

wind shadow thermals. The strength of the wind and the terrain determine the intensity of turbulence. Where the wind is reduced (e.g. on the lee side of a hill) the depth of air warmed by the ground heated in sunshine is less and the maximum temperature reached greater. Glider pilots find that intense thermal upcurrents rise from such places from time to time. A special case is a field of ripe wheat: the temperature within the crop is higher than in a green crop, and a gust of wind bending down the stalks releases a body of very warm air.

wind shear. In general, the rate of change of the wind vector with distance in a direction perpendicular to the wind direction. Usually, the term implies the change of wind with height, and near the ground it is often used to designate the vertical gradient of horizontal velocity.

wind variation. The wind varies in time and place. In the wake of a building, tree or cliff it may fluctuate by 50% or more in a very few seconds and vary in direction by 180°. With the passage of storms and with time of day and season it varies over minutes, hours, days, and months. Even one year may differ markedly from another at the same place. These variations have to be taken into account in defining 'the wind', its average or mean, and in determining the exposure of the measuring instrument.

wing-tip vortices. A pair of counter-rotating vortices (see Vortex) which are formed in the wakes of aircraft and lie along the direction of motion. In general, they result from the curling up of the vortex sheet generated at the trailing edge of a lifting aerofoil which varies in section along its span, but for constant-section wings they form at the wing-tip. The direction of rotation is such as to induce a downward deflection of the air between the two vortices, which is sometimes referred to as downwash.

winter. *See* seasons.

winze. In mining, a nearly vertical opening or tunnel sunk downwards from a drive or level.

Witkar. A car, powered by electric batteries, designed to carry two persons and some luggage, that was introduced experimentally in Amsterdam as an alternative to conventional private cars in the city centre. The Witkars worked among a number of kerbside stations, where their batteries were recharged during resting periods. The cars were used by subscribers, each of whom paid an annual subscription plus a charge for the distance travelled in the cars, and received a magnetic key that unlocked the cars and started their motors. Before starting, the user dialled ahead to reserve a parking place at the desired destination. The scheme was invented by a Dutch engineer, Luud Schimmelpenninck. The cars were withdrawn at the end of the initial period because they were found to be too expensive. *See* white bicycles.

wolfram. A mineral which is a tungstate of iron and manganese in varying proportions, $(Fe, Mn)WO_4$. Wolfram is a major ore mineral of tungsten and occurs with cassiterite and scheelite in contact metamorphic zones and in hydrothermal deposits adjacent to acid igneous rocks, and in placer deposits. Tungsten is used mainly in the manufacture of tungsten carbide (WC) used for cutting edges, dies, and armour-piercing shells, and in alloy steels.

wood. *See* xylem.

woodland hawthorn. Crataegus.

woodpeckers. Picidae.

wood-wasps. *See* Hymenoptera.

woolsorters disease. Anthrax.

work hardening. *See* ductility.

World Wildlife Fund. *See* International Union for Conservation of Nature and Natural Resources.

World Bank. International Bank for Reconstruction and Development.

wrench fault. Strike-slip fault.

Wright, Frank Lloyd (1867–1959). US architect and planner, whose concept of 'organic architecture' was radically different from the high density planning of Le Corbusier. He used styles and materials that complemented the surroundings of buildings. His ideal city was called Usonia (after Butler's *Erewhon*) in which inhabitants were largely self-sufficient each having an acre or so of land to grow food, commuting to nearby factories to augment their incomes, with occasional paid work. Wright aimed to eliminate conventional city life and develop the quality of rural life.

X

xanthism (xanthochroism). A colour variation in which the normal colour of an animal is more or less replaced by yellow pigments.

xanthochroism. Xanthism.

Xanthophyceae. Xanthophyta.

xanthophylls. *See* carotenoids.

Xanthophyta (Yellow-green algae, Xanthophyceae, Heterokontae). A group of mainly freshwater and terrestrial algae which contain carotenoid pigments as well as chlorophyll. Most are unicellular and non-motile, some are colonial or filamentous. A few are colourless and are saprophytic or ingest food particles like Protozoa.

X chromosome. *See* sex chromosome.

xenia. Characteristics (e.g. colour) produced in the endosperm of a seed by the influence of the pollen nucleus which fuses with the primary endosperm nucleus (*See* embryo sac).

xenogamy. A term used in botany for cross fertilisation.

xenolith. An inclusion of pre-existing rock in an igneous rock, literally 'stranger-stone'.

xenoparasite. *See* parasitism.

xerad. Xerophyte.

xerarch succession. Xerosere.

xeromorphy. The possession of features characteristic of a xerophyte.

xerophilous (siccicolous). Drought resistant. *See* Xerophyte.

xerophyte (xerad). A plant which lived in a dry habitat and is able to endure prolonged drought. Some xerophytes (e.g. cactus) store water in succulent tissue for use when none is available from the soil, and have stomates which close during the day. Other features of xerophytes which tend to check transpiration are reduction in leaf size, thickening of the cuticle, sunken or protected stomates and the possession of hairs. In some (e.g. Gorse, *Ulex*) both leaves and stems have modified to resist shrinkage, forming stiff spines containing abundant sclerenchyma. Many xerophytes can endure long periods of wilting. Deciduous trees become xeromorphic when they shed their leaves.

xerosere (xerarch succession). The stages in a plant succession beginning in a dry site and progressing towards moister conditions.

Xiphosura (king crabs, horseshoe crabs). Aquatic arachnida with a body enclosed in a carapace. *Limulus*, the sole living genus, has existed since Triassic times, and grows to 50 cm in length. The four present-day species are shore-living, burrowing animals which feed on worms and molluscs.

X-ray analysis. The diffraction of X rays by crystalline solids is used for their identification and the solution of their crystalline structures.

xylem (wood). Vascular tissue through which water containing dissolved mineral salts is transported in Pterisophyta and seed plants. Xylem also provides mechanical support. It consists of tracheids and/or vessels, fibres and parenchyma, and forms the bulk of the stems and roots in mature woody plants.

xylophagous. Wood eating

Y

yardangs. Ridges formed by wind carrying sand at low level. Sand-blasting of existing grooves in rocks enlarges these into furrows leaving the sharp ridges between them. *See* corrasion.

Y chromosome. *See* sex chromosome.

yeasts. Many species of unicellular fungi, most of which belong to the Ascomycetes and reproduce by budding. The genus *Saccharomyces* is used in brewing and wine making because in low oxygen concentrations it produces zymase, an enzyme system which breaks down sugars into alcohol and carbon dioxide. *Saccharomyces* is also used in baking. Some yeasts are used as a source of protein (*see* novel protein foods) and of vitamins of the B group.

yellow-green algae. Xanthophyta.

Yew. *See* Taxaceae.

yolk sac. An embryonic membrane which encloses and absorbs the yolk in reptiles, birds and many fish. In mammal embryos the yolk sac absorbs nutriment from secretions of the uterus before the placenta has formed. In some marsupials the yolk sac forms a poorly-developed placenta.

Z

Z. Atomic number.

zeatin. One of the cytokinin group of plant hormones.

Zechstein. The upper division of the Permian System in north-west Europe. The Zechstein contains substantial deposits of evaporites.

Zero Population Growth (ZPG). US voluntary organisation, active in the 1960's and early 70's, whose aim was to draw attention to problems associated with a rapid increase in the size of human populations and to urge Americans to aim for a stable population in which birth-rates and death-rates were in balance.

ZETA. Zero Energy Thermonuclear Apparatus. A torus-shaped (i.e. 'doughnut shaped') apparatus used at the UK Atomic Energy Research Establishment, Harwell, to study controlled thermonuclear reactions.

zinc (Zn). Element. A hard, bluish-white metal that occurs as calamine, zincite, and zinc blende. It is used in alloys, especially brass, and in the galvanising of iron. It is an essential micronutrient. AW 65.37; At. No. 30; SG 7.14; mp 419°C; bp 907°C.

zinc blende. Sphalerite.

zineb. Fungicide of the dithiocarbamate group used to control diseases such as potato blight and downy mildews. Can be irritating to the eyes and the skin.

Zn. Zinc.

zonal flow. West-to-east airflow.

zonal index. A measure of the strength of the circulation in a specified large area of the globe (e.g. the temperate zone), often expressed in the form of a pressure difference.

zone fossil. *See* index fossil.

zone of silence. When anomalous audibility occurs at large distances from a sound source, a zone of silence with relatively low audibility is frequently observed closer to the source but beyond the distance reached by the propagating sound waves.

zone time. A local time system in which 24 time zones, each covering 15° of longitude, are designated by letters of the alphabet. Greenwich Mean Time is designated Z-time.

zoning. System of land use planning based on boundaries inside which areas can be used only for specified purposes, e.g. agriculture, dwellings, recreation, industry, etc. Zoning has been used in the USA and Danish planning is based on zoning.

zoobiotic. Applied to an organism which lives parasitically on an animal.

zoocenose. Animal community.

zoochore. A plant whose reproductive structures are dispersed by animals (e.g. burdock, *Arctium*) and blackberry, *Rubus fruticosus*.

APPENDIX

PRINCIPAL ORGANISATIONS CONCERNED WITH THE ENVIRONMENT IN THE UNITED KINGDOM

ADVISORY CENTRE FOR EDUCATION (ACE) 32 Trumpington Street, Cambridge.

AMATEUR ENTOMOLOGISTS' SOCIETY 23 Manor Way, North Harrow, Middx.

ANCIENT MONUMENTS SOCIETY 11 Alexander Street, London W2

ANGLERS' CO-OPERATIVE ASSOCIATION 53 New Oxford Street, London WC1

ARBORICULTURAL ASSOCIATION 38 Blythwood Gardens, Stansted, Essex

ARMAGH FIELD NATURALISTS' SOCIETY Brean Main Street, Richhill, Co. Armagh

ASSOCIATION FOR STUDIES IN THE CONSERVATION OF HISTORIC BUILDINGS Institute of Archaeology, University of London, 31–34 Gordon Square, London WC2

ASSOCIATION FOR THE PRESERVATION OF RURAL SCOTLAND 39 Castle Street, Edinburgh

ASSOCIATION OF AGRICULTURE 78 Buckingham Gate, London SW1

ASSOCIATION OF COUNTY COUNCILS IN SCOTLAND 3 Forres Street, Edinburgh EH3 6BL

ASSOCIATION OF PUBLIC HEALTH INSPECTORS 19 Grosvenor Place, London

ASSOCIATION OF SCHOOL NATURAL HISTORY SOCIETIES c/o Strand School, Elm Park, London SW2

AUTO-CYCLE UNION 31 Belgrave Square, London SW1

BOTANICAL SOCIETY OF EDINBURGH c/o The Royal Botanic Garden, Inverleith Row, Edinburgh EH3 5LR

BOTANICAL SOCIETY OF THE BRITISH ISLES c/o Dept. of Botany, British Museum (Natural History), Cromwell Road, London SW7

BRITISH AGROCHEMICALS ASSOCIATION Alembic House, 93 Albert Embankment, London SE1

BRITISH ARACHNOLOGICAL SOCIETY Peare Tree House, The Green, Blennerhasset, Carlisle CA5 3RE

BBC NATURAL HISTORY UNIT BBC, Broadcasting House, Whiteladies Road, Bristol BS8 2LR

BRITISH BUTTERFLY CONSERVATION SOCIETY LTD. 7 The Drive, Kingston on the Hill, Surrey

BRITISH CYCLING FEDERATION 26 Park Crescent, London W1N 4BL

BRITISH DEER SOCIETY 43 Brunswick Square, Hove BN3 1EE

BRITISH ECOLOGICAL SOCIETY c/o Monks Wood Experimental Station, Nature Conservancy Council, Abbots Ripton, Huntingdon

BRITISH ENTOMOLOGICAL AND NATURAL HISTORY SOCIETY c/o The Alpine Club, 74 South Audley Street, London W1

BRITISH FIELD SPORTS SOCIETY 137 Victoria Street, London SW1

BRITISH INSTITUTE OF RECORDED SOUND 29 Exhibition Road, London SW7

BRITISH MOUNTAINEERING COUNCIL Room 314, 26 Park Crescent, London W1

BRITISH NATURALISTS' ASSOCIATION Hawkshead, Tower Hill, Dorking, Surrey

BRITISH ORNITHOLOGISTS' UNION c/o Bird Room, British Museum (Natural History), Cromwell Road, London SW7

BRITISH PTERIDOLOGICAL SOCIETY 46 Sedley Rise, Loughton, Essex

BRITISH SPELEOLOGICAL ASSOCIATION Duke Street, Settle, Yorks.

BRITISH TRUST FOR CONSERVATION VOLUNTEERS (formerly the Conservation Corps) Zoological Gardens, Regent's Park, London NW1

BRITISH TRUST FOR ENTOMOLOGY 41 Queen's Gate, London SW7

BRITISH TRUST FOR ORNITHOLOGY Beech Grove, Tring, Herts.

BRITISH WATERWAYS BOARD Melbury House, Melbury Terrace, London NW1

BROADS CONSORTIUM The Cedars, Albemarle Road, Norwich NOR 8 1E

CENTRAL COUNCIL FOR RIVERS PROTECTION Fishmonger's Hall, London EC4

CENTRAL COUNCIL OF PHYSICAL RECREATION 26 Park Crescent, London W 1N 4AJ

CENTRAL ELECTRICITY GENERATION BOARD Sudbury House, 15 Newgate Street, London EC1

CHILTERN SOCIETY Cherry Cottage, Stokenchurch, High Wycombe, Bucks.

CIVIC TRUST 18 Carlton House Terrace, London SW1

CIVIC TRUST FOR THE NORTHEAST 34–35 Saddler Street, Durham

CIVIC TRUST FOR THE NORTHWEST 56 Oxford Street, Manchester M1 6EU

CIVIC TRUST FOR WALES 6 Park Place, Cardiff CF1 3DP

COASTAL ANTI-POLLUTION LEAGUE LTD. Alverstoke, Greenway Lane, Bath

COMMITTEE FOR ENVIRONMENTAL CONSERVATION (CoEnCo) c/o CPRE, 4 Hobart Place, London SW1 and c/o RSPB, The Lodge, Sandy, Beds.

COMMONS, OPEN SPACES AND FOOTPATHS PRESERVATION SOCIETY 166 Shaftesbury Avenue, London WC2

COMMUNITY SERVICE VOLUNTEERS Toynbee Hall, 28 Commercial Street, London E1

CONCHOLOGICAL SOCIETY OF GREAT BRITAIN AND IRELAND The Eyrie, 58 Teignmouth Road, London NW2

CONSERVATION SOCIETY 21 Hanyards Lane, Cuffley, Potters Bar, Herts.

CONSERVATION YOUTH ASSOCIATION C.Y.A. Environmental Studies Centre, London Road, Mitcham, Surrey

CONVENTION OF ROYAL BURGHS 51 Castle Street, Edinburgh 2

COUNCIL FOR BRITISH ARCHAEOLOGY 8 St. Andrews Place, Regent's Park, London NW1

COUNCIL FOR ENVIRONMENTAL EDUCATION 26 Bedford Square, London WC1

COUNCIL FOR NATURE Zoological Gardens, Regent's Park, London NW1

COUNCIL FOR SMALL INDUSTRIES IN RURAL AREAS (COSIRA) 11 Cowley Street, London SW1 and 35 Camp Road, Wimbledon Common, London SW19

COUNCIL FOR THE PROTECTION OF RURAL ENGLAND (CPRE) 4 Hobart Place, London SW1

COUNCIL FOR THE PROTECTION OF RURAL WALES Meifod, Montgomery

COUNCIL OF INDUSTRIAL DESIGN 28 Haymarket, London SW1

COUNCIL OF SOCIAL SERVICE FOR WALES AND MONMOUTHSHIRE 2 Cathedral Road, Cardiff CF1 9XR

COUNTRY LANDOWNERS' ASSOCIATION 7 Swallow Street, London W1R 8EN

COUNTRYSIDE COMMISSION John Dower House, Crescent Place, Cheltenham, Glos.

COUNTRYSIDE COMMISSION FOR SCOTLAND Branklyn House, 116 Dundee Road, Perth

COUNTY COUNCILS ASSOCIATION Eaton House, 66A Eaton Square, London SW1

CROFTERS' COMMISSION 9 Ardross Terrace, Inverness

CUMBERLAND COUNTRYSIDE CONFERENCE 6 West Walls, Carlisle

DARTINGTON AMENITY RESEARCH TRUST Shinners Bridge, Dartington, Totnes, Devon

DARTMOOR PRESERVATION ASSOCIATION 23 Wellpark Close, Exeter EX4 1TS

DEPARTMENT FOR AGRICULTURE AND FISHERIES FOR SCOTLAND Room 165, St. Andrews House, Edinburgh EH1 3DA

DEPARTMENT OF THE ENVIRONMENT 2 Marsham Street, London SW1

DISTRICT COUNCILS ASSOCIATION OF SCOTLAND 28 Union Street, Larkhall, Lanarkshire

DUKE OF EDINBURGH'S AWARD SCHEME 2 Old Queen Street, London SW1

EASTERN FEDERATION OF AMENITY SOCIETIES c/o Feilden and Mawson, Ferry Road, Norwich NOR 18S

ENTERPRISE YOUTH 29 Queen Street, Edinburgh EH2 1JX

ENVIRONMENTAL CONSORTIUM 27 Nassau Street, London W1N 8EQ

ENVIRONMENTAL HEALTH OFFICERS ASSOCIATION 19 Grosvenor Place, London SW1

EXMOOR SOCIETY Parish Rooms, Dulverton, Somerset

FARM AND FOOD SOCIETY 37 Tanza Road, London NW3

FARM BUILDINGS ASSOCIATION Twineham Bolney, Haywards Heath, Sussex

FARMERS' UNION OF WALES Llys Amaeth, Queen's Square, Aberystwyth

FAUNA PRESERVATION SOCIETY c/o Zoological Gardens, Regent's Park,. London NW 1

FELL AND ROCK CLIMBING CLUB OF THE ENGLISH LAKE DISTRICT 110 Low Ash Drive, Shipley, Yorks.

FERMANAGH NATURALISTS' FIELD CLUB The Crannog Rakeelan Glebe, Enniskillen, Co. Fermanagh

FIELD STUDIES COUNCIL 9 Devereux Court, Strand, London WC2R 3JR

FOREST SCHOOL CAMPS 3 Pine View, Fairmile Park Road, Cobham, Surrey

FORESTRY COMMISSION 25 Savile Row, London W1X 2AY

FRIENDS OF THE EARTH 9 Poland Street, London W 1V 3DG

FRIENDS OF THE LAKE DISTRICT 27 Greenside, Kendal, Cumbria

GAME CONSERVANCY Fordingbridge, Hants.

GEOGRAPHICAL ASSOCIATION 343 Fulwood Road, Sheffield S10 3BP

GEORGIAN GROUP 2 Chester Street, London SW1

HOME COUNTIES RAMBLING ASSOCIATION 7 Roche Road, Norbury, London SW16

INSTITUTE OF BIOLOGY 41 Queen's Gate, London SW7

INSTITUTE OF LANDSCAPE ARCHITECTS Nash House, 12 Carlton House Terrace, London SW1

INSTITUTE OF PARK AND RECREATION ADMINISTRATION Lower Basildon, Reading, Berks.

INSTITUTE OF PETROLEUM 61 New Cavendish Street, London W1M 8AR

INSTITUTE OF WATER POLLUTION CONTROL Water Pollution Research Laboratory, Elder Way, Stevenage, Herts.

INSTITUTION OF WATER ENGINEERS 6–8 Sackville Street, Piccadilly, London W1X 1DD

INTERNATIONAL PLANNED PARENTHOOD FEDERATION (IPPF) 18–20 Lower Regent Street, London SW1Y 4PW

IRISH MOUNTAINEERING CLUB 157 Stranmillis Road, Belfast BT9 5AH

JOINT COMMITTEE FOR THE CONSERVATION OF BRITISH INSECTS c/o Royal Entomological Society of London, 41 Queen's Gate, London SW7

KEEP BRITAIN TIDY GROUP Cecil Chambers, 78/86 Strand, London WC2

LAND INSTITUTE LTD 29 Belgrave Square, London SW1

LANDS IMPROVEMENT COMPANY 39 Jermyn Street, London SW1

LANDSCAPE RESEARCH GROUP Longmoor, 8 Cunningham Road, Banstead, Surrey

LINNEAN SOCIETY OF LONDON Burlington House, Piccadilly, London W1V 01Q

MALACOLOGICAL SOCIETY OF LONDON Dept. of Biology, Queen Elizabeth College, Campden Hill Road, London W8

MAMMAL SOCIETY c/o Institute of Biology, 41 Queen's Gate, London SW7

MARINE BIOLOGICAL ASSOCIATION OF THE UNITED KINGDOM
The Laboratory, Citadel Hill, Plymouth

MARITIME TRUST 53 Davies Street, London W1Y 1FH

MEN OF THE STONES The Rutland Studio, Tinwell, Stamford, Lincs.

MEN OF THE TREES Hollybank House, Emsworth, Hants. PO10 7UK

MINISTRY OF AGRICULTURE, FISHERIES AND FOOD Whitehall
Place, London SW1

MINISTRY OF AGRICULTURE, GOVERNMENT OF NORTHERN
IRELAND: FORESTRY DIVISION Dundonald House, Upper Newtown-
ards Roads, Belfast BT4 3SB

MINISTRY OF DEVELOPMENT, GOVERNMENT Of NORTHERN
IRELAND: AMENITY LANDS BRANCH Parliament Buildings, Stormont, Belfast BT4 3SS

MINISTRY OF FINANCE, GOVERNMENT OF NORTHERN IRELAND:
ANCIENT MONUMENTS BRANCH Churchill House, Victoria Square,
Belfast BT1 4QW

MINISTRY OF HOME AFFAIRS, GOVERNMENT OF NORTHERN
IRELAND Parliament Buildings, Stormont, Belfast BT4 3SU

MUSEUM AND ART GALLERY SERVICE FOR YORKSHIRE AND
HUMBERSIDE Farnley Hall, Farnley Park, Leeds 12

NATIONAL ASSOCIATION OF BOYS' CLUBS 17 Bedford Square, London
WC1B 3JJ

NATIONAL ASSOCIATION OF PARISH COUNCILS 100 Great Russell
Street, London WC1

NATIONAL COUNCIL OF SOCIAL SERVICE 26 Bedford Square, London
WC1

NATIONAL FARMERS' UNION Agriculture House, Knightsbridge, London
SW1

NATIONAL FEDERATION OF ANGLERS 56 Ward Street, Derby

NATIONAL FEDERATION OF COMMUNITY ASSOCIATIONS 26 Bedford Square, London WC1

NATIONAL FEDERATION OF YOUNG FARMERS' CLUBS YFC Centre,
National Agricultural Centre, Kenilworth, Warwickshire CV8 2LG

NATIONAL HOUSING AND TOWN PLANNING COUNCIL 11 Green
Street, London W1

NATIONAL PLAYING FIELDS ASSOCIATION 57B Catherine Place,
London SW1

NATIONAL RURAL AND ENVIRONMENTAL STUDIES
ASSOCIATION Education Office, County Hall, March, Cambridgeshire and
2 Bates House, Popple Way, Stevenage, Herts.

NATIONAL SOCIETY FOR CLEAN AIR 134 North Street, Brighton BN1
1RG

NATIONAL TRUST FOR PLACES OF HISTORIC INTEREST OR
NATURAL BEAUTY 42 Queen Anne's Gate, London SW1

NATIONAL TRUST FOR SCOTLAND 5 Charlotte Square, Edinburgh EH2 4DU

NATURAL ENVIRONMENT RESEARCH COUNCIL Alhambra House, 29–33 Charing Cross Road, London WC2

NATURE CONSERVANCY COUNCIL 19 Belgrave Square, London SW1

NORTHERN IRELAND BIRD RECORDS COMMITTEE 45 Park Drive, Holywood, Co. Down

NORTHERN IRELAND ORNITHOLOGISTS' CLUB 470 Upper Newtown-ards Road, Belfast 4

NORTHWESTERN NATURALISTS' UNION c/o 91 Stanley Road, Cheadle Hulme, Cheshire SK8 6PL

OFFA'S DYKE ASSOCIATION March House, Wylcwm Street, Knighton, Radnorshire

ORDNANCE SURVEY Romsey Road, Maybush, Southampton

OUTWARD BOUND TRUST Iddesleigh House, Caxton Street, London SW1

PILGRIM TRUST Millbank House, 2 Great Peter Street, London SW1

PURE RIVERS SOCIETY 113 Cluse Court, St. Peter's Street, Islington, London N1

RAMBLERS' ASSOCIATION 1/4 Crawford Mews, London W1H 1PT and 40 Queen Square, Glasgow S1

RED DEER COMMISSION Elm Park, Island Bank Road, Inverness

RIVER THAMES SOCIETY 2 Ruskin Avenue, Kew, Richmond, Surrey

ROUTE NATURALISTS' FIELD CLUB c/o 15 Portrush Road, Coleraine, Northern Ireland

ROYAL AGRICULTURAL SOCIETY OF ENGLAND National Agricul-tural Centre, Kenilworth, Warwickshire CV8 2LG

ROYAL COMMISSION ON ANCIENT AND HISTORICAL MONUMENTS IN WALES AND MONMOUTHSHIRE Edleston House, Queens Road, Aberystwyth, Dyfed

ROYAL COMMISSION ON ENVIRONMENTAL POLLUTION Govern-ment Offices, Great George Street, London SW1

ROYAL COMMISSION ON HISTORICAL MONUMENTS (ENGLAND): NATIONAL MONUMENTS RECORD Fielden House, 10 Great College Street, London SW1

ROYAL COMMISSION ON THE ANCIENT AND HISTORICAL MONUMENTS OF SCOTLAND 52-54 Melville Street, Edinburgh EH3 7HF

ROYAL ENTOMOLOGICAL SOCIETY OF LONDON 41 Queen's Gate, London SW7

ROYAL FORESTRY SOCIETY OF ENGLAND, WALES AND NORTHERN IRELAND 102 High Street, Tring, Herts.

ROYAL GEOGRAPHICAL SOCIETY Kensington Gore, London SW7

ROYAL INSTITUTE OF BRITISH ARCHITECTS 66 Portland Place, London W1

ROYAL SCOTTISH FORESTRY SOCIETY 26 Rutland Square, Edinburgh

EH1 2BU

ROYAL SOCIETY FOR THE PREVENTION OF ACCIDENTS Terminal House, 52 Grosvenor Gardens, London SW1

ROYAL SOCIETY FOR THE PROTECTION OF BIRDS The Lodge, Sandy, Beds. and 17 Regent Terrace, Edinburgh EH7 5BN

ROYAL SOCIETY OF ARTS 6–8 John Adam Street, London WC2N 6EZ

SALMON AND TROUT ASSOCIATION Fishmongers' Hall, London EC4

SALTIRE SOCIETY Gladstone's Lane, 483 Lawnmarket, Edinburgh EH1 2NT

SCHOOL NATURAL SCIENCE SOCIETY 2 Bramley Mansions, Berrylands Road, Surbiton, Surrey

SCOTTISH ANGLERS' ASSOCIATION 117 Hanover Street, Edinburgh 2

SCOTTISH CIVIC TRUST 24 George Square, Glasgow C2

SCOTTISH COUNCIL OF PHYSICAL RECREATION 4 Queensferry Street, Edingburgh EH2 4PB

SCOTTISH COUNTRYSIDE ACTIVITIES COUNCIL 30 Underwood Road, Prestwick, Ayrshire

SCOTTISH DEVELOPMENT DEPARTMENT St. Andrew's House, Edinburgh 1

SCOTTISH HOME AND HEALTH DEPARTMENT St. Andrew's House, Edinburgh 1

SCOTTISH ORNITHOLOGISTS' CLUB Scottish Centre for Ornithology and Bird Protection, 21 Regent Terrace, Edinburgh EH7 5BT

SCOTTISH SPORTING AND FISHERIES LIAISON COMMITTEE c/o British Field Sports Society (Scotland), Haig House, 23 Drumsheugh Gardens, Edinburgh 1

SCOTTISH STANDING CONFERENCE OF VOLUNTARY YOUTH ORGANISATIONS 8 Palmerston Place, Edinburgh EH1 5AA

SCOTTISH WOODLAND OWNERS' ASSOCIATION 6 Chester Street, Edinburgh EH3 7RD

SCOTTISH YOUTH HOSTELS ASSOCIATION 7 Glebe Crescent, Stirling

SCOUT ASSOCIATION 25 Buckingham Palace Road, London SW1

SEABIRD GROUP c/o British Ornithologists' Union, Bird Room, British Museum (Natural History), Cromwell Road, London SW7 and Zoology Department, Aberdeen University, Old Aberdeen

SMALL INDUSTRIES COUNCIL FOR RURAL AREAS OF SCOTLAND 27 Walker Street, Edinburgh EH3 7HZ

SNOWDONIA NATIONAL PARK SOCIETY Dyffryn Mymbyr, Capel Curig, Betws y Coed, Gwynedd

SOCIETY FOR ENVIRONMENTAL EDUCATION St. Luke's College, Exeter

SOCIETY FOR THE PROMOTION OF NATURE RESERVES (SPNR) c/o British Museum (Natural History), Cromwell Road, London SW7 and The Manor House, Alford, Lincs.

SOCIETY OF SUSSEX DOWNSMEN 93 Church Road, Hove, Sussex

SOIL ASSOCIATION Walnut Tree Manor, Haughley, Stowmarket, Suffolk IP14 3RS

SOCIETY FOR THE PROTECTION OF ANCIENT BUILDINGS 55 Great Ormond Street, London WC1N 3JA

SOUTHWESTERN NATURALISTS' UNION Shorton Manor, Shorton, Paignton, Devon

SPORTS COUNCIL 26 Park Crescent, London W1

SPORTS COUNCIL FOR SCOTLAND 4 Queensferry Street, Edinburgh EH2 4PB

STANDING CONFERENCE FOR WALES OF VOLUNTARY YOUTH ORGANISATIONS 2 Cathedral Road, Cardiff CF1 9XR

STANDING CONFERENCE OF NATIONAL VOLUNTARY YOUTH ORGANISATIONS 26 Bedford Square, London WC1B 3HU

STANDING CONFERENCE OF YOUTH ORGANISATIONS IN NORTHERN IRELAND 40 Academy Street, Belfast BT1 2NQ

TOWN AND COUNTRY PLANNING ASSOCIATION 17 Carlton House Terrace, London SW1

TOWN PLANNING INSTITUTE 26 Portland Place, London W1N 4BE

TRANSPORT 2000 9 Catherine Place, London SW1E 6DX

ULSTER COUNCIL OF OUTDOOR ORGANISATIONS 4 Nendrum Gardens, Belfast BT5 5LZ

ULSTER SOCIETY FOR THE PRESERVATION OF THE COUNTRYSIDE West Winds, Carney Hill, Craigavad, Holywood, Co. Down

UNIVERSITIES FEDERATION FOR ANIMAL WELFARE 230 High Street, Potters Bar, Herts.

WATER POLLUTION RESEARCH LABORATORY Elder Way, Stevenage, Herts.

WILDFOWL TRUST Slimbridge, Glos. and Peakirk, Peterborough, Northants.

WILDFOWLERS' ASSOCIATION OF GREAT BRITAIN AND IRELAND Grosvenor House, 104 Watergate Street, Chester and Conservation Centre, Boarstal Duck Decoy, Brill, Aylesbury, Bucks.

WILDLIFE OBSERVATION SOCIETY c/o Royal Society for the Prevention of Cruelty to Animals, 105 Jermyn Street, London SW1

WILDLIFE YOUTH SERVICE OF THE WORLD WILDLIFE FUND Wildlife, Wallington, Surrey

YORKSHIRE FIELD STUDIES Westland, Westfields, Kirkbymoorside, York YO6 6AG

YOUTH HOSTELS ASSOCIATION (ENGLAND AND WALES) 8 St. Stephen's Hill, St. Albans, Herts.

YOUTH HOSTEL ASSOCIATION OF NORTHERN IRELAND LTD. Bryson House, 28 Bedford Street, Belfast BT2 7FE

YOUTH INFORMATION CENTRE Humberstone Drive, Leicester LE5 0RG

ZOOLOGICAL SOCIETY OF LONDON Regent's Park, London NW1